Plant Cuticles

an integrated functional approach

ENVIRONMENTAL PLANT BIOLOGY series

Editor: W.J. Davies
Institute of Environmental and Biological Sciences, Division of Biological Sciences, University of Lancaster, Lancaster LA1 4YQ, UK

Abscisic Acid: physiology and biochemistry

Carbon Partitioning: within and between organisms

Pests and Pathogens: plant responses to foliar attack

Water Deficits: plant responses from cell to community

Photoinhibition of Photosynthesis: from molecular mechanisms to the field

Environment and Plant Metabolism: flexibility and acclimation

Embryogenesis: the generation of a plant

Plant Cuticles: an integrated functional approach

Forthcoming titles include:

Stable Isotopes: integration of biological, ecological and geochemical processes

G. KERSTIENS
Institute of Environmental and Biological Sciences, Division of Biological Sciences, Lancaster University, Lancaster LA1 4YQ, UK

Plant Cuticles

an integrated functional approach

βIOS
SCIENTIFIC
PUBLISHERS

First published 1996

A CIP catalogue record for this book is available from the British Library.

ISBN 1 85996 130 4

BIOS Scientific Publishers Ltd
9 Newtec Place, Magdalen Road, Oxford OX4 1RE, UK
Tel. +44 (0) 1865 726286. Fax +44 (0) 1865 246823
World Wide Web home page: http://www.Bookshop.co.uk/BIOS/

DISTRIBUTORS

Australia and New Zealand
DA Information Services
648 Whitehorse Road, Mitcham
Victoria 3132

India
Viva Books Private Limited
4325/3 Ansari Road, Daryaganj
New Delhi 110002

Singapore and South East Asia
Toppan Company (S) PTE Ltd
38 Liu Fang Road, Jurong
Singapore 2262

USA and Canada
BIOS Scientific Publishers
PO Box 605, Herndon
VA 20172-0605

Typeset by Chandos Electronic Publishing, Stanton Harcourt, UK.
Printed by Biddles Ltd, Guildford, UK.

Contents

Contributors

Abbott, G.D. NRG in Fossil Fuels and Environmental Geochemistry, Drummond Building, University of Newcastle upon Tyne, Newcastle upon Tyne NE3 1TY, UK

Barnes, J.D. Department of Agricultural and Environmental Science, Ridley Building, University of Newcastle upon Tyne, Newcastle upon Tyne NE1 7RU, UK

Baur, P. Institut für Obstbau und Baumschule, Universität Hannover, Am Steinberg 3, D-31157 Sarstedt, Germany

Butler, D.R. International Crops Research Institute for the Semi-Arid Tropics (ICRISAT), Patancheru, Andhra Pradesh 502324, India

Cape, J.N. Institute of Terrestrial Ecology, Bush Estate, Penicuik, Midlothian EH26 0QB, UK

Cardoso-Vilhena, J. Department of Agricultural and Environmental Science, Ridley Building, University of Newcastle, Newcastle upon Tyne NE1 7RU, UK

Carver, T.L.W. Institute of Grassland and Environmental Research, Plas Gogerddan, Aberystwyth, Ceredigion SY23 3EB, UK

Danehower, D.A. Department of Crop Science, North Carolina State University, Raleigh, NC 27695, USA

Edwards, D. Department of Earth Sciences, P.O. Box 914, University of Wales, Cardiff CF1 3YE, UK

Eigenbrode, S.D. Department of Plant, Soil and Entomological Sciences, University of Idaho, Moscow, ID 83843, USA

Espelie, K.E. Department of Entomology, University of Georgia, Athens, GA 30602, USA

Grace, J. Institute of Ecology and Resource Management, University of Edinburgh, Darwin Building, Mayfield Road, Edinburgh EH9 3JU, UK

Huttunen, S. Department of Botany, University of Oulu, P.O. Box 400, FIN-90571 Oulu, Finland

Ingerson, S.M. Institute of Grassland and Environmental Research, Plas Gogerddan, Aberystwyth, Ceredigion SY23 3EB, UK

Jackson, D.M. Agricultural Research Service, United States Department of Agriculture, U.S. Vegetable Laboratory, 2875 Savannah Highway, Charleston, SC 29414, USA

Jeffree, C.E. Science Faculty EM Facility, University of Edinburgh, Daniel Rutherford Building, King's Buildings, Edinburgh EH9 3JH, UK

Kerstiens, G. Institute of Environmental and Biological Sciences, Division of Biological Sciences, Lancaster University, Lancaster LA1 4YQ, UK

Kirsch, T. Julius-von-Sachs-Institut für Biowissenschaften, Lehrstuhl für Botanik II, Universität Würzburg, Mittlerer Dallenbergweg 64, D-97082 Würzburg, Germany

Kolattukudy, P.E. The Ohio State University Neurobiotechnology Center, 206 Rightmire Hall, 1060 Carmack Road, Columbus, OH 43210-1002, USA

Markstädter, C. Julius-von-Sachs-Institut für Biowissenschaften, Lehrstuhl für Botanik II, Universität Würzburg, Mittlerer Dallenbergweg 64, D-97082 Würzburg, Germany

Mendgen, K. Fakultät für Biologie, Universität Konstanz, Phytopathologie, D-78434 Konstanz, Germany

Raven, J.A. Department of Biological Sciences, University of Dundee, Dundee DD1 4HN, UK

Riederer, M. Julius-von-Sachs-Institut für Biowissenschaften, Lehrstuhl für Botanik II, Universität Würzburg, Mittlerer Dallenbergweg 64, D-97082 Würzburg, Germany

Schönherr, J. Institut für Obstbau und Baumschule, Universität Hannover, Am Steinberg 3, D-31157 Sarstedt, Germany

Schreiber, L. Julius-von-Sachs-Institut für Biowissenschaften, Lehrstuhl für Botanik II, Universität Würzburg, Mittlerer Dallenbergweg 64, D-97082 Würzburg, Germany

Thomas, B.J. Institute of Grassland and Environmental Research, Plas Gogerddan, Aberystwyth, Ceredigion SY23 3EB, UK

van Gardingen, P.R. Institute of Ecology and Resource Management, School of Agriculture, University of Edinburgh, West Mains Road, Edinburgh EH9 3JG, UK

Abbreviations

A	absorbance
AS	ammonium sulphate
BET	Brunauer, Emmett and Teller
CAM	crassulacean acid metabolism
CAT	chloramphenicol acetyl transferase
CL	cuticular layer
CM	cuticular membrane(s)
CP	cuticle proper
CRT	canopy reflectance and radiative transfer
CTF	cutinase transcription factor
CW	cell wall
Cyt P_{450}	cytochrome P450
2,4-D	dichlorophenoxyacetic acid
d.c.	donor compartment
DCA	dichloroacetic acid
DCMU	*N*′-(3,4-dichlorophenyl)-*N*,*N*′-dimethyl urea
EA	ethoxylated alcohol
ECL	external cuticular layer
EDTA	ethylenediaminetetraacetic acid
EL	epicuticular lipids
EM	electron microscopy
EW	epicuticular wax(es)
EWF	epicuticular wax film
F_M	maximum chlorophyll fluorescence
F_O	initial level of chlorophyll fluorescence
F_V	variable chlorophyll fluorescence
f.sp.	*formae specialis* (form of the fungus)
FTIR	Fourier transform infrared
g_{min}	minimum conductance
GC–MS	gas chromatography-mass spectroscopy
GE	glucose ester
HS	hexane-soluble fraction
ICL	internal cuticular layer
LAB	lactic acid buffer
LAI	leaf area index
LC	liquid chromatography
LS	leaf surface
LTSEM	low temperature SEM
MWS	methanol–water-soluble fraction
MX	polymer matrix membranes
N-O-Me-DIMBOA	2-hydroxy-4,7-dimethoxy-1,4-benzoxazin-3-one
NMR	nuclear magnetic resonance
ORS	Old Red Sandstone
P	permeance
P_{H20}	permeance to water
PAL	present atmospheric level

PAR	photosynthetically active radiation
PCP	pentachlorophenol
PCW	primary cell wall
PFD	photon flux density
PGT	primary germ tube
PL	pectin lamella
PLS	phospholipid suspension
Py-GC-MS	pyrolysis-GC-MS
R	reflectance
r_a	aerodynamic resistance
r_b	sublayer resistance
r.h.	relative humidity
RI	refractive index
RT-PCR	reverse transcriptase polymerase chain reaction
S/V	surface to volume
SCL	soluble cuticular lipids
SCW	secondary cell wall
SE	sucrose ester
SEM	scanning electron microscopy
SFV	surface flux velocity
SOCO	sorption compartment
SOFU	simulation of foliar uptake
TCA	trichloroacetic acid
TEM	transmission electron microscopy
TI	tobacco introduction
TIC	total ion current
TLC	thin layer chromatography
TLS	trilaminar sheets
UDOS	unilateral desorption from the outer surface
UV	ultraviolet
v_d	deposition velocity
WUE	water-use efficiency

Preface

This book unites the invited reviews, case studies and methodological papers presented at a three-day Discussion Meeting held at the 1996 Annual Conference of the Society for Experimental Biology in Lancaster (UK). Abstracts of the majority of papers and posters presented at the meeting can be found in the *Journal of Experimental Botany* 47 (1996), supplement pp. 50–60 (abstracts P4.1–P4.56).

While the focus of the book is on the cuticle itself, other features of plant surfaces (trichomes and stomata in particular) have received attention if relevant to a particular function. Contributions from the fields of palaeobotany, ultrastructure, molecular biology, physiology and pathology of plants, insect ecology, air pollution research and micrometeorology cover a wide variety of cuticular functions. Some of these functions may be intrinsically linked to each other; for instance, the polymeric matrix of the cuticular membrane contains and supports a wide variety of soluble molecules which act as diffusion barriers and light reflectors, provide a water-repellent cover, impede fungal growth and stimulate feeding and oviposition by insects, to name but a few functions. As much of their efficacy depends not just on the presence of these molecules, but where and how they are situated in relation to each other, it is rather difficult to imagine that evolutionary pressures on certain functions have *not* also affected others. The result may have been a trade-off of one function's fulfilment against another's. Similarly, environmental impacts such as the increase in wettability of leaf surfaces caused by air pollutants are likely to affect all cuticular functions related to its hydrophobicity, such as droplet retention and stimulation of fungal growth. The resulting processes, pollutant deposition and development of an epiphyllous microflora, may themselves affect leaf wettability and create positive or negative feedbacks intersecting with each other.

It was the foremost objective of the meeting to start to explore connections between different processes occurring at the plant surface or within the cuticle. The book reflects this approach. A number of chapters are of particular relevance to plant production, covering the use of adjuvants to speed uptake of agrochemicals across cuticles, and breeding for resistance against pests and pathogens. The case studies describing cuticle-mediated multitrophic interactions between plants, fungal parasites, insect pests and their predators, although carried out with crop species, should also be of interest to ecologists. Last, but not least, the book is meant to remind those researchers (like the editor) whose main interest lies in the cuticle itself, just how excitingly rich is the palette of different functions fulfilled by this structure, and that we may not understand structure–function relationships unless we look at more than one function in our analyses.

The first three chapters in the book provide a framework for the discussion of cuticular functionality. They describe the evolutionary pressures acting on the surface design of the first land plants, and the structure, composition and biosynthesis of cuticles. The next five chapters deal with the cuticle as a barrier for the permeation of water and solutes (including agrochemicals) as well as for the penetration of light and fungi. Chapters 9–13 discuss the isolation and analysis of soluble cuticular compounds, and the role of these compounds as signals for microbes and insects. The next three chapters focus on foliar water films and droplets (whose presence depends strongly on cuticular properties) and their interactions with air pollutants, microbes and insects. A concluding paper takes up the question of natural damage to the cuticle and its ecological importance.

The nomenclature used in relation to the cuticle differs somewhat depending on whether the focus is on its ultrastructure and composition or its barrier or substrate properties, and whether concepts or experiments are concerned. As this may sometimes seem more than a little confusing, there follows a short list of the most important terms. *Cuticular membrane* (CM) is often used today as a synonym for *cuticle*, the region of the epidermis that can be stained with lipophilic dyes and/or isolated from it by dissolving the pectinaceous layer. Originally, 'cuticle' was a synonym for the *cuticle proper*, the thin polysaccharide-free region found at the surface of many plant organs, but is seldom used in this sense today. *Cuticular matrix* refers to the mixture of insoluble polymers (cutin, cutan, polysaccharides, etc.) that make up most of the CM. It is obtained by removing all soluble compounds from isolated cuticles. The term should not be confused with the expression *cutin matrix* which refers to the cutin polymer only.

The term *waxes* (or *soluble cuticular lipids*), in its widest sense, comprises all molecules that can be removed from the CM with organic solvents, but the occurrence of very large and hardly soluble, but not polymeric, molecules such as the estolides of conifers may blur the distinction between waxes and cuticular matrix. The terms *cuticular* and *epicuticular* waxes are often used practically as synonyms. It is usually not clear whether a certain kind of molecule, when extracted and analysed, was situated originally on or in the CM. Where the difference matters, as with regard to permeability, the conceptual distinction between *intracuticular* and *epicuticular* waxes may be made. In most experiments, however, no meaningful distinction between the two categories is possible (see Chapter 9), and *cuticular* waxes is used as a generic term.

In the general discussion that concluded the meeting, there was unanimous agreement that there is a problem referring to the conspicuous platelets, tubules, and so on, seen with the scanning electron microscope as *crystalline* waxes on plant surfaces, not because these waxes are not crystalline, but because their counterparts, the (potentially invisible) *amorphous* epicuticular waxes are very likely also to contain at least some crystalline domains. These should therefore not be called amorphous without qualification. However, the quest for more precise terms is still on!

Gerhard Kerstiens (*Lancaster*)

Acknowledgements

The meeting was generously supported by the Society for Experimental Biology (SEB), the British Ecological Society, BASF AG and Rhône-Poulenc Agriculture Ltd. Thanks are due to the SEB's Environmental Physiology, Plant Metabolism and Plant Development Groups for their sponsorship. I want to extend my sincere thanks to Bill Davies and Paul van Gardingen, without whose patient support and advice this project would never have been realized.

For my parents

Meinen Eltern gewidmet

1

Cuticles of early land plants: a palaeoecophysiological evaluation

D. Edwards, G.D. Abbott and J.A. Raven

1.1 Introduction

The major interest of cuticles for the palaeobotanist has traditionally lain in the information they carry on epidermal cells. Such features have proved reliable taxonomic characteristics, and are particularly important in identification of leaves which often show morphological convergence (e.g. Florin, 1938-1945; Thomas and Bancroft, 1913; for history see Stace, 1965) or in reuniting different organs of the same plant which had been shed in life or separated in traumatic pre-burial events (Oliver and Scott, 1905). Their high fossilization potential means that they are often the only anatomical features preserved in coalified compressions, and also occur as small fragments dispersed in the rock matrix. The latter have led to classifications based on dispersed cuticles (e.g. Roselt and Schneider, 1969) and their use in correlation (e.g. Oldham, 1976) and facies analysis (e.g. Batten *et al.*, 1985). Kerp (1990) touched on the potential of gymnosperm cuticles in palaeoecological and palaeoclimatological reconstruction (e.g. Buscher *et al.*, 1978). Evolutionary and adaptive physiological dimensions are added here particularly as they relate to the earliest land plants, and attention is focused on the construction and chemistry of the cuticle itself as well as epidermal characteristics. Following our summary of the latter (Edwards *et al.*, 1982), there have been some new cuticular descriptions (e.g. *Horneophyton*: Hass, 1991; *Hsüa*: Li, 1992; *Aglaophyton*: Remy and Hass, 1996), with their taxonomic value reinforced by the demonstration that Chinese Emsian fossils assigned to *Drepanophycus spinaeformis* on morphological evidence, have quite different stomatal complexes and have been placed in a new species (Li and Edwards, 1995). More relevant to this account are the detailed descriptions of isolated cuticles with well preserved stomata, which permitted inferences on the mechanics of their functioning (Edwards and Axe, 1992) and calculations on their frequency in relation to atmospheric CO_2 (McElwain and Chaloner, 1995).

In considering the functioning of these early cuticles, and, as a corollary, speculating on the selective pressures that led to their evolution we are faced with new challenges. Applying the maxim that the present is the key to the past it would seem safe to conclude that the presence of a layer, composed of exceedingly resilient non-hydrolysable macromolecules on the surface of the earliest vascular

Plant Cuticles, edited by G. Kerstiens.

plants, with guard cells anatomically indistinguishable from extant forms, would suggest that waterproofing and gaseous exchange were as important approximately 400 million years ago as they are today; the cuticle functioning as a compromise between the requirements of photosynthesis and desiccation avoidance. However, except for anatomical information we have limited evidence on cuticular chemistry (particularly the wax component), on controls on such functions, on the physical environment (including atmosphere and substrate) and on the actual habitats in which the plants grew. Indeed in some cases we have *only* cuticles, usually presumed to derive from non-vascular plants, and *no* information on any other tissues. Such is the evidence on which palaeoecophysiological speculation is based, and which will be briefly reviewed here.

1.2 The record (Figure 1.1)

The oldest illustrated cuticles were isolated by bulk maceration of Ordovician rocks of Libya (Gray *et al.*, 1982). They are in the form of astomatous sheets, anisomorphic in the sense that one surface is smooth (usually interpreted as the outer) while the other is covered by a reticulum of ridges or flanges (sometimes termed muri; e.g. Al-Ameri, 1983). Analogy with vascular plants suggests that these sheets covered the surface of a terrestrial plant with the reticulum marking the junctions between elements of superficial tissues (Figure 1.2). In some instances cellular sheets or sheets of cells are recorded (Pratt *et al.*, 1978). Astomatous cuticles are usually assigned to *Nematothallus* (see Edwards, 1982) erected by Lang (1937) for thallose organisms which he believed comprised a cuticle covering interweaving tubes of two sizes, the larger with unevenly thickened walls. *Nematothallus* is last recorded in the Emsian. Dispersed cuticles with similar internal reticulum and external ornament composed of papillae (*Cosmochlaina*: Edwards, 1986) first occur in the Upper Wenlock Cae Castell Formation (Burgess, 1987) and also persist into the Emsian (Figures 1.4–1.7). The earliest unequivocal cuticles thought to belong to higher plants (better termed rhyniophytoids) are also found dispersed, and derive from sporangia (Edwards and Burgess, 1990). Later examples (Figures 1.11 and 1.12) sometimes have adhering spores. Cuticles were not present on the very poorly preserved earliest specimens of *Cooksonia* (Upper Wenlock of Ireland: Edwards *et al.*, 1983) nor on the southern hemisphere Upper Silurian earliest record of the presumed lycophyte *Baragwanathia* preserved as impressions (Tims and Chambers, 1984). Dispersed cuticles, poorly preserved but lacking stomata, have been recovered from coeval sediments in Wales (Edwards, 1970), and stomata (and hence presumably cuticle) on the surfaces of sterile (see Figure 1.26) and fertile coalified axes from the Pridoli (latest Silurian) of the Welsh Borderland (Edwards, 1996). Similar sterile axes in the basal Devonian have yielded cuticles with stomata (see Figures 1.20–1.23) on boiling in hydrogen peroxide (Edwards and Axe, 1992) and stomatous cuticles have also been recovered from conventional coalified compressions of *Zosterophyllum myretonianum* (Lochkovian, Scotland: Lele and Walton, 1961). Superficial examination suggests that each stoma in *Zosterophyllum* consisted of a slit-like pore surrounded by a single ring-shaped guard cell (Chaloner and Sheerin, 1979), thus leading to comparisons with the non-functioning stomata of certain mosses (Paton and Pearce, 1957). However, the presence of T-shaped thickenings at the poles, and an occasional faint dividing line, as were observed by the original authors, confirm that two guard cells were present and were probably covered by a much thinner layer of cuticle than the surrounding epidermal cells. Descriptions of epidermal features in more recent plants were summarized in Edwards *et al.* (1982) and of stomata in Edwards and Axe (1992).

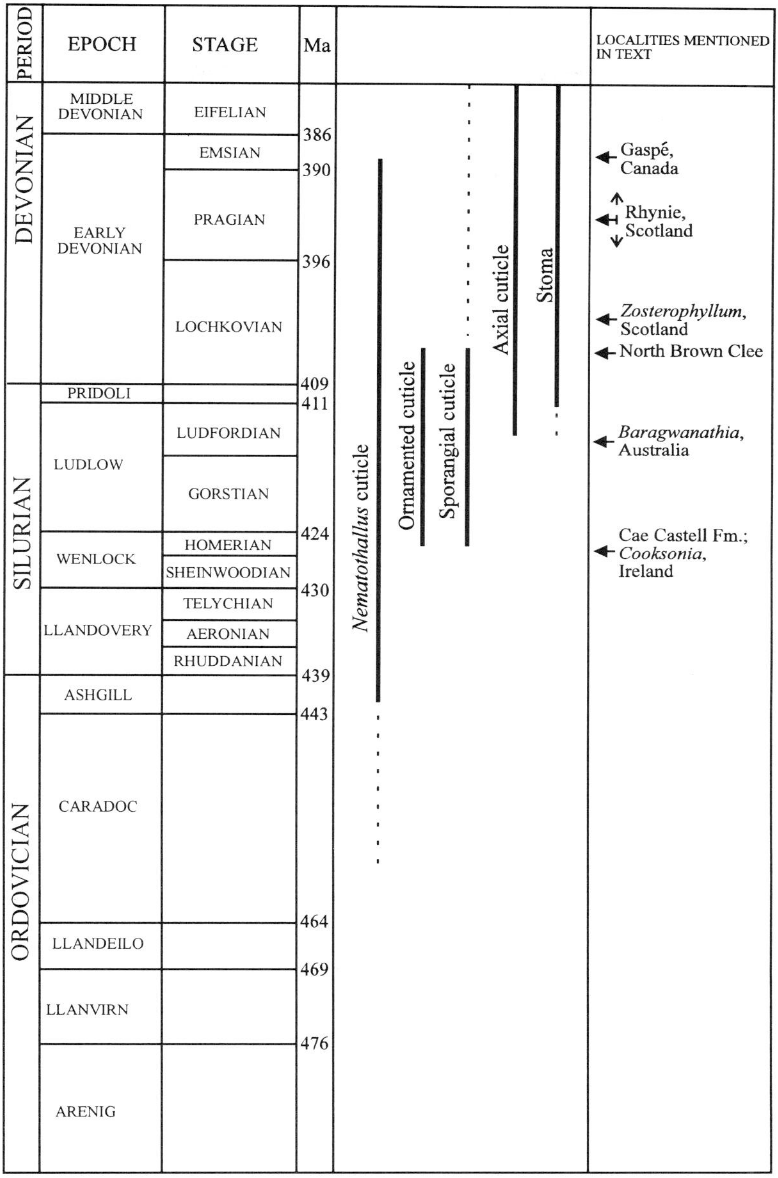

Figure 1.1. *Ordovician-Devonian stratigraphy and the records of early cuticles.*

1.3 Cuticular thickness and structure

Most information again derives from light microscopy of sections of the Rhynie Chert, with thickness values of < 2.5 µm for *Nothia* and < 5.0 µm for *Aglaophyton*. *Asteroxylon*'s aerial parts are reported to be covered by 'thick' cuticle. Rare

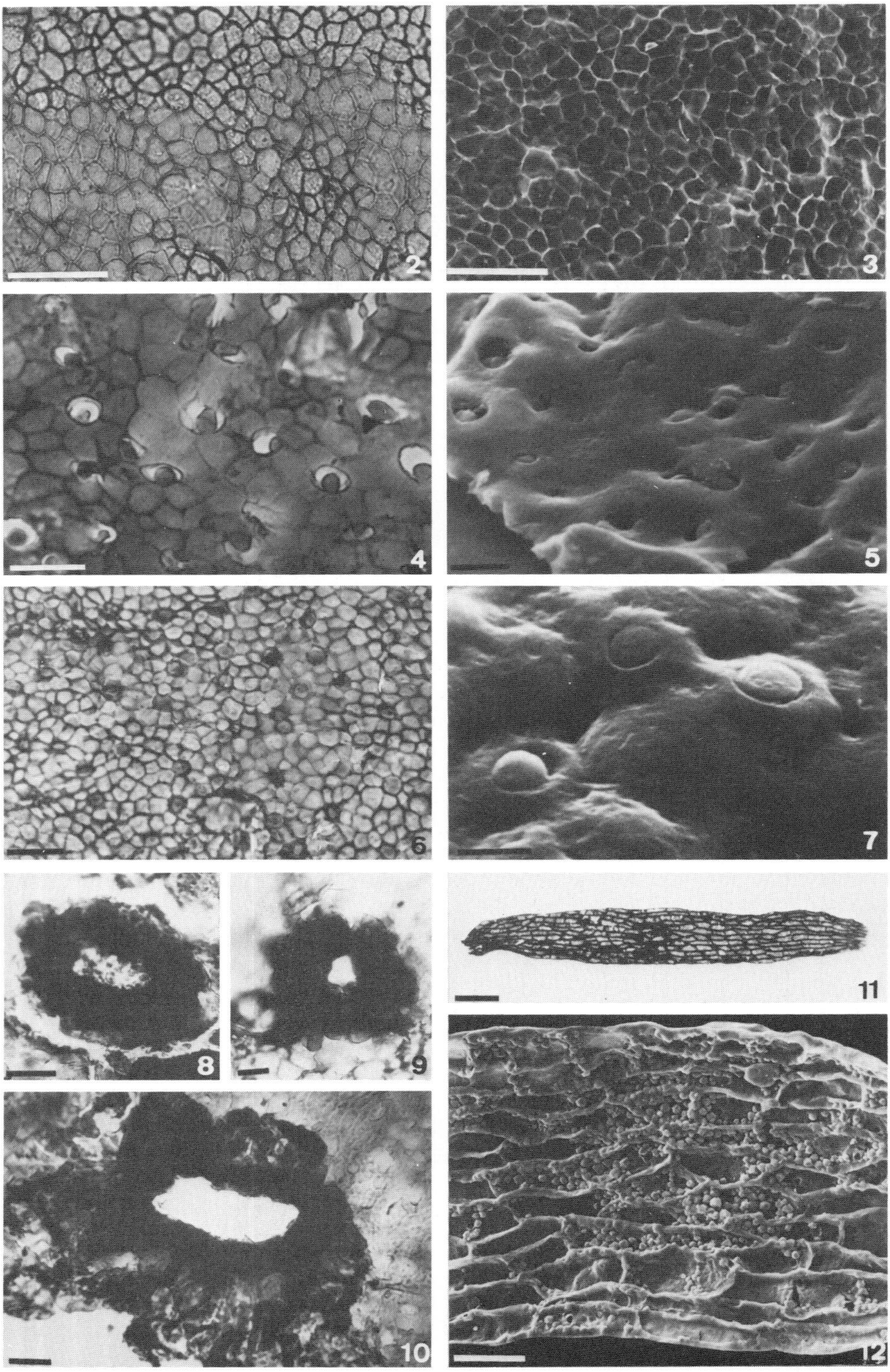
2
3
4
5
6
7
8
9
11
10
12

preparations of the outer or inner surfaces of *Rhynia gwynne-vaughanii* show prominent ridges (Remy and Hass, 1996, Pl. IV, 3 and 4), revealing layering with an opaque layer sandwiched between outer and inner translucent ones, the latter incorporating the outer periclinal wall of the epidermal cells. A similar layering is seen in *Aglaophyton major*, and additional thickenings are observed over the anticlinal walls comprising in section a central opaque rhomboidal band flanked by a thin, translucent layer on either side (Remy and Hass, 1996, Pl. IV, 1 and 2). Additional thickening in the form of papillae or ridges on the centres of epidermal cells in *Sawdonia ornata* present a lenticular outline in section (Rayner, 1983).

Apart from the gross layering in the Rhynie Chert plants, no further structural details are visible even in ultra-thin sections viewed by transmission electron microscopy (TEM) (work in progress in D.E.'s laboratory in Cardiff). In general, variation in cuticular thickness over the surface is evidenced by increased thickness on sporangia and thin or no cuticles on rhizomes. However, Remy and Hass (1996) report increased thickening, possibly by cutin or suberin, on cells adjacent to substomatal cavities, accompanied more proximally by thickened hypodermal cells. Figure 1.28 shows that in *Rhynia gwynne-vaughanii* a pronounced cuticle extends over the apex of an aerial axis, presumably an adaptation for the protection of the exposed meristem against desiccation, biophages, ultraviolet (UV) light, and so on.

1.4 Chemistry

1.4.1 *Higher plants*

Most information on plant history comes from fossils preserved as coalified compressions (*sensu* Schopf, 1975) in which the original tissues are converted into a film of coal. In such fossils original structure is obliterated except for rare occurrences of remains of lignified tissues and more common, but not ubiquitous,

---

Figures 1.2 and 1.3. *LM and SEM of* Nematothallus *cuticles. Lochkovian, North Brown Clee, Welsh Borderland. NMW 96.5G.1. (Scale bars = 50 µm.) (1.2) Note undulating surface. (1.3) Inside surface showing flanges.*
Figures 1.4 and 1.5. *LM and SEM of* Cosmochlaina verrucosa. *Lochkovian, M50, Hereford. NMW 85.20G.1. (Scale bars = 20 µm.) (1.4) Papillae in thinner areas of cuticle often absent. (1.5) Outer surface. (Reproduced from Edwards (1986) Dispersed cuticles of putative non-vascular plants from the Lower Devonian of Britain.* Bot. J. Linn. Soc. ***93****, 259–275, with permission from Academic Press.)*
Figures 1.6 and 1.7. *LM and SEM of* Cosmochlaina maculata. *Lochkovian, M50, Hereford. NMW 85.20G.6. (1.6) Scale bar = 50 µm. (1.7) Outer surface. (Scale bar = 10 µm.) (Reproduced from Edwards (1986) Dispersed cuticles of putative non-vascular plants from the Lower Devonian of Britain.* Bot. J. Linn. Soc. ***93****, 259–275, with permission from Academic Press.)*
Figures 1.8–1.10. *Apertures in* Nematothallus *cuticles by cuticular thickenings. (1.8) NMW 82.33G.18. (Scale bar = 10 µm.) (1.9) NMW 82.33G.14. (Scale bar = 10 µm.) (1.10) NMW 82.33G.17b. (Scale bar = 20 µm.) (Reproduced from Edwards and Rose (1984) Cuticles of* Nematothallus*: a further enigma.* Bot. J. Linn. Soc. ***88****, 35–54, with permission from Academic Press.)*
Figures 1.11 and 1.12. *LM and SEM of sporangial cuticle. Lochkovian, North Brown Clee, Welsh Borderland. NMW 96.5G.2. (1.11) Complete elongate ? valve. (Scale bar = 200 µm.) (1.12) Part of inner surface with entire margins. Granules are pyrite crystals. (Scale bar = 50 µm.)*

occurrences of cuticles. The latter retain their morphological integrity, and indeed, as has been demonstrated earlier, apart from their colour look remarkably similar to cuticle preparations from extant plants. A prerequisite for this preservation is the original presence of molecules that are resistant to chemical and biological degradation (Tegelaar *et al.*, 1991), macromolecules that might be expected to persist in the fossils themselves. Recent research, particularly in the Netherlands, using a variety of pyrolysis, chemical degradation and spectroscopic methods has concentrated on the isolation and chemical characterization of cuticles in fossils and subfossils with the application of similar techniques to extant plants (e.g. Tegelaar *et al.,* 1993). This section will summarize progress on the chemical characterization of macromolecules from fossil plant cuticle, in which cutin is only very rarely detectable (Tegelaar *et al.*, 1993). Cutin is the most abundant biopolyester that forms the structural framework of the cuticle in the living plant (see Chapter 2). This insoluble polymeric material can be depolymerized during fossilization by hydrolysis of the ester linkages which hold the monomer units together. During deposition and subsequent burial, depolymerization is probably catalysed by cutin-hydrolysing enzyme (or enzymes) excreted by microbial organisms. Cutinase has been purified from the extracellular fluid of several species of cutin-grown phytopathogenic fungi (Kolattukudy, 1981; and see Chapters 3 and 8).

Analysis of extant and fossil cuticles (gymnosperm and angiosperm) using flash pyrolysis-gas chromatography–mass spectrometry (Py-GC–MS) and solid-state ^{13}C nuclear magnetic resonance (NMR) as well as Fourier transform infrared (FTIR) spectroscopy has shown that, in addition to cutin, an insoluble, non-hydrolysable highly aliphatic polymer named cutan may also be present (Nip *et al.*, 1986a,b, 1987; see Chapter 2). The major pyrolysis products from cutan include homologous series of *n*-alkanes, *n*-alk-1-enes and *n*-α,ω-alkadienes ranging from C_6 to C_{35} (in contrast to cutin-derived products which are dominated by C_{16} alkanoic, alkenoic and alkadienoic acids; Tegelaar *et al.*, 1991). The cutan-derived products suggest that this polymer is composed of polymethylenic chains (Nip *et al.*, 1986a). The solid-state ^{13}C NMR and FTIR spectra of the cuticular residue, before and after acid treatment, from extant *Agave americana* L. confirm the presence of polymethylenic chains (Tegelaar *et al.*, 1989a). Limited surveys suggest that most extant cuticles contain a mixture of cutin and cutan but some contain one or the other (Tegelaar *et al.*, 1991). Flash pyrolysates from the fossil representatives of some of the extant taxa suggest that cutin is readily degraded (found in significant amounts only in Upper Miocene *Sciadopitys tertiaria*). Instead homologous series of *n*-alkanes, *n*-alk-1-enes and *n*-α,ω-alkadienes ranging from C_6 to C_{35} dominate the pyrolysates of most of the fossil higher plant leaf remains analysed (Tegelaar *et al.*, 1991). There is also evidence for a 'modified mixed' fossil cuticular matrix, for example in the Eocene *Symplocos hallensis*, which is made up of cutan and cutin-derived material with non-saponifiable intermolecular linkages — the cutin-derived material could either have been biosynthesized by the plant itself or be the result of diagenetically induced reactions of reactive groups, such as epoxide, originally present in the cutin (Tegelaar *et al.*, 1993). The fossil cuticle pyrolysates demonstrate the differing responses of the original cuticular components, cutin and cutan, to diagenesis, leading to the conclusions that the possession of cutan would lead to greater chance of preservation in the fossil record, to the overrepresentation of such taxa (Tegelaar *et al.*, 1991), and that cutans are a likely source for the aliphatic components in crude oils (*n*-alkanes; Tegelaar *et al.*, 1989b) and kerogens and coals (*n*-alkyl moieties; Nip *et al.*, 1989).

Each of the flash pyrolysates (G. Ewbank, D. Edwards and G. Abbott, submitted) of coalified axes from selected Lower Devonian vascular plants yields a series of *n*-alkane/*n*-alkene doublets (Figure 1.13). These series show a maximum between *n*-C_{10} and *n*-C_{14} extending to *n*-C_{27} in *Psilophyton dawsonii* (Emsian; Gaspé Bay, Quebec) and to *n*-C_{23} in the slightly older *Zosterophyllum myretonianum* (Lochkovian; Scotland). A similar distribution pattern of *n*-alkanes and *n*-alkenes is observed in the pyrolysate of isolated cuticles of *Sawdonia ornata* (Emsian; Gaspé). All three of the above Py–GC–MS traces suggest that the aliphatic components derive from cutan, there being no evidence at all for cutin monomers. The partial total ion current (TIC) trace from the pyrolysate of Lower Devonian (Lochkovian) representatives of the enigmatic cuticle *Cosmochlaina* (Figure 1.13d) shows a series of *n*-alk-1-enes and *n*-alkanes, with a maximum at C_{15}, ranging from C_8 to C_{25}. C_0–C_1 alkylphenols, C_1–C_2 alkylbenzenes are also prominent pyrolysis products in this trace. The aromatic components may derive from adhering coalified remains of underlying tissue. In *Nematothallus*, the *n*-alkane/*n*-alkene doublets are absent and instead the TIC trace of its pyrolysate is dominated by C_0–C_2 alkylphenols, C_1–C_3 alkylbenzenes and C_0–C_2 alkylnaphthalenes (Figure 1.13e). This implies that the parent plant was probably not closely related to the higher plants, and that in not being chemically homologous may have had different functions (see Section 1.8 on affinities of *Nematothallus*).

1.4.2 *Algae*

Cuticle-orientated deliberations on the ancestry of the higher land plants have centred on attempts to demonstrate the presence of cuticle in green algae, or at least the ability of these algae to synthesize precursors of the cuticular polymers. It must be emphasized that the glycoprotein 'cuticle' of many green, red and brown macroalgae (Craigie *et al.,* 1992; Hanic and Craigie, 1969) is not a cuticle in the higher plant sense. Indeed, certain vascular plants have structures like algal 'cuticles' in their cell walls underneath their *real* cuticle; these structures, like their algal analogues, can have optical effects via thin film interference leading to (usually blue) iridescence (Craigie *et al.*, 1992; Lee, 1991; Vogelmann, 1993).

The synthesis of long polymethylenic chains has recently been shown in *Chlorella* ($< C_{27}$) and *Nanochlorum* (or, more properly, *Nannochloris*) ($< C_{30}$) by pyrolysis, while FTIR and solid state ^{13}C NMR indicate a low degree of branching and large concentrations of polymethylenic chains (Derenne *et al.*, 1992). These polymers were isolated from the outer resilient walls which remain in lyophilized cells and which ultrastructurally have a trilaminar nature (trilaminar sheets (TLS); Atkinson *et al.,* 1972). It is noteworthy, however, that a recent structural characterization of the insoluble residues, obtained by basic and acid hydrolyses, of five common marine green microalgae showed the *Nannochloris* sp. yielded a flash pyrolysate dominated by alkylbenzenes and alkylphenols (Gelin *et al,* 1996). Similar non-hydrolysable polymers termed algaenans have been isolated from the outermost layer of the wall surrounding individual cells of *Scenedesmus* (Derenne *et al.*, 1991) and other unicellular Chlorophyceae (de Leeuw and Largeau, 1993). Such acetolysis-resistant walls with chemical properties similar to those of carotenoid derivatives led to the assumption that sporopollenin or a sporopollenin-like material was present (Burczyk, 1987a,b), but while secondary carotenoids are present in the cytoplasm of *some* of these algae, they are not present in their walls, and hence not part of the non-hydrolysable constituents of chlorophycean TLS (see summary in Derenne *et al.*, 1992). Although it is tempting to consider the outermost lamellate

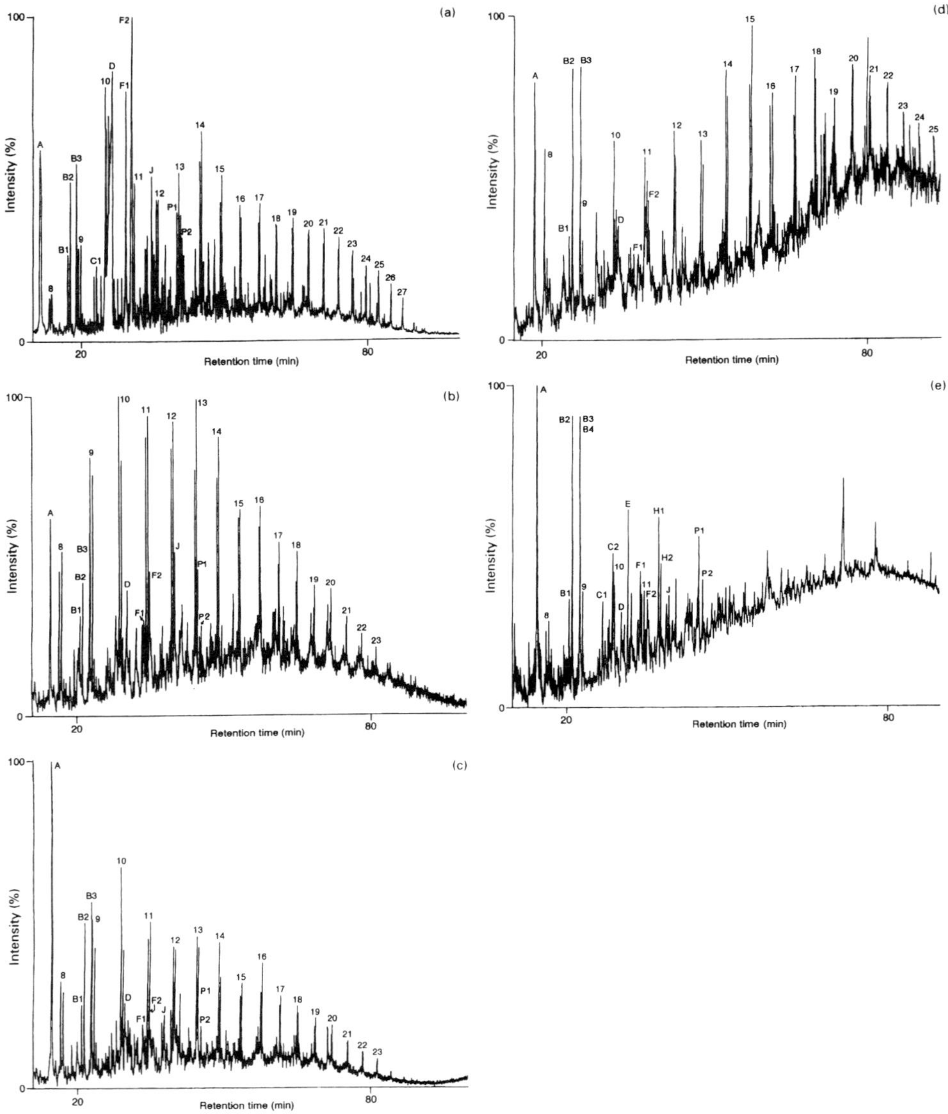

Figure 1.13. *(a) Partial trace for the total ion current (TIC) of* Psilophyton dawsonii *(Emsian; Gaspé Bay, Canada) flash pyrolysate. (b) Partial trace for the TIC of* Zosterophyllum myretonianum *(Lochkovian; Westhall Terrace, Scotland) flash pyrolysate. (c) Partial trace for the TIC of isolated cuticles of* Sawdonia ornata *(Emsian; Gaspé Bay, Canada) flash pyrolysate. (d) Partial trace for the TIC of* Cosmochlaina versiformis *(Lochkovian; North Brown Clee, Welsh Borderland) flash pyrolysate. (e) Partial trace for the TIC of* Nematothallus *(Lochkovian; North Brown Clee, Welsh Borderland) flash pyrolysate. (Numbers indicate total number of carbon atoms for* n*-alk-1-enes and* n*-alkanes.) See Table 1.1 for peak assignments.*

Table 1.1. *Products identified in flash pyrolysates of Lower Devonian plants shown in Figure 1.13*

Peak	Assignment	Abbreviation
A	Toluene	C_1 alkylbenzene
B1	Ethylbenzene	C_2 alkylbenzene
B2	*(m+p)*-xylenes	C_2 alkylbenzene
B3	*o*-xylene	C_2 alkylbenzene
B4	Ethenylbenzene	
C1	C_2-toluene	C_3 alkylbenzene
C2	C_2-toluene	C_3 alkylbenzene
C3	C_2-toluene	C_3 alkylbenzene
D	Phenol	C_0 alkylphenol
E	Indene	C_0 alkylindene
F1	*o*-cresol	C_1 alkylphenol
F2	*(m+p)*-cresols	C_1 alkylphenol
G	2-methoxyphenol (guaiacol)	
H1	Methylindene	C_1 alkylindene
H2	Methylindene	C_1 alkylindene
I1	C_2-phenol	C_2 alkylphenol
I2	C_2-phenol	C_2 alkylphenol
I3	C_2-phenol	C_2 alkylphenol
J	Naphthalene	C_0 alkylnaphthalene
K	4-methyl-2-methoxyphenol (methylguaiacol)	
L	1,2-dihydroxybenzene (catechol)	
M	4-vinylphenol	
N1	C_3-phenol	C_3 alkylphenol
N2	C_3-phenol	C_3 alkylphenol
O	4-ethyl-2-methoxyphenol (guaiacylethane)	
P1	2-metylnaphthalene	C_1 alkylnaphthalene
P2	1-methylnaphthalene	C_1 alkylnaphthalene
Q1	3-methyl-1,2-dihydroxybenzene (3-methylcatechol)	
Q2	4-methyl-1,2-dihydroxybenzene (4-methylcatechol)	
R	4-vinyl-2-methoxyphenol (*p*-vinylguaiacol or guaiacylethene)	
S	2,6-dimethoxyphenol (syringol)	
T	Biphenyl	
U1	Prist-1-ene	
U2	Prist-2-ene	
V	Dibenzofuran	
W	Fluorene	
X1	Phenanthrene	
X2	Anthracene	
Y	Fluoranthene	

layer of the wall of the unicellular chlorophycean algae as homologous (both ultrastructurally and chemically, cf. cutan) with the vascular plant cuticle, it is lacking in multicellular Chlorophyceae and in members of the Charophyta. A positional analogue is the thin superficial layer on the dorsal surface of *Coleochaete* thalli (Graham and Delwiche, 1992; see Figure 1.29). This separates from the thallus following mild acid treatment, shows no reaction to Sudan IV and is partially degraded by putative bacterial cutinase. Its chemical composition is unknown and a TLS has not been demonstrated (see Figure 1.29). However, it is of some interest that members of the group which appear closest to the embryophytes have the ability to synthesize an extracellular layer. Indeed, Graham (1993) speculates that it might function in a similar way to the wax-free cuticles of certain bryophytes in allowing movement of water and carbon dioxide. Wax-free cuticles may have permeability coefficients for CO_2 as high as 7.7 μm s^{-1} (artificially dewaxed cuticles; MacFarlane, 1992), while the plant plasmalemma can have values as low as 15 μm s^{-1} (Sültemeyer and Rinast, 1996).

1.5 Palaeozoic atmosphere

CO_2 and O_2 levels in the Palaeozoic atmosphere are given in Figure 1.14. The model for estimates of CO_2 (Berner, 1993, 1994) takes into account changes in solar radiation and the advent of substantial land vegetation as they affect the rate of weathering of silicates (and uptake of CO_2) in addition to tectonic events and burial of organic material. Thus the graph shows high concentrations of CO_2 in the early Palaeozoic, with maximum (c.17 times PAL (present atmospheric level)) in the Ordovician and with values of c. 12 times PAL at the time of the advent of vascular plants. The oxygen curve from Berner and Canfield (1989) was derived from estimated rates of burial and weathering of organic carbon and pyrite, and has not yet been adjusted for solar radiation. The values for the Ordovician and Silurian are c. 8% below PAL, with a decrease c. 400 million years ago, due to red beds, followed by increase attributed to productivity of land plants. Such levels are sufficiently high to allow the development of an ozone layer from early Palaeozoic times (Levine *et al.,* 1980; Margulis *et al.*, 1976; Wayne, 1992), although Holland (1984) cautions that such calculations omitted effects of tropospheric photochemical or chemical processes. Carver (1980) emphasizes the non-linear relationship between total atmospheric O_2 and total atmospheric O_3. Mora *et al.* (1996) have independently estimated the CO_2 content of the Late Palaeozoic atmosphere using the $^{13}C/^{12}C$ ratio of soil carbonates and organic matter. The agreement with the estimates of Berner (1993, 1994; Figure 1.14) is generally good, although they suggest lower levels (about 3–5 times the present) in the Mid-Upper Devonian. Furthermore, the possibility of stratospheric destruction of O_3 involving photochemical reactions of halocarbons derived from marine biota (Raven, 1995a) should not be ignored. Also relevant to considerations here are the implications for net CO_2 fixation by Rubisco of the atmospheric CO_2:O_2 ratio (Raven, 1993). The absolute CO_2 concentrations, and the CO_2:O_2 ratio, were higher than the present values throughout the Palaeozoic, with the exception of the Upper Carboniferous–Lower Permian. CO_2 limitation of C_3 photosynthesis, and the occurrence of photorespiration, would only have been likely in the Upper Carboniferous–Lower Permian.

1.6 Palaeoenvironments

Information on substrate is essential for any phytoecophysiological analysis, but direct evidence for plants in growth positions is rare in the Siluro-Devonian. The putative lycophyte, *Drepanophycus,* has adventitious roots (the oldest unequivocal record) which ramify through fluvial sediments implying growth on a flood plain (Rayner, 1984: Emsian, Scotland), although nothing is known of the organic component of the substrate. Cores through the autochthonous Rhynie Chert give information on nature of soils and plant succession (Powell, 1994). Soil horizons vary in composition and thickness ranging from those containing little organic material to accumulations of abundant plant debris with scattered detrital grains and

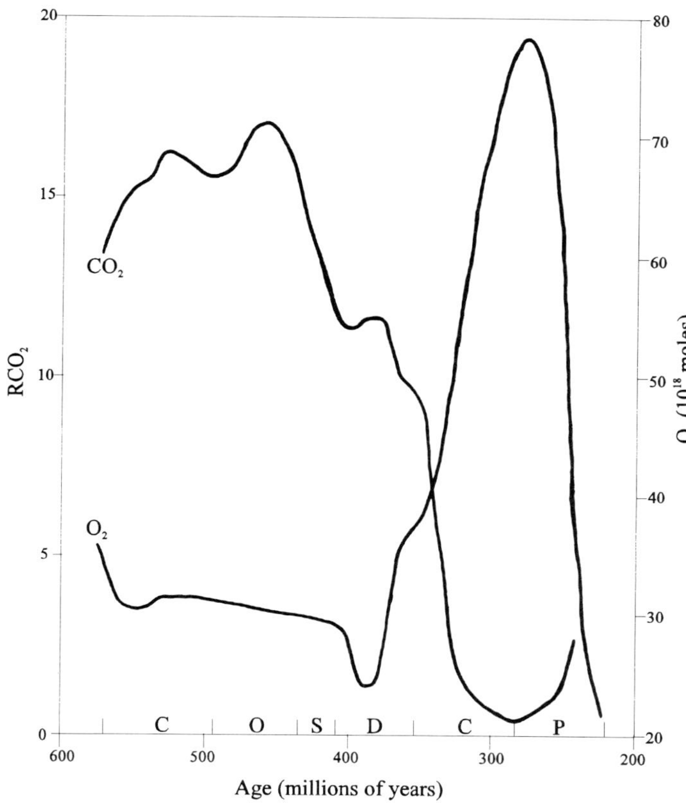

Figure 1.14. *Atmospheric CO_2 (based on Berner, 1993) and O_2 (based on Berner and Canfield, 1989) in the Palaeozoic. RCO_2 is the mass of CO_2 in a past atmosphere divided by that at present. O_2 values represent total amount of atmospheric oxygen.*

evidence (from basal compaction) of accumulation *in situ*. The mineral material may be dominated by sands or sinter with small amounts of clay, but some allochthonous plant assemblages contain high clay content. Plants are recorded in growth on all substrates with some evidence that *Rhynia gwynne-vaughanii* was a pioneer of 'open' mineral surfaces, both sinter and sandy, produced respectively by overflowing from pools fed by the hot springs or overbank flooding on the alluvial plains. In the case of the former, analogy with extant hot springs suggests that the water would be alkaline with higher concentrations of major elements than meteoric water (e.g. calcite is sometimes deposited with chert) thus causing problems of water uptake for the plants. However, there is unequivocal evidence for regrowth after flooding, suggesting that the plants were capable of surviving such water stress and Remy and Hass (1996) provided anatomical information that the sporophytes of *Aglaophyton* were capable of regeneration. The plants themselves lack the peripheral thick-walled tissues characteristic of coeval plants preserved in clastic sediments and relied on turgor for support. This might indicate a constant supply of water, but raises the possibility that tissues were analogous to the succulence seen in certain salt-marsh plants. Remy and Hass (1996) reported extensive thickening, possibly by cutin or suberin, in cells adjacent to substomatal cavities in the basal parts of *Aglaophyton,* which they suggested were adaptations to flooding episodes and/or fungal attacks. They also described more extensive thickenings in walls of hypodermal cells proximal to this region which they thought might have had a supporting function. Evidence for waterlogging is equivocal. Disseminated pyrite in chert and 'peaty' layers suggests anoxic environments at, or immediately after, chert formation.

More general information on habitats is inferred from combined sedimentological and community (as reflected in assemblages) studies on fragments of the aerial parts of plants, broken up and transported in major flooding events and then buried in fluvial or marine sediments. The majority of the plants probably colonized areas close to waterways, for example river banks, levées, mudflats, but some have been traced to more marginal habitats. Thus, for example, Schweitzer (1983) reconstructed a range of coastal-plant communities including sub- and intertidal, saltmarsh and marshland less influenced by the sea in Early Devonian times in Germany. The latter habitat was also postulated for the extensive assemblages of the Emsian of Gaspé, Canada (Andrews *et al.*, 1977). Coastal areas, possibly influenced by the sea, were also postulated as habitats for the earliest axial land plants (see summary in Edwards and Selden, 1993).

Information on the nature of the soils themselves comes from inferences based on modern analogues and direct evidence from palaeosols. Before the advent of the vascular plants with their enhanced productivity and more extensive subterranean parts, terrestrial substrates would have been characterized by limited mineral nutrient resources, low organic content and high physical stresses (for example, erosion, run through, temperature extremes; e.g. Beerbower, 1985; Gray and Boucot, 1977, 1994). While it is now appreciated that 'biologically active' microbial soils had probably existed since Precambrian times (e.g. Campbell, 1979; Retallack, 1990; Retallack and Mindszenty, 1994) neither they nor those under Ordovician–Silurian bryophyte?/lichen? dominated communities would have had the chemical and physical impact of post Silurian vegetation. The extent of the latter was semi-quantified in Berner's recent calculations on atmospheric CO_2, which attributed much of the decline in the Devonian to enhanced weathering of silicate (Berner, 1994).

Retallack, based on detailed observations of palaeosols from the Late Ordovician and Silurian (Retallack, 1985, 1990), concluded that they represented highly oxidized, well drained, sub-humid, sub-tropical soils with ferruginized caliche concretions indicative of a dry season, the older comparable to the moss and lichen cover of desert or alpine boulders today and the younger to modern desert vegetation or alpine or open woodland. For the most part, they were warm, water-stressed, hostile environments, as were the calcrete palaeosols of the Lower Devonian which Allen (1986), from analogy with extant examples, concluded developed in a mean annual temperature of 16–20°C with low but marked seasonal rainfall (100–500 mm). Intuitively these would not seem to be the habitats favouring the growth of homosporous pteridophytes, but recent discoveries in the Rhynie Chert of gametophytes of *Aglaophyton* (*Lyonophyton*), *Horneophyton* (*Langiophyton*) and *Nothia* (*Kidstonophyton*) reveal that these early land plants had isomorphic, diplobiontic life cycles with axial gametophytes which were presumably homoiohydric in view of their possession of cuticles, stomata and water-conducting tissues at maturity (Kenrick, 1994; Remy and Hass, 1991).

However, the earliest stages of their development (germinating spores) apparently occurred in wet or shallow aquatic environments (Remy and Hass, 1996).

1.7 Functions of cuticles in early land plants

From the occurrence of stomata in Silurian and Devonian cuticles, it may be inferred that the cuticle was acting as a barrier to gaseous diffusion even in the earliest vascular plants and hence had a desiccation resistance function. However, its efficacy in this latter role is conjectural, it being unwise to base inferences on cuticle thickness in the absence of constant relationship of the latter with diffusion resistance (thicker cuticles can even have higher permeabilities to water; Becker *et al.,* 1986) and we have no information on presence or distribution of waxes. Indeed the gross morphology of early vascular plants, microphyllous (e.g. *Drepanophycus*) or axial (comparable to stems or petioles), is so different from the seed plants, and particularly broad-leaved angiosperms, on which much experimentation has been undertaken, that the present does not necessarily provide the key to unlock the past in considering functions relating to photosynthesis, transpiration, UV protection, wettability and 'stretchability'.

1.7.1 *The photosynthetic dilemma*

The guard cells themselves are of familiar shape in surface view often with poral thickenings and are shallow compared with the epidermal cells. From the differential colouring of the guard cells and epidermal cells in both permineralized (silicified) Rhynie Chert fossils (Figure 1.25) and certain isolated cuticles, it is concluded that the guard cells contained chloroplasts and thus physiologically functioned as today. Based on distribution of cuticle on guard and surrounding epidermal cells in certain Lower Devonian axial fossils, Edwards and Axe (1992) postulated that opening of the pore followed vaulting outwards of the outer periclinal walls of the guard cells, deformation of the remaining non-poral walls being hampered by cuticularization and the extension of the surrounding epidermal cells below the guard cells (Figure 1.22). Polar thickenings restricted change in length (Figure 1.21). The outer periclinal walls were considered less cuticularized because they often break down during preparation/oxidation, and are often absent from coalified axes (Figures 1.23 and 1.26). It is unlikely that the thinner cuticle on the outer periclinal walls of guard cells would necessarily have aided the sensing of humidity and of gaseous CO_2 since the chemistry and spatial distribution of wax rather than cuticular thickness controls cuticular permeability. A similar opening mechanism is recorded in certain extant ferns (e.g. *Diplazium*; von Guttenberg, 1971) and mosses (Ziegler, 1987), although pronounced cuticular thickenings are not seen in the living plants. That controls on opening and closure were the same seems unlikely in view of the coeval higher atmospheric concentrations of CO_2 (Berner, 1994).

Stomatal frequency and distribution. Reports that stomatal density in early land plants was low (e.g. Edwards, 1993) have recently been confirmed quantitatively by the analyses of McElwain and Chaloner (1995; their Table 1) of *Aglaophyton major* (Pragian?; Rhynie Chert) and *Sawdonia ornata* (Emsian borehole; southern Britain), who related low values (4.5 and 4.3 mm^{-2}) to high atmospheric CO_2 concentrations in the Lower Devonian. Li (1982) reported a similar density

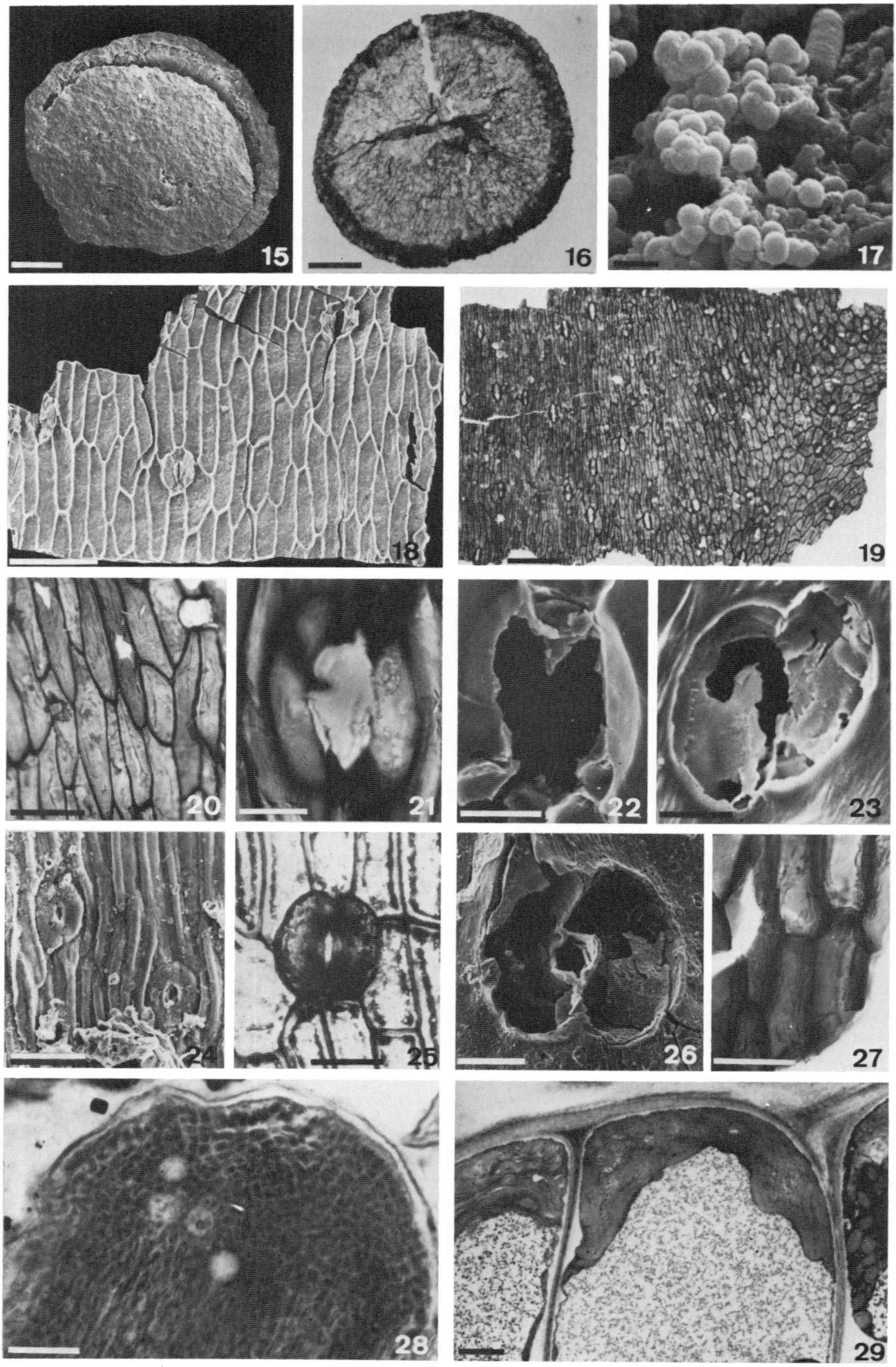
15
16
17
18
19
20
21
22
23
24
25
26
27
28
29

(5 mm^{-2}) in *Hsüa robusta*, but Stubblefield and Banks (1978) calculated 33 mm^{-2} for the stems of *Drepanophycus spinaeformis*. Long-ranging and geographically widespread taxa such as *D. spinaeformis* (Lochkovian–Frasnian) and *Sawdonia* (Emsian–Frasnian) which often occur abundantly in assemblages, offer opportunities for study into intraspecific variation and changes in abundance in relation to the drop of CO_2 concentrations through the Devonian. (Indeed it is not

---

Figures 1.15 and 1.16. *SEM and LM of sporangial cuticle. Pridoli, Ludford Lane, Shropshire. (1.15) Detaching ± circular distal lid. NMW 94.60G.5.(Scale bar = 200 µm.) (1.16) Lid missing. NMW 94.60G.4. (Scale bar = 200 µm.) (Reproduced from Edwards (1996) New insights into early land ecosystems: a glimpse of a Lilliputian world.* Rev. Palaeobot. Palynol. ***90,*** *159–174, with permission from Elsevier Science - NL.)*
Figure 1.17. *Modern bacterial contaminant of sporangium of* Resilitheca. *NMW 94.60G.12. (Scale bar = 1 µm.)*
Figures 1.18 and 1.19. *SEM and LM of cuticle of* Drepanophycus qujingensis *Emsian, Yunnan Province, China. (1.18) Inner surface of cuticle with occasional stomata. CBYn9001 stub 9. (Scale bar = 100 µm.) (1.19) Fragment showing variation in epidermal cell shape. CBYn9001 sl. 2. (Scale bar = 200 µm.) (Reproduced from Li and Edwards (1995) A re-investigation of Halle's* Drepanophycus spinaeformis *Göpp. from the Lower Devonian of Yunnan Province, southern China.* Bot. J. Linn. Soc. ***118,*** *163–192, with permission from Academic Press.)*
Figures 1.20–1.23. *LM and SEM of indeterminate cuticles of axial plants. Lochkovian, Welsh Borderland. (1.20) Typical fragment with single stoma (top right). NMW 90.43G.9. (Scale bar = 40 µm). (1.21) Single stoma showing thickenings at poles and 'cradling epidermal' cells. NMW 90.43G.4. (Scale bar = 10 µm.) (1.22) Inner surface below guard cells showing cradling epidermal cells. NMW 90.43G.8. (Scale bar = 10 µm.) (1.23) Surface view of typical stoma with outer periclinal walls of guard cells absent. NMW 90.43G.7. (Scale bar = 10 µm.) (Reproduced from Edwards and Axe (1992) Stomata and mechanics of stomatal functioning in some early land plants.* Cour. Forsch.-Inst. Senckenberg ***147,*** *59–73, with permission from Senckenbergische Naturforschende Gesellschaft.)*
Figure 1.24. *SEM of atypical example with two stomata in close proximity on axis of* ?Sporogonites. *NMW 96.5G.3. (Scale bar = 50 µm.)*
Figure 1.25. *Paradermal section through* Aglaophyton major *showing densely staining guard cells. ?Siegenian, Rhynie Chert, Aberdeen. AGL.EP.1. (Scale bar = 50 µm.) (Reproduced from Edwards (1993) Tansley Review No. 53. Cells and tissues in the vegetative sporophytes of early land plants.* New Phytol. ***125,*** *225–247, with permission from the* New Phytologist.*)*
Figure 1.26. *SEM of Silurian stoma on sterile axis with outer periclinal walls of guard cells absent. Pridoli, Ludford Lane. NMW 96.5G.4. (Scale bar = 10 µm.) (Reproduced from Edwards (1993) Tansley Review No. 53. Cells and tissues in the vegetative sporophytes of early land plants.* New Phytol. ***125,*** *225–247, with permission from the* New Phytologist.*)*
Figure 1.27. *LM fragment of thick-walled cuticle from* Tortilicaulis *sporangium. Lochkovian, Welsh Borderland. NMW 96.5G.5. (Scale bar = 5 µm.)*
Figure 1.28. *LS apex of* Rhynia gwynne-vaughanii. *Siegenian, Rhynie Chert, Aberdeen. Note covering of slightly detached cuticle. No. 2425. (Scale bar = 100 µm.) (Reproduced from Edwards (1994) Towards an understanding of pattern and process in the growth of early vascular plants. In:* Shape and Form in Plants and Fungi, Linn. Soc. Symp. Ser. No. 16 *(eds Ingram and Hudson). Academic Press, London, pp. 39–59, with permission from Academic Press.)*
Figure 1.29. *TS thallus of* Coleochaete. *(Negative courtesy of Prof. Linda Graham. Scale bar = 2 µm.)*

clear whether the Stubblefield and Banks (1978) high value refers to Upper Devonian examples.) However, such studies require accurate identification. C.-S. Li and D. Edwards are currently studying a possible lycophyte from the Emsian of China which is almost indistinguishable from *D. spinaeformis*, and was so originally identified, but differs in epidermal configurations relating to stomata. Preliminary studies show relatively high frequencies on the stems, and low on the leaves, comparable to coeval *D. spinaeformis* values. The cuticle shows two types of epidermal patterning: elongate cells in which stomatal frequency was 3–21 mm^{-2} and rarer shorter types where frequency was 10–13 mm^{-2} (Figures 1.18 and 1.19). Variation in stomatal indices suggests that the latter relate to variation in stomatal differentiation rather than lack of expansion of surrounding epidermal cells, but far more measurements are required (Li and Edwards, 1995). On-going quantification of sterile coalified axes of small dimensions and cuticles in D.E.'s laboratory in Cardiff (0.5–1.7 mm^2) reveal that most are astomatous, some possess a single stoma, while a few show clustering of two or more stomata (maximum five). Only rarely are stomata scattered over the whole surface. An unusual Silurian example possesses 13 stomata on an axis 4 mm long. Stomatal indices cannot be calculated, because epidermal cell outlines cannot be detected on the coalified surfaces. Our studies also show that stomata are frequently associated with sporangia either on the sporangium wall (Figures 1.30 and 1.33) or on the subtending axes immediately below (Figures 1.31, 1.35, 1.37 and 1.38). The bivalved specimen illustrated in Figure 1.30 shows a proximal, oblique band, while a similarly organized and also as

---

Figures 1.30–1.39. *SEMs of sporangia. Lochkovian, North Brown Clee, Welsh Borderland.*

Figure 1.30. *Unnamed reniform sporangium with oblique band of stomata (marked by spots). NMW 94.60G.11. (Scale bar = 500 µm.) (Reproduced from Edwards (1996) New insights into early land ecosystems: a glimpse of a Lilliputian world.* Rev. Palaeobot. Palynol. ***90****, 159–174, with permission from Elsevier Science – NL.)*

Figure 1.31. Cooksonia pertoni *with irregularly fractured distal surface. Arrows indicate stomata. NMW 94.60G.17. (Scale bar = 500 µm.) (Reprinted with permission from* Nature *(Edwards* et al. *(1992) A vascular conducting strand in the early land plant* Cooksonia. ***357****, 683–685. (Copyright (1992) Macmillan Magazines Limited.)*

Figure 1.32. *Unnamed terminal sporangia each splitting distally into three valves. NMW 96.5G.6. (Scale bar = 100 µm.)*

Figure 1.33. *One valve of unnamed sporangium, with numerous stomata (black spots). NMW 96.5G.7. (Scale bar = 500 µm.)*

Figure 1.34. *Arrowed stoma in Figure 1.33 enlarged. (Scale bar = 10 µm.)*

Figure 1.35. *Sporangium of* Cooksonia pertoni *shape with prominent cuticle. Arrows indicate stomata. NMW 96.5G.8. (Scale bar = 100 µm.)*

Figures 1.36, 1.39, 1.40–1.42. *Sporangia of* Tortilicaulis. *(1.36) Intact sporangium with adhering tubes. NMW 96.5G.9. (Scale bar = 100 µm.) (1.39) Fractured sporangium showing extremely thick outer periclinal walls of sporangial wall and* in situ *spores. NMW 93.98G.13. (Scale bar = 50 µm.) (Reproduced from Edwards and Fanning (1994) Lower Devonian coalified sporangia from Shropshire:* Salopella *Edwards & Richardson and* Tortilicaulis *Edwards.* Bot. J. Linn. Soc. ***116****, 89–110, with permission from Academic Press.) (1.40) Close-up of tube on axis of sporangium in Figure 1.36. (Scale bar = 20 µm.) (1.41) Close-up of tube and amorphous matrix on sporangium in Figure 1.36. (Scale bar = 20 µm.) (1.42) Irregularities on surface of cuticle of sporangium in Figure 1.36. (Scale bar = 5 µm.)*

Figures 1.37, 1.38. *Unnamed sporangium with stomata (arrowed on 1.38) near base. NMW 96.5G.10. (1.37) Scale bar = 200 µm. (1.38) Scale bar = 50 µm.*

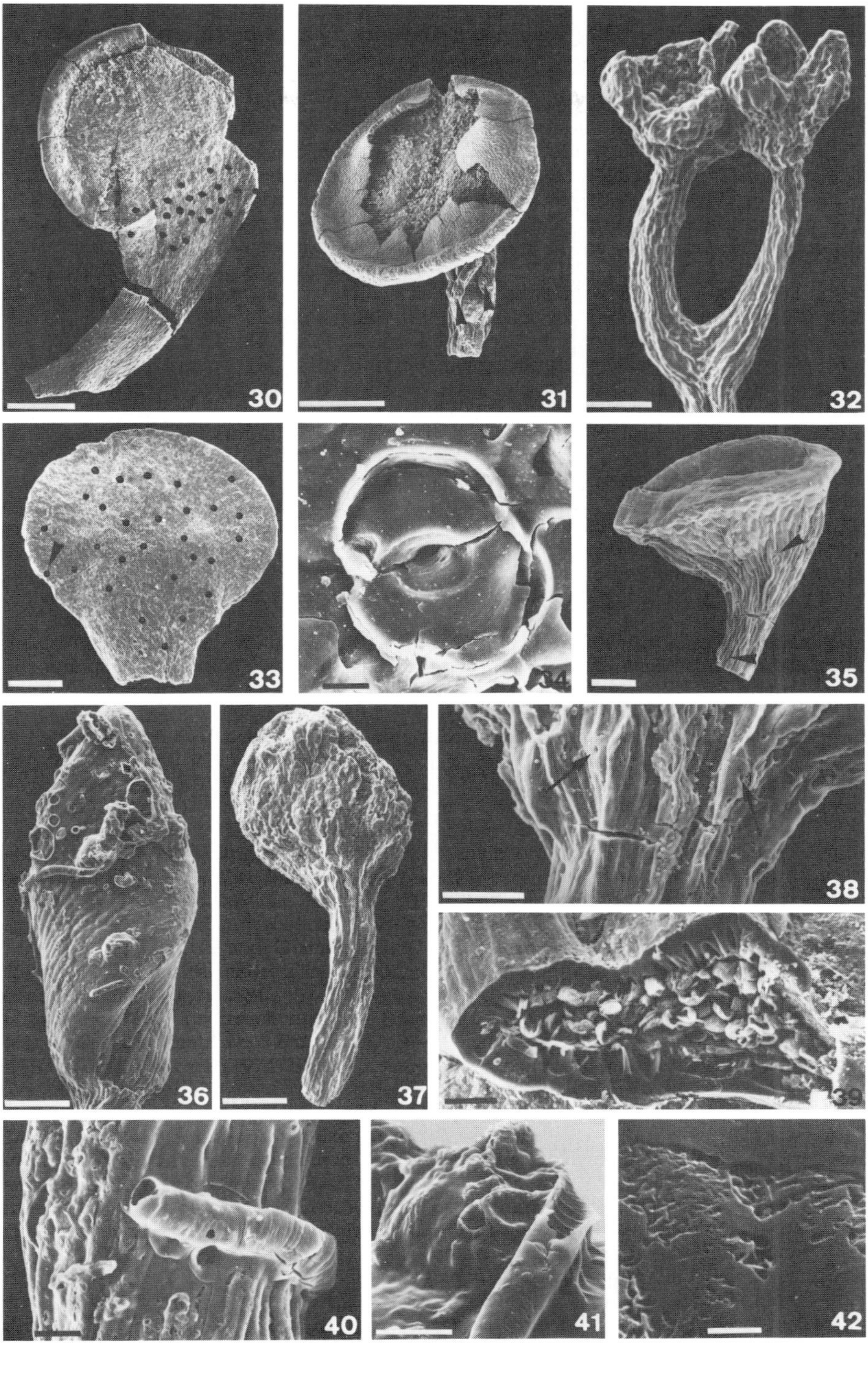
30
31
32
33
34
35
36
37
38
39
40
41
42

yet unnamed sporangium (but with different *in situ* spores) has stomata scattered over the whole valve surface (Figure 1.33; 7.25 mm^{-2}). The latter condition is also recorded in *Hsüa robusta* (Li, 1982).

Explanations. The demonstration that increasing atmospheric CO_2 over the last 200 years has been accompanied by a decrease in stomatal density and indices (Woodward, 1987), was followed by both experimental testing of the relationship (Radoglou and Jarvis, 1990; Woodward and Bazzaz, 1988) and measurements of extant (Beerling and Chaloner, 1993a; Paoletti and Gellini, 1993; Peñuelas and Matamala, 1990), sub-fossil (Beerling and Chaloner, 1993b,c, 1994; Beerling *et al.*, 1992; van de Water *et al.,* 1994) and fossil leaves (McElwain and Chaloner, 1995; Wolfe, 1990). Such activity against a background of concern over increasing CO_2 and global warming has tended to overshadow the roles of other environmental parameters, for example, water and nutrient availability (Hsaio and Fischer, 1975) and light intensity (Gay and Hurd, 1975), and to the underestimation of variation in extant plants (Poole *et al.*, 1996 and references therein). McElwain and Chaloner (1975) related low frequencies in the Lower Devonian to high carbon dioxide, and eliminated the possibility that variation could be morphology-dependent by using axial modern analogues, comparing *Aglaophyton* with *Juncus* and *Sawdonia* with *Psilotum*. While it is accepted that in the Lower Devonian low stomatal frequencies correlate with postulated high CO_2, the influences of water stress, nutrient availability and light intensity are more difficult to assess.

Water stress. (a) Rhynie chert plants: although communities grew on an alluvial plain on probably well-drained substrates (but see above), it is possible that the groundwater was sometimes supersaturated in silica and rich in other minerals. Thus water with such low chemical potential would not readily have been available to plants such as *Aglaophyton* (which has a low stomatal frequency). (b) Plants preserved in clastic Old Red Sandstone (ORS) sediments (deposited 411–363 million years ago): the dry periods typical of a monsoonal climate would have produced water shortages even in the vicinity of waterways, and it has been suggested that the peripheral structural tissues as are found in *Sawdonia* would have provided support during such periods (Edwards, 1993, 1994; Rayner, 1983). (c) The highest stomatal frequencies recorded in the Lower Devonian are in *Drepanophycus qujingensis,* which grew in a presumably moist depositional environment (Li and Edwards, 1995). (d) In both Rhynie Chert and ORS axial plants the presumed photosynthesizing tissues are deep-seated. In the former, cells with granular contents (chlorenchyma?) and limited intercellular spaces occur below one to a few layers of closely packed hypodermal cells and are connected to stomata by channels rather than well-developed substomatal cavities. The most detailed descriptions are for *Aglaophyton* (Remy and Hass, 1996). Here the guard cells have common walls with hypodermal as well as epidermal cells. The walls of hypodermal cells adjacent to the narrow canal are thickened and the authors suggest that this might be related to support or an 'additional closing mechanism'. Thus internal exposed surfaces may not be extensive. In 'mature' axes of *Aglaophyton* and *Rhynia* there is a further zone (1–4 cells thick) between the hypodermis and cortex with large intercellular spaces, which is the region in which the arbuscules of the mycorrhizal fungi are located (Remy and Hass, 1996). Such information is not available for the ORS plants, but the more or less continuous sterome with no obvious intercellular spaces suggests a similar organization. Indeed the presence of extensive sub-stomatal cavities in such a peripheral system might well have detracted from its structural role.

Raven (1993) has suggested that photosynthesis might be light-limited in these erect, axial, homoiohydric plants at the prevailing high CO_2 partial pressures. The deep-seated (presumed) chlorenchyma, contrasting with the positioning seen in McElwain and Chaloner's modern analogue *Psilotum*, might restrict light absorption by reducing the projected area of chlorenchyma relative to that of the axis as a whole, a problem which could, in principle, be overcome by radiation focusing by overlying cells (Lee, 1986; Vogelmann, 1993). Restriction of light absorption by chlorenchyma as a result of attenuation by overlying tissues need not be a significant problem (Vogelmann, 1993). How the deep-seated chlorenchyma, with (probably) significant diffusion resistance to gas exchange through overlying tissues, impacts on the radiation balance of the tissue remains to be modelled. It is possible that more energy could be dissipated as sensible heat rather than as latent heat of evaporation, thus decreasing the already low water cost of photosynthesis in a high CO_2 atmosphere.

In conclusion, while it cannot be denied that low frequency of stomata in the Lower Devonian coincides with postulated high CO_2, the above deliberations indicate that the former might not be caused *only* by the latter, and that water relationships might also have been involved.

Further conjecture. On-going research on small, leafless, cuticularized vegetative axes from uppermost Silurian and Lower Devonian rocks shows them to be virtually astomatous. Progress in demonstration of vascular tissues is disappointing, possibly because it consisted of a small xylem strand containing small-diameter conducting elements (Edwards *et al.*, 1992; *Cooksonia*), so that their tracheophyte status is only an assumption; thus fertile examples are termed rhyniophytoid. While it is possible that fragments derive from the tips of much larger plants, compression fossils from a number of localities in a variety of matrices suggest that the branching sporophytes rarely exceeded 10 cm in height.

Such evidence suggests that the earliest vascular plants were small, with few stomata and a small fraction of the axis occupied by conducting tissue. Even a small fraction of xylem in the axis (1% of the volume) with conducting elements only 10 μm in diameter would be adequate to supply water to a stomata-less or a stomata-containing organism with a small water potential (pressure) gradient (Raven, 1993; Roth *et al.,* 1994). This conclusion applies even at the highest likely photosynthetic rates with the highest likely transpiration: photosynthesis ratios in the late Silurian (Raven, 1993).

However, even without stomata these 10 cm long axes with conducting tissue could achieve photosynthetic rates of some 6 μmol CO_2 m^{-2} axis surface area s^{-1} in the uppermost Silurian–Lower Devonian atmosphere. Could the origin of stomata be related to the supply of water and hence nutrients, to particular parts of the axis? This possibility could find support in the location of stomata preferentially on sporangia if the sporangia need higher N, P or Fe per unit C than vegetative tissue. Stomata would permit a greater water loss per surface area, and hence a greater delivery of xylem-borne N, P or Fe, than in a similar astomatous area (noting that rapid water movement *per se* is not needed to supply nutrients to above-ground parts of the plant; Smith, 1991). The presence of stomata would, of course, permit photosynthesis to occur at a higher rate on an external surface area basis (Raven, 1993). At any event, the presence of stomata would yield similar C:N:P:Fe ratios in the tissue as in astomatous tissue unless the stomatal tissue sacrifices a low water cost of CO_2 fixation as a way of increasing the C:N, C:P or C:Fe ratio. This would involve a lower CO_2 fixation rate in the stomatal tissue than could be achieved at the expressed rate of transpiration. An alternative way of increasing the C:N, and so on, ratios in sporangia would be symplasmic movement of N, P, Fe, and so on, from transpirational termini on the sterile axis to the sporangium. Symplasmic transport of organic C (as well as of N, P and Fe) from sterile regions to the sporangia is common in extant organisms, where *inter alia* the lower photosynthetic capacity of sporangia reduces oxygen production there with a corresponding smaller risk of free radical damage to the germ-line (Raven *et al.*, 1994).

It is important to note that this suggestion has echoes of that of Church (1919) who conjectured that stomata evolved in relation to nutrient transport in the transpiration stream. However, Church (1919) denied that stomata were involved in CO_2 exchange, since he believed that CO_2 could diffuse as readily through water as through an equivalent thickness of air, although this was already known to some workers to be untrue (Brown and Escombe, 1900).

Nutrient availability. By Siluro–Devonian times, higher plants, initially with rhizomes and rhizoids and eventually roots, and associated humous and soil microbial populations, would have begun to influence weathering rates and processes (references from Berner, 1992, 1994; Cawley *et al.,* 1969; Yapp and Poths, 1994). While these effects would have increased the concentration of nutrient solutes in soils they may not have reached the levels found in fertile non-agricultural soils today. This poses a problem for nutrient solute supply to early vascular plants in view of the low diffusion coefficient for many solutes (especially phosphate) in many soils. Raven (1993) points out that transpirational water flux through such soil to the plant surface would have been grossly inadequate for most nutrient solutes, and especially for phosphate, and with the (low) transpiratory water loss per CO_2 fixed in an atmosphere 10 or more times the present CO_2 mol fraction. Thus, movement of nutrients to the plant by mass flow with water in the soil is inadequate to supply the early vascular plant's requirements; it is not clear if this was part of the argument of Church (1919). This inadequacy of the transpiratory water flux through the *soil* to supply plant nutrients does not contradict earlier statements (penultimate paragraph in previous section) about the role of transpiratory water flow in the *xylem,* since vascular plant sporophytes (and, apparently, endohydric moss gametophytes) exclude some solutes (e.g. Ca^{2+}, Na^+) from the xylem while accumulating others (e.g. K^+, N and soluble N and P species) to values in the transpiration stream in the xylem which are higher than those in the soil solution (de Boer and Wegner, 1996; Raven, 1994). How this was achieved in early vascular plants (and, indeed, in extant endohydric mosses) in the absence of the anatomical correlate of this discrimination, i.e. the endodermis, is not clear (Raven, 1994). It is likely that early rhyniophytes (and rhyniophytoids) could make suberin, the solute-excluding component in the endodermal cell walls, as part of wound-repair processes (see Kevan *et al.,* 1975).

As we have seen, the diffusion of solutes from sites of release in weathering or the breakdown of organic remains in the soil to the plant surface is slow; solute uptake rates per unit plant biomass are greatly increased by allocation of part of the below-ground biomass to evaginations with a high surface area per unit volume, which grow into a larger volume of soil than can be exploited by diffusion to a rhizome even when this is only a millimeter or so in diameter. Rhizoids fit this job description, and can exploit soil nutrients by continued elongation, growth, accumulation of nutrients within the cytoplasm, and diffusion or (better) mass flow in cytoplasmic streaming as a means of moving the solutes to the rest of the plants. 'Rhizoids' occur on the Rhynie plants; however, comparison with extant plants suggests that such evaginations are less effective (for incompletely understood reasons) than are mycorrhizas (e.g. Harrison and Buuren, 1995). It has even been suggested that the initial colonization of the land was possible only through a symbiotic partnership between semi-aquatic algae and fungi (Pirozynski and Malloch, 1975). There seem to be no examples of nutritional mutualism between extant Charophyceae and fungi. However, the semi-terrestrial *Geosiphon* is a photosynthetic symbiosis of the cyanobacterium *Nostoc* and a zygomycetous fungus closely related to the Glomades and Edogonales which are the arbuscular mycorrhizas of embryophytes (Gehring *et al.,* 1995; Schüssler *et al.*, 1995). The demonstration of arbuscular mycorrhizas in *Aglaophyton* and *Rhynia gwynne-vaughanii* (Remy and Hass, 1996; cf. Nicolson, 1975) in the Rhynie Chert is the first direct demonstration of such a relationship (Taylor *et al.*, 1995).

1.7.2 *Structural and mechanical roles of cuticle*

In the Rhynie Chert plants, where cortical tissues are predominantly thin-walled and support is via turgor, a peripheral inextensible layer in mature axes would increase bending resistance in the longitudinal plane by improving the elastic modulus (T.G. Morgan, personal communication), but its effects would be negligible in axial plants which possess a more or less continuous sterome (Edwards, 1994). In this respect a thicker cuticle would have greater effect. The additional prominent cuticular ridges on *Aglaophyton* and *Rhynia* (Remy and Hass, 1996) were interpreted as giving extra strength to the cuticle and preventing deformation of the epidermal cells during desiccation. In sporangia, where cuticles are often much thicker, a support function might be important during spore development in maintaining shape, but would be a hindrance to the liberation of spores. Our on-going studies on the earliest land plants, suggest that even in the Lower Devonian a wide range of dehiscence mechanisms existed, sometimes involving modifications in the cells of the sporangium wall (*Uskiella*; Shute and Edwards, 1989), although the mechanisms of dehiscence remain poorly understood, and it is uncertain whether or not the actual splitting of the wall involves physical tearing or more controlled chemical lysis of the cuticle. Thus, for example, the majority of sporangial cuticles isolated from basal Devonian sediments clearly derive from elongate sporangia which probably split into two or more valves (Figures 1.11 and 1.12). The consistent shape of these cuticular fragments suggests that dehiscence was predetermined. The entirety of the margins show no evidence for cuticle tearing and raises the possibility that the cuticle was thinner in those areas. In one Silurian example, known only from cuticles and spores, an ellipsoidal sporangium, resembling a depressed funnel maybe incomplete distally, possesses an incurved cuticular rim, or less rarely shows a circular area of cuticle in place (Figures 1.15 and 1.16). Such circles of cuticle are also recovered from the macerations suggesting that spore liberation occurred after separation of the distal cuticle (+ possibly sporangial wall) with its irregular margin indicating tearing following the generation of radial or ring stress. In Lower Devonian *Cooksonia pertoni* sporangia which are of similar shape, a heavy cuticle is present in some specimens (Figure 1.35), but is less obvious in those which show separation between cells of the outermost layer of cells of the sporangium wall. In this taxon spore liberation is thought to be achieved by imprecise breakdown of the sporangium wall and cuticle disintegration in the central distal region of the discoidal sporangium (Figure 1.31), its overall shape being maintained by marginal thickening. Finally, a new Lower Devonian taxon (Figure 1.32) shows dehiscence into three valves distally associated with differential wrinkling in the distal and proximal (non-split) regions. Is it possible here that increased cuticular water loss and hence shrinkage of the basal part of the sporangial wall created the tensions that resulted in the predetermined distal splitting? Such deliberations relating to dehiscence are highly speculative and more specimens are needed before it can be deduced whether or not the sporangial cuticle was an obstacle to be overcome in spore liberation or more actively involved in dehiscence mechanisms.

As regards *Nematothallus*-like cuticles, Edwards and Rose (1984) illustrated a number with distinctive thickenings, some of which surrounded perforations (Figures 1.8–1.10), and in a highly speculative discussion on their functions, suggested that they might have developed either to facilitate dispersal of the plant's own propagules or those of pathogens colonizing the organism.

1.7.3 *Protection*

Uncertainties as to the chemical composition of the cuticle are perhaps less important here; the cuticle clearly forms a physical barrier between plant tissues and environment, which could retard entry of pathogens and herbivores. Conversely it acts against the easy liberation of propagules, not the least the plant's own spores (but see above). However, the presence of waxes and the surface properties of the cuticles *are* relevant in considering defence against flightless arthropod microherbivores, particularly in the erect, axial branching systems, where a 'greasy pole' scenario (see Section 10.2.1) might have protected the sporangia with their concentrations of nutrients in the distal parts of the plant. This considerably predates the evolution of a slippery pedicel or peduncle in angiosperms (see Juniper, 1995) as an anti-robbing device against coleopteran non-pollinators. Evidence for the existence of such predators and pathogens in the mid-Palaeozoic is in the form of actual fossils or anatomical 'abnormalities' in the plant suggestive of response to attack. Such a palaeopathological explanation was offered for cuticular developments associated with perforations in *Nematothallus* cuticles and regeneration of tissues sometimes associated with staining in Rhynie plants (Rolfe, 1985) and *Psilophyton dawsonii* (Banks and Colthart, 1993).

Bacteria. Kidston and Lang's (1921) discussion of possible fossil records of bacteria is typical of the dilemma in the recognition of bacteria. They record irregularly lobed granular masses in a matrix with decaying algae which show superficial similarities with extant colonies in similar habitats (both freshwater and marine) today. Remy and Hass (1996) related regeneration of water-conducting cells in *Aglaophyton* and *Rhynia* in phloem or cortical tissues to infection by bacteria and/or fungi. For the most part, size and morphological simplicity make it impossible to demonstrate unequivocal bacteria, and, even if possible, whether they were pathogens or saprobes. An added problem is that of the recognition of modern contamination. To avoid this, particularly for geochemical studies, we temporarily store all our isolated material in dilute hydrochloric acid. However, in a recent study of Devonian coprolites (faecal pellets), where we were specifically searching for bacterial decomposers, a specimen left in water for a few days displayed structures similar to those illustrated in Figure 1.17. We found no bacteria-like remains in the remainder of the material. All our isolated cuticles look remarkably 'clean', showing no signs of bacterial attack.

Fungi. The classic studies of Kidston and Lang (1921) included descriptions of a variety of Phycomycetes, none of which were identified as parasites although the authors stated that initial parasitism could have progressed into saprophytism. Recent, more comprehensive, surveys, which have transformed the early fossil record of the fungi, contain two records of possible relevance. The first involves the existence of chytrids as potential pathogens (Hass *et al.,* 1994). As they are predominantly aquatic fungi it might seem more likely that they were saprophytes on immersed decaying terrestrial higher plants. However, Remy and Hass' remarkable account of the development of the gametophytes of *Aglaophyton* includes germinating spores always found in wet or shallow-water environments, as evidenced by presence of algae or aquatic fungi which were parasitized by chytrids very early on, when the gametophyte was a globular mass of cells emerging from the spore and lacking a cuticle (Remy and Hass, 1996). They also recorded concentrations of chytrids in the apical and sub-apical regions of aerial axes, with infection through stomata and subsequent spread beneath the cuticle. The authors postulated that access must have been easier in this area. The excellent preservation of the apices and lack of host response suggests that infection probably occurred soon after the organs were immersed in water, but pathogenicity cannot be eliminated. Direct evidence for the mode of entry is also absent in the second possible example (Taylor *et al.,* 1995) which presents the earliest unequivocal arbuscular mycorrhiza in *Aglaophyton major*. Infection sites indicative of multiple invasion are located between rhizoids where the cuticle is thin, but not through the rhizoids themselves. There is no evidence of entry through the stomata present on all surfaces of the rhizome. In *Rhynia gwynne-vaughanii,* Kidston and Lang (1917) noted that lateral projections, some with apical stomata, contained cavities associated with necrosed tissue and staining plus some regeneration, but failed to demonstrate any associated fungi.

Sporangia of* Tortilicaulis *and banded tubes. Evidence for fungi associated with compression fossils of higher plants is very rare, and the mode of nutrition of such fungi would be impossible to determine unequivocally. Palynological preparations from Wenlock–Lower Devonian rocks often contain isolated tubes with thickenings in rings or helices on the inner walls (Burgess and Edwards, 1991). Similar banded tubes have been recorded in *Nematothallus* and the tetrad-containing plant which provides evidence for the kind of land vegetation in the Ordovician (Edwards *et al.,* 1995a), while isolated examples are often found associated with a variety of coalified fragments. That illustrated in Figures 1.36 and 1.41 is unusual, in that it is attached to a sporangium of *Tortilicaulis* (a rhyniophytoid) by an amorphous substance which extends over the surface of the sporangium, and is reminiscent of the extracellular matrix by which fungal spores adhere to hydrophobic surfaces of plants (e.g. conidia of *Colletotrichum graminicola*: Mercure *et al.*,1994; uredospores and germlings of *Uromyces viciae-fabae*: Clement *et al.*, 1993; Section 8.3). Further tubes and 'cement' occur on the axes itself. Thus it is tempting to assume that the tubular structure is part of a possible pathogen, and while impossible to determine whether the amorphous material was merely a cement or the location of enzymes associated with breakdown of the cuticles, the irregularities on the adjacent cuticle, which normally has a smooth surface, might be evidence for the latter (Figure 1.42). The outer periclinal wall of the superficial cells of the sporangium is exceptionally thick in *Tortilicaulis* (Figure 1.39) and in fractured section showed no indications of the cuticle, which is recovered following oxidation of the sporangium with nitric acid

(Figure 1.27). It is of considerable thickness compared with axial cuticles and, while this might be related to a structural role associated with dehiscence (see Section 1.7.2), it is also possible that it is an adaptation to increase resistance to invasion both by fungal pathogens and herbivores.

Herbivores. The existence of herbivores in the Lower Devonian remains contentious (Banks and Colthart, 1993; Edwards *et al.*, 1995b). Body fossils indicate that carnivores (trigonotarbids and centipedes) and detritivores (collembolans and mites) were present in early terrestrial ecosystems (Gray and Boucot, 1994; Jeram *et al.*, 1990) with arthropleurids (related to millipedes) as possible herbivores or detritivores (Shear and Selden, 1995). Evidence for spore feeding and, if from living plants, the necessity to protect sporangia, comes from spore-dominated coprolites (Edwards *et al.*, 1995b), and for herbivory from sterile axes showing their regeneration (Banks and Colthart, 1993) and possible chemical responses (Kevan *et al.*, 1975; Scott *et al.*, 1985). Banks and Colthart argued forcefully that the localized periderm development in certain axes of *Psilophyton* was produced when surface cells were chewed, a case where cuticle would probably not be very effective in protection, and also reported regeneration associated with more deep-seated injuries related to superficial multiple puncturing where a thick cuticle would act as a deterrent to piercing (although apparently unsuccessful in this case!). Kevan *et al.* (1975) discussed the possibility that lesions in Rhynie Chert plants represented sites of attack by sap-feeding arthropods.

UV radiation. Natural waters invariably attenuate the (damaging) UV-B radiation (280–320 nm) to a greater extent than photosynthetically active radiation (PAR 400–700 nm; Kirk, 1994a,b). Terrestrial organisms are thus exposed to greater incident UV-B fluxes relative to PAR fluxes with a given solar radiation output and atmospheric attenuation. Thus, regardless of whether the ratio of UV-B:PAR incident on natural water surfaces and terrestrial vegetation was higher in the Siluro–Devonian than is the case today, the terrestrial organisms would have a relatively greater exposure to UV-B. The attenuation of UV-B relative to PAR by components of primarily aquatic phototrophs involves extracellular phenolic compounds and intracellular (nitrogenous) mycosporine-like amino acids (Garcia-Pichel, 1994; Raven, 1991a, 1995b). Most of the data on which these conclusions are based are for marine algae, and so are not directly relevant to the immediate ancestors of terrestrial plants. Certainly extant embryophytes appear to lack both the extracellular phenolic materials of the type found in algae and mycosporine-like amino acids within cells.

UV-B attenuation relative to PAR attenuation in extant terrestrial embryophytes can involve, from the outside inwards, the wax layer on the cuticle, the cuticle, phenolics (including lignin; see Lowry *et al.*, 1980) in cell walls and intracellular (mainly vacuolar) phenolics such as flavonoids (see Chapter 7; Bornman *et al.*, 1996; Krauss *et al.*, 1996; Raven, 1991b; Vogelmann, 1993). The cuticle and its wax layer are thus part of a spatial series of UV-B screens; it is not readily decided if the cuticle plus wax were relatively more or less important as UV-B attenuators in the earliest vascular plants than they are in extant examples. It is clear that in certain extant plants the wax layer functions to attenuate high, potentially damaging, fluxes of PAR (Robinson *et al.*, 1993), although it is not certain if there is differential UV-B and PAR attenuation in these cases. It is possible that the non-cuticular cell wall and vacuolar UV-B screens show a more effective differential attenuation of UV-B relative to PAR than do cuticles and waxes, so that they might be the evolutionarily selected major attenuators of UV-B if maximizing PAR absorption while minimizing UV-B absorption by the genetic and photosynthetic apparati. This would effectively substitute (non-nitrogenous) soluble intracellular UV-B screens, such as flavonoids, for the mycosporine-like amino acids in the descendants of the remote marine ancestors of the higher plants (noting that extant charophyceans show little evidence of possessing flavonoids; Harborne, 1988). However, there is the intrinsic advantage of extracellular UV-B attenuators (e.g. cuticles and wax) over intracellular ones, in that they protect all of the intracellular components, not just those internal to the outermost vacuoles containing the screening material (see Raven, 1991a, 1995b). Indeed, in Section 1.8 it is shown that phenolics were present in the thallus cuticle of *Nematothallus*; these could have had a role in extracellular UV-B screening.

1.8 *Nematothallus*: a salutary tale

Lang (1937) placed the astomatous cuticles which he believed covered wefts of tubes of two sizes in an informal group, the Nematophytales, because he considered them neither algae nor higher plants, a distinction reiterated by Strother (1993) who called them the Paraphyta. Strother (1988) had earlier described a new species with only the wefts of smaller tubes present. Indeed Lang's type species, *N. pseudo-vasculosus*, has no cuticle. Speculation on the nature of the underlying tissue that produced the reticulum (Strother, 1988) includes tubes aligned at right angles to the surface, a conventional epidermis (Lang, 1945) or pseudoparenchyma/paraplec-tenchyma, the latter currently favoured in our Cardiff studies. Such organization, and hence affinities, clearly will affect the functions of the cuticle as will its chemistry.

The recent discovery that the cuticle of *Nematothallus* does not have the same composition as those of coeval axial higher plants raises major queries as to its function, queries exacerbated by our continuing ignorance of its affinities. Although Edwards *et al.* (1982) speculated that on the basis of their greater resilience to decay (perhaps related to their greater thickness) than coeval vascular plants, and their different appearance on rock surfaces might be evidence of differences in chemical composition, their functions were considered as if part of a photosynthesizing plant, albeit a thalloid one. Thus a water repellency role was discussed. Our recent researches show that the surface of *Nematothallus* was extremely irregular (note the lack of uniform focus in Figure 1.2), suggestive of pockets for accumulating water, and it is tempting to speculate that the 'cuticular' layer itself had a water absorption and storage facility much as in extant lichens. Although contemporary oxygen levels, and hence ozone levels, were probably higher (Berner and Canfield, 1989) than was thought earlier (Berkner and Marshall, 1965), it is still possible that the *Nematothallus* cuticles acted as a UV shield.

Acknowledgements

We thank Greg Ewbank, Catherine Rogerson and Elena Mendez for the use of unpublished data. D.E. is grateful to Charles Wellman and Lindsey Axe for their unselfish help in the preparation of the manuscript. Aspects of her research were financed by a number of NERC grants.

References

Al-Ameri, T.K. (1983) Acid-resistant microfossils used in the determination of Palaeozoic palaeoenvironments in Libya. *Palaeogeogr. Palaeoclimatol. Palaeoecol.* **44,** 103–116.

Allen, J.R.L. (1986) Old Red Sandstone calcretes. In: *Palaeosols, their Recognition and Interpretation* (ed. V.P. Wright). Blackwell Scientific Publications, Oxford, pp. 58–82.

Andrews, H.N, Kasper, A.E., Forbes, W.H. *et al.* (1977) Early Devonian flora of the Trout Valley Formation of northern Marine. *Rev. Palaeobot. Palynol.* **23,** 255–285.

Atkinson, A.W. Jr, Gunning, B.E.S. and John, P.C.L. (1972) Sporopollenin in the cell wall of *Chlorella* and other algae: ultrastructure, chemistry and incorporation of ^{14}C-acetate, studied in synchronous cultures. *Planta* **107,** 1–32.

Banks, H.P. and Colthart, B.J. (1993) Plant-animal-fungal interactions in Early Devonian trimerophytes from Gaspé, Canada. *Am. J. Bot.* **80,** 992–1001.

Batten, D.J., Creber, G.F. and Zhiyan, Z. (1985) Fossil plants and other organic debris in Cretaceous sediments from deep sea drilling project leg 80: their paleoenvironmental significance and source potential. *Init. Rep. Deep Sea Drill. Proj.* **80**, 629–641.

Becker, M., Kerstiens, G. and Schönherr, J. (1986) Water permeability of plant cuticles: permeance, diffusion and partition coefficients. *Trends Ecol. Evol.* **1,** 54–60.

Beerbower, R. (1985) Early development of continental ecosystems. In: *Geological Factors and the Evolution of Plants* (ed. B.H. Tiffney). Yale University Press, New Haven, pp. 47–91.

Beerling, D.J. and Chaloner, W.G. (1993a) The impact of atmospheric CO_2 and temperature change on stomatal density: observations from *Quercus robur* Lammas leaves. *Ann. Bot.* **71,** 231–235.

Beerling, D.J. and Chaloner, W.G. (1993b) Stomatal density responses of Egyptian *Olea europaea* L. leaves to CO_2 change since 1327 BC. *Ann. Bot.* **71,** 431–435.

Beerling, D.J. and Chaloner, W.G. (1993c) Evolutionary responses of stomatal density to global CO_2 change. *Biol. J. Linn. Soc.* **48,** 343–353.

Beerling, D.J. and Chaloner, W.G. (1994) Atmospheric CO_2 changes since the last glacial maximum: evidence from the stomatal density record of fossil leaves. *Rev. Palaeobot. Palynol.* **81,** 11–17.

Beerling, D.J., Chaloner, W.G., Huntley, B *et al.* (1992) Variations in the stomatal density of *Salix herbacea* L. under the changing atmospheric CO_2 concentrations of late- and post-glacial time. *Phil. Trans. R. Soc. Lond. B.* **336,** 215–224.

Berkner, L.V. and Marshall, L.C. (1965) On the origin and rise of oxygen concentration in the Earth's atmosphere. *J. Atmos. Sci.* **22,** 225–261.

Berner, R.A. (1992) Weathering, plants and the long-term carbon cycle. *Geochim. Cosmochim. Acta* **56,** 3225–3231.

Berner, R.A. (1993) Paleozoic atmospheric CO_2: importance of solar radiation and plant evolution. *Science* **261,** 68–70.

Berner, R.A. (1994) 3GEOCARBII: A revised model of atmospheric CO_2 over phanerozoic time. *Am. J. Sci.* **294,** 56–91.

Berner, R.A. and Canfield, D.E. (1989) A new model for atmospheric oxygen over Phanerozoic time. *Am. J. Sci.* **289,** 333–361.

de Boer AH and Wegner LH. (1996) Molecular properties of ion channels in xylem parenchyma cells suggest controlled salt release to the xylem. *J. Exp. Bot.* **47** (suppl.), 62.

Bornman, J.F., Reuber, S. and Weissenbock, G. (1996) Protective pigments. *J. Exp. Bot.* **47** (suppl.), 17.

Brown, H.T. and Escombe, F. (1900) Static diffusion of gases and liquids in relation to the assimilation of carbon and translocation in plants. *Phil. Trans. R. Soc. Lond. B.* **193,** 223–291.

Burczyk, J. (1987a) Biogenetic relationships between keto-carotenoids and sporopollenins in green algae. *Phytochemistry* **26,** 113–119.

Burczyk, J. (1987b) Cell wall carotenoids in green algae which form sporopollenins. *Phytochemistry* **26,** 121–128.

Burgess, N.D. (1987) Micro- and megafossils of land plants from the Silurian and Lower Devonian of the Anglo-Welsh Basin. Unpublished Ph.D. thesis, University of Wales (Cardiff).

Burgess, N.D. and Edwards, D. (1991) Classification of uppermost Ordovician to Lower Devonian tubular and filamentous macerals from the Anglo-Welsh Basin. *Bot. J. Linn. Soc.* **106,** 41–66.

Buscher, R., Hass, H. and Remy, R. (1978) Möglichkeiten und Grenzen der Deutung von Ökologie und Klima mit Hilfe disperser Kutikulen aus dem Autun des Nahe-Rammes (Pfalz). *Argumenta Palaeobotanica* **5,** 149–160.

Campbell, S.E. (1979) Soil stabilization by a prokaryotic desert crust: implications for Precambrian land biota. *Orig. Life* **9**, 335-348.

Carver, J.H. (1980) Oxygen and ozone evolution in palaeoatmospheres. In: *Biogeochemistry of Ancient and Modern Environments* (eds P.A. Trudinger, M.R. Walter and B.J. Ralph). Springer-Verlag, Berlin, pp. 55-71.

Cawley, J.L., Burruss, R.C. and Holland, H.D. (1969) Chemical weathering in central Iceland: an analog of pre-Silurian weathering. *Science* **165**, 391-392.

Chaloner, W.G. and Sheerin, A. (1979) Devonian macrofloras. In: *The Devonian System. Special Papers in Palaeontology*, Vol. 23 (eds M.R. House, C.T. Scrutton and M.G. Bassett). Oxford University Press, Oxford, pp. 145-161.

Church, A.H. (1919) Thalassiophyta and the subaerial transmigration. *Oxford Bot. Mem. No. 3*, 1-95. Oxford University Press, Oxford.

Clement, J.A., Martin, S.G., Porter, R. *et al.* (1993) Germination and the role of extracellular matrix in adhesion of urediniospores of *Uromyces viciae-fabae* to synthetic surfaces. *Mycol. Res.* **97**, 585-593.

Craigie, J.S., Correae, J.A. and Gordon, M.E. (1992) Cuticles from *Chondrus crispus* (Rhodophyta). *J. Phycol.* **28**, 777-786.

de Leeuw, J.W. and Largeau, C. (1993) A review of macromolecular organic compounds that comprise living organisms and their role in kerogen, coal and petroleum formation. In: *Organic Geochemistry: Principles and Applications* (eds M.H. Engel and S.A. Macko). Plenum Publishing Group, New York, pp. 23-72.

Derenne, S., Largeau, C., Casadevall, E. *et al.* (1991) Chemical evidence of kerogen formation in source rocks and oil shales via selective preservation of thin resistant outer walls of microalgae: origin of ultralaminae. *Geochim. Biophys. Acta* **55**, 1041-1050.

Derenne, S., Largeau, C., Berkaloff, C. *et al.* (1992) Non-hydrolysable macromolecular constituents from outer walls of *Chlorella fusca* and *Nanochlorum eucaryotum*. *Phytochemisty* **31**, 1923-1929.

Edwards, D. (1970) Fertile Rhyniophytina from the Lower Devonian of Britain. *Palaeontology* **13**, 451-461.

Edwards, D. (1982) Fragmentary non-vascular plant microfossils from the late Silurian of Wales. *Bot. J. Linn. Soc.* **84**, 223-256.

Edwards, D. (1986) Dispersed cuticles of putative non-vascular plants from the Lower Devonian of Britain. *Bot. J. Linn. Soc.* **93**, 259-275.

Edwards, D. (1993) Tansley review No. 53. Cells and tissues in the vegetative sporophytes of early land plants. *New Phytol.* **125**, 225-247.

Edwards, D. (1994) Towards an understanding of pattern and process in the growth of early vascular plants. In: *Shape and Form in Plants and Fungi,* Linn. Soc. Symp. Ser. No. 16. (eds D.S. Ingram and A. Hudson). Academic Press, London, pp. 39-59.

Edwards, D. (1996) New insights into early land ecosystems: a glimpse of a Lilliputian world. *Rev. Palaeobot. Palynol.* **90**, 159-174.

Edwards, D. and Axe, L. (1992) Stomata and mechanics of stomatal functioning in some early land plants. *Cour. Forschungsinst. Senckenb.* **147**, 59-73.

Edwards, D. and Burgess, N.D. (1990) Terrestrialization — plants. In: *Palaeobiology — a Synthesis* (eds D.E.G. Briggs and P.R. Crowther). Blackwell Scientific Publications, Oxford, pp. 60-64.

Edwards, D. and Rose, V. (1984) Cuticles of *Nematothallus*: a further enigma. *Bot. J. Linn. Soc.* **88**, 35-54.

Edwards, D. and Selden, P.A. (1993) The development of early terrestrial ecosystems. *Bot. J. Scot.* **46**, 337-366.

Edwards, D., Edwards, D.S. and Rayner, R. (1982) The cuticle of early vascular plants and its evolutionary significance. In: *The Plant Cuticle,* Linn. Soc. Symp. Ser. No. 10. (eds D.F. Cutler, K.L. Alvin and C.E. Price). Academic Press, London, pp. 341-361.

Edwards, D., Feehan, J. and Smith, D.G. (1983) A late Wenlock flora from Co. Tipperary, Ireland. *Bot. J. Linn. Soc.* **86**, 19-36.

Edwards, D., Davies, K.L. and Axe. L. (1992) A vascular conducting strand in the early land plant *Cooksonia. Nature* **357**, 683-685.

Edwards, D., Duckett, J.G. and Richardson, J.B. (1995a) Hepatic characters in the earliest land plants. *Nature* **374**, 635-636.

Edwards, D., Selden, P.A., Richardson, J.B. *et al.* (1995b) Coprolites as evidence for plant-animal interaction in Siluro-Devonian terrestrial ecosystems. *Nature* **377**, 329-331.

Florin, R. (1938-45) Die Koniferen des Oberkarbons und des unteren Perms. *Palaeontogr. Abt. B.* **85**, 2-729.

Garcia-Pichel, F. (1994) A model for internal self-shading in planktonic organisms and its implications for the usefulness of ultraviolet sunscreens. *Limnol. Oceanogr.* **39,** 1704–1777.

Gay, A.P. and Hurd, R.G. (1975) The influence of light on stomatal density in the tomato. *New Phytol.* **75,** 37–46.

Gehring, H., Schüssler, A. and Kluge, M. (1995) Molecular characterization of the taxonomy of *Geosiphon pyriforme*. In: *Abstracts of the International Symposium of the Sonderforschungsbereich 199: Ecophysiology of Plants: Molecules, Membranes and Mechanisms*, Ebernburg, Nahe, March 1995 (eds R. Ratajczak and C. Ullrich), pp. 26–27.

Gelin, F., Boogers, I., Noordeloos, A.A.M., Sinninghe Damsté, J.S., Hatcher, P.G. and de Leeuw, J.W. (1996) Novel resistant microalgal polyethers: an important sink of organic carbon in the marine environment? *Geochim. Cosmochim. Acta* **60,** 1275–1280.

Graham, L.E. (1993) *Origin of Land Plants*. John Wiley & Sons, New York.

Graham, L.E. and Delwiche, C.F. (1992) The occurrence and phylogenetic significance of a surface layer on thalli of *Coleochaete* (Charophyceae). *Am. J. Bot.* **79,** 101a.

Gray, J. and Boucot, A.J. (1977) Early vascular land plants: proof and conjecture. *Lethaia* **10,** 145–174.

Gray, J. and Boucot, A.J. (1994) Early Silurian nonmarine animal remains and the nature of the early continental ecosystem. *Acta Palaeontol. Pol.* **38,** 303–328.

Gray, J., Massa, D. and Boucot, A.J. (1982) Caradocian land plant microfossils from Libya. *Geology* **10,** 197–201.

Hanic, L.A. and Craigie, J.S. (1969) Studies on the algal cuticle. *J. Phycol.* **5,** 89–102.

Harborne, J.B. (ed.) (1988) *The Flavonoids: Advances in Research Since 1980*. Chapman & Hall, London.

Harrison, M.J. and van Buuren, M.L. (1995) A phosphate transporter from the mycorrhizal fungus *Glomus versiforme*. *Nature* **378,** 626–629.

Hass, H. (1991) Die Epidermis von *Horneophyton lignieri* (Kidston & Lang) Barghoorn & Darrah. *Neues Jahrb. Geol. Paläontol. Abh.* **183,** 61–85.

Hass, H., Taylor, T.N. and Remy, W. (1994) Fungi from the Lower Devonian Rhynie Chert: Mycoparasitism. *Am. J. Bot.* **81:** 29–37.

Holland, H.D. (1984) *The Chemical Evolution of the Atmosphere and Oceans*. Princeton University Press, Princeton, NJ.

Hsaio, T.C. and Fischer, R.A. (1975) Microscopic measurements. In: *Measurements of Stomatal Aperture and Diffusion Resistance* (ed. E.T. Kanesmasu). College of Agricultural Research Centre, Washington State University, Pullman, USA, pp. 2–5.

Jeram, A.J., Selden, P.A. and Edwards, D. (1990) Land animals in the Silurian: arachnids and myriapods from Shropshire, England. *Science* **250,** 658–661.

Juniper, B.E. (1995) Waxes on plant surfaces and their interactions with insects. In: *Waxes: Chemistry, Molecular Biology and Functions* (ed. R.J. Hamilton). The Oily Press, Dundee, pp. 157–175.

Kenrick, P. (1994) Alternation of generations in land plants: new phylogenetic and palaeobotanical evidence. *Biol. Rev.* **69,** 293–330.

Kerp, H. (1990) The study of fossil gymnosperms by means of cuticular analysis. *Palaios* **5,** 548–569.

Kevan, P.G., Chaloner, W.G. and Savile, D.B.O. (1975) Interrelationships of early terrestrial arthropods and plants. *Palaeontology* **18,** 391–417.

Kidston, R. and Lang, W.H. (1917) On Old Red Sandstone plants showing structure, from the Rhynie Chert Bed, Aberdeenshire. Part 1. *Rhynia gwynne-vaughani,* Kidston and Lang. *Trans R. Soc. Edin.* **51,** 761–784.

Kidston, R. and Lang, W.H. (1921) On Old Red Sandstone plants showing structure, from the Rhynie Chert Bed, Aberdeenshire. Part V. The Thallophyta occurring in the peat-bed; the succession of the plants throughout a vertical section of the bed, and the conditions of accumulation and preservation of the deposit. *Trans. R. Soc. Edinb.* **52,** 855–902.

Kirk, J.T.O. (1994a) *Light and Photosynthesis in Aquatic Ecosystems*, 2nd Edn. Cambridge University Press, Cambridge.

Kirk, J.T.O. (1994b) Optics of UV-B in natural waters. *Arch. Hydrobiol. Beihefte* **43,** 1–16.

Kolattukudy, P.E. (1981) Structure, biosynthesis and biodegradation of cutin and suberin. *Annu. Rev. Plant Physiol.* **32,** 539–567.

Krauss, P.S., Markstädter, C. and Riederer, M. (1996) Attenuation of ultraviolet radiation by plant cuticles. *J. Exp. Bot.* **47** (suppl.), 58.

Lang, W.H. (1937) On the plant-remains from the Downtonian of England and Wales. *Phil. Trans. R. Soc. Lond. B.* **227,** 245-291.

Lang, W.H. (1945) *Pachytheca* and some anomalous early plants (*Prototaxites, Nematothallus, Parka, Foerstia, Orvillea* n.g.). *J. Linn. Soc. Lond. Bot.* **52,** 535-552.

Lee, D.W. (1986) Unusual strategies of light absorption in rain-forest herbs. In: *On the Economy of Plant Form and Function* (ed. T.J. Givnish). Cambridge University Press, Cambridge, pp. 105-131.

Lee, D.W. (1991) Ultrastructural basis and function of iridescent blue colour of fruits in *Elaeocarpus. Nature* **349,** 260-262.

Lele, K.M. and Walton, J. (1961) Contributions to the knowledge of *Zosterophyllum myretonianum* Penhallow from the Lower Old Red Sandstone of Angus. *Trans. R. Soc. Edinb.* **64,** 469-475.

Levine, J.S., Boughner, R. and Smith, K.A. (1980) Ozone, ultraviolet flux and temperature of the paleoatmosphere. *Orig. Life* **10,** 199-213.

Li, Cheng-Sen. (1982) *Hsüa robusta*, a new land plant from the Lower Devonian of Yunnan, China. *Acta Phytotaxon. Sin.* **8,** 331-342 (in Chinese with English abstract).

Li, Cheng-Sen. (1992) *Hsüa robusta*, an Early Devonian plant from Yunnan Province, China and its bearing on some structures of early land plants. *Rev. Palaeobot. Palynol.* **71,** 121-147.

Li, Cheng-Sen and Edwards, D. (1995) A re-investigation of Halle's *Drepanophycus spinaeformis* Göpp. from the Lower Devonian of Yunnan Province, southern China. *Bot. J. Linn. Soc.* **118,** 163-192.

Lowry, B., Lee, D. and Hébant, C. (1980) The origin of land plants: a new look at an old problem. *Taxon* **29,** 183-197.

MacFarlane, J.J. (1992) Permeability of the cuticle of *Vallisneria spiralis* to carbon dioxide and oxygen. *Aquat. Bot.* **43,** 129-135.

Margulis, L., Walker, J.C.G. and Rambler, M. (1976). Reassessment of roles of oxygen and ultraviolet light in Precambrian evolution. *Nature* **264,** 620-624.

McElwain, J.C. and Chaloner, W.G. (1995) Stomatal density and index of fossil plants track atmospheric carbon dioxide in the Palaeozoic. *Ann. Bot.* **76,** 389-395.

Mercure, E.W., Leite, B. and Nicholson, R.L. (1994) Adhesion of ungerminated conidia of *Colletotrichum graminicola* to artificial hydrophobic surfaces. *Physiol. Mol. Plant Pathol.* **45,** 421-440.

Mora, C.I., Driese, S.G. and Colarusso, L.A. (1996) Middle to late Paleozoic atmospheric CO_2 levels from soil carbonate and organic matter. *Science* **271,** 1105-1107.

Nicolson, T.H. (1975) Evolution of vesicular-arbuscular mycorrhizas. In: *Endomycorrhizas* (eds E. Saunders, B. Mosse and P.B. Tinker). Academic Press, London, pp. 25-34.

Nip, M., Tegelaar, E.W., de Leeuw, J.W. *et al.* (1986a) A new non-saponifiable highly aliphatic and resistant biopolymer in plant cuticles. Evidence from pyrolysis and ^{13}C-NMR analysis of present-day and fossil plants. *Naturwissenschaften* **73,** 579-585.

Nip, M., Tegelaar, E.W., Brinkhuis, H. *et al.* (1986b) Analysis of modern and fossil plant cuticles by Curie point Py-GC and Curie point Py-GC-MS: recognition of a new highly aliphatic and resistant biopolymer. *Org. Geochem.* **10,** 769-778.

Nip, M., de Leeuw, J.W., Holloway, P.J. *et al.* (1987) Comparison of flash pyrolysis, differential scanning calorimetry, ^{13}C NMR and IR spectroscopy in the analysis of a highly aliphatic biopolymer from plant cuticles. *J. Anal. Appl. Pyrolysis* **11,** 287-295.

Nip, M., de Leeuw, J.W., Schenck, P.A. *et al.* (1989) A flash pyrolysis and petrographic study of cutinite from the Indiana paper coal. *Geochim. Cosmochim. Acta* **53,** 671-683.

Oldham, T.C.B. (1976) Flora of the Wealden plant debris beds of England. *Palaeontology* **19,** 437-502.

Oliver, F.W. and Scott, D.H. (1905) On the structure of the Palaeozoic seed *Lagenostoma lomaxi*, with a statement of the evidence upon which it is referred to *Lyginodendron. Phil. Trans. R. Soc. Lond. B* **197,** 193-247.

Paoletti, E. and Gellini, R. (1993) Stomatal density variations in beech and holm oak leaves collected over the last 200 years. *Acta Oecol.* **14,** 173-178.

Paton, J.A. and Pearce, J.V. (1957) The occurrence, structure and functions of the stomata in British bryophytes. *Trans. Br. Bryol. Soc.* **3,** 228-259.

Peñuelas, J. and Matamala, R. (1990) Changes in N and S leaf content, stomatal density and specific leaf area of 14 plant species during the last three centuries of CO_2 increase. *J. Exp. Bot.* **41,** 1119-1124.

Pirozynski, K.A. and Malloch, D.W. (1975) The origin of land plants: a matter of mycotrophism. *BioSystems* **6**, 153–164.
Poole, I., Weyers, J.D.B., Lawson, T. *et al.* (1996) Variations in stomatal density and index: implications for palaeoclimatic reconstructions. *Plant Cell Environ.* **19**, 705–712.
Powell, C.L. (1994) The palaeoenvironments of the Rhynie Cherts. Unpublished PhD thesis. University of Aberdeen.
Pratt, L.M., Phillips, T.L. and Dennison, J.M. (1978) Evidence of non-vascular land plants from the early Silurian (Llandoverian) of Virginia, U.S.A. *Rev. Palaeobot. Palynol.* **25**, 121–149.
Radoglou, K.M. and Jarvis, P.G. (1990) Effects of CO_2 enrichment on four poplar clones. II. Leaf surface properties. *Ann. Bot.* **65**, 627–632.
Raven, J.A. (1991a) Responses of aquatic photosynthetic organisms to increased solar UV-B. *J. Photochem. Photobiol. B.* **9**, 239–244.
Raven, J.A. (1991b) Long-term functioning of enucleate sieve elements: possible mechanisms of damage avoidance and damage repair. *Plant Cell Environ.* **14**, 139–146.
Raven, J.A. (1993) The evolution of vascular plants in relation to quantitative functioning of dead water-conducting cells and stomata. *Biol. Rev.* **68**, 337–363.
Raven, J.A. (1994) Physiological analysis of aspects of the functioning of vascular tissue in early plants. *Bot. J. Scot.* **47**, 49–64.
Raven, J.A. (1995a) The early evolution of land plants: aquatic ancestors and atmospheric interactions. *Bot. J. Scot.* **47**, 151–175.
Raven, J.A. (1995b) Costs and benefits of low intracellular osmolarity in cells of freshwater algae. *Functional Ecol.* **9**, 701–707.
Raven, J.A., Johnston, A.M., Parsons, R. *et al.* (1994) The influence of natural and experimental high O_2 concentrations on O_2-evolving phototrophs. *Biol. Rev.* **69**, 61–94.
Rayner, R.J. (1983) New observations on *Sawdonia ornata* from Scotland. *Trans. R. Soc. Edinb. Earth Sci.* **74**, 79–93.
Rayner, R.J. (1984) New finds of *Drepanophycus spinaeformis* Göppert from the Lower Devonian of Scotland. *Trans. R. Soc. Edinb. Earth Sci.* **75**, 353–363.
Remy, W. and Hass, H. (1991) Gametophyten und sporophyten im Unterdevon – Fakten und Spekulationen. *Argumenta Palaeobot.* **8**, 193–223.
Remy, W. and Hass, H. (1996) New information on gametophytes and sporophytes of *Aglaophyton major* and inferences about possible environmental adaptations. *Rev. Palaeobot. Palynol.* **90**, 175–193.
Retallack, G.J. (1985) Fossil soils as grounds for interpreting the advent of large plants and animals on land. *Phil. Trans. R. Soc. Lond. B* **309**, 105–142.
Retallack, G.J. (1990) *Soils of the Past. An Introduction to Paleopedology*. Unwin Hyman, Boston.
Retallack, G.J. and Mindszenty, A. (1994) Well preserved late Precambrian paleosols from northwest Scotland. *J. Sed. Res. A* **64**, 264–281.
Robinson, S.A., Lovelock, C.E. and Osmond, C.B. (1993) Wax as a mechanism for protection against photoinhibition — a study of *Cotyledon orbiculata*. *Bot. Acta* **106**, 307–312.
Rolfe, W.D.I. (1985) Early terrestrial arthropods: a fragmentary record. *Phil. Trans. R. Soc. Lond. B* **309**, 207–218.
Roselt, G. and Schneider, W. (1969) Cuticulae dispersae, ihre Merkmale, Nomenklatur und Klassifikation. *Paläont. Abh.* **B3**, 1–128.
Roth, A., Mosbrugger, V. and Neugebauer, H.J. (1994) Efficiency and evolution of water-transport systems in higher plants: a modelling approach. I. The earliest land plants. *Phil. Trans. R. Soc. Lond. B* **345**, 137–152.
Schopf, J.M. (1975) Modes of fossil preservation. *Rev. Palaeobot. Palynol.* **20**, 27–53.
Schüssler, A., Schepf, E., Mollenhauer, D. *et al.* (1995) Structural and functional characterisation of the endosymbiosis *Geosiphon pyriforme*. In: *Abstracts of the International Symposium of the Sonderforschungsbereich 199: Ecophysiology of Plants: Molecules, Membranes and Mechanisms*, Ebrenburg, Nahe, March 1995 (eds R. Ratajczak and C. Ullrich), pp. 24–25.
Schweitzer, H.-J. (1983) Die Unterdevonflora des Rheinlandes. *Palaeontogr. Abt. B* **189**, 1–138.
Scott, A.C., Chaloner, W.G. and Paterson, S. (1985) Evidence of pteridophyte-arthropod interactions in the fossil record. *Proc. R. Soc. Edinb. B* **86**, 133–140.
Shear, W. and Selden, P. (1995) *Eoarthropleura* (Arthropoda, Arthropleurida) from the Silurian of Britain and the Devonian of North America. *Neues Jahrb. Geol. Paläontol. Abh.* **196**, 347–375.
Shute, C.H. and Edwards, D. (1989) A new rhyniopsid with novel sporangium organization from the Lower Devonian of South Wales. *Bot. J. Linn. Soc.* **100**, 111–137.

Smith, J.A.C. (1991) Ion transport and the transpiration stream. *Bot. Acta* **104**, 416–421.

Stace, C.A. (1965) Cuticular characters as an aid to plant taxonomy. *Bull. Br. Mus. (Nat. Hist.) Bot.* **4**, 3–78.

Strother, P.K. (1988) New species of *Nematothallus* from the Silurian Bloomsburg Formation of Pennsylvania. *J. Paleontol.* **62**, 967–982.

Strother, P.K. (1993) Clarification of the genus *Nematothallus* Lang. *J. Paleontol.* **67**, 1090–1094.

Stubblefield, S. and Banks, H.P. (1978) The cuticle of *Drepanophycus spinaeformis*, a long-ranging Devonian lycopod from New York and eastern Canada. *Am. J. Bot.* **65**, 110–118.

Sültemeyer, D. and Rinast, K.A. (1996) The CO_2 conductivity of plasma membranes from high- and low-C_1 cells of *Chlamydomonas reinhardtii*. In: *Abstracts of 1st European Phycological Congress*, Cologne, July 1996 (ed. M. Melkonian).

Taylor, T.N., Remy, W., Hass, H. *et al.* (1995) Fossil arbuscular mycorrhizae from the Early Devonian. *Mycologia* **87**, 560–573.

Tegelaar, E.W., de Leeuw, J.W., Derenne, S. *et al.* (1989a) A reappraisal of kerogen formation. *Geochim. Cosmochim. Acta* **53**, 3103–3106.

Tegelaar, E.W., Matthezing, R.M., Jansen, J.B.H. *et al.* (1989b) Possible origin of n-alkanes in high-wax crude oils. *Nature* **342**, 529–531.

Tegelaar, E.W., Kerp, H., Visscher, H. *et al.* (1991) Bias of the paleobotanical record as a consequence of variations in the chemical composition of higher vascular plant cuticles. *Paleobiology* **17**, 133–144.

Tegelaar, E.W., Wattendorff, J. and de Leeuw, J.W. (1993) Possible effects of chemical heterogeneity in higher land plant cuticles on the preservation of its ultrastructure upon fossilization. *Rev. Palaeobot. Palynol.* **77**, 149–170.

Thomas, H.H. and Bancroft, N. (1913) On the cuticles of some recent and fossil Cycadian fronds. *Trans. Linn. Soc. Lond.* **8**, 155–204.

Tims, J.D. and Chambers, T.C. (1984) Rhyniophytina and Trimerophytina from the early land flora of Victoria, Australia. *Palaeontology* **27**, 265–279.

van de Water, P.K., Leavitt, S.W. and Betancourt, J.L. (1994) Trends in stomatal density and $^{13}C/^{12}C$ ratios of *Pinus flexilis* needles during last glacial-interglacial cycle. *Science* **264**, 239–243.

von Guttenberg, H. (1971) Bewegungsgewebe und Perzeptionsorgane. In: *Handbuch der Pflanzenanatomie,* Vol. 5 (ed. K. Linsbauer). Gebrüder Borntraeger, Berlin, pp. 203–219.

Vogelmann, T.C. (1993) Plant tissue optics. *Annu. Rev. Plant Physiol. Plant Mol. Biol.* **44**, 231–251.

Wayne, R.P. (1992) Atmospheric chemistry: the evolution of our atmosphere. *J. Photochem. Photobiol. A.* **62**, 379–396.

Wolfe, J.A. (1990) Palaeobotanical evidence for a marked temperature increase following the Cretaceous/Tertiary boundary. *Nature* **343**, 153–156.

Woodward, F.I. (1987) Stomatal numbers are sensitive to increases in CO_2 from pre-industrial levels. *Nature* **327**, 617–618.

Woodward, F.I. and Bazzaz, F.A. (1988) The responses of stomatal density to CO_2 partial pressure. *J. Exp. Bot.* **39**, 1771–1781.

Yapp, C.J. and Poths, H. (1994) Productivity of pre-vascular continental biota inferred from the $Fe(CO_3)OH$ content of goethite. *Nature* **368**, 49–51.

Ziegler, H. (1987) The evolution of stomata. In: *Stomatal Function* (eds E. Zeiger, G.D. Farquhar and I.R. Cowan). Stanford University Press, Stanford, CA, pp. 29–57.

2

Structure and ontogeny of plant cuticles

C.E. Jeffree

2.1 Introduction

All primary aerial surfaces of vascular plants and some bryophytes are covered by a thin, superficial film, the cuticle, which is composed of soluble and polymeric lipids. The functions attributed to the cuticle are the protection and waterproofing of the plant surface (Holloway, 1994). The latter is often simply taken to mean defence against external water (e.g. rain and dew) by virtue of water-repellency, but its function as a permeability barrier against water loss via the epidermal surface is the central role of the cuticle in all vascular land plants, and together with stomatal regulation of gas exchange enables homeostasis of tissue hydration to be achieved under a wide range of hydrological conditions (Raven, 1984). Although the epicuticular waxes primarily govern the wettability of the plant surface (Holloway, 1970), they do not function as the primary permeability barrier, which is now believed to be governed by soluble cuticular lipids (SCL) embedded in the cuticular polymer matrix (Riederer, 1991; Schönherr and Mérida, 1981; Schreiber *et al.*, 1996). The aim of this chapter is to review currently available data on the structure, ontogeny and composition of the cuticular membrane, and to integrate the observed variability in cuticle types into a structural and developmental model.

2.2 The cuticular membrane

2.2.1 *Nomenclature*

The predominant structural model is of a bilayer *cuticular membrane* (CM), the two layers distinguishable by their ontogeny, ultrastructure and chemical composition. The most distal portion, which forms a continuous layer outside the epidermal cell wall (Lee and Priestley, 1924), is composed predominantly of soluble and polymerized aliphatic lipids, while the inner layer, which forms by impregnation of the cell wall, also contains substantial amounts of various embedded cell wall polysaccharides. In addition, the CM typically bears a superficial layer of soluble epicuticular wax, which may appear amorphous or which may develop obviously crystalline protrusions. The CM is subtended, particularly during the early stages of its ontogeny, by a layer of pectin-rich cell wall polysaccharide, the so-called pectin lamella. This bilayer model of cuticle structure is evident in the descriptions of various authors (Chafe and Wardrop, 1973; Crisp, 1963; Hülsbruch,

Plant Cuticles, edited by G. Kerstiens.

1966a,b; Jarvis and Wardrop, 1974; Lee and Priestley, 1924; Sargent, 1976a,b,c; Sitte and Rennier, 1963; von Mohl, 1847 (progenitor of the concept that the cuticle is a layered structure); Ziegenspeck, 1928) and has been summarized in a wide variety of published diagrams (Hadley, 1981; Holloway, 1971, 1982a, 1993, 1994; Jeffree, 1986; Juniper and Jeffree, 1983; Martin and Juniper, 1970; Roelofsen, 1952; Schönherr *et al.*, 1991; Wattendorff and Holloway, 1984). This chapter employs the widely accepted nomenclature employed by Wattendorff and Holloway (1980) and Holloway (1982a), with additions where appropriate. Briefly, this nomenclature derives from von Mohl (1847) who referred to the superficial layer of the CM as the *cuticle proper* (CP), defining it as composed of saponifiable lipids that leave no cellulosic residue. Esau (1953) used the term 'cuticularized layer' to describe this outermost layer of the cuticle. Beneath the CP is the *cuticular layer* (of the cell wall) (CL) (von Mohl, 1847), or 'cutinized layer' in Esau's terminology (Esau, 1953), which is usually regarded as the cutin-impregnated outer part of the epidermal cell wall (CW) (Roelofsen, 1952, 1959). However, Sitte and Rennier (1963) and Hülsbruch (1966a,b) distinguished a cellulose-free outer zone in the CL of leaves of *Ficus elastica* and *Ilex integra*, thus making it difficult to distinguish this zone from the CP by von Mohl's definition. Although Hülsbruch introduced the term '*cuticularsaum*' to describe this layer, neither this term, nor Esau's nomenclature, which is readily confused with that of von Mohl, have become widely adopted.

2.2.2 *The pectin lamella*

In Sargent's view (1976a,c) the cuticular membrane of *Libertia elegans* develops centripetally, produced continuously at its junction with the underlying pectic layer. No doubt influenced by the recent observations of Sitte and Rennier and Hülsbruch, Sargent (1976a,c) saw the secondary cuticle (= CL) not as a cutin-impregnated region of primary cell wall (PCW) but as an external secretion. Fibrils extending into the secondary cuticle were interpreted as extensions from the pectin lamella, which bonded the CM to the cell wall. Especially in the early stages of cuticle development the CM may be bonded to the periclinal walls of the epidermal cells by a layer which stains intensely with ruthenium red and other cationic dyes, is hydroxylamine/ferric chloride positive (see e.g. Tenberge, 1992) and stains with Schiff reagent following periodate oxidation (Holloway, 1982a), suggesting that it is pectin-rich, and it may be possible to isolate the CM from the epidermis solely by treatment with pectinase (Baker and Procopiou, 1975; Norris and Bukovac, 1968; Orgell, 1955). The pectin lamella is often considered an extension of the middle lamella (e.g. Sargent, 1976a). Jeffree (1986) perpetuated this myth, but consideration of the origins of the middle lamella, from the cell plate, shows it to be a distinct structure. The middle lamella forms inside the mother PCW, and is therefore separated by it from any sub-cuticular pectin lamella, as shown in Figure 2.1b–d. The pectin lamella layer may be consumed in the progressive cutinization of the PCW, especially in species with thick cuticles (Tenberge, 1992), and there may be no ultrastructural correlate of the structure detected by light microscopy (Wattendorff and Holloway, 1980). Although the base of the cuticle may give reactions with ruthenium red and $KMnO_4$ consistent with the presence of pectin, an alternative explanation of this reactivity is that it results from high concentrations of carboxyl groups in the immature cutin. Later in development, many cuticles become harder to isolate using pectinases alone. Mixtures of cellulase and pectinase are often effective in separating the cuticle when it is cross-linked to the cell wall with cellulose (Schieferstein and Loomis, 1959) but more powerful cellulolytic (cellulose lytic) reagents such as zinc chloride/HCl (Holloway and Baker, 1968) or cuprammonium complex (Baker *et al.*,

1964; Huelin, 1959; Huelin and Gallop, 1951) may be required, especially with the thicker CMs (Baker and Parsons, 1971; Holloway and Baker, 1968), implying that the CM becomes more strongly cross-linked to the CW by embedment of cellulose microfibrils. In conifers such as *Picea abies* progressive lignification of the epidermal CW occurs over a period as long as 2 years, with correspondingly increasing difficulty in isolating the cuticle enzymatically (Lendzian *et al.*, 1986).

2.2.3 *Structural polymers of the cuticular membrane (CM)*

Cutin. Until very recently, it was believed that the dominant structural polymer in the plant cuticle is the lipid polyester, cutin. Cutin is a polar, cross-linked polymer, in principle of very great molecular weight, constructed of inter-esterified aliphatic hydroxy acids, with chain-lengths of C_{16} and C_{18}. The detailed composition of the component monomers of cutin, and of the SCLs, has been comprehensively reviewed elsewhere (Baker, 1982; Holloway, 1982b). The presence of cutin renders the cuticular membrane comparatively unreactive and acid-resistant, and the polymeric lipid membrane may be isolated from the epidermis of leaves by retting with microorganisms (Skoss, 1955), by polysaccharide hydrolases such as pectinase and cellulase (Lendzian *et al.*, 1986; Orgell, 1955), by various acid treatments which dissolve pectin (ethylenediaminetetraacetic acid (EDTA) (Letham, 1958; Schneider, 1960); ammonium oxalate/oxalic acid (Roelofsen, 1952)) and/or cellulose (e.g. zinc chloride/HCl (Holloway and Baker, 1968); 6 M HCl (Schmidt and Schönherr, 1982); 10% chromic + 10% nitric acid (Jeffrey, 1917)). The enzyme-isolated CM contains 20–80% of the lipid polymer cutin by weight (Holloway, 1994), contains quantities of embedded SCLs, predominantly C_{16} and C_{18} fatty acids, and is encrusted with epicuticular waxes, of which the predominant constituents are long-chain primary alcohols (C_{26}, C_{28}, C_{30}), secondary alcohols (C_{29}), alkanes (C_{29}, C_{31}), β-diketones (C_{31}, C_{33}), alkanals and alkanones. Polysaccharide microfibrils, their identity poorly defined, but probably including pectin, crystalline cellulose microfibrils, and hemicelluloses (e.g. xylan and xyloglucan) are also embedded in the CL.

Cutan. It now appears possible that not one, as previously thought, but two major lipid polymers may be present in plant cuticles. The polyester polymer cutin is readily solubilized or saponified, by alkaline hydrolysis, which cleaves the ester linkages. However, a very substantial insoluble residue often remains after saponification of the CM. Recently it has been shown that, when subjected to Curie-point (770°C) pyrolysis-gas chromatography-mass spectroscopy (CP-pyrolysis-GC-MS), the alkali-resistant fraction yields homologous series of *n*-alkanes, alkenes and α, ω-alkadienes, unlike the fatty acid/hydroxy fatty acid profile of fragments generated by pyrolysis of polyester cutin (Nip *et al.*, 1986a,b, 1987; Tegelaar, 1990; Tegelaar *et al.*, 1989). The characteristic CP-pyrolysis-GC-MS signature of this 'cutan' (Tegelaar, 1990) has been interpreted as consistent with that of a polymethylene polymer, analogous in composition to polythene, and on this basis cutan has been reported in a number of recent and fossil species (Table 2.1). Although the cuticles of some species appear completely to lack cutan, notably those of tomato fruit, *Citrus* and *Erica* leaves, cutan may be the principal polymer in some others, notably in *Beta vulgaris* in which cutin is not detectable (Holloway and Baker, 1970; Table 2.1), and the mixed cutin/cutan type of CM may turn out to be very widespread. The two polymers may occur in any ratio (Tegelaar, 1990), and differ in their relative abundance at different stages of cuticle development. However, it should be noted that the structure of cutan has yet to be confirmed by conventional chemical analysis, and its biosynthetic origin is unknown.

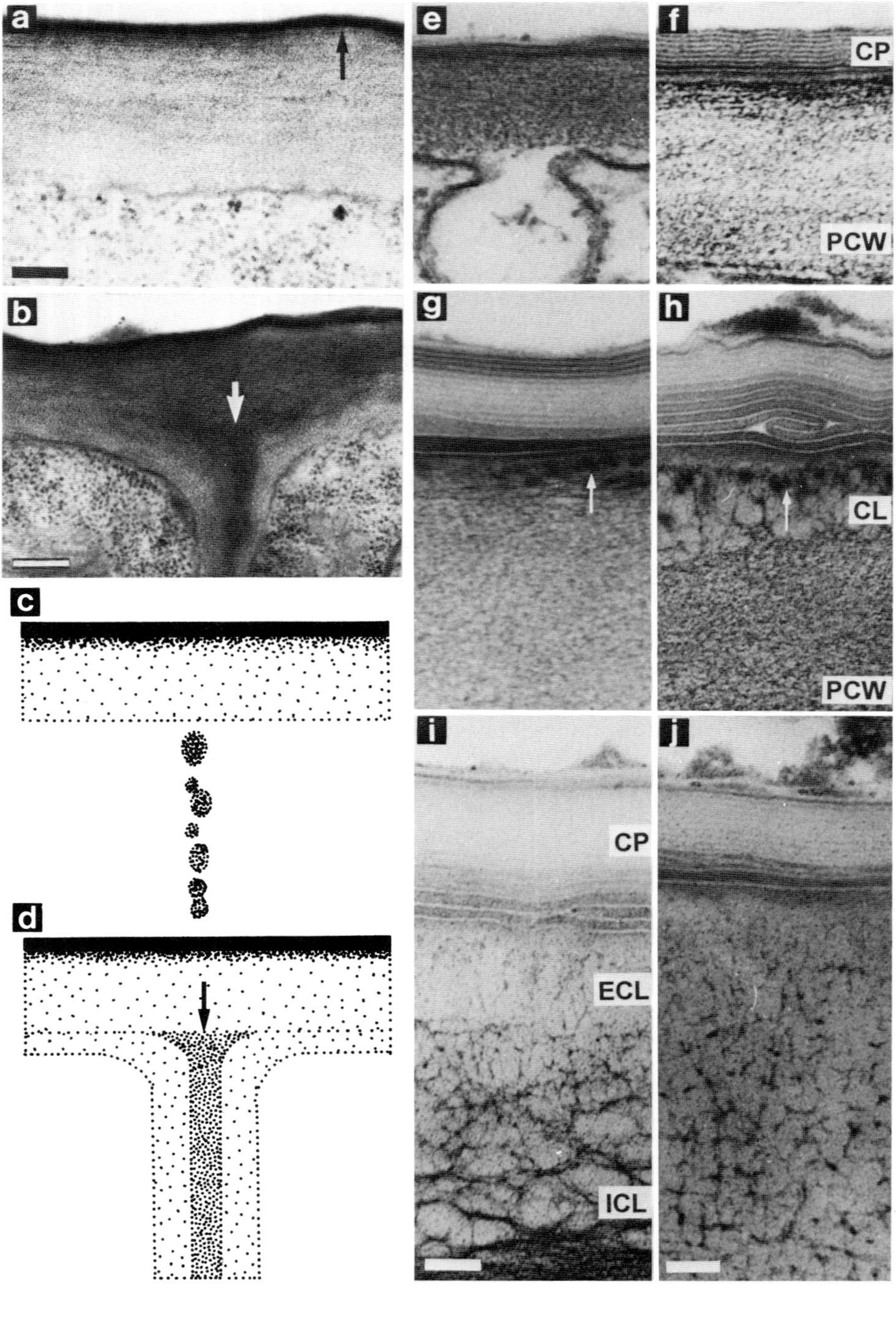
a
b
c
d
e
f
CP
PCW
g
h
CL
PCW
i
j
CP
ECL
ICL

Figure 2.1. Pisum sativum *(Leguminosae) apical dome: outer epidermal CWs of protodermal cells. (a) Periclinal PCW with a uniformly electron-dense procuticle (arrow). (b) PCW and procuticle overlying an anticlinal PCW. The middle lamella terminates beneath the mother PCW (arrow) (cf. c, d). (c, d) Diagrams of (c) cell plate formation near an outer epidermal cell wall, with covering procuticle; (d) the mother PCW separates the middle lamella between daughter cell PCW from the procuticle and underlying pectic lamella. (e–j)* Clivia miniata *(Amaryllidaceae) leaf adaxial. (e) Leaf base: a simple CP, with 1–2 pairs of lamellae; (f) 1 cm from base: CP with 8–9 pairs of lamellae, underlain by electron-dense reticulum; (g) 2 cm from base: CP lamellae are of greater and more variable thickness. The innermost and outermost lamellae are denser than in the central region (cf. Figure 2.3b). Electron-dense globules are accumulating beneath the CP (arrow). (h) 3 cm from base: a reticulate CL has formed, electron-dense globules remaining in its outer layer (arrow). (i) 4 cm from base (cell-expansion phase): the CL has a finely reticulate external layer (ECL) and a coarsely reticulate inner layer (ICL). (j) 5 cm from base: at late cell-expansion phase, the reticulum of the ECL extends as far as the CP. (Osmium tetroxide; uranyl acetate/lead citrate × 75 000; bar = 0.1 μm.) ((e–j) Reproduced from Riederer and Schönherr (1988) Development of plant cuticles fine structure and cutin composition of* Clivia miniata *Reg. leaves.* Planta ***174****, 127–138, with permission from Springer-Verlag.)*

2.3 The cuticle proper: is it the primary cuticle?

2.3.1 *The procuticle*

The CP is present on leaves at a very early stage in their development (Jarvis and Wardrop, 1974; Sargent, 1976a,b; Wattendorff and Holloway, 1980 and see Figure 2.3b; Riederer and Schönherr, 1988; Schmidt and Schönherr, 1982 and see Figure 2.1e and f; Tenberge, 1992 and see Figure 2.6a). Sargent (1976a,b) consequently proposed that the CP be called the 'Primary cuticle'. There is a certain consistency about this nomenclature, since the terms primary and secondary are also applied to ontogenetically distinct phases of the cell wall. However, the CP cannot be regarded as the primary cuticle. At the very earliest stages of epidermal development, rapidly dividing cells in the shoot apices are covered by a highly water-repellent and -impermeable osmiophilic film, which forms a dense, amorphous, superficial layer ~ 20 nm thick in transmission electron microscope (TEM) sections (Figure 2.1a and b). In a number of species, including *Cuscuta gronovii* and *Utricularia sandersonii* (Figure 2.2a–e), an amorphous procuticle ~ 22 nm thick is present on the earliest protodermal surface (Heide-Jørgensen, 1991). During primordial development the procuticle increases in thickness to ~ 30 nm, slightly reducing in density to develop a globular transition cuticle (Figure 2.2b) before transforming into a lamellate layer ~ 35–40 nm thick containing 2–3 pairs of electron-dense and lucent lamellae (Figure 2.2c), identifiable as the first manifestation of the lamellate CP (Figures 2.2d and e).

The procuticle and its relationship to the 'internal cuticle' lining intercellular spaces. The cuticle, as conventionally understood, is solely a product of the epidermis. It develops on epidermal cell surfaces which have been topologically external since their earliest beginnings in the zygotic embryo (Scott, 1966). However, a cuticle can develop deep between the anticlinal walls of adjacent

Table 2.1. *Occurrence of the polymer types cutin and cutan in the CM of fossil and extant plant species*

Species (family)	Refs*
Cutin-only	
Citrus limon (Rutaceae)	1
Erica carnea (Ericaceae)	1
Lycopersicon esculentum, fruit (Solanaceae)	1
Salicornia europaea (Chenopodiaceae)	2
Cutan-only	
Beta vulgaris (Chenopodiaceae)	1
Mixed cutin/cutan	
Ginkgo biloba (Gynkgoaceae)	1
Ginkgo huttonii (Gynkgoaceae, **Jurassic**)	1
Abietities sp. (Pinaceae, **Cretaceous**)	1
Sciadopytis verticillata (Taxodiaceae)	2
Amentotaxus argotaenia (Taxaceae)	2
Picea abies (Karst) (Pinaceae)	4
Symplocos hallensis (Theaceae, **Eocene**)	1, 3
Symplocos paniculata (Theaceae)	1, 3
Gossypium sp. (Malvaceae)	2
Malus pumila (Rosaceae)	2
Prunus laurocerasus (Rosaceae)	2
Hydrangea macrophylla (Saxifragaceae)	2
Limonium vulgare (Plumbaginaceae)	2
Acer platanoides (Aceraceae)	2
Quercus robur (Fagaceae)	2
Quercus sp. (Fagaceae, **Miocene**)	1
Zostera marina (Zosteraceae)	2
Clivia miniata (Amaryllidaceae)	1
Agave americana (Agavaceae)	1

*References: (1) Nip *et al.* (1986a, b); (2) Tegelaar (1990); (3) Tegelaar *et al.* (1993); (4) Kögel-Knabner *et al.* (1994).

epidermal cells to form ‘spandrels’ (see Figure 2.9c), which delineate and record the epidermal cell pattern in the isolated cuticle, and also forms on new epidermal cell wall surfaces which arise schizogenously, as between the opposed guard cell surfaces (Figures 2.3a and 2.4d), and on *epidermal cell surfaces* lining the substomatal cavity. The cuticle on the guard cell surfaces inside the substomatal cavity of cotton leaves is 150–200 nm thick, about 75% of that on the leaf surface (Wullschleger and Oosterhuis, 1989; see Table 2.3).

2.3.2 *Internal cuticle*

Circumstantial evidence for a fatty, superficial layer on mesophyll cell surfaces derives from the observation that their exposed surfaces are difficult to wet with pure water but are readily wetted with oils, organic solvents or water contaminated with a wetting agent (Lewis, 1945). Observations of condensation/distillation of water droplets with high contact-angles onto cell surfaces bordering air spaces in leaves provide additional evidence (Jeffree *et al.*, 1987; Sheriff, 1977; Willison and Pearce, 1983; Willison *et al.*, 1984). Droplets could not form under such circumstances on a wet hydrophilic surface. Precisely contrary evidence was

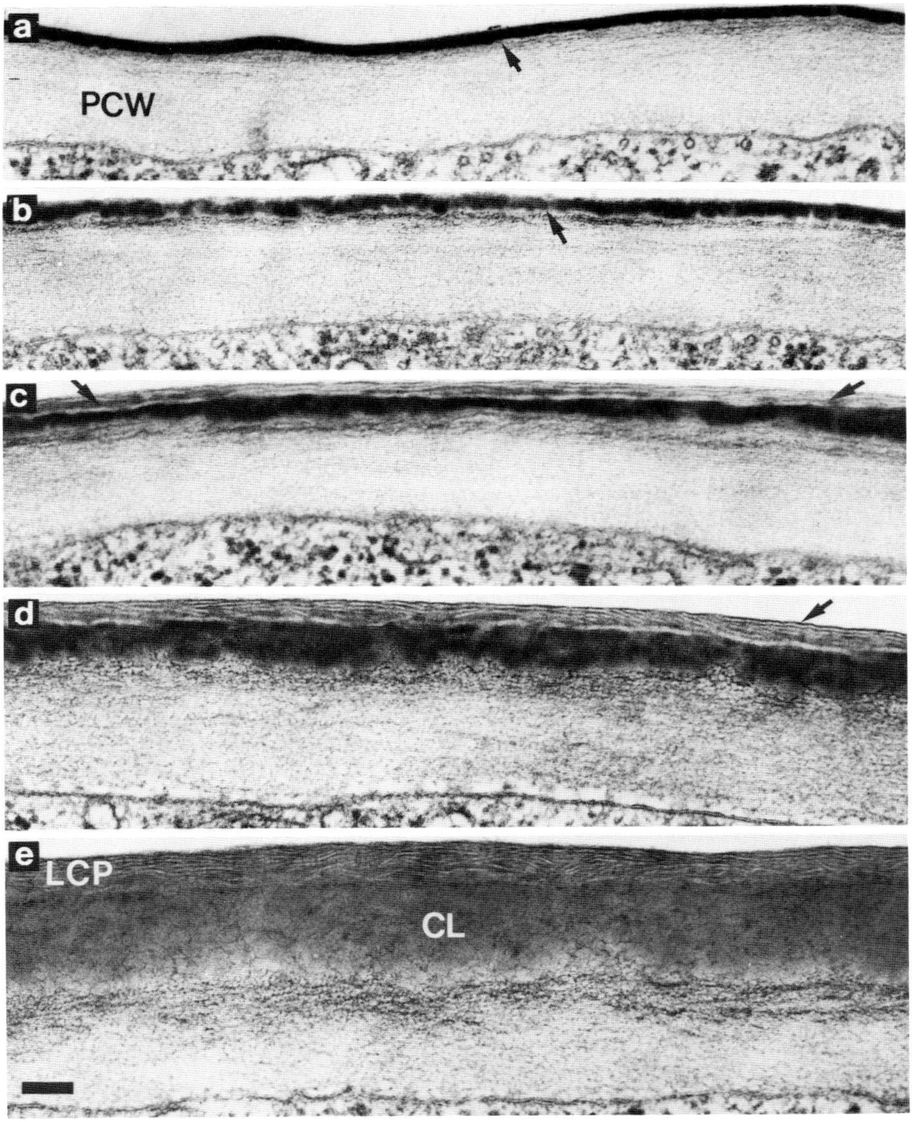

Figure 2.2. *(a–e)* Utricularia laterifolia *(Droseraceae) bladder primordium. Stages in CM development, showing (a) a uniformly electron-dense procuticle (arrow); (b) transition zone with globular, electron-dense material in a translucent matrix (arrow). The adjacent cell wall microfibrils are electron-dense. (c) The procuticle is transformed into a lamellate CP (arrows), with an underlying globular electron-dense layer. (d) A reticulate CL is constructed beneath the CP, which has increased further in thickness. (e) The reticulate CL is now as thick as the original PCW, but the thickness of the uncutinized cell wall is undiminished. (LCP, lamellate cuticle proper. × 64 400; bar = 0.1 µm. ((a–e) Reproduced from Heide-Jørgensen (1991) Cuticle development and ultrastructure - evidence for a procuticle of high osmium affinity.* Planta ***183****, 511–519, with permission from Springer-Verlag.)*

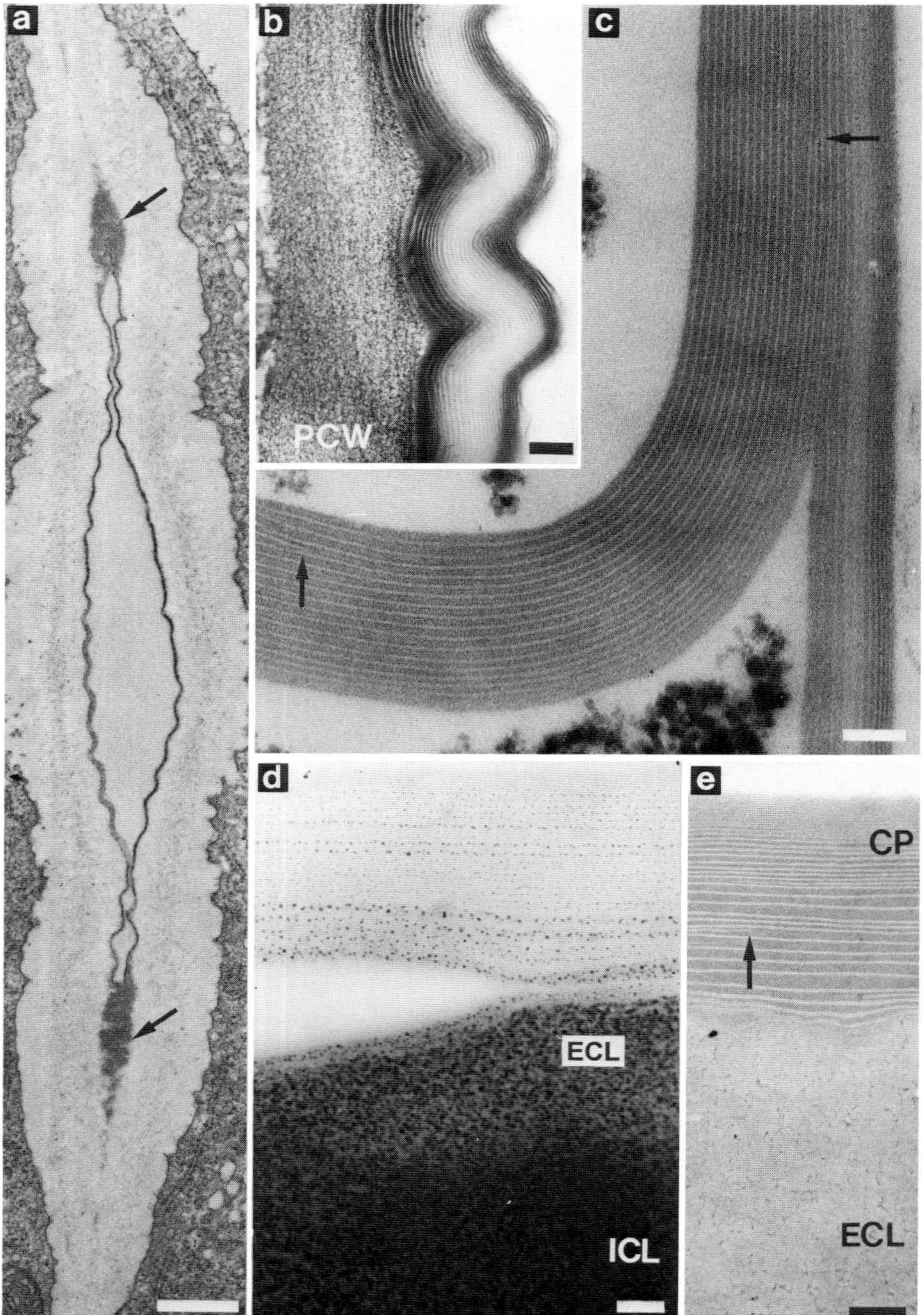

Figure 2.3. *(a)* Phaseolus vulgaris *(Leguminosae) primary leaf, abaxial. Guard cell separation viewed in paradermal section. Cell separation is limited by electron-dense end-granules (arrows). The new external PCW surfaces are covered with an amorphous*

obtained by Willison and Pearce (1983), who reported the cell wall microfibrils to be exposed at the mesophyll cell surface in freeze-fractured rye leaves, except in the vicinity of cell–cell junctions, where they were obscured by an overlay of amorphous material *on which frozen water droplets were visible*. Mesophyll cell surfaces can be dyed with basic but not with acidic chromophores, indicating the presence of fatty acids (Lewis, 1945), and Scott (1948) reported histochemical reactions at the surfaces of mesophyll cells in leaves of over 50 vascular plant species consistent with the presence of a suberized or cutinized surface. Scott (1966) interpreted this lipid deposition on the CW surface as resulting from wounding of the plasmodesmata consequent upon cell–cell separation. However, in both higher plants and bryophytes, cell separation occurs along predetermined regions of the middle lamella where characteristic deposits of an osmiophilic, electron-dense material occur (Roland, 1978; Sack and Paolillo, 1983a,b and see Figure 2.4d; Jeffree *et al.*, 1986; Kollöffel and Linssen, 1984). Following cell separation, an amorphous osmiophilic layer (Figure 2.2a), ultrastructurally similar to the procuticle described by Heide-Jørgensen (1991) remains on the cell surface (Kaufman *et al.*, 1970). It is tempting to think that there may be some relationship between the procuticle and the hydrophobic 'internal cuticle'. However, neither has been isolated, nor shown to contain cutin. In the latter respect, it is relevant to note that the capacity to convert hydroxy-fatty acids into the polymer cutin is believed to be unique to the epidermal cells and lacked by mesophyll (Kolattukudy, 1984).

2.4 Ontogeny, composition and ultrastructure of the cuticle proper

2.4.1 *Structure and ontogeny of the CP*

In many species, perhaps in the majority, the CP of the mature CM is of lamellate construction (see Table 2.3). The lamellate region is typically about 50–150 nm,

---

electron-dense layer ~ 20 nm thick, equivalent to the procuticle (cf. Figure 2.2a). (× 21 800; bar = 0.5 µm. Ga/Os, U/Pb). (b–e) Agave americana *(Agavaceae) leaf. (b) Wrinkled, lamellate CP on a young leaf near the apical cone has greater area than the underlying PCW. Superficial layers of the CP are more electron dense than the central region. (Ga/Os,* en bloc *UAc, U/Pb, × 52 000; bar = 0.1 µm). (c) CP detached from CL during processing has split within an electron-lucent lamella. Electron-dense lamellae show two dense subunits. Arrows show terminations of electron-lucent lamellae within electron-dense lamellae. Periodicity of lamellae decreases from inner to outer surfaces. (Ga/Os, U/Pb, × 84 000; bar = 0.1 µm). (d, e) CP, ECL and part of ICL of a mid-aged leaf, stained by the H_2SO_4-I_2/KI-Ag proteinate reaction (Holloway* et al.*, 1981) to localize epoxide groups in the electron-dense lamellae, and in ICL > ECL. As in (c), splitting of the CP has occurred within an electron-dense lamella. (Ga/Os, H_2SO_4-I_2/KI-AgP, × 58 800; bar = 0.1 µm). (e) CP and ECL fixed in Ga/Os/ruthenium red, no section stain. Electron-dense lamellae of CP show enhanced density. Periodicity of lamellae decreases towards exterior, and lamellae may merge. Reticulations in ECL extend to the base of the CP. Faint lamellae occur in ECL. Terminations of electron-lucent lamellae within electron-dense at arrows (× 69 000; bar = 0.1 µm). (Ga, glutaraldehyde; Os, osmium tetroxide; U, uranyl acetate; Pb, lead citrate.) ((b–e) Reproduced from Wattendorff and Holloway (1980) Studies on the ultrastructure and histochemistry of plant cuticles: the cuticular membrane of* Agave americana *L.* in situ. Ann. Bot. *46, 13–28, with permission from Academic Press.)*

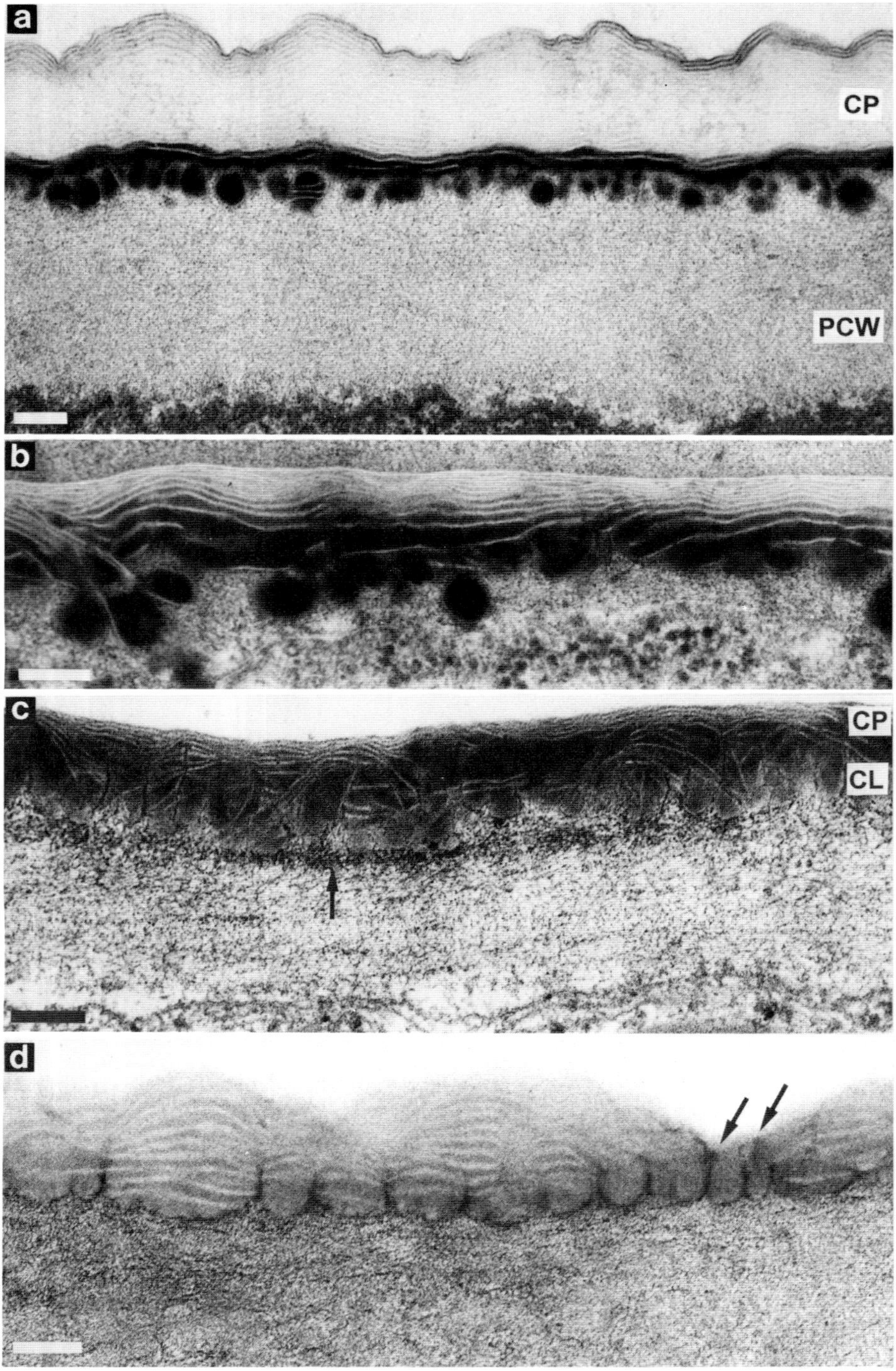
a
CP
PCW
b
c
CP
CL
d

exceptionally > 1 µm in thickness, as in *Hakea suaveolens* (Heide-Jørgensen, 1978). The thickness of the lucent lamellae is approximately 5-8 nm (Chafe and Wardrop, 1973), generally rather greater than for suberin lamellae (3-6 nm, Heide-Jørgensen, 1980), although the ranges overlap. Furthermore the lucent lamellae are more constant in thickness, at about 5 nm, than the electron-dense lamellae, which may exceed 15 nm in width, and this variability may give rise to graded periodicity in the lamellae, with the closest spacing at the external surface as in Figure 2.3c and e.

Ontogeny of the CP. The earliest appearance of the lamellate CP in *Phormium tenax* (Jarvis and Wardrop, 1974) is strikingly similar to that in leaves of *Clivia miniata*, which grow from a basal meristem, and in which a lamellate CP was the earliest cuticular structure observed in sections taken from the leaf base (Riederer and Schönherr, 1988). At the stage shown in Figure 2.5a, which may be analogous to that in the transition from the procuticle (Figure 2.1a), the CP consisted only of two pairs of lucent/dense lamellae, and contained no detectable ester cutin. At 1–2 cm from the base, the CP had now increased to about 70 nm thickness, had extended to 8–10 pairs of periclinal lamellae, and contained ester cutin, although the non-saponifiable fraction at this stage was of greater mass. In the first 5 cm of the leaf base of *Clivia*, leaf area increases by a factor of about nine, but the CP lamellae give no appearance of being stretched by this, maintaining the number of lamellae through to maturity. Indeed the thickness of the lamellae, and thus of the CP, is at its greatest during the period of most rapid expansion up to about 6 cm from leaf base.

Construction of the CP lamellae. In *Utricularia sandersonii*, the flocculent appearance of the procuticle, prior to the appearance of lamellae (Figure 2.2b), persists in the sublamellar layers of the cell wall, initially as a very electron-dense zone that increases in thickness to about 100 nm while further lamellae are added

Figure 2.4. *(a)* Hakea suaveolens *(Proteaceae): 2.1 cm long developing leaf. Construction of a thick (~1 µm) lamellate CP, the inner layers are thickened and electron dense. Electron-dense globules coated with electron-lucent shells appear to fuse with the base of the CP (× 72 000; bar = 0.1 µm). (b)* Cassytha pubescens *(Lauraceae): suberized inner cell wall of an epidermal idioblast. Lamellate structure analogous to a CP is formed* within *the CW. As in (a) electron-dense globules, each coated with electron-lucent shell, appear to fuse with the basal lamellae. The electron-lucent coat contributes to the electron-lucent lamella, and the core to the electron-dense lamella. Within the most recently formed lamellae, adjacent dense globules are still visible, enabling the number contributing to each layer to be appreciated (× 95 000; bar = 0.1 µm). ((a,b) by H.S. Heide-Jørgensen (unpublished).) (c)* Cuscuta gronovii *(Convolvulaceae): paradermally lamellate CP. The CL with random lamellae may be interpreted as composed of massed globules similar to those seen in (a, b), and Figure 2.8. The CL is subtended by an electron-dense fibrillar layer of PCW (arrow) (× 92 000; bar = 0.1 µm). (d)* Funaria hygrometrica *(Musci): mature schizogenous CM of the capsule stomatal pore. CM is lamellate, interrupted by cell-wall polysaccharide fibril bundles, which may reach to the surface of the cuticle (arrows) (× 89 000; bar = 0.1 µm). ((c) Reproduced from Heide-Jørgensen (1991) Cuticle development and ultrastructure – evidence for a procuticle of high osmium affinity.* Planta ***183,*** *511–519, with permission from Springer-Verlag. (d) Reproduced from* Sack and Paolillo *(1983) Stomatal pore and cuticle formation in* Funaria. Protoplasma ***116,*** *1–13, with permission from Springer-Verlag.)*

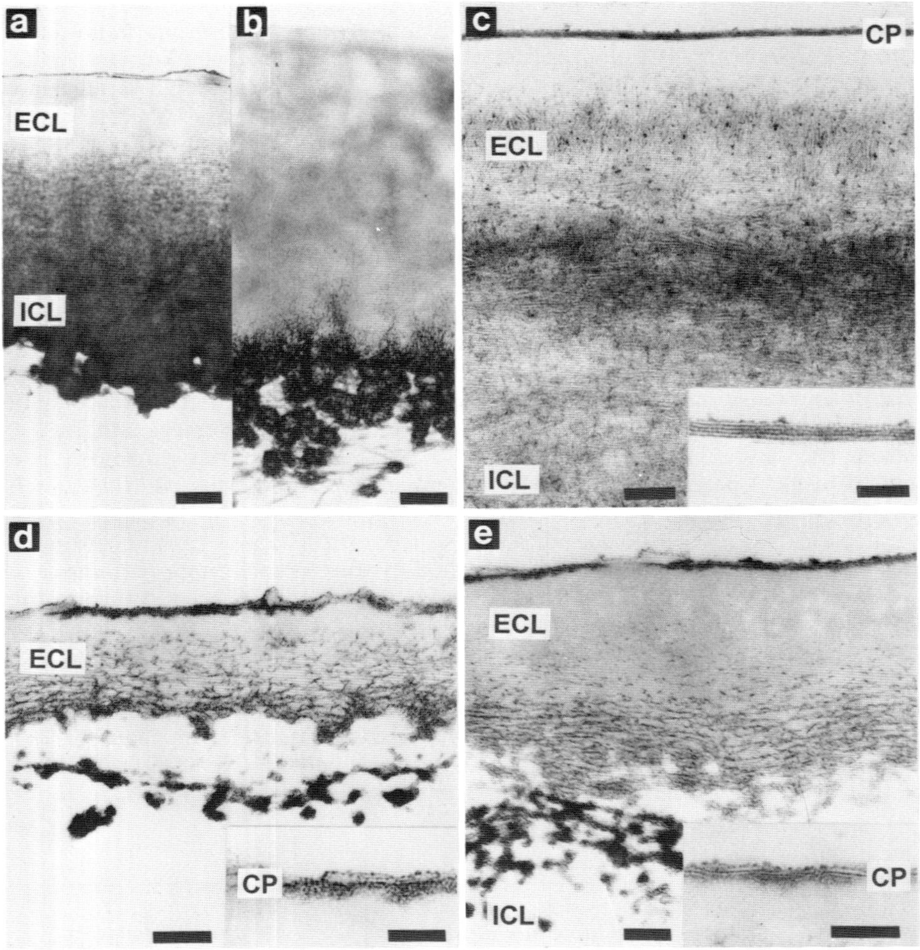

Figure 2.5. *(a–e)* Clivia miniata *(Amaryllidaceae) leaf, adaxial: sections of enzyme-isolated CM and* $CHCl_3$*/methanol-extracted, acid hydrolysed (6 M HCl, 120°C, 24 h) polymer matrix membranes (MX) (Ga/*$KMnO_4$*; U/Pb). (a) CM of young leaf, 6.5 cm from base, showing dense (polar) ICL and electron-lucent (non-polar) ECL. (b) CM of a mature leaf, 12.5 cm from base, with polar cutinized cystolites in the inner ICL. (a, b × 12 200; bar = 0.5 µm.) (c) MX, showing lamellate CP (inset). CL layers are both reticulate and polylamellate. (× 36 000; bar = 0.2 µm. Inset × 63 000; bar = 0.1 µm.) (d, e) MX de-esterified in* BF_3*-methanol. (d) 6.5 cm from base: the ECL is alkali-resistant (non-ester cutin). Reticulate fibrils occur in the inner two-thirds of the ECL. (× 38 000; bar = 0.2 µm. Inset × 80 000; bar = 0.1 µm.) (e) 12.5 cm from base: non-ester cutin layer has doubled in thickness. The outer two-thirds is now non- or weakly reticulate. ICL is damaged, but polar, globular residue remains, probably polysaccharide-rich. Insets to (d, e) show lamellate structure remains in the CP after solvent extraction, acid and alkaline hydrolysis. (× 30 000; bar = 0.2 µm. Inset × 90 000; bar = 0.1 µm.) (Ga, glutaraldehyde; U, uranyl acetate; Pb, lead citrate.) ((a–e) Reproduced from Schmidt and Schönherr (1982) Development of plant cuticles – occurrence and role of non-ester bonds in cutin of* Clivia miniata *Reg. leaves.* Planta ***156****, 380–384, with permission from Springer-Verlag.)*

(Figure 2.2c and d), sometimes in a less precisely periclinal orientation to those first secreted. Ultimately the lamellate region of the cuticle proper is about 80 nm thick, and is subtended by a reticulate CL > 200 nm thick which has lost much of its electron density compared with the initial stages (Figure 2.2e). Hitherto unpublished micrographs by H.S. Heide-Jørgensen show the unusually thick lamellate layer under construction in *Hakea suaveolens* cuticle (Figure 2.4a). The basal CP lamellae are densely osmiophilic, and osmiophilic globules, enclosed in an electron-lucent envelope about 5 nm thick, appear to fuse with the basal layer. In Figure 2.4b (H.S. Heide-Jørgensen, unpublished data) the electron-lucent envelopes appear to unwrap from the dense droplets as they approach the lamellate layer in the wall of a suberized epidermal idioblast in *Cassytha pubescens* (see also de Vries, 1968a). These images provide glimpses of a possible mechanism of construction of the lamellate CP, and stimulate many further questions; for example, the lucent lamellae in the CP are widely interpreted as containing SCL (Fisher and Bayer, 1972; Hallam, 1964; Heide-Jørgensen, 1978; Schönherr and Mérida, 1981; Sitte, 1975; and see Section 2.4.2). If composed of SCL, then the position of the lucent lamellae outside such hydrophobic globules, interfaced with the aqueous environment of the apoplast would be surprising. However, close examination shows the globules to have a thin (~2.5 nm), dense border, which may represent a shell of polar material, possibly protein or glycoprotein. The manner in which these globules move through the cell wall matrix (if indeed they do so) is obscure. Are they tangled up with the cell wall matrix, polysaccharide microfibrils running through them, do they part the matrix in transit, remaining polysaccharide free, or are they immobilized within the polysaccharide matrix, growing by the accumulation of material until fusion occurs? Finally the composition of the electron-dense and electron-lucent components of this system, and the detailed mechanism of the construction process (self-assembly or enzyme mediated assembly?) are currently obscure, and offer fertile ground for fundamental research.

2.4.2 *Composition of the CP*

Lamellation in *Hedera helix* CP is apparently removed by extraction with lipid solvents (Viougeas *et al.*, 1995), suggesting that the lamellar structure may be attributable to alternation of non-polar waxy layers with more polar cutiniferous layers, as is often proposed (Fisher and Bayer, 1972; Hallam, 1964). By contrast, Mérida *et al.* (1981) found the lamellae to be resistant to extraction in chloroform, indeed subsequently staining more prominently. This would be consistent with the exposure of more polar material to staining by the removal of obscuring SCL. Schönherr and Mérida (1981) found the CP of *Allium cepa* to yield a scant 1 μg cm^{-2} of wax, equivalent at best to about 25% of the amount required to fit the alternating polymer/wax model. This does not entirely negate the possibility that the equivalent chemical species are involved in this lamellar structure. A possible model would be that the waxes involved in the lucent phase of the CP are primary alcohols esterified to an ester cutin, but set against this is the key observation of the extreme sensitivity of cuticular permeability to solvent extraction (Schönherr and Mérida, 1981), which indicates that the permeability barrier is soluble.

Cutan is the principal polymer in *Beta vulgaris* (Tegelaar, 1990) in which cutin may not be detectable (Holloway and Baker, 1970; see Table 2.1). The cuticle of *B. vulgaris* has a CP with a pronounced periclinal lamellate structure about 100–200 nm thick, composed of 5–15 lamellae, but lacks significant development of the CL. Schmidt and Schönherr (1982) observed that following exhaustive extraction of enzyme-isolated CM in chloroform/methanol (to remove SCL) and 6 M HCl, 120°C for 24 h (to hydrolyse proteins and polysaccharides), the lamellate structure of the CP of *Clivia* was distorted but not destroyed by treatment with 20% BF_3-methanol, 80°C for 24 h, which cleaves ester bonds in cutin. Its appearance also predates the appearance of identifiable ester cutin in the membrane (Riederer and Schönherr, 1988). Histochemical reactivity of the opaque lamellae in *Clivia* and *Agave* is, however, consistent with the conclusion that polar molecules or substituted groups are present. Thus the lamellar structure is reported by several authors to be enhanced by treatment with ruthenium red (Krüger *et al.*, 1996; Sargent, 1976a; Wattendorff and Holloway, 1980; Figure 2.3e), and with $KMnO_4$ or $Ba(MnO_4)_2$ (Hoch, 1979; Mérida *et al.*, 1981; Wattendorff and Holloway, 1982), which points to (but cannot be held to confirm) the presence of carboxyl groups. Further the dense lamellae of *Agave* react positively with iodine-silver proteinate (Wattendorff and Holloway, 1980; Figure 2.3d), which is believed specifically to identify epoxy groups in the C_{18} type cutins by the formation of epoxide iodohydrin derivatives with iodide that reduce silver proteinate to metallic silver (Holloway et al., 1981; Wattendorff and Holloway, 1980, 1982). These findings have a number of implications. Thus, while the electron-lucent lamellae of the CP may contain some SCL, the indication is once again that its structure may not be wholly dependent either on soluble waxes or on cutin. In addition, the CP in these species fail von Mohl's definition of the CP, that it should be completely saponifiable, and the possibility is now raised that either of two polymers, cutin and cutan, or various mixtures of them, may equally be involved in forming the electron-dense lamellae.

2.5 Ontogeny, composition and structure of the cuticular layer

Sargent (1976a, c) regarded the CL as structurally and developmentally distinct from that of the CP, and, observing that it was formed in a second secretory phase after the expansion of the epidermis was complete, proposed the term '*secondary cuticle*'. The development of the CP also appears to precede that of the CL in other species that have been investigated (*Eryngium rostratum*, Chafe and Wardrop, 1973; *Phormium tenax*, Jarvis and Wardrop, 1974; *Agave americana*, Wattendorff and Holloway, 1980 and Figure 2.3b; *Utricularia laterifolia, Athanasia parviflora, Cuscuta gronovii*, Heide-Jørgensen, 1991 and Figure 2.2c; *Clivia miniata*, Schmidt and Schönherr, 1982; Riederer and Schönherr, 1988 and Figure 2.1f; *Picea abies*, Tenberge, 1989, 1992 and Figure 2.6a; *Hedera helix*, Viougeas *et al.*, 1995). In *Clivia*, above about 3 cm from the leaf base, in the region of most rapid area expansion, the CL begins rapidly to increase in thickness (Schmidt and Schönherr, 1982). At least initially, this increase results from cutin deposition within the superficial layers of the primary cell wall, immediately beneath the CP. The transition of the CL with the CW is marked by globular cutin-rich cystoliths (Mérida *et al.*, 1981), which seem to arise by accretion, the accumulation of cutin on nuclei within the interstices of the CW, until the globules so formed fuse laterally to form a continuous matrix. In consequence, the structure of the CL is permeated with electron-dense, fibrillar CW material spread apart by the growing cutin globules. Very similar events occur in *Picea abies*, the reticulate CL similarly appearing immediately beneath a clearly demarcated CL, which is, however, apparently amorphous. In spruce, the reticulum is PaTAgP-positive, indicating the presence of polysaccharides, and is labelled by cellulase-gold sol (Figure 2.7b), indicating the presence of a glucan polysaccharide — cellulose or xyloglucan — but a mannanase-gold probe also locates these structures (Figure 2.7f; Tenberge, 1989, 1992). The reticulum also stains strongly with $KMnO_4$ (Figure 2.7e), hydroxylamine-ferric chloride, reactions which are abolished by extraction with ammonium oxalate, indicating that pectin may be present (Tenberge, 1989, 1992). In *Clivia*, following the cessation of cell expansion at about 5-6 cm, the absolute amount of ester cutin in the membrane continues to increase, reaching a peak at 10–15 cm (Riederer and Schönherr, 1988; Schmidt and Schönherr, 1982).

2.5.1 *ICL/ECL: layering in the CL*

The CL of *Clivia* is clearly itself a layered structure. Mérida *et al.* (1981) showed that the inner and outer regions of the CL showed differential staining responses; the internal cuticular layer (ICL) staining strongly with $KMnO_4$, the external cuticular layer (ECL) only weakly. The ICL and ECL are differentiated here *sensu* Wattendorff and Holloway (1980) and Schmidt and Schönherr (1982), and not *sensu* Mérida and Ogura (1987), who included the CP in the ECL. The distinction is evident from the late-expansion zone at about 6 cm from the leaf base, and also at maturity (Figure 2.5a and b). The ECL may be noticeably denser than the ICL when stained in the conventional way with osmium, uranium and lead as in *Nicotiana glauca* (Mérida and Ogura, 1987). Two CL layers may be observed in many other species, as in *Agave americana* (Wattendorff and Holloway, 1980); *Hedera helix* (Holloway *et al.*, 1981; Viougeas *et al.*, 1995; see also Figures 2.6e and 2.9a and b), but the distinction was not recognized in earlier investigations of *Avena sativa* (O'Brien, 1967), *Eryngium* or *Apium* (Chafe and Wardrop, 1973); *Phormium tenax* (Jarvis and Wardrop, 1974); *Agave americana* (Wattendorff, 1974) or *Libertia elegans* (Sargent, 1976a,b,c). Schmidt and Schönherr (1982), found that the outermost 0.5 μm of the CM in expanding leaves of *Clivia* withstood saponification in BF_3 (Figure 2.5d and e), and consequently referred to the residue as non-ester cutin. As the rate of cell expansion declines, the quantity of non-ester cutin in the CL of the *Clivia* leaf cuticle increases, at first in parallel with the deposition of saponifiable ester cutin, but more rapidly in the later stages, so that non-ester cutin ultimately becomes the more abundant polymer, comprising about 60% of the polymer matrix cuticle (Riederer and Schönherr, 1988; Schmidt and Schönherr, 1982; Tegelaar, 1990). The non-ester cutin observed by Schmidt and Schönherr (1982), is presumably conterminous with the polymethylenic fraction referred to as cutan by Tegelaar (1990). This significant deposit of cutan does not, as might be anticipated, begin to accumulate in the innermost layers of the ICL, but immediately beneath the CP, indicating that the cutan precursors pass through the CL *en route*. The resistant layer was in excess of 500 nm thick in mature leaves (Figure 2.5e). In *Picea abies* also, both the amorphous CP (100 nm) and the outer third of the CL (350 nm) appear to be alkali-insoluble, and can be detached as a unit from the underlying alkali-soluble CL (Tenberge, 1992). Recent evidence using CP/MS ^{13}C nuclear magnetic resonance (NMR) spectroscopy and analytical pyrolysis suggests, however, that the non-saponifiable residue of spruce cutin may be more closely related to lignin than to cutan (Kögel-Knabner *et al.*, 1994). Two striking observations derive from the images shown in Figure 2.5d and e. First, despite having been subjected to acid hydrolysis in 6 M HCl at 120°C for 24 h, the saponification-resistant ECL is permeated with electron-dense reticulum in its basal region, indicating that in this region the polysaccharide is profoundly inaccessible to aqueous reagents, and in turn signifying that the prospects for a low-resistance water and solute transport route, using the fibrils as high permeability 'channels',

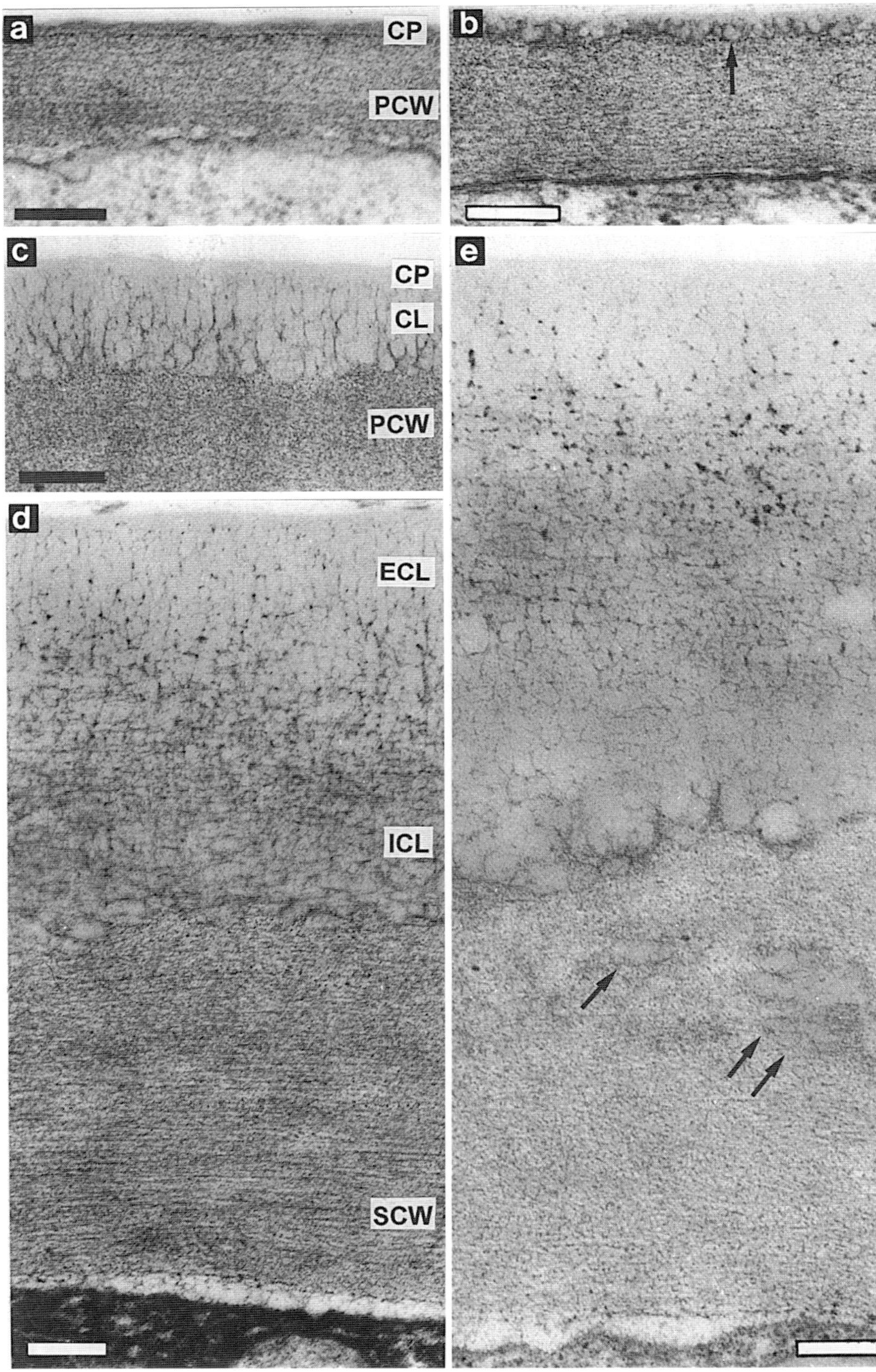
a
CP
PCW
b
c
CP
CL
PCW
e
d
ECL
ICL
SCW

are not good. It has been suggested that the aliphatic polymethylene in the cutan of Agave is not simply contaminated with a polysaccharide moiety but covalently bound to it (Nip *et al.*, 1986a,b; Tegelaar, 1990; Tegelaar *et al.*, 1989). Second, in *Clivia,* the alkali-extracted membrane is almost clear of reticulum in the superficial layer (70–80 nm) of the immature cuticle at the 6.5-cm position and absent from the superficial 300–400 nm of the mature ECL, despite the fact that the reticulum extended to the base of the CP in the earlier stages of development. Thus, in *Clivia*, though apparently not in *Agave* or *Picea*, cutan deposition in the CL may be by intrusive accumulation of precursor material between the cutinized CL and the CP. It cannot, however, be ruled out that the 'cutanization' of the ECL occurs not by additional deposition of a novel polymer, but by modification *in situ* of the previously deposited cutin, possibly by intensive cross-linking via the epoxide groups to form ether linkages. This hypothesis is strengthened by the observation that as the non-ester cutin increases, ultimately to become the major polymer of the *Clivia* CM, the ester cutin declines (Riederer and Schönherr, 1988; Schmidt and Schönherr 1982). The apparent lack of a polysaccharide reticulum in the ECL of mature *Clivia* leaves recalls the findings of Sitte and Rennier (1963) that the CL in *Ficus elastica* may be cellulose-free, and of Hülsbruch (1966a,b) that there may be both cellulosic and cellulose-free regions, as in *Ilex integra*. Sargent (1976a,b) interpreted the whole CL (her 'secondary cuticle') as interposed between the CP and the primary cell wall in several species of the Iridaceae, probably for the same reason. However, by contrast with the previous examples, in *Agave americana*, which also has a mixed-polymer cuticle (Tegelaar, 1990), the reticulate region extends fully as far as the CP (Wattendorff and Holloway, 1980; Figure 2.3e), and the same applies in *Picea abies* (Figure 2.6d). In both *Agave* and *Clivia*, the outer CL shows a rapid transition from periclinal lamellate layering to anticlinally or randomly oriented lamellae, which nevertheless have polysaccharide fibrils ramifying between (Wattendorff and Holloway, 1982). Thus Figure 2.5c of the mature *Clivia* cuticle shows lamellation throughout a bilayered CL, much of which is also reticulate (Schmidt and Schönherr, 1982). It may therefore be concluded that the CL lamellae are not a manifestation of the presence of cutan, and that they can occur equally, whether the polymer matrix is predominantly cutan or predominantly cutin.

2.5.2 *Further development of the CL*

In *Picea abies*, an amorphous CP (~100 nm) is present on the protodermal cells in-bud (Phase I, Tenberge, 1992; Figure 2.6a). At this stage, the leaves bear no

---

Figure 2.6. *(a-e)* Picea abies *(Pinaceae): stages in the development of the CM.* *(a)* Phase I: *a thin (~ 24 nm) amorphous CP (L1, Tenberge, 1992) is formed outside the 240-nm thick PCW. (× 62 200; bar = 0.2 μm.) (b)* Phase II: *globular deposits beneath the CP initiate construction of a CL (arrow). (× 61 100; bar = 0.2 μm.) (c)* Phase III: *striated reticulation in the CL (L2, Tenberge, 1992). (× 59 600; bar = 0.2 μm.) (d) The CL, now 1 μm thick, shows outer parallel-striated (L2s) and inner cross-linked reticulate (L2r) regions above non-cutinized SCW thickened to 1 μm. (× 51 600; bar = 0.2 μm.) (e) The CM, 1.6 μm thick over 1.3 μm of SCW, shows pale L2s contrasted against denser L2r and treelike L2t layers. Cutinized cystoliths of L2r structure are forming detached from the main CM (arrows). (× 53 400; bar = 0.2 μm.) ((a-e) Reproduced from Tenberge (1992) Ultrastructure and development of the outer epidermal wall of spruce (*Picea abies*) needles.* Can. J. Bot. ***70,*** *1467–1487, with permission from the NRC Research Press.)*

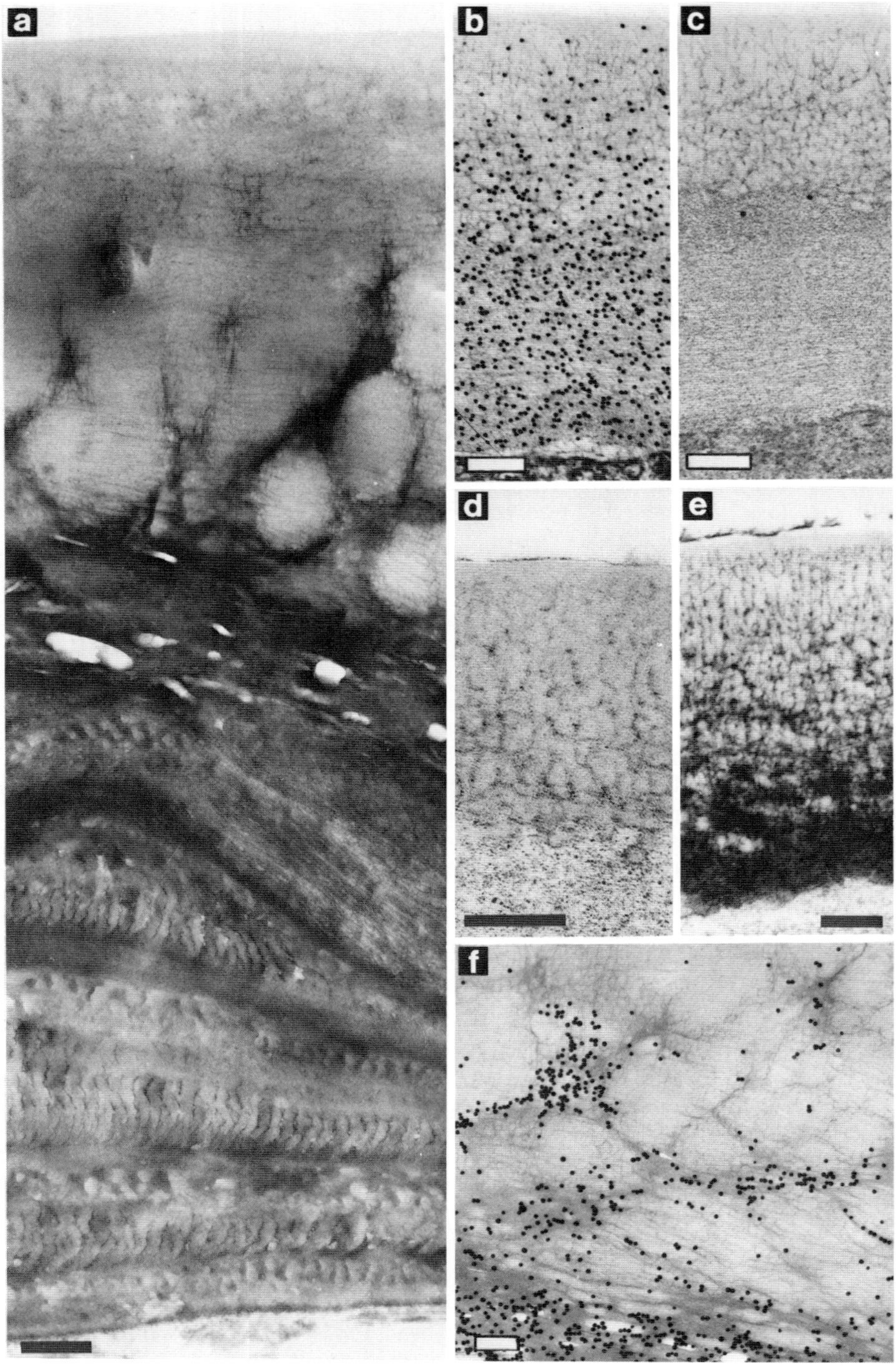
a
b
c
d
e
f

crystalline epicuticular wax, which first appears on rapidly expanding leaves at about the time of the division of the guard cell mother-cells (Jeffree, 1974). Construction of the CL starts in April (early Phase II), the first accumulations of the CL entrapping polysaccharide fibrils (Figure 2.6b). As in *Clivia* (Riederer and Schönherr, 1988; Schmidt and Schönherr, 1982), the CL of spruce is thus formed not by encrustation but by impregnation of the CW with cutin. The first cutin-embedded fibrils of the CL were identified by Tenberge (1992) as carboxylated and alkali-soluble, but *not cellulose*, the first stage of development of the CL coinciding with the disappearance of the pectin lamella as a distinct substratum (Figure 2.6b and c). Contrary to Sargent's interpretation that the lamellate CP constructs the epicuticular waxes and is in the process consumed, the CP in spruce remains constant in appearance and thickness throughout leaf expansion and CM development and the synthesis of the crystalline epicuticular layer.

Tenberge (1992) recognized three sub-layers in the mature CL, differentiated chiefly by the form of the fibrils ramifying through them. In the outermost layer (L2s ~350 nm) the fibrils are fine, radially aligned, and lack cross-links, appearing as dark granules when the CL is sectioned paradermally. The central sub-layer (L2r ~1100 nm) contains a three-dimensional reticulum of cross-linking fibrils with the same appearance in whichever orientation it is sectioned. In the lowest layer (L2t ~1800 nm) the fibrillar polysaccharide material is treelike, with substantial bundles of polysaccharide microfibrils, continuous with the secondary CW, between globular units of cutinous matrix penetrated by an L2r-like three-dimensional network. The underside of the isolated CM often shows the rounded, pillow-like profiles of these cutin globules.

The differences in the structure of the layers of the CL, which arise successively during development, result from the fact that they are formed by cutin impregnation of successively deeper layers, first of the primary, and later of the secondary cell wall. Particularly where the CL develops to substantial depth within a thick secondary cell wall, as in the heavy CM layers of xerophytic species, or in spandrels, accumulation of cutin may occur at locations isolated from the main body of the CM, to form cystoliths (see Figures 2.5b and 2.9c). Morphologically and chemically (judging solely by their reactivity with electron microscopy (EM) stains and fixatives) cystoliths are comparable to the innermost layers of the ICL. It has

---

Figure 2.7. *(a–f)* Picea abies *(Pinaceae): (a) CM at maturity (*Phase V*). The innermost CL (L2t, Tenberge, 1992) shows substantial branching bundles of polysaccharide microfibrils, separating large cutin globules. The lignified SCW is 2.3 μm thick, containing an outer calcium crystal-zone, and up to 2 μm of lignified helicoidal SCW. (× 47 100; bar = 0.2 μm.) (b–e) Reactivity of the Phase III CM. (b) Ga/Os, cellulase-gold solution, U/Pb, showing correlation of gold granules with the electron-dense reticulum. (× 35 800; bar = 0.2 μm.) (c) Control for (b), Ga/Os, carboxymethyl cellulose, cellulase-gold solution, U/Pb. Reactivity of reticulum with gold solution is abolished. (× 40 000; bar = 0.2 μm.) (d) Ga/Os, PA-T-AgP; silver granules locate to the reticulum, indicating the presence of* vic *(1,2)-diols, probably of hemicelluloses or pectins. (× 69 500; bar = 0.2 μm.) (e) GA/KMnO$_4$, U/Pb: strong staining of the inner ICL and reticulum. (× 41 000; bar = 0.2 μm.) (f) Phase V: Ga/Os, mannanase-gold solution, U/Pb. Labelling of the treelike reticulum at the inner ICL (L2t, Tenberge, 1992). (× 25 300; bar = 0.2 μm.) (GA, glutaraldehyde; Os, osmium tetraoxide; U, uranyl acetate; Pb, lead citrate.) ((a–f) Reproduced from Tenberge (1992) Ultrastructure and development of the outer epidermal wall of spruce (*Picea abies*) needles.* Can. J. Bot. ***70****, 1467–1487, with permission from the NRC Research Press.)*

been suggested that cutin cystoliths are packets of cutin or cutin precursors *en route* from the cell membrane to the CL (Mahlberg and Kim, 1992). This seems improbable, however, since they are usually to be seen embedded in a tangle of polysaccharide microfibrils, and it would be necessary to explain how this intractable material might be swept aside from their path. More plausibly, cystoliths represent sites of accumulation of cutin within the CW which have not yet enlarged sufficiently to fuse laterally, and with the base of the CL. That development of the CL proceeds in this way, by accumulation of cutin and waxes upon nuclei in the interstices of the cell wall, is evident from the increasingly flocculate morphology of the CL with distance from the CP. The CL layers closest to the CP contain fewest polysaccharide microfibrils, but as construction of the CL proceeds, more of the cutinization occurs within the structure of the cell wall, the polysaccharide fibrils being swept aside as the globular masses of CL form. The multiglobular profile of the inner CL is apparent from many examples of the morphology of the profile seen in TEM, and of the inner surface morphology of isolated CMs seen in scanning electron microscopy (SEM). The completion of the construction of the cuticular layer continues over about 3 months following bud-burst, being completed by late June (Tenberge, 1992). At this time, rapid lignification of the secondary cell wall may halt the process of cutinization, but it seems more probable that the precursors are not simply blocked (since they would have to accumulate elsewhere, and this is not observed), but that the synthesis of precursors is down-regulated. The development of the CL may continue over more than one season in spruces close to altitudinal or latitudinal tree-lines (Baig and Tranquillini, 1976), although Grace (1990) found no evidence that Scots pine *krummholz* had thinner cuticles. Judged by the evidence of the increasing difficulty over the first 1 or 2 years of isolating the cuticles using enzymes (Lendzian *et al.,* 1986), further development of the lignified secondary walls may occur progressively over much of a needle's lifetime. There is, in general, little data on the post-majority changes in the CM of long-lived leaves.

2.6 Expansion of the plant cuticle during epidermal cell growth

2.6.1 *Wrinkled and blebbed cuticles in general*

The cuticular membrane covering cells is often ornamented with papillae or ribbing, as in the ribbed surfaces of the papillate petal epidermal cells of *Tropaeolum speciosum*, and the surfaces of glandular trichomes in the nectary of *Lonicera* (Jeffree, 1986). Such wrinkling is often accomplished by simple folding caused by area expansion of the CM, as described above, but may also be due to massive local thickening of the CL (see Figure 2.9d). The thickness of the CP may be reduced over such ornaments, or may remain at constant thickness, possibly depending on whether the development occurs during or after completion of the main phase of leaf expansion, as in the trichomes of *Hakea leucoptera* (Sargent, 1976a).

2.6.2 *Wrinkled cuticles and area expansion of the epidermis*

The plant cuticle undergoes very large changes in area as cells expand during organ growth. Over organs which undergo very rapid area expansion, the cuticle may initially be very loose and wrinkled, a CM of even thickness being thrown up into pronounced folds which increase the surface area by a factor of two to three compared with that of the underlying epidermis (Figure 2.3b). This form of pre-expansion wrinkling was described by Considine and Knox (1979) and Rosenquist and Morrison (1988) on the grape ovary, which undergoes about 10^4-fold volume expansion in the 40 days post anthesis, representing about a 22-fold increase in linear dimensions of the epidermis. As growth proceeds, the wrinkles flatten out and eventually disappear. Although the surface area of the wrinkled cuticle is insufficient to accommodate all of the post-anthesis extension, wrinkling may help to prevent cuticle splitting during the most rapid phase. The thickness both of the polylamellate CP, and of the lamellae themselves, is maintained in *Hedera helix* during substantial area expansion of the underlying epidermal cell wall (Viougeas *et al.*, 1995). Indeed, in *Clivia miniata*, the thickness of the lamellae appears to be greatest during cell expansion (Riederer and Schönherr, 1988), not least as might be the anticipated consequence of stretching. In *Clivia*, to maintain both constant thickness of the CP and of the lamellae, new material equivalent to at least nine times the original mass would have to be incorporated. It is not sufficient to explain such expansion merely in terms of passive properties of elasticity, creep and plasticity. Since the lamellate structure appears to be preserved, growth may occur by intussusception, the intercalation of new material between and within the layers deposited earlier. If so, it must be presumed that the necessary precursors can traverse the lamellae. However, the periodicity of the lamellae in the CP of *Agave* diminishes towards the exterior (see Figure 2.3c and e) and this is discernible in some other species (e.g. *Hakea*, Figure 2.8), offering the possibility that, as the outer lamellae are stretched, they are simply replaced by new lamellae beneath. However, the mechanisms underlying these processes are far from clear, and the available observations pose more questions than answers; for example it is not known how the epidermal cells of *Vitis* ovaries (or of *Agave americana*; Figure 2.3b) produce their wrinkled cuticles. If the cuticle polymer covering a plant cell or organ during growth forms large macromolecular domains, it may be necessary to explain the process of its expansion in terms of active mechanisms of bond cleavage and reformation, as discussed by Kolattukudy (1984), and it must be presumed that the enzyme systems necessary to accomplish this are in place in the cuticle or in the underlying cell wall. By analogy with the transglycosylation mechanisms involved in cell wall loosening, it is postulated that transacylation mechanisms involving a hydroxy acid hydroxy-acyltransferase may be involved in hydroxy acid chain cleavage and reconnection. Hoffmann-Benning and Kende (1994) and Hoffmann-Benning *et al.* (1994), showed that the tension-bending of rice coleoptiles was relieved when treated in cutinase, indicating that the cuticle is under tension during coleoptile elongation, and can provide a mechanical constraint on organ growth. The mechanical properties of the cuticle during growth have commercial significance, especially in the fruit industry, where cuticular cracking causes blemishing of top fruit. However, the phenomenon of cuticle loosening is also of fundamental importance to understanding of the rate-limitation of plant growth by the outer epidermal cell wall, of which the cuticle is an integral component.

2.7 Cuticular microchannels

Extrusion of wax to the surface via pores was first proposed by de Bary (1871, 1884), who nevertheless failed to observe them. A century later, Hall and Donaldson (1962, 1963) and von Wettstein-Knowles (1974) again postulated the involvement of microchannels in epicuticular wax production by an extrusion process. The form

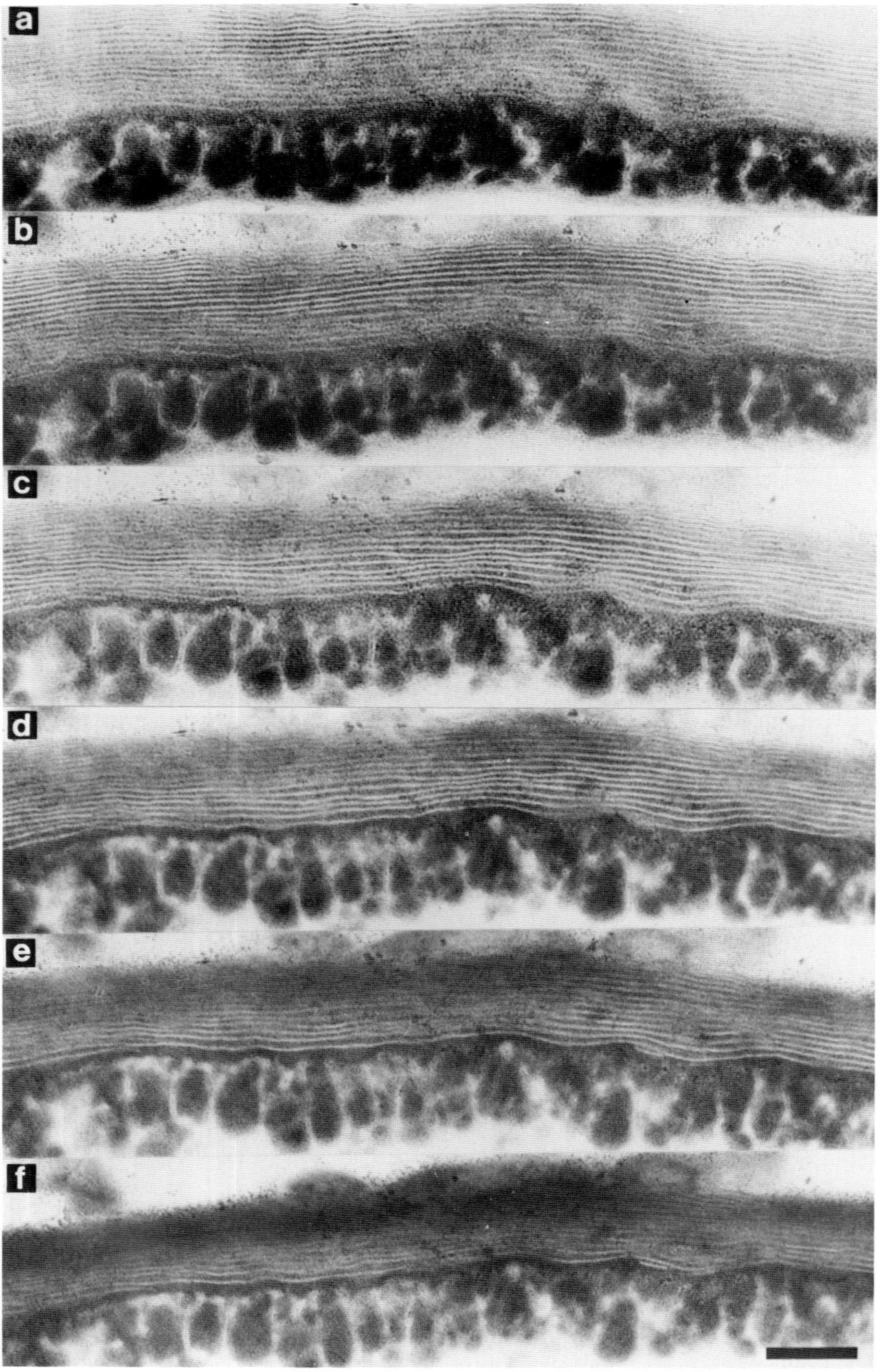
a
b
c
d
e
f

of the wax rods on leaf sheaths of *Saccharum*, nodes of *Arundinaria* and other species with '*Strelitzia*-type' waxes (Frölich and Barthlott, 1988; Meusel *et al.*, 1994) is suggestive of extruded structures, and the issue of wax extrusion through cuticular pores has again been raised by Machado and Barros (1995) in *Syagurus coronata*, the ouricuri palm of Brazil, which bears *Strelitzia*-type waxes and massive epidermal wax crusts. Hallam (1967) suggested that wax might migrate to the surface through anastomosing channels between the cuticular lamellae of *Eucalyptus cinerea*. Wax microchannels were reported in freeze-etched cuticles of *Trifolium repens* (Hall, 1967), and Miller (1986) reported the 'existence of discrete, natural cuticular pores concomitant with anticlinally-oriented transcuticular canals' in leaves of more than 50 species, but his observations were made with the light microscope, the resolution of which is inadequate to confirm the identity of such features. Electron microscopy has consistently failed to reveal transcuticular pores in the sense of open transcuticular channels via which wax, or other materials, might pass freely. In the petiole of *Vaccinium reticulatum*, and in sections of leaf cuticles of *Plantago major* and *Ardisia crenata* (Fisher and Bayer, 1972), the CL is penetrated by a 'channel-like reticulum which emanates from the primary cell wall', consisting of cutin-embedded microfibrillar bundles (Singh and Hemmes, 1978). Pectic radiating strands in the reticulate CL of *Spartocytisus filipes* were referred to as 'microchannels' by Lyshede (1978) and in *Hakea suaveolens* by Heide-Jørgensen (1978). These structures, which correspond with massively enlarged forms of the treelike reticulum in the spruce cuticle (Tenberge, 1989, 1992; Figure 2.9c), are light in TEM negatives, of greater electron-density than the matrix. They are thus inconsistent with the concept of open pores or channels.

The reticulum of polysaccharide microfibrils ramifying through the CL offers a potential pathway for diffusion (but not mass flow) of water and solutes through the CL. Although the reticulum demonstrably traverses the entire CL in *Agave* (Wattendorff and Holloway, 1980), and also in *Picea abies* (Tenberge, 1989, 1992) it stops short of the ECL in *Clivia*, which has notably low permeability to water (Schmidt *et al.*, 1981; Schmidt and Schönherr, 1982). The ultrastructure of the pathway for diffusion of water and other molecules across a lamellate CP is probably at intermolecular scale. It has never been observed, and may permanently elude observation by electron microscopy. Penetration of the CP by aqueous solutes is very slow, as judged by the staining reactions with osmium tetroxide, for example, where the inner and outer lamellae are more densely stained than those of the central region (Figure 2.3b). Periclinal penetration of permanganate is two to four times faster in the electron-dense lamellae than in the electron lucent ones. In the CL, penetration by the fibrillar pathway is some 10 times or more faster than via the cuticular polymer matrix (Wattendorff and Holloway, 1984). The lamellate cuticle covering the guard cells of *Funaria hygrometrica* (Sack and Paolillo, 1983a,b) is penetrated to the surface by polysaccharide microfibrillar bundles (Figure 2.4d). Sound evidence for the existence of discrete pores, capable of functioning as pathways for fluid mass flow, in cuticles of higher plants is practically confined to

---

Figure 2.8. *(a–f)* Hakea suaveolens *(Proteaceae): developing CM in a 2-cm leaf. The same section has been photographed with 5° tilt between successive photographs, showing the effect of angle relative to the beam on the contrast of the lamellae. The apparent mean periodicity of the lamellae is about: (a) 8.2 nm; (b) 7.4 nm; (c) 7 nm; (d) 6.6 nm; (e) 5.7 nm; (f) 5.1 nm. Electron-dense globules beneath the CP are coated with an electron-lucent shell (cf. Figure 2.4a-c). (By H.S. Heide-Jørgensen, unpublished.)*

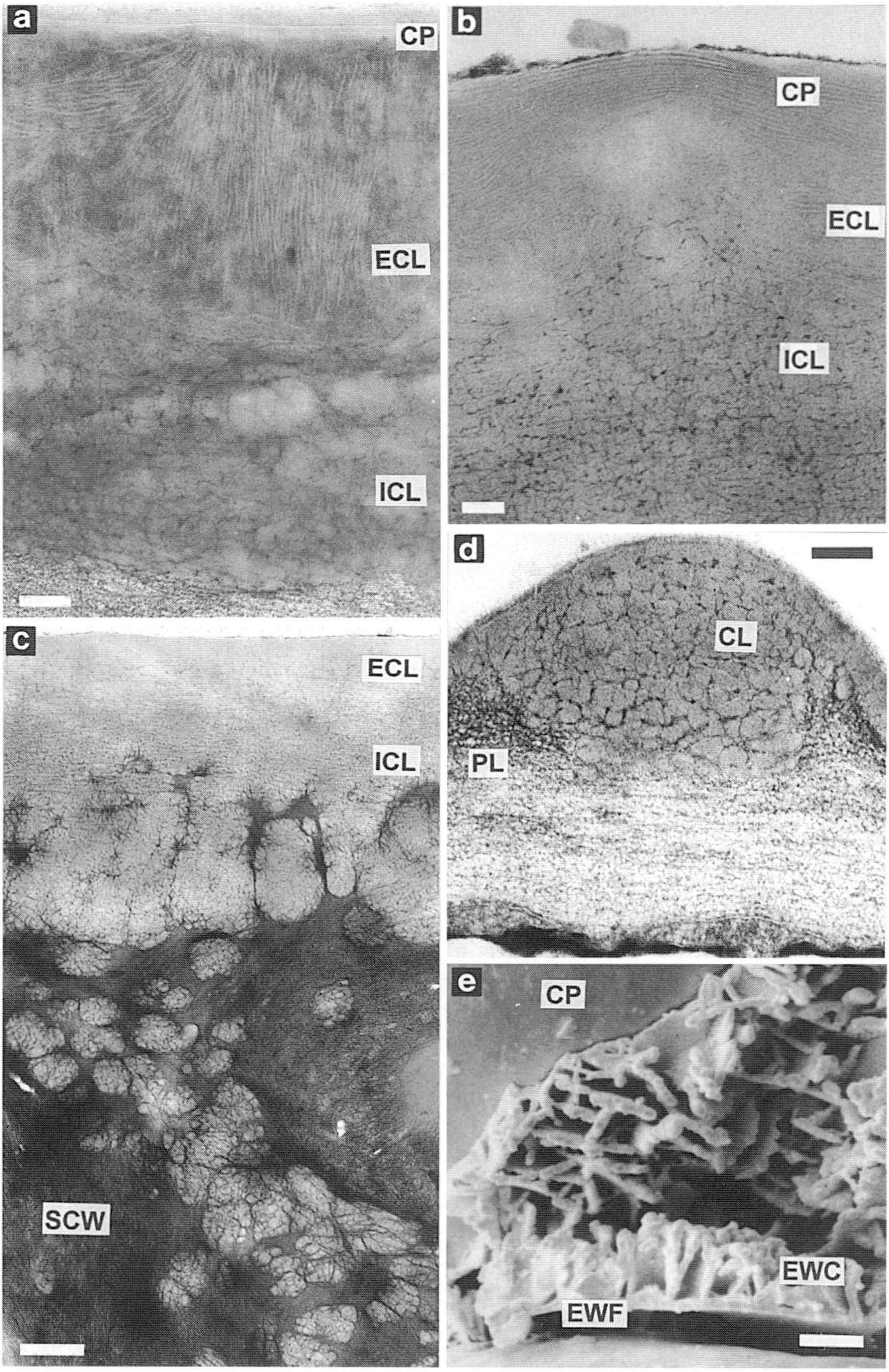
a
CP
ECL
ICL
b
CP
ECL
ICL
c
ECL
ICL
SCW
d
CL
PL
e
CP
EWC
EWF

secretory cells, including the receptive stigmatic papillae. In glandular hairs of many species, secretion product accumulates between the primary cell wall and the CM, which is therefore detached. Release of secretions may be via pores or ruptures in these cuticles. In *Abutilon venosum*, nectar droplets pass through a cuticle 0.4 μm thick perforated with pores 0.15–0.25 μm across at the tip of each nectary hair (Findlay and Mercer, 1971 and see Table 2.3). These pores seemed to have valve-like action, periodically releasing nectar accumulated between the cuticle and the cell wall. Gaps in the cuticular membrane are reported for other gland types. In many cases the evidence points to the mechanical formation of breaks or discontinuities at weak points in the CM. Thus the thin cuticles of *Drosophyllum lusitanica* (Joel and Juniper, 1982; Joel *et al.*, 1983), which are discontinuous at maturity, rupture mechanically at predetermined points of weakness under tension caused by area expansion of the underlying PCW, which reduces in thickness by a factor of >2. Discontinuous cuticles are found on the stigmatic papillae of *Secale cereale* (Heslop-Harrison and Heslop-Harrison, 1982; see Figure 2.11d and e) and in the remarkable chambered cuticles of *Crocus chrysanthus* (Heslop-Harrison, 1977; Heslop-Harrison and Heslop-Harrison, 1975, 1982; see Figure 2.11c).

2.8 Epicuticular wax

The epicuticular waxes (EW), which form a waxy bloom on the surfaces of many plants, are conspicuous and readily accessible to investigation, and their properties, structure and chemistry are consequently much studied (Baker, 1982; Jeffree, 1986;

Figure 2.9. *(a)* Pyrus communis *(Rosaceae), leaf, adaxial (Type 2). CP weakly lamellate; lamellae in all orientations in non-reticulate ECL. Reticulate ICL lacks lamellae. (Ga/Os, Pb; × 68 000; bar = 0.1 μm.) (b)* Hedera helix *(Araliaceae), leaf, adaxial (Type 2). CP strongly lamellate. Boundary between CP and ECL indistinct. Strongly contrasted lamellae occur in ECL, which also contains electron-lucent non-lamellate zones with or without reticulum. ICL strongly reticulate contains few lamellae. (× 56 000; bar = 0.1 μm.) (c)* Picea sitchensis *(Pinaceae) (Type 4): no distinguishable CP, fine reticulum reaching to the surface of the CM. Outer third of CM is lightly reticulate, equivalent to ECL. Inner third shows coarse globular cutinization with coarse reticulum (cf. Tenberge, 1992, layer L2t, Figure 2.7a). Loose aggregates of cutin cystoliths form intercellular cutinized layers within lignified SCW. (Ga/Os, Pb; × 17 300; bar = 0.5μm.) (d)* Utricularia sandersonii *(Lentibulariaceae): cuticular wart on a stolon, formed by massive amplification of the cutinization of the CL. (PL, pectin lamella; × 84 000; bar = 0.1 μm.) (e)* Quercus pubescens *(Fagaceae), leaf adaxial: low-temperature SEM of an epicuticular wax fragment retained over the anticlinal cell walls following low-temperature (-196°C) stripping of the wax layer using a cryo-adhesive (*n-pentane*). The wax crystallites can be seen to emerge from a continuous epicuticular wax film (EWF; see Figure 2.12). The underlying CP contains no macroscopic pores. (× 4250; bar = 2 μm.) (Ga, glutaraldehyde; Os, osmium tetroxide; Pb, lead citrate.) ((a, c) Reproduced from Holloway (1982a) Structure and histochemistry of plant cuticular membranes: an overview. In:* The Plant Cuticle *(eds Cutler* et al.*). Academic Press, London, pp. 1-32, with permission from Academic Press. (b) (Reproduced from Viougeas* et al. *(1995) Structural-changes and permeability of ivy (*Hedera helix *L.) leaf cuticules in relation to leaf development and after selective chemical treatments.* New Phytol. ***130,*** *337-348, with permission from the* New Phytologist. *(d) Reproduced from Heide-Jørgensen (1991) Cuticle development and ultrastructure - evidence for a procuticle of high osmium affinity.* Planta ***183,*** *511-519, with permission from Springer-Verlag. (e) By C.E. Jeffree (unpublished).)*

see Chapters 3 and 9). At bud-break spruce needles already have a CP and a CL, and epicuticular waxes are crystallizing as tubes on top of the CM, which must therefore be permeable to *n*-nonacosan-10-ol and other constituents of the EW. No structure corresponding to a pathway for wax precursors is visible, however, and the mechanism by which the EW is delivered to the plant surface remains unconfirmed. The recent detection of acyl transfer proteins in the epicuticular wax of broccoli (*Brassica oleracea*; Pyee and Kolattukudy, 1995; Pyee *et al.*, 1994) appears to provide support for the hypothesis of Hallam (1982) that a protein carrier might be involved. That the structure of epicuticular wax crystallites is determined by the crystal morphology of dominant chemical constituents is well established both from correlative and genetic observations (Hallam, 1967; von Wettstein-Knowles, 1974), and by experimental analysis (Jeffree, 1974, 1986; Jeffree *et al.*, 1975, 1976; Jetter and Riederer, 1994, 1995; Lister and Thair, 1981) in many species. However, both the morphology of the *Strelitzia*-type waxes found in many species (Frölich and Barthlott, 1988; Meusel *et al.*, 1994) and the non-random arrangement of the structures in these and other groups that show conspicuous wax patterning suggest that other, as yet unidentified, mechanisms must also be involved. The existence of an amorphous wax layer underlying the EW crystallites was proposed by Jeffree *et al.* (1975, 1976), and in some instances (e.g. on the leaves of *Beta vulgaris*, *Clematis vitalba* and *Phaseolus vulgaris*, which have smooth surfaces with little discernible ultrastructure) has been deduced from quantitative analysis of wax deposits (Baker, 1982), but the existence of a discrete superficial wax layer has previously eluded direct confirmation. However, Figure 2.9e shows a low-temperature SEM image of the surface of a *Quercus pubescens* leaf from which the wax was mechanically stripped at -196°C, using pentane as a cryo-adhesive. At the edge of the stripped layer, a flake of EW has lifted, showing that the EW crystallites do not arise independently from the CP, but from a continuous covering layer of apparently amorphous wax, about 0.2–0.3 μm thick. That the layer is in fact EW, and not a stratum of CP, splitting along its lamellae, is evidenced by the lack of a corresponding step in the substrate, the layer splitting cleanly at a uniform level. No pores are visible in the underlying cuticle.

2.9 Structural variation in plant cuticles

2.9.1 *Structural types*

Holloway (1982a) surveyed the range of cuticle structural types in vascular plants, and the terminology employed to describe their main features, using a combination of his own observations of cuticle structure and the data available in the literature, and tentatively identified six types of cuticle structure (Table 2.2). The allocation of species to cuticle types by Holloway (1982a) is summarized in Table 2.3 (**bold**), together with additional species by this author (Table 2.3, *italics*). The structural variety of cuticles could be viewed either as evidence of a corresponding variety of fundamentally divergent approaches to cuticle biosynthesis, or alternatively as a manifestation of the versatility of the same structural and developmental model. Current evidence indicates that particular cuticle types are not confined to, or generally characteristic of, particular taxa. In the Solanaceae, for example, the fruit cuticle of *Lycopersicon esculentum* is Type 4, reticulate throughout, lacking a defined CP. However, recent evidence (Krüger *et al.*, 1996; Mérida and Ogura, 1987; Figure 2.10a and b) confirms the presence of a lamellate CP in the leaf cuticles of *Nicotiana glaucum* and *Nicotiana tabacum*, by contrast with the assignment of the latter as Type 4 by Holloway (1982a), based on the data of Akers *et al.* (1978).

Table 2.2. *Summary of the six cuticle structural types identified by Holloway (1982a)*

Cuticle type	Cuticle morphology
1	Lamellate CP Reticulate CL
2	Faintly lamellate CP, gradually merging with reticulate CL
3	Amorphous CP, mainly reticulate CL
4	All regions reticulate
5	All regions lamellate
6	Mainly amorphous

The CL in many conifers is also strongly reticulate, and in common with *Abies balsamea* (Chabot and Chabot, 1977), *Pinus sylvestris* (Walles *et al.,* 1973) and *Pinus nigra* (Campbell, 1972), where no CP was detected, the leaf cuticle of *Picea sitchensis* is classified as Type 4 (Holloway, 1982a; Figure 2.9c). However, more recent reports show that Norway spruce (*Picea abies*; Tenberge, 1989, 1992; Figures 2.6, 2.7), red spruce (*Picea rubens*; Percy *et al.,* 1992; Figure 2.10e and f), and white pine (*Pinus strobus*; McQuattie and Rebbeck, 1994) all have a distinct CP, placing them in Type 3. The CP in *Pinus strobus* is amorphous (McQuattie and Rebbeck, 1994), and is never lamellate in *Picea abies* (K.B. Tenberge, personal communication). However, *Cunninghamia lanceolata* has a lamellate CP (Sargent, 1976a) and is thus Type 1, and the absence of CP lamellae is not therefore characteristic of the Coniferae, or even necessarily of the genus *Picea*. The data shown in Table 2.3 only represents 119 species from 94 genera, a trivial and arguably unrepresentative fraction of the world's species. Nevertheless, all six cuticle types are recorded in each of the superorders Dileniidae, Rosidae and Asteridae, and it is possible that comparable diversity would also be detected at the Order and Family levels if sufficient observations were available.

2.9.2 *Fugitive lamellae*

Confirmation of lamellate CP structure is made difficult by variation in the appearance of cuticles during their ontogeny, and by their response to technical variations in specimen preparation procedure. Hülsbruch (1966a,b) thought that the CP of *Ilex integra* might be sloughed after synthesis of the CL. Sargent (1976a) reported that the lamellar structure of *Sisyrinchium filifolium* was absent in the mature cuticle, visible only where the CL (secondary cuticle) is < 2 μm thick, and that 'a primary cuticle is not generally found beneath a fully-differentiated wax cover' (Sargent, 1976a). In TEM sections, contrast between lamellae depends not on intrinsic electron optical properties of the materials, but on their capacity to stain differentially. There may be a parallel between the progressive loss of the staining capacity of the supposedly cutanized CL, and the loss of differential reactivity of the CP lamellae reported by Sargent (1976a). In *Tamarix pentandra* it was difficult to detect a CP over cuticular papillae (Sargent, 1976a) (see also *Utricularia sandersonii*, Heide-Jørgensen, 1991; Figure 2.9d). McQuattie and Rebbeck (1994) reported the loss of the CP lamellae when *Liriodendron tulipifera* leaves were exposed to twice ambient concentrations of ozone.

Table 2.3. *Summary of cuticle types (as defined by Holloway, 1982a; see Table 2.2) and dimensions (species are arranged in order of taxonomic relatedness;* **bold numbers** *refer to the classification by Holloway (1982a);* italic numbers *refer to recent or additional data)*

Species	Order	Family	Organ	Type	ECL/ ICL	PL	CM (μm)	CP (μm)	CL (μm)	Reference
Bryophyta, Musci										
Funaria hygrometrica		Funariales	Cap	*1*	~	~	0.2	0.13	0.07	Sack and Paolillo (1983a)
Pteridophyta, Filicopsida										
Psilotum nudum			S	*1*	~	~	0.38	0.2	0.18	Whittier and Peterson (1995)
Dryopteris filix mas	Filicales	Aspidiaceae	L	**3**	~	~	~	~	~	Holloway (1982a)
Phyllitis scolopendrium	Filicales	Aspleniaceae	L	**3**	~	~	~	~	~	Holloway (1982a)
Adiantum capillus-veneris	Filicales	Adiantaceae	L	*1*	~	~	1.0	~	~	Archer and Cole (1986)
Gymnospermae, Coniferopsida										
Abies balsamea	Coniferae	Pinaceae	L	*4*	-	-	1.2	~	~	Chabot and Chabot (1977)
Chamaegigas intrepidus	Coniferae	Pinaceae	L	*3*	~	~	~	~	~	Lehmann and Schultz (1976)
Cryptomeria japonica	Coniferae	Pinaceae	L	*4*	~	~	2.0	0.0	2.0	Sargent (1976a)
Cunninghamia lanceolata	Coniferae	Pinaceae	L	**1**	~	~	~	~	~	Sargent (1976a)
Picea abies	Coniferae	Pinaceae	L	*3*	+	-	3.6	0.05	3.56	Tenberge (1992)
Picea rubens	Coniferae	Pinaceae	L	*3*	+	-	0.4	0.05	0.33	Percy *et al.* (1992)
Picea sitchensis	Coniferae	Pinaceae	L	*4*	+	-	2.2	~	~	Holloway (1982a)
Pinus nigra	Coniferae	Pinaceae	L	*4*	~	~	4.7	~	~	Campbell (1972)
Pinus strobus	Coniferae	Pinaceae	L*	*3*	~	~	0.15	0.05	0.1	McQuattie and Rebbeck (1994)
Pinus strobus	Coniferae	Pinaceae	L†	*3*	~	~	0.5	0.05	0.45	McQuattie and Rebbeck (1994)
Pinus sylvestris	Coniferae	Pinaceae	L	*4*	~	~	~	~	~	Walles *et al.* (1973)
Pseudotsuga menziesii	Coniferae	Pinaceae	L	**1**	~	~	~	0.15	~	Sargent (1976a)

Angiospermae, Dicotyledonae, Magnoliidae										
Liriodendron tulipifera	Magnoliales	Magnoliaceae	L†	*1*	~	~	0.2	0.05	0.15	McQuattie and Rebbeck (1994)
Liriodendron tulipifera	Magnoliales	Magnoliaceae	L*	*3*	+	~	0.15	0.05	0.1	McQuattie and Rebbeck (1994)
Magnolia grandiflora	Magnoliales	Magnoliaceae	LAd	*4*	~	~	1.8	~	~	Postek (1981)
Caltha palustris	Ranunculales	Ranunculaceae	Petal	*4*	-	-	0.6	0.0	0.58	Whatley (1984)
Sarracenia psittacina	Sarraceniales	Sarraceniaceae	Pitcher	*1*	~	~	0.12	0.04	0.08	Joel and Heide-Jørgensen (1985)
Sarracenia psittacina	Sarraceniales	Sarraceniaceae	Pitcher	*1*	~	~	0.04	0.03	0.02	Joel and Heide-Jørgensen (1985)
Sarracenia purpurea	Sarraceniales	Sarraceniaceae	Pitcher	*1*	+	+	0.77	0.04	0.73	Joel and Heide-Jørgensen (1985)
Angiospermae, Dicotyledonae, Caryophillidae										
Dianthus caryophyllus	Caryophyllales	Caryophyllaceae	L	**3**	~	~	~	~	~	Reed (1979)
Silene dioica	Caryophyllales	Caryophyllaceae	Stigma	*1*	-	-	0.33	0.08	0.26	H-H‡ (1975, 1977, 1982)
Stellaria media	Caryophyllales	Caryophyllaceae	L	**6**	~	~	~	~	~	Holloway (1982a)
Anabasis articulata	Caryophyllales	Chenopodiaceae	S	*1*	+	+	1.02	0.13	0.89	Lyshede (1977a)
Anabasis articulata	Caryophyllales	Chenopodiaceae	S	**2**	+	+	1.12	0.18	0.94	Lyshede (1977a)
Atriplex semibaccata	Caryophyllales	Chenopodiaceae	L	**5**	-	-	0.1	0.1	0.0	Kolattukudy (1980)
Beta vulgaris	Caryophyllales	Chenopodiaceae	L	**5**	-	-	0.06	0.06	0.0	Holloway (1982a)
Spinacia oleracea	Caryophyllales	Chenopodiaceae	L	**2**	-	+	0.24	0.04	0.21	Holloway *et al.* (1981)
Rumex conglomeratus	Polygonales	Polygonaceae	G	*3*	+	-	0.65	0.15	0.5	Chafe and Wardrop (1973)
Angiospermae, Dicotyledonae, Dilleniidae										
Gordonia axilaris	Theales	Theaceae	L	**5***(1)*	~	+	2.0	0.16	1.84	Sargent (1976a,c)
Abutilon striatum	Malvales	Malvaceae	P	**2**	~	~	~	~	~	Chafe and Wardrop (1973)
Abutilon venosum	Malvales	Malvaceae	N	*poCh*	-	-	0.4	~	~	Findlay and Mercer (1971)

(continued)

Species	Order	Family	Organ	Type	ECL/ ICL	PL	CM (μm)	CP (μm)	CL (μm)	Reference
Gossypium hirsutum	Malvales	Malvaceae	L	*3*	~	~	0.16	~	~	Ryser and Holloway (1985)
Gossypium hirsutum	Malvales	Malvaceae	SC	*3*	~	~	10.1	0.06	10.0	Ryser and Holloway (1985)
Gossypium hirsutum	Malvales	Malvaceae	Fibre	*P*	~	~	0.02	~	~	Ryser and Holloway (1985)
Cannabis sativa	Urticales	Moraceae	G	*4*	~	+	0.16	0.0	0.16	Kim and Mahlberg (1995)
Cannabis sativa	Urticales	Moraceae	G	*3*	-	-	0.2	0.02	0.18	Mahlberg and Kim (1992)
Cannabis sativa	Urticales	Moraceae	T	*4*	-	+	0.19	0.0	0.19	Mahlberg and Kim (1991)
Cannabis sativa	Urticales	Moraceae	T	**6**	~	~	~	~	~	Hammond and Mahlberg (1978)
Cannabis sativa	Urticales	Moraceae	T	*4*	-	-	0.59	0.0	0.59	Hammond and Mahlberg (1978)
Ficus elastica	Urticales	Moraceae	L	**2**	~	~	~	~	~	Holloway (1982a)
Ficus lyrata	Urticales	Moraceae	L	*3*	~	~	2.0	~	~	Davis (1987)
Humulus lupulus	Urticales	Moraceae	P	**1**	~	~	~	~	~	Chafe and Wardrop (1973)
Tamarix pentandra	Violales	Tamaricaceae	L	**1**	~	~	~	0.2	~	Sargent (1976a)
Tamarix pentandra	Violales	Tamaricaceae	EP	4	~	~	~	0.0	~	Sargent (1976a)
Brassica napus	Capparales	Cruciferae	L	**6**	~	~	~	~	~	Armstrong and Whitecross (1976)
B. oleracea cv. *gemmifera*	Capparales	Cruciferae	L	**6**	~	~	~	~	~	Reed (1979)
B. oleracea cv. *botrytis*	Capparales	Cruciferae	L	**6**	~	~	~	~	~	Holloway (1982a)
B. oleracea cv. *botrytis*	Capparales	Cruciferae	Stigma	4	-	-	0.13	0.0	0.13	Elleman *et al.* (1988)
Vaccinium reticulatum	Ericales	Ericaceae	P	3	+	+	2.1	0.0[§]	2.1	Singh and Hemmes (1978)
Ardisia crenata	Primulales	Myrsinaceae	LAb	*1*	+	+	0.8	0.16	0.64	Fisher and Bayer (1972)

Ardisia crenata	Primulales	Myrsinaceae	LAd	*B*	+	+	1.6	0.25	1.35	Fisher and Bayer (1972)
Ardisia crenata	Primulales	Myrsinaceae	L	**2**	+	+	1.6	0.25	1.35	Fisher and Bayer (1972)
Angiospermae, Dicotyledonae, Rosidae										
Dionaea muscipula	Rosales	Droseraceae	G	*3 po*	~	~	0.1	~	~	Joel and Juniper (1982)
Dionaea muscipula	Rosales	Droseraceae	T	**2**	~	~	~	~	~	Sievers (1968)
Dionaea muscipula	Rosales	Droseraceae	T	*1*	+	+	0.56	0.1	0.47	Sievers (1968)
Drosera sp.	Rosales	Droseraceae	G	*4 po*	-	-	0.13	~	~	Chafe and Wardrop (1973)
Drosophyllum lusitanica	Rosales	Droseraceae	G	*3 po*	-	-	0.3	0.8	0.22	Joel and Juniper (1982)
Fragaria ananassa	Rosales	Rosaceae	L	**3**	~	~	~	~	~	Holloway (1982a)
Fragaria ananassa	Rosales	Rosaceae	L	**6**	~	~	~	~	~	Holloway *et al.* (1981)
Malus pumila	Rosales	Rosaceae	F	*1*	~	~	~	~	~	Linskens and Gelissen (1966)
Malus pumila	Rosales	Rosaceae	F	*1*	~	~	~	~	~	Hilkenbäumer (1958)
Malus pumila	Rosales	Rosaceae	L	*1*	+	+	0.38	0.04	0.34	Hoch (1979)
Malus pumila	Rosales	Rosaceae	LAd	**2**	~	~	~	~	~	Hoch (1975, 1979)
Malus pumila	Rosales	Rosaceae	LAb	**3**	+	-	0.98	0.05	0.93	Holloway (1982a)
Malus pumila	Rosales	Rosaceae	LAb	*1*	+	-	0.98	0.05	0.93	Holloway (1982a)
Malus pumila	Rosales	Rosaceae	F	**4**(*3*)	+	-	10.0	0.6	9.4	de Vries (1968b)
Malus pumila	Rosales	Rosaceae	LAb	**6**	~	~	~	~	~	Holloway *et al.* (1981)
Prunus laurocerasus	Rosales	Rosaceae	L	**4**	~	~	~	~	~	Holloway (1982a)
Prunus persica	Rosales	Rosaceae	L	**3**	~	~	~	~	~	Schneider and Dargent (1977)
Pyrus communis	Rosales	Rosaceae	LAd	**2**	+	+	1.1	0.04	1.06	Holloway (1982a)
Pyrus communis	Rosales	Rosaceae	LAb	**3**	~	~	~	~	~	Holloway (1982a)
Hydrangea macrophylla	Rosales	Saxifragaceae	L	**4**	~	~	~	~	~	Holloway (1982a)
Ribes nigrum	Rosales	Saxifragaceae	L	**3**	~	~	~	~	~	Holloway (1982a)
Ribes nigrum	Rosales	Saxifragaceae	LAb	*6*	~	~	~	~	~	Holloway *et al.* (1981)
Ribes nigrum	Rosales	Saxifragaceae	LAd	*6*	~	~	~	~	~	Holloway *et al.* (1981)
Saxifraga granulata	Rosales	Saxifragaceae	L	**6**	-	-	~	~	~	Lehmann and Schulz (1976)
Cicer arietinum	Fabales	Leguminoseae	H	*D*	~	~	~	~	~	Schnepf (1969)
Genista aetnensis	Fabales	Leguminosae	L	*4*	-	-	0.85	0.0	0.85	Lyshede (1982)

(continued)

Species	Order	Family	Organ	Type	ECL/ ICL	PL	CM (μm)	CP (μm)	CL (μm)	Reference
Phaseolus vulgaris	Fabales	Leguminoseae	L	**2**	~	~	~	~	~	Holloway (1982a)
Phaseolus vulgaris	Fabales	Leguminoseae	ssc	*Pc*	~	~	0.02	~	~	Jeffree (unpublished data)
Pisum sativum	Fabales	Leguminoseae	Ap	*Pc*	~	~	0.02	~	~	Jeffree (1986)
Spartocytisus filipes	Fabales	Leguminoseae	S	**1**	~	~	~	~	~	Lyshede (1978)
Spartocytisus filipes	Fabales	Leguminoseae	S	*3*	~	~	17.0	0.14	17.0	Lyshede (1978)
Spartocytisus filipes	Fabales	Leguminoseae	Mc	*4*	~	~	0.11	~	0.11	Lyshede (1977b)
Spartocytisus filipes	Fabales	Leguminoseae	Mc	**4**	~	~	~	~	~	Lyshede (1977b)
Spartocytisus filipes	Fabales	Leguminoseae	T	**6**	~	~	~	~	~	Lyshede (1976)
Vicia faba	Fabales	Leguminoseae	GC	*6*	-	-	0.07	~	~	Pallas and Mollenhauer (1972)
Vicia faba	Fabales	Leguminoseae	L	**6***(4)*	-	-	0.1	0.0	0.1	McKeen (1974)
Myriophyllum verticillatum	Haloragales	Haloragaceae	Ap	*Po*	-	-	0.02	0.02	0.0	Hallam (1982)
Oenothera organensis	Myrtales	Onagraceae	Stigma	*Cb*	-	-	~	~	~	H-H‡ (1985)
Eucalyptus cinerea	Myrtales	Myrtaceae	L	**5**	~	~	0.59	0.1	0.49	Hallam (1964)
Eucalyptus papuana	Myrtales	Myrtaceae	L	*1*	~	~	0.1	0.05	0.05	Hallam (1970)
Eucalyptus papuana	Myrtales	Myrtaceae	L	**2**	~	~	0.1	0.05	0.05	Hallam (1970)
Eucalyptus perriniana	Myrtales	Myrtaceae	L	*1*	~	~	0.32	0.18	0.14	Hallam (1982)
Eucalyptus perriniana	Myrtales	Myrtaceae	L	**3**	~	~	0.5	~	~	Hallam (1970)
Hakea leucoptera	Proteales	Proteaceae	L	**1**	+	-	0.15	~	~	Sargent (1976a)
Hakea leucoptera	Proteales	Proteaceae	T	**4**	+	-	~	~	~	Sargent (1976a)
Hakea suaveolens	Proteales	Proteaceae	L	*1*	~	~	11.5	1.5	10.0	Heide-Jørgensen (1978)
Hakea suaveolens	Proteales	Proteaceae	T	*1*	~	~	1.49	0.06	1.43	Heide-Jørgensen (1980)
Ilex aquifolium	Celastrales	Aquifoliaceae	L	**4**	~	~	~	~	~	Holloway (1982a)
Ilex aquifolium	Celastrales	Aquifoliaceae	L	*4*	~	~	10.0	~	~	Sargent (1976a)
Ilex integra	Celastrales	Aquifoliaceae	L	*4*	~	~	~	~	~	Hülsbruch (1966a,b)
Ricinus communis	Euphorbiales	Euphorbiaceae	L	*6*	~	~	~	~	~	Holloway *et al.* (1981)
Pistacia palaestina	Sapindales	Anacardiaceae	L	**1**	~	~	0.25	0.17	0.08	Sargent (1976a)
Citrus limon	Sapindales	Rutaceae	L	**4**	~	~	~	~	~	Holloway (1982a)
Citrus paradisi	Sapindales	Rutaceae	Peel	*4*	~	~	1.9	~	~	Espelie *et al.* (1980)
Citrus paradisi	Sapindales	Rutaceae	SC	*4*	+	~	1.0	~	~	Espelie *et al.* (1980)

Citrus paradisi	Sapindales	Rutaceae	Juice-sac	*?*	~	~	0.1	~	~	Espelie *et al.* (1980)
Citrus sinensis	Sapindales	Rutaceae	F	**4**	~	~	~	~	~	Thomson and Platt-Alloia (1976)
Tropaeolum majus	Geraniales	Tropaeolaceae	Stigma	**6**(*4*)	-	-	0.24	0.0	0.24	Shayk *et al.* (1977)
Hedera helix	Umbellales	Araliaceae	L	**1**	+	+	2.5	0.1	2.4	Viougeas *et al.* (1995)
Hedera helix	Umbellales	Araliaceae	L	**2**	+	-	1.96	0.26	1.7	Holloway (1982a)
Apium graveolens	Umbellales	Umbelliferae	P	**1**	+	+	0.27	0.05	0.22	Chafe and Wardrop (1973)
Apium graveolens	Umbellales	Umbelliferae	P	**1**	+	?	0.85	0.13	0.72	Juniper and Cox (1973)
Apium sativum	Umbellales	Umbelliferae	P	**1**	-	?	0.66	0.09	0.58	Hallam and Juniper (1971)
Eryngium rostratum	Umbellales	Umbelliferae	P	**2**(*1*)	+	+	0.8	0.13	0.67	Chafe and Wardrop (1973)
Angiospermae, Dicotyledonae, Asteridae										
Ligustrum ovalifolium	Gentianales	Oleaceae	L	**3**	-	-	0.77	~	~	Hallam and Juniper (1971)
Cuscuta campestris	Polemoniales	Convolvulaceae	S	*1*	+	+	0.07	0.03	0.04	Jeffree (1986)
Cuscuta gronovii	Polemoniales	Convolvulaceae	S	*1, 5*	+	+	0.26	0.05	0.21	Heide-Jørgensen (1991)
Cuscuta odorata	Polemoniales	Convolvulaceae	S	**3**	~	~	~	~	~	Weinert and Barckhaus (1975)
Lycopersicon esculentum	Polemoniales	Solanaceae	F	**4**	-	-	9.0	0.0	9.0	Wilson and Sterling (1976)
Lycopersicon esculentum	Polemoniales	Solanaceae	F	**4**	-	-	2.0	0.0	2.0	Rijkenberg *et al.* (1980)
Nicotiana glauca	Polemoniales	Solanaceae	L	*1*	+	+	0.25	0.05	0.2	Mérida and Ogura (1987)
Nicotiana tabacum	Polemoniales	Solanaceae	L	*1*	-	-	0.26	0.04	0.23	Krüger *et al.* (1996)
Nicotiana tabacum	Polemoniales	Solanaceae	T	**4**	-	-	0.3	0.0	0.3	Akers *et al.* (1978)
Plantago major	Plantaginales	Plantaginaceae	L	*1*	+	+	0.23	0.05	0.18	Fisher and Bayer (1972)
Plantago major	Plantaginales	Plantaginaceae	L	**3**	~	~	~	~	~	Crisp (1965)
Plantago major	Plantaginales	Plantaginaceae	L	**3**	~	~	~	~	~	Holloway (1982a)
Eccremocarpus scaber	Scrophulariales	Bignoniaceae	Td	*4wr*	-	-	0.25	~	~	Junker (1977)
Utricularia laterifolia	Scrophulariales	Lentibulariaceae	LAd	*1*	-	-	0.31	0.08	0.24	Heide-Jørgensen (1991)
Utricularia laterifolia	Scrophulariales	Lentibulariaceae	LAd	*Pc*	-	-	0.03	~	~	Heide-Jørgensen (1991)
Utricularia sandersonii	Scrophulariales	Lentibulariaceae	S	*4*	-	-	0.52	0.0	0.52	Heide-Jørgensen (1991)

(continued)

Species	Order	Family	Organ	Type	ECL/ ICL	PL	CM (μm)	CP (μm)	CL (μm)	Reference
Coffea arabica	Rubiales	Rubiaceae	L	*4*	-	-	2.0	0.0	2.0	Holloway *et al.* (1981)
Lonicera japonica	Dipsacales	Caprifoliaceae	N	*D wr*	~	~	0.4	~	~	Fahn and Rachmilevitz (1970)
Lonicera japonica	Dipsacales	Caprifoliaceae	N	*D wr*	~	~	3.0	~	~	Fahn and Rachmilevitz (1970)
Lactuca sativa	Asterales	Compositae	L	**2**	-	-	0.25	0.04	0.21	Srivastava *et al.* (1977)
Lactuca sativa	Asterales	Compositae	L	**6**	~	~	~	~	~	Holloway *et al.* (1981)
Taraxacum officinale	Asterales	Compositae	L	**5**	~	~	0.18	0.03	0.15	Holloway (1982a)
Elodea canadensis	Hydrocharitales	Hydrocharitaceae	L	**6**	-	-	~	~	~	Crisp (1965)
Potamogeton crispus	Najadales	Potamogetonaceae	L	**6**	-	-	0.04	~	~	Holloway (1982a)
Zostera capensis	Najadales	Zosteraceae	L	**6**	-	-	0.57	~	~	Barnabus *et al.* (1977)
Posidonia australis	Najadales	Posidoniaceae	L	**6**	-	-	~	~	~	Kuo (1978)
Angiospermae, Monocotyledonae, Commelinidae										
Avena fatua	Poales	Gramineae	C	4	-	-	1.1	-	1.1	O'Brien (1967)
Avena fatua	Poales	Gramineae	L	**6**	~	~	~	~	~	Holloway (1982a)
Avena fatua	Poales	Gramineae	St	*Pc*	-	-	0.03	0.0	0.03	Kaufman *et al.* (1970)
Hordeum vulgare	Poales	Gramineae	L	**6**	~	~	~	~	~	Sargent and Gay (1977)
Secale cereale	Poales	Gramineae	Stigma	*1po*	-	-	0.03	0.01	0.02	H-H‡(1982)
Triticum aestivum	Poales	Gramineae	O	4	-	-	0.87	0.0	0.87	Morrison (1975)
Zea mays	Poales	Gramineae	L	**6**	-	-	~	~	~	Holloway (1982a)
Lemna minor	Arales	Lemnaceae	L	**6**	~	~	~	~	~	Holloway (1982a)
Angiospermae, Monocotyledonae, Liliidae										
Bobartia gracilis	Liliales	Iridaceae	L	**1**	+	~	2.5	0.5	2.0	Sargent (1976c)
Iris germanica	Liliales	Iridaceae	L	**1**	+	-	0.8	0.1	0.7	Holloway *et al.* (1981)
Libertia elegans	Liliales	Iridaceae	L	**1**	+	?	0.7	0.15	0.55	Sargent (1976a,b)
Libertia ixioides	Liliales	Iridaceae	L	**1**	+	?	0.7	0.15	0.55	Sargent (1976 a,b)
Sisyrinchium filifolium	Liliales	Iridaceae	L	4	~	~	2.5	~	~	Sargent (1976a)
Allium cepa	Liliales	Liliaceae	Bs	*1*	+	-	0.76	0.16	0.6	Schönherr and Mérida (1981)

Aloe arborescens	Liliales	Liliaceae	L	4	-	-	2.2	0.0	2.2	Kluge *et al.* (1979)
Crocus chrysanthus	Liliales	Liliaceae	Stigma	*1*	-	+	0.9	0.1	0.8	H-H‡ (1975, 1977, 1982)
Gasteria planifolia	Liliales	Liliaceae	L	**1**	~	~	~	~	~	Holloway (1982a)
Gasteria verrucosa	Liliales	Liliaceae	L	*1*	~	~	~	0.18	~	Heumann (1990)
Lilium candidum	Liliales	Liliaceae	L	**3**	-	-	0.5	0.15	0.35	Maier (1968)
Muscari atlanticum	Liliales	Liliaceae	L	**3**	~	~	~	~	~	Holloway (1982a)
Tulipa gesneriana	Liliales	Liliaceae	L	**3**	~	~	~	~	~	Holloway (1982a)
Tulipa gesneriana	Liliales	Liliaceae	L	*6*	~	~	~	~	~	Holloway *et al.* (1981)
Clivia miniata	Liliales	Amaryllidaceae	L	**1**	+	+	~	0.21	~	Holloway (1982a)
Clivia miniata	Liliales	Amaryllidaceae	L	*1*	+	+	4.5	0.2	4.3	Holloway (1982a)
Clivia miniata	Liliales	Amaryllidaceae	LAd	**1**	+	~	6.7	0.13	6.57	Mérida *et al.* (1981)
Clivia nobilis	Liliales	Amaryllidaceae	L	**1**	~	~	~	~	~	Holloway (1982a)
Agave americana	Liliales	Agavaceae	L	**1**	+	+	~	~	~	Crisp (1965)
Agave americana	Liliales	Agavaceae	L	**1**	+	+	~	~	~	Wattendorff and Holloway (1980)
Agave americana	Liliales	Agavaceae	L	*1*	+	+	4.0	0.45	3.55	Wattendorff and Holloway (1984)
Phormium tenax	Liliales	Agavaceae	L	**1**	+	-	0.64	0.12	0.52	Jarvis and Wardrop (1974)
Phormium tenax	Liliales	Agavaceae	EP	*1*	-	-	4.3	0.2	4.1	Jarvis and Wardrop (1974)

Abbreviations: A, shoot apex; Ab, abaxial; Ad, adaxial; Bs, bulb scale; C, coleoptile; Ch, chambered CM; CM, cuticular membrane; CP, cuticle proper; CL, cuticular layer; ECL, external CL; ICL, internal CL; PL, pectin lamella; GC, guard cell; D, detached CM; EP, epidermal papillae; F, fruit; G, gland; H, hydathode; L, leaf; Mc, mucilaginous cell; O, ovule; N, nectary; P, petiole; Pc, procuticle; po, porous CM; S, stem; ssc, sub-stomatal cavity; SC, seed coat; St, stoma; T, trichome; Td, tendril; wr, wrinkled CM; ECL/ICL, CL subdivided into ECL and ICL; ~, no data available; +, present; -, absent. *Plants grown in twice-ambient ozone. †Plants grown in charcoal-filtered air. ‡H-H, Heslop-Harrison and Heslop-Harrison, or Heslop-Harrison. §A lamellate CP reported, but not measurable from published figures.

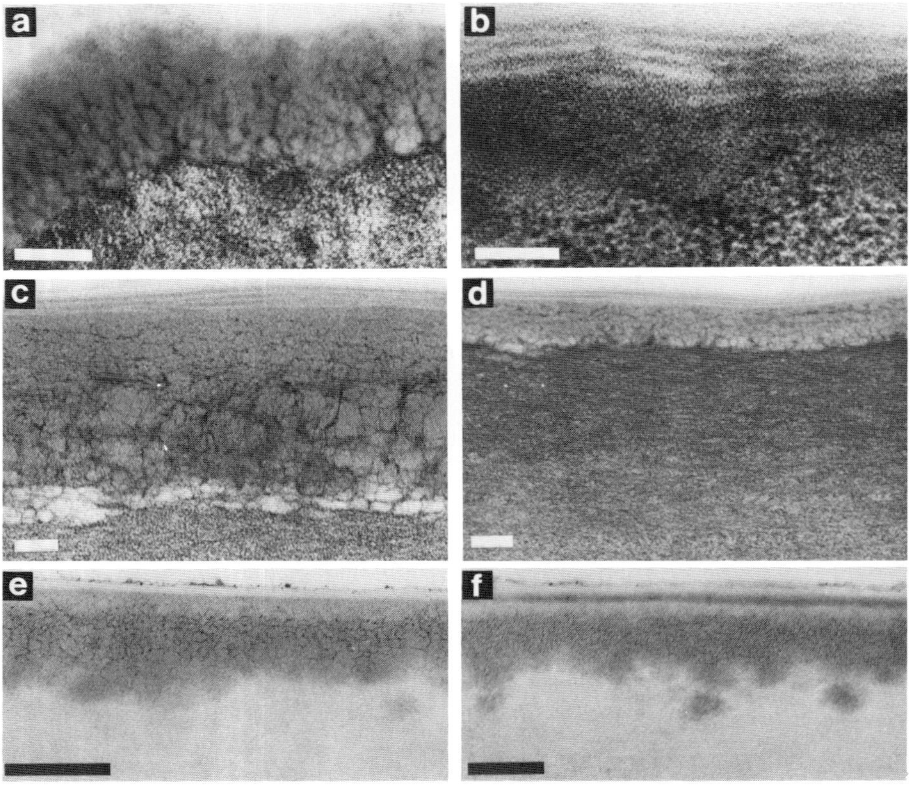

Figure 2.10. *(a, b)* Nicotiana tabacum *(Solanaceae), leaf, adaxial. (a) Reticulate structure only seen in an oblique section. (× 100 000; bar = 0.1 μm.) (b) Lamellate CP seen in perfect transverse section. (× 220 000; bar = 50 nm. Ga/Os, U/Pb.) (c, d)* Liriodendron tulipifera *(Magnoliaceae), leaf, adaxial. (c) CM from leaf grown in carbon-filtered air, showing 0.5-μm CL with finely reticulate ECL and coarse ICL. (d) CM from leaf exposed to twice-ambient ozone, showing reduced development of the CL, which corresponds in type to the ECL produced in clean air. ((c, d) × 54 900; bar = 0.1 μm.) (e, f)* Picea rubens *(Pinaceae), leaf. (e) Charcoal-filtered air. Distinct amorphous CP. Finely reticulate ECL, coarsely reticulate ICL. (× 32 200; bar = 0.5 μm.) (f) 250 p.p.b.* O_3 *and pH 4.2 acid fog. A dense amorphous layer is interposed between the CP and ECL. (× 20 100; bar = 0.5 μm.) (Ga, glutaraldehyde; Os, osmium tetroxide; U, uranyl acetate; Pb, lead citrate.) ((a, b) Reproduced from Krüger* et al. *(1996) Cuticular membrane fine structure of* Nicotiana tabacum *L. leaves.* Ann. Bot. ***77****, 11–16, with permission from Academic Press. (c, d) Reproduced from McQuattie and Rebbeck (1994) Effect of ozone and elevated carbon dioxide on cuticular membrane ultrastructure of yellow poplar (*Liriodendron tulipifera*). In:* Air Pollution and the Leaf Cuticle *(eds Percy* et al.*), NATO ASI Series G. Vol. 36. Springer-Verlag, Heidelberg, pp. 175–182, with permission from Springer-Verlag. (e, f) Reproduced from Percy* et al. *(1992) Effects of ozone and acidic fog on red spruce needle epicuticular wax production, chemical-composition, cuticular membrane ultrastructure and needle wettability.* New Phytol. ***122****, 71–80, with permission from the* New Phytologist.*)*

Detachment of the CP. The absence of a lamellate CP in the extremely thick CM of *Ilex aquifolium* was attributed by Sargent (1976a) to possible damage during preparation. A number of examples of splitting of the CP along the lamellae are shown in this paper. The CP may split along the lucent lamellae during handling for EM (Wattendorff and Holloway, 1980; Figure 2.3c and d), and in some instances these lamellae may conceivably represent a potential 'dehiscence layer', by which the CP is sloughed. Although Hülsbruch's observation in *Ilex* is unconfirmed, detachment of the lamellate region of the cuticle from the cutinized layer is documented in the stigmatic papillae of *Silene* (Heslop-Harrison and Heslop-Harrison, 1982; Figure 2.11b). Sargent (1976a) stated that in those instances where the lamellate structure was not identifiable, the cuticle preparations were of dubious quality, suggesting that their disappearance might sometimes have been artefactual. By contrast, she cites Wardrop's view (personal communication) that the lamellae are artefacts of fixation. The observation that lamellae are generally absent beneath a fully differentiated wax layer (Sargent, 1976a) appears to be contradicted by the evidence of several other workers (e.g. Hallam, 1964; Tenberge, 1992; Wattendorff and Holloway, 1980), and her statement that the majority of plant species examined have a lamellate CP with a reticulate CL (Sargent, 1976a) represents the mainstream of current opinion.

Tilt angle and lamellar contrast. Since the thickness of EM sections is up to 20 times that of the lucent lamellae of the CP (~ 5 nm), a clear view of the lamellate structure depends on the cuticle being cut at right angles, so that the lamellae are parallel to the beam (Heide-Jørgensen, 1980; Joel and Heide-Jørgensen, 1985). The effect of tilt angle on both the contrast of the lamellae and their apparent thickness, is shown in Figure 2.8, which shows the CP of a developing *Hakea suaveolens* leaf (H.S. Heide-Jørgensen, unpublished observations). Krüger *et al.* (1996) found that lamellae were only visible in the *Nicotiana tabacum* cuticle when cut in perfect transverse section, reticulate CL structure being seen to greater advantage in slightly oblique sections, and this effect may account for some records of apparently amorphous or absent CP layers.

Staining and lamellar contrast. The cuticle of *Apium graveolens* was observed by Hallam and Juniper (1971) to have a faintly lamellate CP with a reticulate CL, thus of Type 2. By contrast, Juniper and Cox (1973) provided evidence of a conspicuously lamellate CP and a reticulate CL (Type 1), while Chafe and Wardrop (1972) showed no evidence of a CP in *A. graveolens* (Type 4). In *Malus pumila*, Holloway (1982a) categorized the leaf adaxial cuticle as Type 3 (outer region amorphous, inner region reticulate), but Hoch (1979) detected lamellae in the CP of this species, placing it in Type 1. Holloway (1982a) detected lamellae in the *Ficus elastica* leaf adaxial cuticle, classifying it as Type 2, but lamellae were not shown in the CP of *F. lyrata* (Davis, 1987), suggesting that they may easily be overlooked. The contrast of the lamellae varies with the adopted staining protocol. Frequently, the superficial and basal two to three electron-dense CP lamellae are stained more strongly than those in the centre (Figures 2.1g-j; 2.3b), perhaps due to the exclusion of osmium from the central region. The contrast of both cutin and suberin lamellae can be enhanced by incorporation of $KMnO_4$ or ruthenium red (Wattendorff and Holloway, 1980; Figure 2.3e) in the primary and secondary fixation procedures, or by treatment of sections with hydrogen peroxide prior to staining with lead citrate and uranyl acetate (Heumann, 1990). It must also be presumed that fixative and stain penetration times, concentrations and temperatures, and tissue block size are

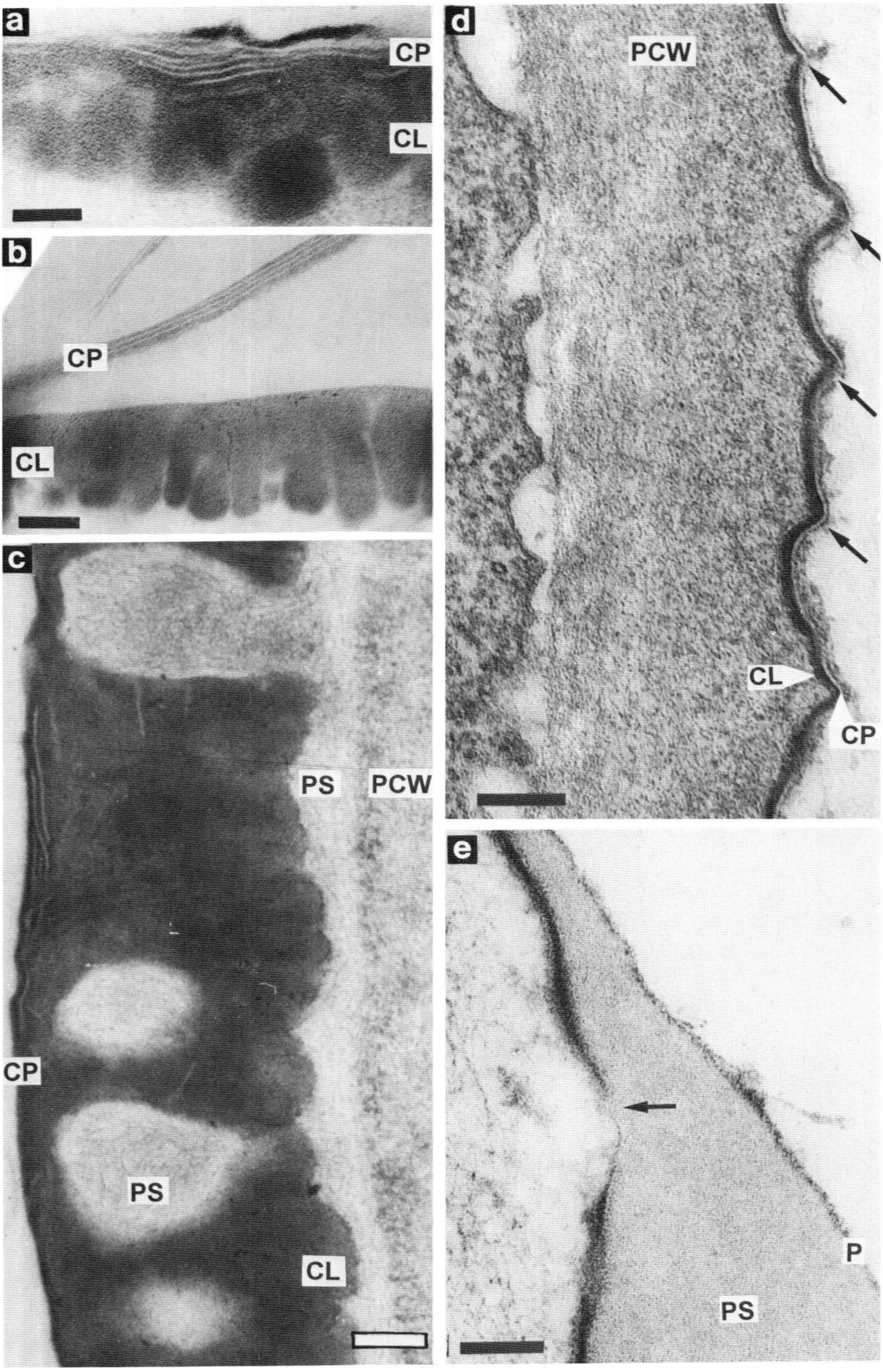
a
CP
CL
b
CP
CL
c
PS
PCW
CP
PS
CL
d
PCW
CL
CP
e
P
PS

important factors in obtaining reproducibility of results. It is therefore conceivable that cuticles of Types 4 and 6 (known examples of the latter represent few species and few observations of each species) may be revealed as lamellate under optimal preparative conditions. In testing this hypothesis it would be advisable to use not simply the standard glutaraldehyde-osmium fixation, but a range of fixation and contrasting techniques, such as those proposed by Heumann (1990), to obtain maximal enhancement of the contrast of the lamellae, which may sometimes be low and close to the limit of detectability.

2.9.3 *Intermediates between cuticle types*

Holloway (1982a) distinguished three basic lamellate types, Type 1 (lamellate CP-reticulate CL), Type 2 (faintly lamellate CP, gradually merging with a reticulate CL), and Type 5, in which lamellate structure, periclinally orientated or otherwise, is present throughout the CM. Type 5 is represented by two cuticles of contrasting character, that of *Gordonia axillaris* containing lamellae both in the CP and throughout a very substantial CL (Sargent, 1976a), while that of *Beta vulgaris* (Holloway, 1982a, his Figure 18) is very thin, and may be interpreted as being composed entirely of a lamellate CP without significant CL development. These three cuticle types, 1, 2 and 5, probably represent just three out of many possible stages in a continuous spectrum of intermediates between, on the one hand, the presence of lamellae in the CP only, and the occurrence of lamellae throughout the CP and CL grading from exterior to interior, and on the other hand, the presence of reticulum at any level in the CL grading from interior to exterior (note that reticulum has never been shown to appear in the CP). The CL of *Cuscuta gronovii* (Heide-Jørgensen, 1991; Figure 2.4c) is composed of chaotically packed electron-dense globules separated by lucent lamellae, with fibrillar material ramifying between. In this species, and also in *Taraxacum officinale* (Holloway, 1982a), the CL lamellae, which become more anticlinally orientated and chaotic in the inner layers of the CL, are continuous with those of the CP, indicating that they are manifestations of the same physicochemical structure. Lamellae occupy a substantial component of the ECL in the thicker cuticles of *Pyrus communis* (Holloway, 1982a; Figure 2.9a) and also of *Hedera helix* (Viougeas *et al.*, 1995; Figure 2.9b), which also have a reticulate ICL. In *Clivia* and *Agave*, periclinal CP lamellae grade into the more chaotically orientated lamellae which occur throughout the substantial thickness of otherwise

---

Figure 2.11. *(a, b)* Silene dioica *(Caryophyllaceae): stigmatic papilla. (a) CP lamellate. The globular CL is comparable with early CL development in* Hakea suaveolens *(cf. Figures 2.4a and c; 2.8). (× 95 000; bar = 0.1 µm.) (b) CP separated from the CL at maturity. The rodlet-CL forms a continuous layer only at the surface. (× 80 000; bar = 0.1 µm.) (c)* Crocus chrysanthus *(Iridaceae): stigmatic papilla. Chambered cuticle sectioned near the base of a papilla. Ovoid chambers contain protein-polysaccharide secretion (PS), underlying the entire CM, which separates from PCW to permit passage of pollen tubes. CP lamellae extend into reticulate CL. (× 48 000; bar = 0.2 µm.) (d, e)* Secale cereale *(Graminae): stigmatic papilla. (d) Pleated pre-secretory cuticle. CP is a single pair of lamellae overlying a simple discontinuous CL. Slit-like discontinuities (arrows) open by mechanical rupture of CP. (e) Open slit at maturity (arrow), releasing a mucilaginous polysaccharide secretion (PS) covered with a protein pellicle (P). (× 115 000; bar = 0.1 µm.) ((a–e) Reproduced from Heslop-Harrison and Heslop-Harrison (1982) The specialized cuticles of the receptive surfaces of angiosperm stigmas.* The Plant Cuticle *(eds Cutler* et al.*). Academic Press, London, pp. 99–119, with permission from Academic Press.)*

reticulate CL (Schmidt and Schönherr, 1982; Wattendorff and Holloway, 1984), and although classified as Type 1 (Holloway, 1982a), it would be equally logical to allocate them to Type 5. The CL in these species is a hybrid lamellate/reticulate structure. Typically, the gradient in development of the lamellate CL is from the CP downwards towards the CW. However, in the heavily thickened CM of the xeromorphic leaves of *Hakea suaveolens*, even the cutin cystoliths, accreting to the base of the CL may show lamellae (Heide-Jørgensen, 1978; Figure 2.4a, arrow). The observed variation in the presence and extent of lamellae in the CP and CL forms a continuous spectrum within Types 1, 2 and 5 in Holloway's classification, and it is difficult to identify or assign boundaries to these types.

2.9.4 *Subdivision of the CL: ICL/ECL*

The significance of the subdivision of the CL into exterior and interior regions has generally been appreciated only since the publication of the paper by Schmidt and Schönherr (1982), describing the appearance of non-ester cutin in *Clivia* (Figure 2.5d and e). The observation that the inner CL region is more ($KMnO_4$) or less (OsO_4) electron dense than the outer CL must now be regarded as widespread. Often, as in *Clivia,* the apparent termination of the reticulate region corresponds with this dark/light boundary, but histochemical observations nevertheless often reveal the presence of a reticulum of polysaccharide fibrillae reaching right up to the base of the cuticle proper, as in *Agave* (Wattendorff and Holloway, 1980; Figure 2.3e) and spruce (Tenberge, 1992; Figure 2.6d and e) but not in *Clivia* (Schmidt and Schönherr, 1982; Figure 2.5a–c). The ECL and ICL layers are probably to be regarded as ontogenetically related and may converge by maturation of the CL to become indistinguishable in the fully mature CM.

2.10 A structural and developmental model of the plant cuticle

2.10.1 *Summary of cuticle ontogeny*

Summarizing the process of development of plant cuticles, it is evident that an amorphous procuticle (Figure 2.12a), present as early as the protodermis or tunica of the apical dome, is replaced early in organ differentiation by a CP which is most often of lamellate ultrastructure (Figure 2.12b and c). At this stage a pectinaceous lamella bonds the CP to the PCW surface. Development of the cutinized layers proceeds by sequential impregnation of successive cell wall layers with cutin by growth and lateral fusion of globular cutin cystoliths within first the primary, and later successive layers of secondary cell wall (Figures 2.12e and f; 2.13a). During this phase and at maturity, the CL may contain either paradermal or a more random lamellate structure closest to the CP, but lamellae may extend to any depth within the CL. Soluble epicuticular lipids accumulate and crystallize at the CP surface throughout the process of cutinization (Figure 2.12e and f), and the outer CL layers may be developed by maturation of the polymer matrix, involving more intensive cross-linking of the existing cutin, or by further progressive impregnation of the ECL with highly aliphatic polymers and SCL. Further elaboration of the CM, especially where cutinization occurs in cells with walls that have undergone heavy secondary thickening (Figure 2.13b), may involve the accumulation and accretion of CL-like material at locations detached from the main CL deposit. These deposits may be recognizable as cystoliths, which may grow from the basic size of the pre-cutin

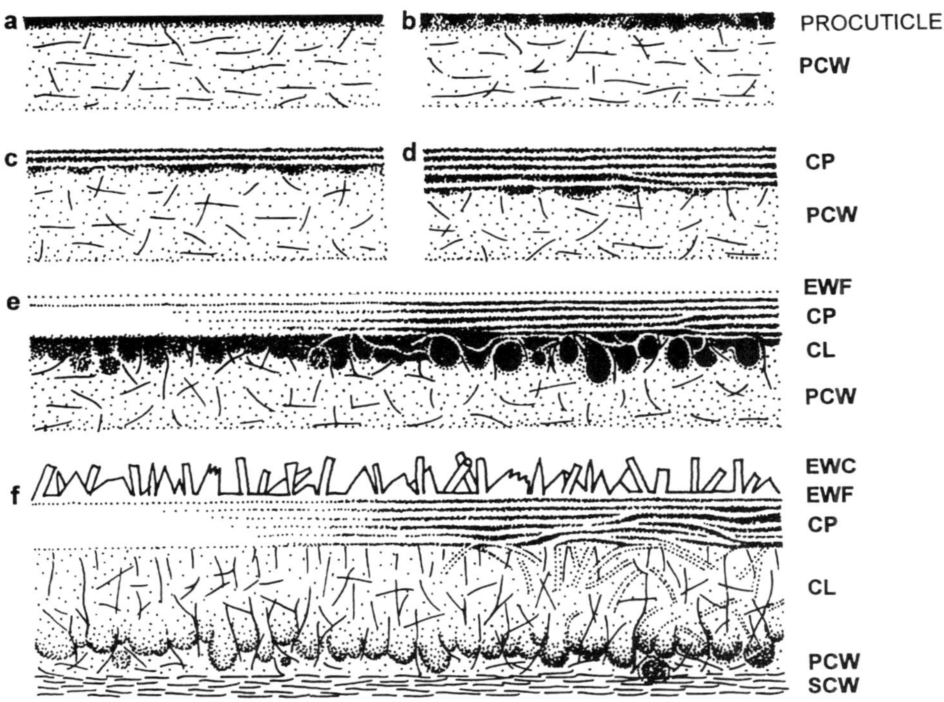

Figure 2.12. *(a-f) Summary diagrams of the development of the CP of the plant cuticle, and the early development of the CL within the PCW up to the complete cessation of cell expansion. (a) Uniformly electron-dense procuticle. (b) Procuticle: CP transition phase. (c) The procuticle has been converted to a lamellate CP. (d) Further construction of lamellate CP may involve globular lipids as in (e). (e) The CP/CL transition. Globular lipids coated with electron-lucent shells construct the CP and the CP/CL transition zone. Lamellae may become less regular. An apparently amorphous EWF is present on the CP. (f) Incorporation of PCW in the CL. Reticulations are light and predominantly radial, reaching as far as the CP. Lamellae may occur to any depth within the CL. Epicuticular wax crystals begin to form before cell expansion has ceased.*

globules visible beneath most developing and mature cuticles, to several hundred nanometers in diameter. The later stages of development of the CW may involve, deposition of inorganic material (silica and calcium salts) in and beneath the CL (Davis, 1987; Postek, 1981; Tenberge 1989, 1992), and progressive lignification of the underlying SCW (as exemplified by *Picea*; Tenberge, 1989, 1992).

2.10.2 *Structural variation and cuticle development*

When considering the variability of cuticle structure against this summary, it should be recognized that the six types identified by Holloway (1982a; Table 2.2) were suggested as a provisional pragmatic grouping, in order to obtain a descriptive handle on the available data. The type boundaries are not hard (nor intended to be), nor can they be accounted for by fundamental differences in cuticle chemistry, taxonomic relationships or plant function. The spectrum of variation seen in

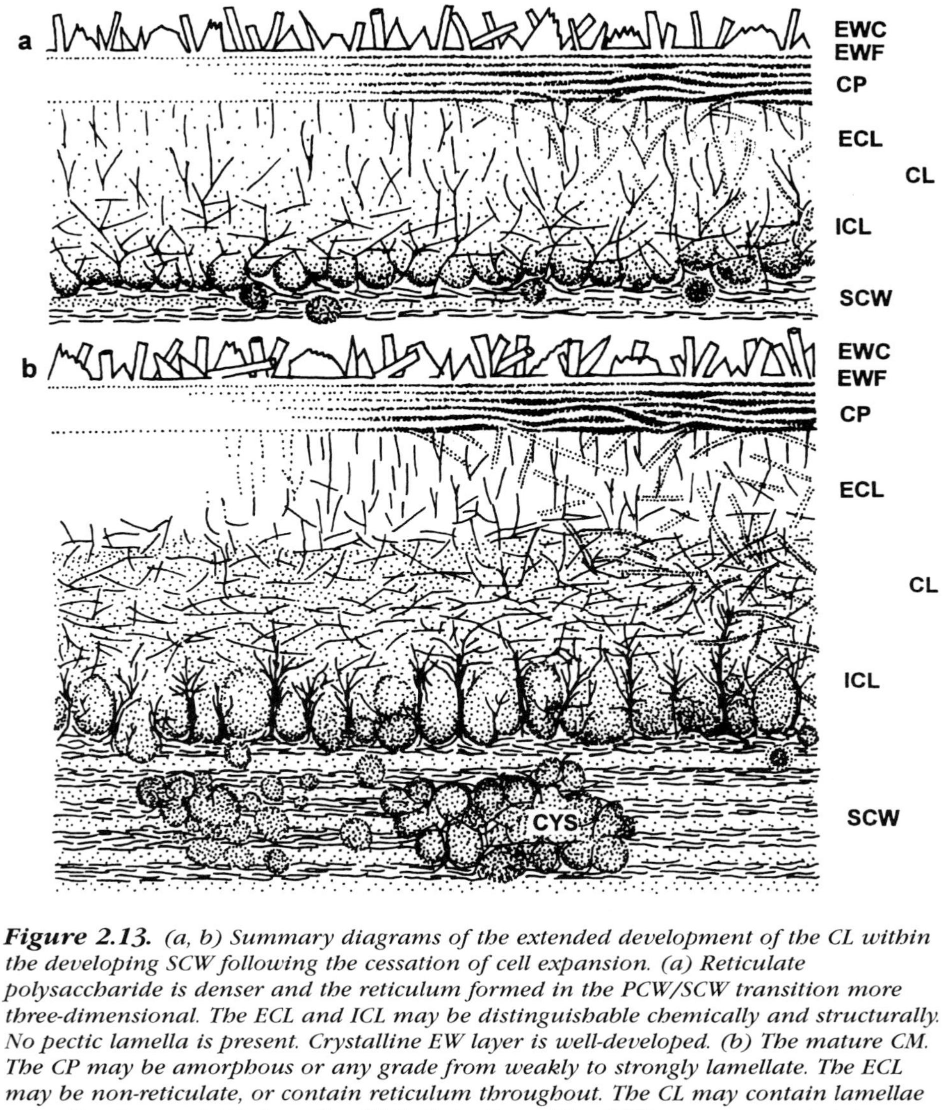

Figure 2.13. *(a, b) Summary diagrams of the extended development of the CL within the developing SCW following the cessation of cell expansion. (a) Reticulate polysaccharide is denser and the reticulum formed in the PCW/SCW transition more three-dimensional. The ECL and ICL may be distinguishable chemically and structurally. No pectic lamella is present. Crystalline EW layer is well-developed. (b) The mature CM. The CP may be amorphous or any grade from weakly to strongly lamellate. The ECL may be non-reticulate, or contain reticulum throughout. The CL may contain lamellae extending to any depth from the CP. Cutinization of the SCW may occur at points detached from the main CM, forming cystoliths (CYS).*

cuticles at maturity appears to be continuous, individual types often straddling type boundaries, making allocation to a type difficult. Although structural variation is apparent in the cuticles of bryophytes and vascular plants, it is also evident that they contain elements of a common set of the fundamental ultrastructural features. Despite the potential dangers inherent in this, and despite the cautionary notes against oversimplification and generalization of Holloway (1982a), and others, I therefore suggest a solution that emphasizes the unity of the structure and ontogeny of plant CM as dominant over their structural differences. It is common to find parallels between the structural patterns of mature cuticles, and stages in the development of the thicker and more elaborate cuticles of *Agave*, *Clivia* and *Picea* on which, currently, the predominant developmental and ultrastructural data is based. Probably, the major differences between cuticle structural types can be ascribed to the relative extent of the development of the CM and of the underlying CW, rather than to fundamental differences in mechanism, biochemistry or approach. Thus, for example, the simple lamellate CM of *Beta vulgaris* (Holloway, 1982a) is structurally analogous to those of *Clivia* (Figure 2.1f) and *Agave* (Figure 2.3b) at the earliest stages of development that have been documented (Riederer and Schönherr, 1988; Wattendorff and Holloway, 1980). In the specialized cuticle of the stigma papillae of *Silene dioica* (Heslop-Harrison and Heslop-Harrison, 1982; Figures 2.11a and b) a more elaborate, but nevertheless delicate, form is observed, with a combination of lamellate CP and a CL composed of osmiophilic globules or rods that bear a striking resemblance to the morphology of the (ultimately) complex cuticle of *Hakea suaveolens* early in its ontogeny, at the point where the CL is just beginning its ultimately extensive development (see Figures 2.4a and 2.8). The cuticle of *Silene* stigmas thus appears as if suspended in development. Equally, the CM of *Secale cereale* stigmas, which could be categorized as Type 6 in Holloway's scheme, may alternatively be interpreted as consisting of a simple CP with a single pair of lamellae, subtended by a thin, discontinuous CL, placing it as a simplest-possible expression of Type 1. The CM of the glandular trichomes of *Cannabis sativa* (Mahlberg and Kim, 1991, 1992) and of *Symplocos paniculata* (Tegelaar, 1990) are similar to that of *Picea abies* at comparable thicknesses in Stage 2 (Tenberge, 1992; Figure 2.6c). Again, the cuticle of *Iris germanica*, which at ~ 0.8 μm is much thinner than that of *Clivia,* has CL development which strongly resembles that of *Clivia* at 5–6 cm from the base, before the latter cuticle goes on to elaborate differentially the ECL and ICL. The hypothesis is therefore erected that the spectrum of observed cuticle structural variation falls within the scope of this structural and developmental model.

Acknowledgements

Grateful thanks are extended to Dr Peter Holloway, Dr Henning S. Heide-Jørgensen, Professor John Heslop-Harrison, FRS, and Dr Yolande Heslop-Harrison, Dr Klaus Tenberge, Professor Dr Jörg Schönherr, Professor Dr Markus Riederer, Professor Dr Joachim Wattendorff, Dr Fred Sack, Dr Carolyn McQuattie, Professor A. Chamel and Dr Hester Krüger, and their colleagues, for many valuable discussions and for permission to use their electron micrographs to illustrate this review.

References

Akers, C.P., Weybrew, J.A. and Long, R.C. (1978) Ultrastructure of the glandular trichomes of leaves of *Nicotiana tabacum* L., cv. Xanthi. *Am. J. Bot.* **65,** 282–292.

Archer, K.J. and Cole, A.L.J. (1986) Cuticle, cell-wall ultrastructure and disease resistance in maidenhair fern. *New Phytol.* **103,** 341–348.

Armstrong, D.J. and Whitecross, M.I. (1976) Temperature effects on formation and fine structure of *Brassica napus* leaf waxes. *Aust. J. Bot.* **24,** 309–318.

Baig, M.N. and Tranquillini, W. (1976) Studies on the upper timberline: morphology and anatomy of Norway spruce (*Picea abies*) and stone pine (*Pinus cembra*) needles from various habitat conditions. *Can. J. Bot.* **54,** 1622–1632.

Baker, E.A. (1982) Chemistry and morphology of plant epicuticular waxes. In: *The Plant Cuticle* (eds D.J. Cutler, K.L. Alvin and C.E. Price). Academic Press, London, pp. 139–165.

Baker, E.A. and Parsons, E. (1971) Scanning electron microscopy of plant cuticles. *J. Microsc.* **94**, 39–49.

Baker, E.A. and Procopiou, J. (1975) The cuticles of *Citrus* species. Composition of the intracuticular lipids of leaves and fruits. *J. Sci. Food Agric.* **26**, 1347–1352.

Baker, E.A., Batt, R.F. and Martin, J.T. (1964) Studies on plant cuticle. VII. The nature and determination of cutin. *Ann. Appl. Biol.* **53**, 59–65.

Barnabus, A.D., Butler, V. and Steinke, T.D. (1977) *Zostera capensis*, Setchell. I. Observations on the fine structure of the leaf epidermis. *Z. Pflanzenphysiol.* **85**, 417–427.

Campbell, R. (1972) Electron microscopy of the epidermis and the cuticle of the needles of *Pinus nigra* var. *maritima* in relation to infection by *Lophodermella sulcigena. Ann. Bot.* **36**, 307–314.

Chabot, J.F. and Chabot, B.F. (1977) Ultrastructure of the epidermis and stomatal complex of balsam fir (*Abies balsamea*). *Can. J. Bot.* **55**, 1064–1075.

Chafe, S.C. and Wardrop, A.B. (1972) Fine structural observations on the epidermis I. The epidermal cell wall. *Planta* **107**, 269–278.

Chafe, S.C. and Wardrop, A.B. (1973) Fine structural observations on the epidermis. II. The cuticle. *Planta* **109**, 39–48.

Considine, J.A. and Knox, R.B. (1979) Development and histochemistry of the pistil of the grape, *Vitis vinifera. Ann. Bot.* **43**, 11–22.

Crisp, C.E. (1965) The biopolymer cutin. Ph.D. Thesis, California University.

Crisp, D.J. (1963) Waterproofing mechanisms in animals and plants. In: *Waterproofing and Water-Repellency* (ed. J.L. Moilliet). Elsevier Publishing Co., Amsterdam, pp. 416–481.

Davis, R.W. (1987) Ultrastructure and analytical microscopy of silicon in the leaf cuticle of *Ficus lyrata* (Warb). *Bot. Gaz.* **148**, 318–323.

de Bary, A. (1871) Über die Wachsüberzüge der Epidermis. *Bot. Ztg* **29**, 129–176.

de Bary, A. (1884) Comparative anatomy of the vegetative organs of the phanerogams and ferns. *Bot. Ztg* **29**, 145–154.

de Vries, H.A.M.A. (1968a) Development of the structure of the russetted apple skin. *Acta Bot. Neerl.* **17**, 405–415.

de Vries, H.A.M.A. (1968b) Development of the structure of the normal, smooth cuticle of the apple 'Golden delicious'. *Acta Bot. Neerl.* **17**, 229–241.

Elleman, C.J., Willson, C.E., Sarker, R.H. *et al.* (1988) Interaction between pollen tube and stigmatic cell wall following pollination in *Brassica oleracea. New Phytol.* **109**, 111–117.

Esau, K. (1953) *Plant Anatomy*. Chapman and Hall, London.

Espelie, K.E., Davis, R.W. and Kolattukudy, P.E. (1980) Composition, ultrastructure and function of the cutin- and suberin-containing layers in the leaf, fruit peel, juice-sac and inner seed coat of grapefruit (*Citrus paradisi* Macfed.). *Planta* **149**, 498–511.

Fahn, A. and Rachmilevitz, T. (1970) Ultrastructure and nectar secretion in *Lonicera japonica. Bot. J. Linn. Soc.* **63**, 51–56.

Findlay, N. and Mercer, F.V. (1971) Nectar production in *Abutilon*. I. Movement of nectar through the cuticle. *Aust. J. Biol. Sci.* **24**, 647–656.

Fisher, D.A. and Bayer, D.E. (1972) Thin sections of plant cuticles, demonstrating channels and wax platelets. *Can. J. Bot.* **50**, 1509–1511.

Frölich, D. and Barthlott, W. (1988) Mikromorphologie der epicuticularen Wachse und das System der Monokotylen. *Tropische und subtropische Pflanzenwelt* **63**, 7–132.

Grace, J. (1990) Cuticular water loss unlikely to explain tree-line in Scotland. *Oecologia* **84**, 64–68.

Hadley, N.F. (1981) Cuticular lipids of terrestrial plants and arthropods — a comparison of their structure, composition, and waterproofing function. *Bio. Rev. Cambridge Phil. Soc.* **56**, 23–47.

Hall, D.M. (1967) Wax microchannels in the epidermis of white clover. *Science* **158**, 505–506.

Hall, D.M. and Donaldson, L.A. (1962) Secretion from pores of surface wax on plant leaves. *Nature* **194**, 1196.

Hall, D.M. and Donaldson, L.A. (1963) The ultrastructure of wax deposits on plant leaf surfaces. I. Growth of wax on leaves of *Trifolium repens. J. Ultrastruct. Res.* **9**, 259–267.

Hallam, N.D. (1964) Sectioning and electron microscopy of Eucalypt leaf waxes. *Aust. J. Biol. Sci.* **17**, 587–590.

Hallam, N.D. (1967) An electron microscope study of the leaf waxes of the genus *Eucalyptus* L'Heritier. Ph.D. Thesis, University of Melbourne.

Hallam, N.D. (1970) Leaf wax fine structure and ontogeny in *Eucalyptus* demonstrated by means of a specialized fixation technique. *J. Microsc.* **92**, 137–144.

Hallam, N.D. (1982) Fine structure of the leaf cuticle and the origin of leaf waxes. In: *The Plant Cuticle* (eds D.F. Cutler, K.L. Alvin and C.E. Price). Academic Press, London, pp. 197–214.

Hallam, N.D. and Juniper, B.E. (1971) The anatomy of the leaf surface. In: *The Ecology of Leaf Surface Micro-organisms* (eds T.F. Preece and C.H. Dickinson). Academic Press, London, pp. 3–37.

Hammond, C.T. and Mahlberg, P.G. (1978) Ultrastructural development of capitate glandular hairs of *Cannabis sativa* L. (Cannabaceae). *Am. J. Bot.* **65**, 140–151.

Heide-Jørgensen, H.S. (1978) The xeromorphic leaves of *Hakea suaveolens* R.Br. II. Structure of epidermal cells, cuticle development and ectodesmata. *Bot. Tidsskr.* **72**, 227–244.

Heide-Jørgensen, H.S. (1980) The xeromorphic leaves of *Hakea suaveolens* R.Br. III. Ontogeny, structure and function of the T-shaped trichomes. *Bot. Tidsskr.* **75**, 181–198.

Heide-Jørgensen, H.S. (1991) Cuticle development and ultrastructure — evidence for a procuticle of high osmium affinity. *Planta* **183**, 511–519.

Heslop-Harrison, Y. (1977) The pollen–stigma interaction: pollen-tube penetration in *Crocus. Ann. Bot.* **41**, 913–922.

Heslop-Harrison, J. and Heslop-Harrison, Y. (1975) Fine structure of the stigmatic papilla of *Crocus. Micron* **6**, 45–52.

Heslop-Harrison, J. and Heslop-Harrison, Y. (1982) The specialized cuticles of the receptive surfaces of angiosperm stigmas. In: *The Plant Cuticle* (eds D.F. Cutler, K.L. Alvin and C.E. Price). Academic Press, London, pp. 99–119.

Heslop-Harrison, J. and Heslop-Harrison, Y. (1985) The secretory system of the stigma of *Oenothera organensis* Munz: some developmental and quantitative features. *Israel J. Bot.* **34**, 187–204.

Heumann, H-G. (1990) A simple method for improved vizualization of the lamellated structure of cutinized and suberized plant cell walls by electron microscopy. *Stain Technol.* **65**, 183–187.

Hilkenbäumer, F. (1958) Elektronenmikroscopische Untersuchungen über den Aufbau kräftig entwickelter Cuticulae von Apfelfrüchten. *Z. Naturf. C-Biosci.* **13B**, 666–668.

Hoch, H.C. (1975) Ultrastructural alterations observed in isolated apple leaf cuticles. *Can. J. Bot.* **53**, 2006–2013.

Hoch, H.C. (1979) Penetration of chemicals into the *Malus* leaf cuticle. *Planta* **147**, 186–195.

Hoffmann-Benning, S. and Kende, H. (1994) Cuticle biosynthesis in rapidly growing internodes of deep-water rice. *Plant Physiol.* **104**, 719–723.

Hoffmann-Benning, S., Klomparens, K.L. and Kende, H. (1994) Characterization of growth-related osmiophilic particles in corn coleoptiles and deep-water rice internodes. *Ann. Bot.* **74**, 563–572.

Holloway, P.J. (1970) Surface factors affecting the wetting of leaves. *Pestic. Sci.* **1**, 156–162.

Holloway, P.J. (1971) The chemical and physical characteristics of leaf surfaces. In: *The Ecology of Leaf Surface Micro-organisms* (eds T.F. Preece and C.H. Dickinson). Academic Press, London, pp. 39–53.

Holloway, P.J. (1982a) Structure and histochemistry of plant cuticular membranes: an overview. In: *The Plant Cuticle* (eds D.F. Cutler, K.L. Alvin and C.E. Price). Academic Press, London, pp. 1–32.

Holloway, P.J. (1982b) The chemical constitution of plant cutins. In: *The Plant Cuticle* (eds D.F. Cutler, K.L. Alvin and C.E. Price). Academic Press, London, pp. 45–85.

Holloway, P.J. (1993) Structure and chemistry of plant cuticles. *Pestic. Sci.* **37**, 203–206.

Holloway, P.J. (1994) Plant cuticles: Physicochemical characteristics and biosynthesis. In: *Air Pollution and the Leaf Cuticle (eds* K.E. Percy, J.N. Cape, R. Jagels and C.J. Simpson). NATO ASI Series G, Vol. 36. Springer Verlag, Berlin, pp. 1–14.

Holloway, P.J. and Baker, E.A. (1968) Isolation of plant cuticles with zinc chloride-hydrochloric acid solution. *Plant Physiol.* **11**, 1878–1879.

Holloway, P.J. and Baker, E.A. (1970) The cuticles of some angiosperm leaves and fruits. *Ann. Appl. Biol.* **66**, 145–154.

Holloway, P.J., Brown, G.A. and Wattendorff, J. (1981) Ultrahistochemical detection of epoxides in plant cuticular membranes. *J. Exp. Bot.* **32**, 1051–1066.

Huelin, F.E. (1959) Studies in the natural coating of apples. IV. The name of cutin. *Aust. J. Biol. Sci.* **12**, 175–180.

Huelin, F.E. and Gallop, R.A. (1951) Studies in the natural coating of apples. I. Preparation and properties of fractions. *Aust. J. Sci. Res.* **B 4**, 526–532.

Hülsbruch, M. (1966a) Cutin als konstruktive Wandsubstanz nicht nur Akkruste oder Inkruste. *Ber. Dtsch. Bot. Ges.* **79**, 87–91.

Hülsbruch, M. (1966b) Zur Radialstreifung cutinisierter Epidermisaussenwände. *Z. Pflanzenphysiol.* **55**, 181–197.

Jarvis, L.R. and Wardrop, A.B. (1974) The development of the cuticle in *Phormium tenax*. *Planta* **119,** 101–112.

Jeffree, C.E. (1974) Some aspects of the fine structure of the cuticle and epicuticular waxes of Sitka spruce, *Picea sitchensis*, (Bong.) Carr. PhD Thesis, University of Aberdeen.

Jeffree, C.E. (1986) The cuticle, epicuticular waxes and trichomes of plants, with reference to their structure, function and evolution. In: *Insects and Plant Surfaces* (eds B.E. Juniper and R. Southwood). Edward Arnold, London, pp. 23–64.

Jeffree, C.E., Baker, E.A. and Holloway, P.J. (1975) Ultrastructure and recrystallisation of plant epicuticular waxes. *New Phytol.* **75,** 539–549.

Jeffree, C.E., Baker, E.A. and Holloway, P.J. (1976) Origins of the fine structure of plant epicuticular waxes. In: *Microbiology of Aerial Plant Surfaces* (eds C.H. Dickinson and T.F. Preece). Academic Press, London, pp. 119–158.

Jeffree, C.E., Dale, J.E. and Fry, S.C. (1986) The genesis of intercellular spaces in developing leaves of *Phaseolus vulgaris*. *Protoplasma* **132,** 90–98.

Jeffree, C.E., Read, N.D., Smith, J.A.C. *et al.* (1987) Water droplets and ice deposits in leaf intercellular spaces: redistribution of water during cryofixation for scanning electron microscopy. *Planta* **172,** 20–37.

Jeffrey, E.C. (1917) *The Anatomy of Woody Plants.* University of Chicago Press, Chicago, USA.

Jetter, R. and Riederer, M. (1994). Epicuticular crystals of nonacosan-10-ol — in-vitro reconstitution and factors influencing crystal habits. *Planta* **195,** 257–270.

Jetter, R. and Riederer, M. (1995). In-vitro reconstitution of epicuticular wax crystals — formation of tubular aggregates by long-chain secondary alkanediols. *Bot. Acta* **108,** 111–120.

Joel, D.M. and Heide-Jørgensen, H.S. (1985) Ultrastructure and development of the pitcher epithelium of *Sarracenia*. *Israel J. Bot.* **34,** 331–349.

Joel, D.M. and Juniper, B.E. (1982) Cuticular gaps in carnivorous plant glands. In: *The Plant Cuticle* (eds D.F. Cutler, K.L. Alvin and C.E. Price). Academic Press, London, pp. 121–130.

Joel, D.M., Rea, P.A. and Juniper, B.E. (1983) The cuticle of *Dionaea muscipula* Ellis (Venus's flytrap) in relation to stimulation, secretion and absorption. *Protoplasma* **114,** 44–51.

Juniper, B.E. and Cox, G.C. (1973) The anatomy of the leaf surface: The first line of defence. *Pestic. Sci.* **4,** 543–561.

Juniper, B.E. and Jeffree, C.E. (1983) *Plant Surfaces*. Edward Arnold, London.

Junker, S. (1977) Ultrastructure of tactile papillae on tendrils of *Eccremocarpus scaber* R. et P. *New Phytol.* **78,** 607–610.

Kaufman, P.B., Petering, L.B., Yocum, C.S. *et al.* (1970) Ultrastructural studies in stomata development in internodes of *Avena sativa*. *Am. J. Bot.* **57,** 33–49.

Kim, E. and Mahlberg, P.G. (1995) Glandular cuticle formation in *Cannabis* (Cannabaceae). *Am. J. Bot.* **82,** 1207–1214.

Kluge, M., Knapp, I., Kramer, D. *et al.* (1979) Crassulacean acid metabolism (CAM) in leaves of *Aloe arborescens* Mill. Comparative studies of the carbon metabolism of chlorenchym and central hydrenchym. *Planta* **145,** 357–363.

Kögel-Knabner, I., de Leeuw, J.W., Tegelaar, E.W. *et al.* (1994) A lignin-like polymer in the cuticle of spruce needles — implications for the humification of spruce litter. *Org. Geochem.* **21,** 1219–1228.

Kolattukudy, P.E. (1980) Biopolyester membranes of plants: cutin and suberin. *Science* **208,** 990–1000.

Kolattukudy, P.E. (1984) Biochemistry and function of cutin and suberin. *Can. J. Bot.* **62,** 2918–2933.

Kollöffel, C. and Linssen, P.W.T. (1984) The formation of intercellular spaces in the cotyledons of developing and germinating pea seeds. *Protoplasma* **120,** 12–19.

Krüger, H., van Rensburg, L. and Peacock, J. (1996) Cuticular membrane fine structure of *Nicotiana tabacum* L. leaves. *Ann. Bot.* **77,** 11–16.

Kuo, J. (1978) Morphology, anatomy and histochemistry of the Australian sea grasses of the genus *Posidonia* König (Posidonaceae). I. Leaf blade and leaf sheath of *Posidonia australis* Hook. f. *Aquatic Bot.* **5,** 171–190.

Lee, B. and Priestley, J.H. (1924) The plant cuticle. I. Its structure, distribution, and function. *Ann. Bot.* **38,** 525–545.

Lehmann, H. and Schulz, D. (1976) *Die Pflanzenzelle. Struktur und Funktion*. Verlag Eugen Ulmer, Stuttgart.

Lendzian, K.J., Nakajima, A. and Ziegler, H. (1986) Isolation of cuticular membranes from various conifer needles. *Trees* **1,** 47–53.

Letham, D.S. (1958) Maceration of plant-tissues with ethylene-diamine-tetra-acetic acid. *Nature* **181,** 135–136.

Lewis, F.J. (1945) Physical conditions of the surface of the mesophyll cell walls of the leaf. *Nature* **156,** 407–409.

Linskens, H.F. and Gelissen, A. (1966) Die Natur der Rauchschaligkeit bei Früchten der Apfelsorte 'Golden Delicious'. *Phytopathol. Z.* **57,** 1–7.

Lister, G.R. and Thair, B.W. (1981) *In vitro* studies on the fine structure of epicuticular leaf wax from *Pseudotsuga menziesii*. *Can. J. Bot.* **59,** 640–648.

Lyshede,O.B. (1976) Structure and function of trichomes in *Spartocytisus filipes*. *Bot. Notiser.* **129,** 395–404.

Lyshede, O.B. (1977a) Structure of the epidermal and subepidermal cells of some desert plants of Israel. *Anabasis articulata* and *Calligonum comosum*. *Israel J. Bot.* **26,** 1–10.

Lyshede, O.B. (1977b) Studies on the mucilaginous cells in the leaf of *Spartocytisus filipes* W.B. *Planta* **133,** 255–260.

Lyshede, O.B. (1978) Studies on outer epidermal cell walls with microchannels in a xerophytic species. *New Phytol.* **80,** 421–426.

Lyshede, O.B. (1982) Structure of the outer epidermal wall in xerophytes. In: *The Plant Cuticle* (eds D.F. Cutler, K.L. Alvin and C.E. Price). Academic Press, London, pp. 87–98.

Machado, R.D. and Barros, C.F. (1995) Epidermis and epicuticular waxes of *Syagrus coronata* leaflets. *Can. J. Bot.* **73,** 1947–1952.

Mahlberg, P.G. and Kim, E. (1991) Cuticle development on glandular trichomes of *Cannabis sativa* (Cannabaceae). *Am. J. Bot.* **78,** 1113–1122.

Mahlberg, P.G. and Kim, E. (1992) Secretory vesicle formation in glandular trichomes of *Cannabis sativa* (Cannabaceae). *Am. J. Bot.* **79,** 166–173.

Maier, U. (1968) Dendritenartige Strukturen in der Cuticularschicht von *Lilium candidum*. *Protoplasma* **65,** 243–246.

Martin, J.T. and Juniper, B.E. (1970) *The Cuticles of Plants*. Edward Arnold, London.

McKeen, W.E. (1974) Mode of penetration of epidermal cell walls of *Vicia faba* by *Botrytis cinerea*. *Phytopathol.* **64,** 461–467.

McQuattie, C.J. and Rebbeck, J. (1994) Effect of ozone and elevated carbon dioxide on cuticular membrane ultrastructure of yellow poplar (*Liriodendron tulipifera*). In: *Air Pollution and the Leaf Cuticle,* NATO ASI Series G, Vol. 36 (eds K.E. Percy, J.N. Cape, R. Jagels and C.J. Simpson). Springer Verlag, Heidelberg, pp. 175–182.

Mérida, T. and Ogura, M. (1987) Fine structure of the cuticular membrane of *Nicotiana glauca* Grah. leaves. *Acta Cient. Venez.* **38,** 77–83.

Mérida, T., Schönherr, J. and Schmidt, H.W. (1981) Fine structure of plant cuticles in relation to water permeability: The fine structure of *Clivia miniata* Reg. leaves. *Planta* **152,** 259–267.

Meusel, I., Leistner, E. and Barthlott, W. (1994) Chemistry and micromorphology of compound epicuticular wax crystalloids (*Strelitzia* type). *Plant Syst. Evol.* **193,** 115–123.

Miller, R.H. (1986) The prevalence of pores and canals in leaf cuticular membranes. 2. supplemental studies. *Ann. Bot.* **57,** 419–434.

Morrison, I.N. (1975) Ultrastructure of the cuticular membranes of the developing wheat grain. *Can. J. Bot.* **53,** 2077–2087.

Nip, M., Tegelaar, E.W., de Leeuw, J.W. *et al.* (1986a) A new non-saponifiable highly-aliphatic and resistant biopolymer in plant cuticles. Evidence from pyrolysis and ^{13}C-NMR analysis of present-day and fossil plants. *Naturwissenschaften* **73,** 579–585.

Nip, M., Tegelaar, E.W., Brinkhuis, H. *et al.* (1986b) Analysis of modern and fossil plant cuticles by Curie-point py-gc and Curie-point py-gc-ms - recognition of a new, highly aliphatic and resistant bio-polymer. *Org. Geochem.* **10,** 769–778.

Nip, M., de Leeuw, J.W., Holloway, P.J. *et al.* (1987) Comparison of flash pyrolysis, differential scanning calorimetry, ^{13}C NMR and IR spectroscopy in the analysis of a highly aliphatic biopolymer from plant cuticles. *J. Anal. Appl. Pyrol.* **11,** 287–295.

Norris, R.F. and Bukovac, M.J. (1968) The structure of pear leaf cuticles with special reference to cuticular penetration. *Am. J. Bot.* **55,** 975–983.

O'Brien, T.P. (1967) Observations on the fine structure of the oat coleoptile. I. The epidermal cells of the extreme apex. *Protoplasma* **63,** 385–415.

Orgell, W.H. (1955) The isolation of plant cuticle with pectic enzymes. *Plant Physiol.* **30,** 78–80.

Pallas, J.E. and Mollenhauer, H.H. (1972) Electron microscopic evidence for plasmodesmata in dicotyledonous guard cells. *Science* **175,** 1275–1276.

Percy, K.E., Jensen, K.F. and McQuattie, C.J. (1992) Effects of ozone and acidic fog on red spruce needle epicuticular wax production, chemical-composition, cuticular membrane ultrastructure and needle wettability. *New Phytol.* **122,** 71–80.

Postek, M.T. (1981) The occurrence of silica in the leaves of *Magnolia grandiflora* L. *Bot. Gaz.* **142,** 124–134.

Pyee, J. and Kolattukudy, P.E. (1995) The gene for the major cuticular wax-associated protein and 3 homologous genes from broccoli (*Brassica oleracea*) and their expression patterns. *Plant J.* **7,** 49–59.

Pyee, J., Yu, H.S. and Kolattukudy, P.E. (1994) Identification of a lipid transfer protein as the major protein in the surface wax of broccoli (*Brassica oleracea*) leaves. *Arch. Biochem. Biophys.* **311,** 460–468.

Raven, J.A. (1984) Physiological correlates of the morphology of early vascular plants. *Bot. J. Linn. Soc.* **88,** 105–126.

Reed, D.W. (1979) Ultrastructural studies on plant cuticles: Environmental effects, permeability, electron microscope preparation and specific staining. Ph.D. Thesis, Cornell University.

Riederer, M. (1991) Cuticle as barrier between terrestrial plants and the atmosphere — significance of growth-structure for cuticular permeability. *Naturwissenschaften* **78,** 201–208.

Riederer, M. and Schönherr, J. (1988) Development of plant cuticles fine structure and cutin composition of *Clivia miniata* Reg. leaves. *Planta* **174,** 127–138.

Rijkenberg, F.H.J., de Leeuw, G.T.N. and Verhoeff, K. (1980) Light and electron microscopy studies on the infection of tomato fruits by *Botrytis cinerea*. *Can. J. Bot.* **58,** 1394–1404.

Roelofsen, P.A. (1952) On the submicroscopic structure of cuticular cell walls. *Acta Bot. Neerl.* **1,** 99–114.

Roelofsen, P.A. (1959) The cell plant cell-wall. In: *Handbuch der Pflanzenanatomie (Encyclopedia of Plant Anatomy)* (eds W. Zimmermann and P.G. Ozenda). Gebrüder Borntraeger, Berlin-Nikolassee.

Roland, J.-C. (1978) Cell wall differentiation and stages involved with intercellular gas space opening. *J. Cell Sci.* **32,** 325–336.

Rosenquist, J.K. and Morrison, J.C. (1988) The development of the cuticle and epicuticular wax of the grape berry. *Vitis* **27,** 63–70.

Ryser, U. and Holloway, P.J. (1985) Ultrastructure and chemistry of soluble and polymeric lipids in cell walls from seed coats and fibres of *Gossypium* species. *Planta* **163,** 151–163.

Sack, F.D. and Paolillo, Jr, D.J. (1983a) Stomatal pore and cuticle formation in *Funaria*. *Protoplasma* **116,** 1–13.

Sack, F.D. and Paolillo Jr, D.J. (1983b) Structure and development of walls in *Funaria stomata*. *Am. J. Bot.* **70,** 1019–1030.

Sargent, C. (1976a) Studies on the ultrastructure and development of the plant cuticle. Ph.D. Thesis, London University.

Sargent, C. (1976b) *In situ* assembly of cuticular wax. *Planta* **129,** 123–126.

Sargent, C. (1976c) The occurrence of a secondary cuticle in *Libertia elegans* (Iridaceae). *Ann. Bot.* **40,** 355–359.

Sargent, C. and Gay, J.L. (1977) Barley epidermal apoplast structure and modification by powdery mildew contact. *Physiol. Plant Pathol.* **11,** 195–205.

Schieferstein, R.H. and Loomis, W.E. (1959) Development of the cuticular layers in angiosperm leaves. *Am. J. Bot.* **46,** 625–635.

Schmidt, H.W. and Schönherr, J. (1982) Development of plant cuticles — occurrence and role of non-ester bonds in cutin of *Clivia miniata* Reg. leaves. *Planta* **156,** 380–384.

Schmidt, H.W., Mérida, T. and Schönherr, J. (1981) Water permeability and fine-structure of cuticular membranes isolated enzymatically from leaves of *Clivia miniata* Reg. *Z. Pflanzenphysiol.* **105,** 41–51.

Schneider, A. and Dargent, R. (1977) Localisation et compartement du mycélium de *Taphrina deformans* dans le mésophylle et sans le cuticule des feuilles de pêcher (*Prunus persica*). *Can. J. Bot.* **55,** 2485–2495.

Schneider, W. (1960) Beiträge zum chemischen Aufbau der Apfelschale. Ph.D. Thesis, Julius-Maximillian University, Würzburg.

Schnepf, E. (1969) Sekretion und Exkretion bei Pflanzen. *Protoplasmatologia* **8,** 1–181.

Schönherr, J. and Mérida, T. (1981) Water permeability of plant cuticular membranes: the effects of humidity and temperature on the permeability of non-isolated cuticles of onion bulb scales. *Plant Cell Environ.* **4,** 349–354.

Schönherr, J., Riederer, M., Schreiber, L. *et al.* (1991) Foliar uptake of pesticides and its activation by adjuvants: theories and methods for optimization. In: *Pesticide Chemistry. Advances in International Research, Development and Legislation* (ed. H. Frehse). VCH, Weinheim, Germany, pp. 237-253.

Schreiber, L., Kirsch, T. and Riederer, M. (1996) Transport-properties of cuticular waxes of *Fagus sylvatica* L. and *Picea abies* (L.) Karst — estimation of size-selectivity and tortuosity from diffusion coefficients of aliphatic molecules. *Planta* **198**, 104-109.

Scott, F.M. (1948) Internal suberization of plant tubers. *Science* **108**, 654-655.

Scott, F.M. (1966) Cell wall surface of the higher plants. *Nature* **210**, 1015-1017.

Shayk, M., Kolattukudy, P.E. and Davis, R. (1977) Production of a novel extracellular cutinase by the pollen and the chemical composition and ultrastructure of the stigma cuticle of nasturtium (*Tropaeolum majus*). *Pl. Physiol.* **60**, 907-915.

Sheriff, D.W. (1977) Evaporation sites and distillation in leaves. *Ann. Bot.* **41**, 1081-1082.

Sievers, A. (1968) Zur Epidermisaussenwand der Fühlborsten von *Dionaea muscipula*. *Planta* **83**, 49-52.

Singh, A.P. and Hemmes, D.E. (1978) Ultrastructure of the cuticle of Ohelo (*Vaccinium reticulatum*). *Am. J. Bot.* **65**, 919-921.

Sitte, P. (1975) Fine structure of fruit hair cell wall in *Clematis*. *Ber. Dtsch. Bot. Ges.* **86**, 551-561.

Sitte, P. and Rennier, R. (1963) Untersuchungen an cuticularen Zellwandschichten. *Planta* **60**, 19-40.

Skoss, J.D. (1955) Structure and composition of plant cuticle in relation to environmental factors and permeability. *Bot. Gaz.* **117**, 55-72.

Srivastava, L.M., Sawhney, V.K. and Bonnettemaker, M. (1977) Cell growth, wall deposition, and correlated fine structure of colchicine-treated lettuce hypocotyl cells. *Can. J. Bot.* **55**, 902-917.

Tegelaar, E.W. (1990) Resistant biomacromolecules in morphologically characterized constituents of kerogen: a key to the relationship between biomass and fossil fuels. Ph.D. Thesis, University of Utrecht.

Tegelaar, E.W., de Leeuw, J.W., Largeau, C. *et al.* (1989) Scope and limitations of several pyrolysis methods in the structural elucidation of a macromolecular plant constituent in the leaf cuticle of *Agave americana* L. *J. Anal. Appl. Pyrol.* **15**, 29-54.

Tegelaar, E.W., Wattendorff, J. and de Leeuw, J.W. (1993) Possible effects of chemical heterogeneity in higher land plant cuticles on the preservation of its ultrastructure upon fossilization. *Rev. Palaeobot. Palynol.* **77**, 149-170.

Tenberge, K.B. (1989) Entwicklung und Aufbau der Epidermiszellwände von Fichtennadeln (*Picea abies* (L.) Karsten) — Lichtmikroskopische, elektronenmikroskopische und cytochemische Untersuchungen. Ph.D. Thesis, Westfälische Wilhelms-Universität Münster.

Tenberge, K.B. (1992) Ultrastructure and development of the outer epidermal wall of spruce (*Picea abies*) needles. *Can. J. Bot.* **70**, 1467-1487.

Thomson, W.W. and Platt-Alloia, K.A. (1976) Ultrastructure of the epidermis of developing, ripening, and scenescing navel oranges. *Hilgardia* **44**, 61-82.

Viougeas, M.A., Rohr, R. and Chamel, A.R. (1995) Structural-changes and permeability of ivy (*Hedera helix* L.) leaf cuticles in relation to leaf development and after selective chemical treatments. *New Phytol.* **130**, 337-348.

von Mohl, H. (1847) Untersuchungen der Frage: Bildet die Cellulose die Grundlage sämmtlicher vegetabilischen Membranen? *Bot. Ztg.* **5**, 497-505.

von Wettstein-Knowles, P. (1974) Ultrastructure and origin of epicuticular wax tubes. *J. Ultrastruct. Res.* **46**, 483-498.

Walles, B., Nyman, B. and Aldén, T. (1973) On the ultrastructure of needles of *Pinus sylvestris* L. *Studia Forestalia Suecica* **106**, 1-26.

Wattendorff, J. (1974) Ultrahistochemical reactions of the suberized cell walls in *Acorus, Acacia* and *Larix*. *Z. Pflanzenphysiol.* **73**, 214-225.

Wattendorff, J. and Holloway, P.J. (1980) Studies on the ultrastructure and histochemistry of plant cuticles: the cuticular membrane of *Agave americana* L. *in situ*. *Ann. Bot.* **46**, 13-28.

Wattendorff, J. and Holloway, P.J. (1982) Studies on the ultrastructure and histochemistry of plant cuticles — isolated cuticular membrane preparations of *Agave americana* L. and the effects of various extraction procedures. *Ann. Bot.* **49**, 769-804.

Wattendorff, J. and Holloway, P.J. (1984) Periclinal penetration of potassium permanganate into mature cuticular membranes of *Agave* and *Clivia* leaves: new implications for plant cuticle development. *Planta* **161**, 1-11.

Weinert, H. and Barckhaus, R.H. (1975) Fortified synthesis of cutin at the contact-zones between *Cuscuta odorata* and *Pelargonium zonale*. *Cytobios* **13**, 17–22.
Whatley, J.M. (1984) The ultrastructure of plastids in the petals of *Caltha palustris* L. *New Phytol.* **97**, 227–231.
Whittier, D.P. and Peterson, R.L. (1995) The cuticle on *Psilotum gametophytes*. *Can. J. Bot.* **73**, 1283–1288.
Willison, J.H.M. and Pearce, R.S. (1983) A comparative study of the structure of cell wall surfaces: air spaces in leaves are exceptional in having exposed microfibrils. *Can. J. Bot.* **61**, 2153–2158.
Willison, J.H.M., Pearce, R.S. and Odense, P.H. (1984) Cryo-SEM observation of water in leaf intercellular spaces. *Proc. 42nd Annu. Meet. Electron Microsc. Soc. Am.* 728–729.
Wilson, L.A. and Sterling, C. (1976) Studies on the cuticle of tomato fruit. I. Fine structure of the cuticle. *Z. Pflanzenphysiol.* **77**, 359–371.
Wullschleger, S.D. and Oosterhuis, D.M. (1989) The occurrence of an internal cuticle in cotton (*Gossypium hirsutum* L.) leaf stomates. *Environ. Exp. Bot.* **29**, 229–235.
Ziegenspeck, H. (1928) Über das Ergusswachstum des Kutins bei *Aloe*-Arten. *Bot. Arch.* **21**, 1–8.

3

Biosynthetic pathways of cutin and waxes, and their sensitivity to environmental stresses

P.E. Kolattukudy

3.1 Introduction

The outer envelope of land organisms consists of a polymeric structural material and associated soluble lipid materials. This structural component is proteinaceous in vertebrates and a carbohydrate polymer, chitin, in insects. However, in plants, the structural component is mainly a polyester called cutin derived from cellular lipids, except in barks and underground parts where the outer layer of cells suberize by depositing a polyester domain on an aromatic matrix attached to the cell wall and serve the barrier function. The soluble lipids associated with the polymers are composed of a complex mixture of lipids collectively called waxes that are distinctly different from cellular lipids. Early experiments demonstrated that cuticular wax biosynthesis is a specialized function of the epidermal cells (Kolattukudy, 1968a) and this conclusion has been confirmed by subsequent findings (Cassagne and Lessire, 1974, 1978; Kolattukudy and Buckner, 1972). Thus, it would appear that this biosynthetic process would be sensitive to environmental factors. In fact, in the 1960s, I found that incorporation of labelled acetate in pea leaves was extremely sensitive to growth temperature and light conditions under which pea plants were grown (unpublished data). I shall review the biosynthesis of the most common major components of the plant cuticle and briefly touch upon some aspects of interaction of physical and biological stress on the cuticle.

3.2 Biosynthesis of cuticular waxes

3.2.1 *Composition*

The most common major components of cuticular waxes are hydrocarbons and their oxygenated derivatives such as secondary alcohols and ketones, very long-chain fatty acids, aldehydes and alcohols, and wax esters composed of very long-chain fatty alcohols and fatty acids (Kolattukudy, 1980). Very long-chain β-diketones

Plant Cuticles, edited by G. Kerstiens.

and pentacyclic triterpenes are sometimes major components of waxes particularly on stems and fruits of some plants (Table 3.1). The hydrocarbons and derivatives and β-diketones have odd numbers of carbon atoms in the range of 27 to 33, whereas the other classes of compounds have even numbers of carbon atoms in the range of C_{26} to C_{32}. In each class of the aliphatic compounds found in the cuticular waxes, one or a few chain lengths often predominate. Suberin-associated waxes contain aliphatic components similar to those found in cuticular waxes but often there is a wider range of chain length distribution with less dominance by individual components (Kolattukudy and Espelie, 1985, 1989).

As if to provide chemical and metabolic stability, the cuticular and suberin-associated waxes, which function at the surface exposed to atmospheric oxygen and other components as well as microbes, are mostly saturated and unusually hard for microbes to absorb and metabolize (Kolattukudy, 1976).

3.2.2 *Biosynthesis of very long-chain fatty acids*

One of the characteristic features of the cuticular waxes is that they are usually composed of aliphatic chains much longer than those found in cellular lipids. Such very long-chain fatty acids are produced by elongation of the cellular fatty acids generated by the fatty acid synthase. In the 1960s, it was shown that exogenous saturated fatty acids are readily elongated by tissue slices (Kolattukudy, 1966), whereas unsaturated acid (oleic acid) was not elongated and incorporated into wax, demonstrating specificity of the elongating systems, which is reflected in the wax composition. Such a specificity was later found in microsomal elongating systems (Agrawal *et al.*, 1984). The presence of multiple elongating enzyme systems is suggested by the presence of different ranges of chain lengths present in different classes of wax components that occur in the same plant; for example, in pea waxes hydrocarbons and derivatives constitute the longest with the most dominant component with 31 carbon atoms followed by fatty alcohols dominated by components with 26 and 28 carbon atoms, and fatty acids with 16 to 22 carbon atoms (Kolattukudy, 1970a). During the 1970s the occurrence of multiple-chain elongating systems was suggested on the basis of the differential inhibition of incorporation of labelled precursors into the various chain length classes by inhibitors such as trichloroacetates, thiocarbamates, arsenite, cyanide and β-

Table 3.1. *Waxes: common major components*

	Chain length	
	Range	Major
Hydrocarbons		
n-Alkanes	C_{21}–C_{35}	C_{27},C_{29},C_{31}
Branched		
Iso	C_{21}–C_{35}	C_{27},C_{29},C_{31}
Anteiso	C_{22}–C_{36}	C_{26},C_{28},C_{30},C_{32}
Secondary alcohols and ketones	C_{21}–C_{35}	C_{27},C_{29},C_{31}
Fatty alcohols	C_{22}–C_{34}	C_{26}, C_{28}
Fatty acids	C_{16}–C_{34}	
Aldehydes	C_{23}–C_{35}	C_{26},C_{28}; C_{30},C_{32}
Wax esters (β-diketones, terpenes, flavanoids, etc.)		

mercaptoethanol (Kolattukudy, 1980; von Wettstein-Knowles, 1993). Such tentative conclusions were supported by analyses of wax composition of mutants, which revealed accumulation of aliphatic chains caused by mutations in genes whose product would have elongated such chains. Such wax mutations in *Brassica*, maize, pea and *Arabidopsis* have been analysed (Coe *et al.*, 1990; Holloway *et al.*, 1977a,b; Kunst *et al.*, 1992; von Wettstein-Knowles, 1987).

Since the unique components of the cuticle are made in the epidermis (Kolattukudy, 1968a), the early attempts to seek cell-free systems catalysing chain elongation were made with excised epidermis (Kolattukudy and Buckner, 1972). Thus, the chain-elongating activity in the pea epidermis was found to be associated with membranes as also found in cotyledons and germinating peas (Bolton and Harwood, 1977; Harwood and Stumpf, 1971; Macey and Stumpf, 1968). The characteristics of the elongating systems showed that endogenous acyl chains are elongated with malonyl-CoA and reduced pyridine nucleotides as the substrates. Efforts to solubilize and purify elongating systems have made some progress. Solubilization with detergent followed by protein fractionation separated elongating activities with preference for C_{18} from that which prefers C_{20} (Lessire *et al.*, 1985). Attempts to purify the C_{18} elongating system yielded a preparation containing multiple proteins (Bessoule *et al.*, 1989; Lessire *et al.*, 1989). Similarly, attempts to purify elongating systems from *Lunaria annua* also yielded a preparation containing many proteins. The only fatty acid elongating enzyme that has been purified to homogeneity and characterized is mycocerosic acid synthase of Mycobacteria which is a multifunctional enzyme that catalyses elongation of n-C_{20} fatty acyl-CoA with methylmalonyl-CoA as the elongating agent (Rainwater and Kolattukudy, 1985). The organization of the domains in this enzyme has been elucidated by cloning and sequencing of the gene (Mathur and Kolattukudy, 1992) and it is quite similar to the open-reading frames identified in the genes involved in the macrocyclic lactone synthesis in *Saccharopolyspora erythraea* (Donadio *et al.*, 1991). Whether multifunctional enzymes are involved in the epidermal chain elongation is unknown. Until the enzyme(s) involved is purified no definitive conclusion can be drawn regarding the nature and specific subcellular-location of the elongating enzyme(s).

3.2.3 *Fatty acyl-CoA reduction*

Fatty acyl-CoA reduction was first demonstrated in 1970 with cell-free preparations from *Euglena gracilis,* which produces wax esters as a major form of energy storage (Kolattukudy, 1970b). This membrane-associated reductase produces alcohol with no detectable accumulation of free aldehyde. However, aldehyde intermediate could be demonstrated by chemically trapping it. Since then membrane-associated acyl-CoA reductase has been obtained from a variety of systems from both the plant and animal kingdoms (Blomquist and Ries, 1979; Khan and Kolattukudy, 1973; Kolattukudy, 1971; Kolattukudy and Rogers, 1986; Kolattukudy *et al.*, 1981; Moore and Snyder, 1982; Snyder and Malone, 1970; Wang, *et al.*, 1972). In these cases, where metabolic end-products were wax esters or ether lipids derived from fatty alcohols, the detectable product of acyl-CoA reductase was fatty alcohol with little accumulation of fatty aldehyde intermediate, as originally observed with the *Euglena* system. In plant waxes free fatty aldehydes are found and as described in a later section fatty aldehyde has been shown to be the immediate precursor of hydrocarbons. In fact, the search for acyl-CoA reductase had already demonstrated enzymatic synthesis of aldehyde by an acyl-CoA reductase. Thus, fractionation of an acetone powder extract of young broccoli leaves yielded a protein fraction that catalysed reduction of acyl-CoA to aldehyde (Kolattukudy, 1971). More recently fatty acyl-CoA reduction to aldehyde has been demonstrated with microsomal preparations from plant and insect systems that produce hydrocarbons (Dennis and Kolattukudy, 1991; Yoder *et al.*, 1992). This reductase is probably involved in providing the aldehyde substrate for hydrocarbon synthesis. Thus, there appear to be two acyl-CoA reductases, one that generates aldehyde for hydrocarbon production and the other that generates alcohol for wax ester production.

The two fatty acyl-CoA reductases have been recently purified. Microsomal preparation from *Botryococcus braunii,* which produces large amounts of hydrocarbons, has been solubilized with octyl glucoside and purified to near homogeneity by Blue A agarose and palmitoyl-CoA agarose affinity chromatography (Wang and Kolattukudy, 1995a). This 35 kDa protein required NADH as the preferred reductant. Surprisingly the N-terminal amino acid sequence of this protein showed high homology to the N-terminus of fatty acid reductase from bioluminescent bacteria. Recently the *Botryococcus* gene, which encodes the reductase, has been cloned (J. Vioque and P.E. Kolattukudy, unpublished data). Microsomal preparation from the uropygial gland of duck, which produces large amounts of wax esters, was solubilized and purified to homogeneity using Blue A agarose and palmitoyl-CoA agarose affinity chromatography followed by gel filtration. This 56 kDa enzyme showed specificity for NADPH (Wang and Kolattukudy, 1995b).

More recently two reductases have been resolved and purified from the same tissue that produces hydrocarbons and wax esters. Microsomal preparations from rapidly expanding pea leaves were solubilized and two reductases were separated (J. Vioque and P.E. Kolattukudy, unpublished data). The alcohol-generating acyl-CoA reductase appeared to be a 56 kDa protein and showed preference for NADPH. The aldehyde-generating reductase is a 35 kDa protein and showed preference for NADH.

3.2.4 *Wax ester synthesis*

Wax ester synthesis was originally thought to involve direct esterification of fatty alcohol with free fatty acid by presumed reversal of the hydrolytic reaction because requirement for activation of fatty acid could not be demonstrated in enzyme preparations from animals (Friedberg and Greene, 1967). Suspecting that active thioesterases present in the cell-free preparations could prevent the detection of the involvement of a CoA ester in the esterification reaction, protein fractionation was used to test for a fatty acyl-CoA:fatty alcohol transacylase activity. Such an activity was detected in acetone powder extracts prepared from rapidly expanding broccoli leaves (Kolattukudy, 1967a). This reaction showed a near neutral pH optimum in contrast to the acidic pH optimum observed for esterification with free fatty acids. In addition, evidence was obtained that acyl moiety could be enzymatically transferred to fatty alcohol from phospholipids. Although it is likely that esterification catalysed by the acyl-CoA:fatty alcohol transacylase would be the major route for the biosynthesis of wax esters such issues have not been explored experimentally.

3.2.5 *Biosynthesis of hydrocarbons and their oxygenated derivatives*

Experimental exploration of the mechanism of biosynthesis of hydrocarbons began to succeed in the mid 1960s when the use of rapidly expanding leaves was found to incorporate exogenous radioactive precursors into hydrocarbons (Kolattukudy, 1965). Early efforts were to test for the validity of the then prevailing hypothesis that hydrocarbons are synthesized by a head-to-head condensation between two fatty acids, with decarboxylation of one to generate a ketone which upon reduction would yield a secondary alcohol that could then undergo dehydration and reduction to produce a hydrocarbon. This hypothesis was attractive as the postulated intermediates, the ketone and secondary alcohol, were known to occur in plant waxes. However, the labelling pattern of hydrocarbons produced from exogenous fatty acids specifically labelled with ^{14}C and/or ^{3}H showed that such a head-to-head condensation mechanism is not involved in hydrocarbon biosynthesis in plants (Kolattukudy, 1966, 1967b). On the basis of the experimental results it was proposed in 1966 that in broccoli leaves the product of fatty acid synthase is elongated to a C_{30} fatty acid by an elongating system and this fatty acid is decarboxylated to yield C_{29} hydrocarbon, that is the predominant hydrocarbon in this plant. Even though this hypothesis was questioned in the 1960s by researchers working with insects, bacteria and plants, experimental evidence which accumulated with plant systems showed the validity of the elongation decarboxylation mechanism as summarized elsewhere (Kolattukudy, 1980). The strongest evidence was demonstration of direct conversion of C_{30} fatty acid and C_{32} fatty acid into alkanes with one carbon less than the precursor acid (Kolattukudy *et al.*, 1972). In *Vicia faba* flower petals, which generate C_{27}, C_{29} and C_{31} alkanes, exogenous labelled C_{28} fatty acid gave rise to C_{27} alkane as well as C_{29} and C_{31} alkanes derived from elongated fatty acids, exogenous labelled C_{30} acid gave rise to labelled C_{29} alkane as well as labelled C_{31} alkane from the elongated fatty acid, whereas exogenous labelled C_{32} fatty acid gave only labelled C_{31} alkane (Kolattukudy *et al.*, 1974). Analyses of mutant plants with alterations in wax composition also gave results consistent with the elongation decarboxylation mechanism as summarized elsewhere (von Wettstein-Knowles, 1993). Experimental evidence similar to that obtained with plants has been obtained with insects (Blomquist *et al.*, 1993).

Cuticular waxes of some plants such as tobacco contain a large proportion of 2-(*iso*) or 3-(*anteiso*) methyl branched hydrocarbons (Tulloch, 1976). They could be formed by substitution of 2-methylpropionyl-CoA or 2-methylbutanoyl-CoA derived from valine and isoleucine, respectively, for acetyl-CoA as the starter for fatty acid synthesis to yield *iso*-branched and *anteiso*-branched C_{16} and C_{17} fatty acids, respectively. These fatty acids could then be elongated to very long-chain acids and subsequently converted to hydrocarbons yielding *iso*-branched alkanes with an odd number of carbon atoms and *anteiso* alkanes with an even number of carbon atoms. This hypothesis was strongly supported by the demonstration that labelled valine fed to rapidly growing tobacco leaves specifically labelled *iso*-branched fatty acids including very long-chain acids with an even number of carbon atoms and the naturally occurring *iso*-alkanes with an odd number of carbon atoms, whereas labelled isoleucine gave rise to labelled *anteiso*-fatty acids including very long-chain acids with an odd number of carbon atoms and the naturally occurring *anteiso*-alkanes with an even number of carbon atoms (Kolattukudy, 1968b).

Elucidation of the enzymology of hydrocarbon biosynthesis has progressed slowly. Even though a cell-free system that produced hydrocarbons from fatty acids was described in 1974 (Khan and Kolattukudy, 1974) elucidation of the mechanism of hydrocarbon formation from fatty acids remained a mystery for another decade. The cell-free system generated hydrocarbons with two carbon atoms less than the fatty acid precursor, thus suggesting that the conversion of the fatty acid to the hydrocarbons occurring in the cell-free preparation involved multiple steps probably including steps that would not be involved *in vivo*. Obviously, conversion of a fatty acid with an even number of carbon atoms to a hydrocarbon with an even number of carbon atoms would not be consistent with the composition of naturally occurring hydrocarbons that have an odd number of carbon atoms.

It was postulated that hydrocarbon formation in the cell-free system involved formation of a compound from the fatty acid involving the loss of one carbon atom and subsequent formation of hydrocarbon from this compound by the loss of another carbon atom. Time-course of formation of compounds from the labelled fatty acid was examined in search of the postulated precursor of alkane. Fatty aldehyde with one carbon less than the parent acid was found to be the only compound that could fit the bill for being the immediate precursor of alkane (Bognar *et al.*, 1984). While this result was puzzling because no organic chemical reaction was known directly to convert an aldehyde to an alkane with one carbon less than the aldehyde, it was reported in 1980 that ruthenium–porphyrin complex could catalyse decarbonylation of aldehyde to alkanes (Domazetis *et al.*, 1981). An analogous reaction catalysed by a decarbonylase enzyme was then shown to produce alkane and CO from an aldehyde (Cheesbrough and Kolattukudy, 1984). Once this biochemical reaction was elucidated, it explained earlier observations that inhibition of hydrocarbon synthesis caused by mutations and chemicals had caused accumulation of aldehydes with one carbon more than the alkane. This is an interesting case of how elucidation of a biochemical process had to wait for progress in organic chemistry.

A heavy particulate fraction (sp.gr. 1.30) from rapidly expanding pea leaves was shown to catalyse conversion of C_{18} aldehyde to C_{17} alkane and CO with retention of the aldehydic hydrogen in the alkane (Cheesbrough and Kolattukudy, 1984). This enzyme was severely inhibited by metal ion chelators, suggesting the involvement of metal ions in the enzymatic decarbonylation. A similar decarbonylation was found to be catalysed by a particulate preparation from the uropygial glands of eared grebe that produces hydrocarbons as the major secretion products (Cheesbrough and Kolattukudy, 1988). In this case the aldehydic hydrogen was not retained but the solvent provided the hydrogen. The decarbonylation reaction was demonstrated to be reversible. More recently decarbonylation was demonstrated to be involved in the hydrocarbon synthesis in a colonial microalga, *Botryococcus braunii*, which produces up to 32% of its dry weight in hydrocarbons (Dennis and Kolattukudy, 1991). In this case, with crude microsomes CO_2 was the major product. However, upon solubilization and purification, aldehyde was found to be converted to hydrocarbon and CO (Dennis and Kolattukudy, 1992). The absorption spectrum of the purified enzyme showed bands at 410 mm, 425 mm, 580 mm and 620 mm suggesting the presence of a porphyrin, although a corinoid structure could not be ruled out. Electron microprobe analysis of the purified enzyme showed the presence of Co and purification of the enzyme from cultures grown in the presence of $^{57}CoCl_2$ showed association of ^{57}Co with the decarbonylase activity. From such results it was concluded that the decarbonylase contained Co-porphyrin and this conclusion was supported by the observation that Co protoporphyrin IX, but not other metal porphyrins, catalysed decarbonylation of aldehyde at elevated temperature. Decarbonylation of aldehydes to hydrocarbons and CO was also demonstrated with microsomal preparations from diapausing fleshfly pupa (Yoder *et al.*, 1992). With adult housefly microsomes aldehyde was confirmed to be the immediate precursor of hydrocarbon. However, the process required NADPH and O_2 and was suggested to involve cytochrome (*Cyt*) P_{450} yielding CO_2 instead of CO (Reed *et al.*, 1994). We have also found that adult housefly and adult fleshfly microsomes require NADPH and O_2 yielding CO_2 (F. Schneider and P.E. Kolattukudy, unpublished data). It appears that the decarbonylation of the aldehyde in the adult fly is tightly coupled to oxidation of the CO to CO_2, possibly because the coupled oxidative removal of CO might facilitate the overall reaction. In our experience, the adult fly microsomes are much more effective in converting aldehyde to alkane than the microsomes from pupa or the other biological systems mentioned above. The present understanding of the biosynthetic reactions involved in the biosynthesis of cuticular hydrocarbons and derivatives is summarized in Figure 3.1.

Conventional biochemical approaches to solubilize, purify and characterize the membrane-localized enzymes involved in cuticular wax biosynthesis present a formidable task. Molecular biological and genetic approaches might be more suited for the next level of investigations into the biosynthetic mechanisms involved. Since the epidermis is the site of cuticular wax biosynthesis, one approach to seek the enzymes involved is to examine the transcripts of the genes expressed uniquely in the epidermis. Such an effort in my laboratory with a cDNA library prepared from mRNA isolated from the excised epidermis of the growing leaves of *Senecio odorus* yielded several clones that represented genes uniquely expressed in the epidermis (J. Pyee and P.E. Kolattukudy, unpublished data). One of them turned out to be homologous to lipid transfer protein, while another one termed EPI 23 gave an open-reading frame that showed no homology to any other genes in the data banks. Recently it was found that the *cer*1 gene tagged by insertion of the maize I/d Spm elements and cloned from *Arabidopsis* is highly homologous to the epidermis-specific EPI 23 we had previously cloned from *Kleinia odora* (*S. odorus*) (Aarts *et al.*, 1995). Wax from *cer*1 mutant of *Arabidopsis* was found to be high in aldehydes but lacked alkanes, suggesting that CER1 protein is involved in conversion of aldehydes to alkanes (Hannoufa *et al.*, 1993; Jenks *et al.*, 1995; Lemieux *et al.*, 1994; McNevin *et al.*, 1993). Thus, the epidermis-specific EPI 23 originally cloned from *K. odora* and CER1 from *Arabidopsis* are thought to be the aldehyde decarbonylase. This protein contains histidine-rich motifs containing long, hydrophobic domains between the histidine motifs as found in other proteins containing histidine-rich motifs. These domains can span a membrane twice, making this protein an integral membrane protein. The histidine-rich motifs imply non-haem iron-containing active site (Shanklin *et al.*, 1994). One possibility that cannot be overlooked is that this protein functions with the true Co-containing decarbonylase, possibly to remove oxidatively the CO generated and, thus, is involved in alkane synthesis from aldehyde. Recently a maize mutant blocked in wax synthesis was studied and the mutated gene was cloned and sequenced. The sequence was found to be highly homologous to the same epidermis-specific EPI 23 we had cloned from *K. odora* mentioned above. In this case, however, the composition of the wax in the mutant showed that the mutation decreased the formation of alcohols and aldehydes not alkanes (J.D. Hansen, J. Pyee, Y. Xia *et al.*, submitted). Therefore, it was suggested that gl1 mutation had global effects on wax biosynthesis. EPI 23, CER1 and GL1 are all members of a family of proteins with five to seven transmembrane domains. It is possible that this class of proteins are involved in intercompartmental transport of precursors, intermediates and final products involved in cuticular wax synthesis and deposition. Successive elongation steps, reduction, esterification and decarbonylation probably involve different membrane compartments, the endoplasmic reticulum, the golgi, the plasma membrane and possibly extracellular compartments. If the integral membrane proteins of the *epi23-cer1*-gl1 family are involved in vesicular trafficking among such compartments and final deposition of the wax, differential effects on the synthesis of different classes of cuticular lipids could arise, depending on the membrane compartments involved in their synthesis.

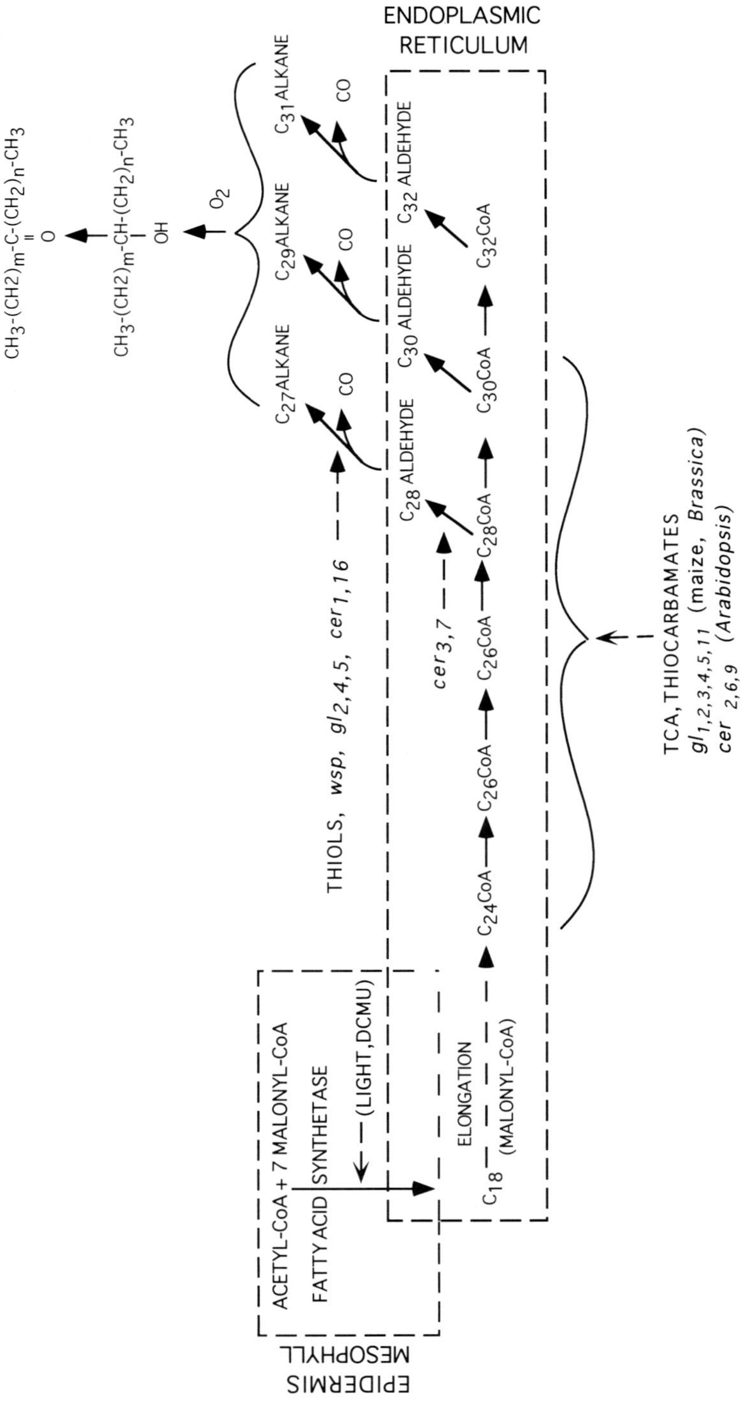

Figure 3.1. *Proposed pathway for the synthesis of alkanes and derivatives. (DCMU,* N′-*(3,4-dichlorophenyl)*-N,N′-*dimethyl urea; TCA, trichloroacetate;* gl_{1}*-5 and* gl_{11}*, mutants of* Brassica oleracea*;* wsp*, mutant of* Pisum sativum*;* $cer_{1\text{-}3,6,7,9,16}$ *mutants of* Arabidopsis.)

Such a global function could also explain the observation that *cer1* mutation affects both epicuticular wax biosynthesis and pollen fertility (Aarts *et al.*, 1995) because vesicular trafficking similar to that involved in the biosynthesis and deposition of cuticular wax is probably involved in the synthesis and deposition of pollen wall components. The high degree of conservation of this family of proteins, from *Arabidopsis* and related *Brassica* species, unrelated compositae *Senecio odorus* and the distantly related monocot species rice and maize, also is consistent with the postulated global role. Obviously, further work is needed to establish the precise functions of the products of the genes mutated.

3.2.6 *Coupling of reactions involved in the biosynthesis of wax components*

The presence of various chain length classes in the different types of wax components probably reflects channelling of the elongated acyl chains into different classes of lipids by coupling the various elongation systems to different terminal steps that generate the different end-products; for example, elongation up to C_{22} could be generating CoA esters that are then used to form wax esters. Another elongating system could elongate acyl chains up to C_{28} or so with C_{26} and C_{28} being the major products, that are linked to alcohol generating fatty acyl-CoA reductase. Early experiments with microsomes from *E. gracilis* strongly suggested that acyl-CoA synthesis and its reduction are probably tightly coupled into wax ester-synthesizing enzyme, so that labelled malonyl-CoA was incorporated into the wax esters without any detectable intermediate fatty acids or fatty alcohols (Khan and Kolattukudy, 1973). Another elongating system could be responsible for production of C_{30} or C_{32} acyl chains and this elongating system could be coupled to the aldehyde generating acyl-CoA reductase and decarbonylase yielding the C_{29} or C_{31} hydrocarbons. The chain length distribution of the end-products would then depend on the specificity of the chain elongating system to which the terminal steps are coupled; the enzymes catalysing the terminal steps would not have to have strict chain length specificity. In fact, acyl-CoA reductase from plant tissues that produce C_{26} and longer alcohols were found to be capable of reducing C_{16} CoA ester (Kolattukudy, 1971). Cell-free synthesis of hydrocarbons could accept C_{16} to C_{32} fatty acids with very little preference for chain lengths (Bognar *et al.*, 1984; Kolattukudy, 1987). Thus, aldehyde decarbonylase preparation from pea leaves and *Botryococcus braunii* that generate C_{29} alkane and C_{27}–C_{31} olefins, respectively, are routinely assayed with C_{18} aldehyde because the decarbonylases do not seem to show a clear preference for chain length or unsaturation. On the other hand, even early experiments strongly suggested strict specificity of elongating systems. In broccoli, which produces saturated alkane, $C_{18:0}$ was elongated and converted to wax components but $C_{18:1}$ was not elongated (Kolattukudy, 1966). More recent studies have further confirmed this conclusion (Agrawal *et al.*, 1984). The elongating reactions are coupled to two reductases and one transacylase system. In tissues that produce only wax esters, such as jojoba seeds and uropygial gland (Wang and Kolattukudy, 1995b), a 56 kDa fatty alcohol-generating acyl-CoA reductase is coupled to the esterifying system. In cells that produce only hydrocarbons such as *B. braunii*, the elongating system is coupled to a 35 kDa aldehyde-generating acyl-CoA reductase and to a decarbonylase that converts the aldehyde to a hydrocarbon. In the epidermis of higher plants, such as pea, both reductases and their coupled partners function to produce wax esters and hydrocarbons. The coupled membrane localized processes involved in cuticular wax synthesis are indicated in Figure 3.2.

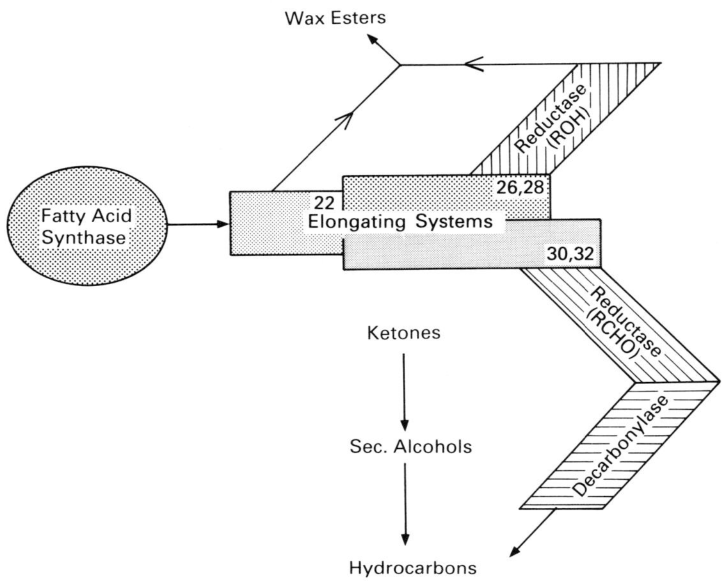

Figure 3.2. *Coupling of elongation systems with the alcohol-generating acyl-CoA reductase (ROH) and aldehyde-generating reductase (RCHO) and their coupling to wax ester synthesis and hydrocarbon synthesis, respectively.*

3.2.7 *Biosynthesis of oxygenated components*

With the finding that ketones and secondary alcohols are not intermediates involved in the biosynthesis of hydrocarbons, the origin of these oxygenated compounds became obscure. That the secondary alcohols are most probably derived from the alkane by hydroxylation and the ketone is derived from the secondary alcohol was shown by the demonstration that exogenous labelled alkane was converted to the secondary alcohols and ketones, whereas exogenous secondary alcohol was converted to the ketone in broccoli leaf slices (Kolattukudy and Liu, 1970). Analysis of the products and the position of hydroxylation of the exogenous alkane suggested that these conversions of the exogenous substrates reflected the natural positional specificity of hydroxylation and oxidation of the secondary alcohols (Kolattukudy *et al.*, 1973a). The hydroxylation of exogenous substrate was inhibited by phenanthroline and this inhibition was reversed by Fe^{+2} suggesting the involvement of a mixed function oxidase in this hydroxylation. Whether hydroxylation can occur at the fatty acid or aldehyde level *in vivo* is not certain. In any case, the hydroxylation must be position-specific to explain the variety of positions at which secondary alcohols are found in plant waxes. *Arabidopsis* mutants blocked in alkane synthesis did not produce secondary alcohols and ketones, supporting the conclusion drawn from early labelling studies (Jenks *et al.*, 1995; Lemieux *et al.*, 1994; McNevin *et al.*, 1993).

3.2.8 *β-Diketones*

In monocots such as barley very long-chain β-diketones occur as major components (von Wettstein-Knowles, 1972). Currently available evidence, mostly from the

Carlsberg group, suggests that this class of molecules are produced by a series of reactions analogous to those involved in hydrocarbon synthesis, except that during elongation a couple of keto groups are left unreduced as known in the case of polyketide synthases (Mikkelsen, 1979; Mikkelsen and von Wettstein-Knowles, 1978; Figure 3.3). β-diketones, like the hydrocarbons, can also be hydroxylated. Biosynthetically related to this process is the formation of methyl ketones, and alkan-2-ols found in barley and some other plant waxes (Mikkelsen, 1984; von Wettstein-Knowles, 1976, 1985, 1993). In this case the β-ketoacids are presumably decarboxylated to the methyl ketones which are reduced to alkan-2-ols (Mikkelsen and von Wettstein-Knowles, 1984) that can be esterified using acyl-CoA as substrates (Mikkelsen, 1984). Analyses of the composition of wax from mutant plants, and incorporation of exogenous labelled acetate and other fatty acids, with or without inhibitors, supported the above conclusions concerning the biosynthesis of β-diketones and alkan-2-ols (von Wettstein-Knowles, 1993).

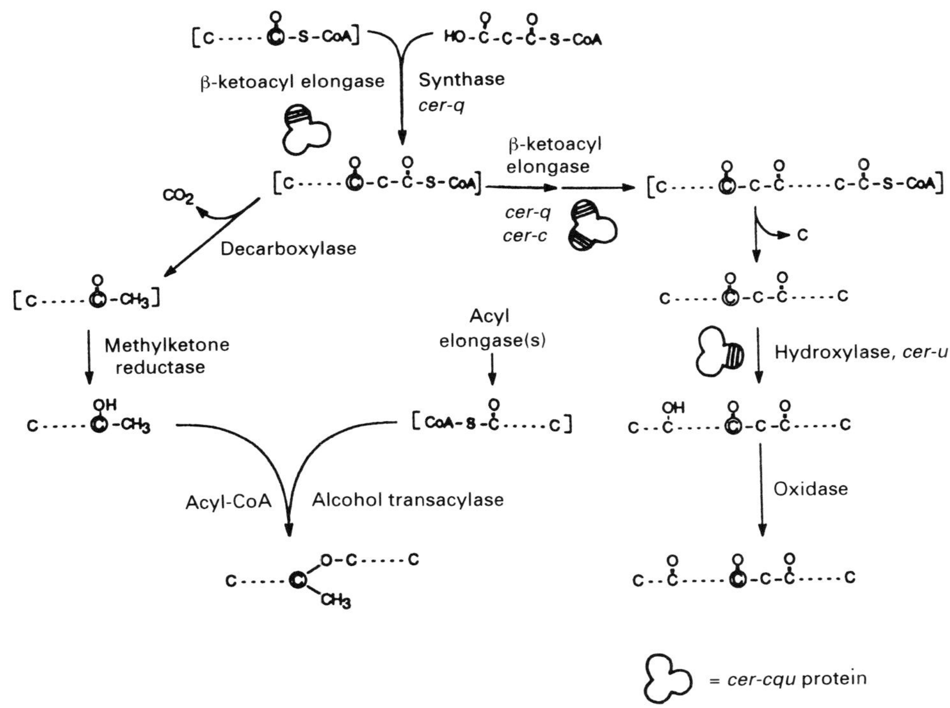

Figure 3.3. *The β-ketoacyl elongase pathways leading to the synthesis of alkan-2-ols and β-diketones.* (cer-cqu, *multifunctional protein determined by the* cer-cqu *gene in barley.) (Reproduced from von Wettstein-Knowles (1993) Waxes, cutin and suberin. In:* Lipid Metabolism in Plants *(ed. Moore). CRC Press, FL, pp. 127–166, with permission from CRC Press.)*

3.3 Biosynthesis of cutin

3.3.1 *Structure and composition of cutin*

Cutin is composed mainly of two families of hydroxy and hydroxy epoxy fatty acids derived from the most common cellular fatty acids: one derived from C_{16} saturated fatty acid and the other derived from C_{18} unsaturated fatty acids. The chief members of the two families of cutin monomers are shown in Figure 3.4. Some plants have mainly C_{16} family of monomers, whereas others have a mixture of both C_{16} and C_{18} family of monomers. The composition can depend on the anatomical region of the plant; for example, flower petals and stigma of apple have mostly C_{16} family of monomers, leaves have some additional C_{18} family of monomers, whereas in the fruit C_{18} and C_{16} families of monomers are major components (Espelie *et al.*, 1979). These monomers are interesterified to form cutin, the insoluble polymer. Chemical modification of the polymer that tags the free functional groups followed by depolymerization and analysis of the products indicated that most of the primary hydroxyl groups are esterified, giving a linear polymer and about half of the mid-chain secondary hydroxyl groups are esterified indicating branching and cross-linkages (Deas and Holloway, 1977; Kolattukudy, 1977). Nuclear magnetic resonance (NMR) studies also indicated that cutin is a moderately flexible netting with motional constraints at cross-link sites (Zlotnik-Mazori and Stark, 1988). More than half the methylenes were in the rigid category, with about 36% in the mobile category. This picture, obtained with lime cutin which probably has less cross-linking capability than the cutins with a higher content of midchain hydroxyl groups such as apple cutin, could be overstating the flexibility of plant cutins.

3.3.2 *Depolymerization-resistant core of cutin (cutan?)*

It has been recognized for a long time that exhaustive treatments of cutin which cleave ester bonds, such as hydrolysis, hydrogenolysis with $LiAlH_4$, or transesterification leave behind insoluble residues (Crisp, 1965; Kolattukudy and Walton, 1973). These residues are thought to represent polymers in which cuticular components are held together by non-ester bonds. The observation that treatment of this material with HI generates soluble materials suggested the presence of ether

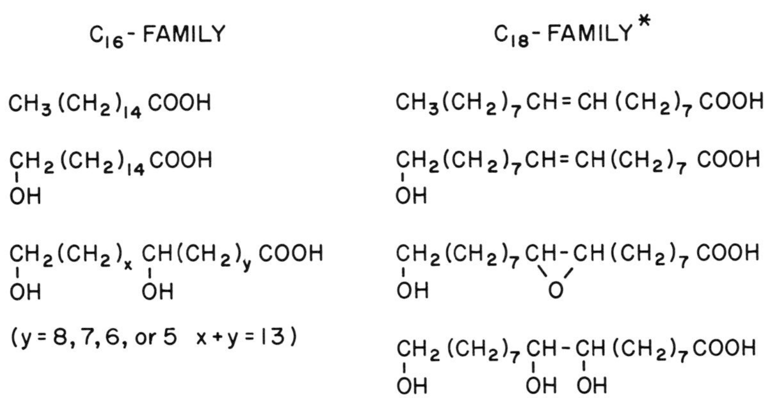

Figure 3.4. *Structure of the major monomers of cutin. *Δ^{12} unsaturated analogues also occur.*

bonds. Some insoluble material remains after all of these treatments. In view of the finding that cuticle contains phenolics (Riley and Kolattukudy, 1975), and peroxidases are expressed in the epidermal cells (Mohan *et al.*, 1993), it is probable that some of the cuticular components are peroxidatively coupled generating C–C bonds resistant to all of the cleavage reactions. In fact, cross-polarization/magic angle spinning NMR of insoluble material remaining after the exhaustive hydrogenolysis with $LiAlH_4$ of cutin from the fruits of apple, pepper and tomato revealed the presence of methylene chains (P.E. Kolattukudy and G.E. Maciel, unpublished data). Since the cuticular components are interdispersed with some cell wall components, cutin preparations may contain some carbohydrate-type materials that could be part of the final insoluble residue. The relative amounts of the non-ester bound components may be species-dependent. The presence of non-ester bound polymeric materials found in fossilized cuticles was called 'cutan' and was considered to be 'cutin-derived' (Tegelaar *et al.*, 1993). Recently, such residues and non-ester bound polymeric material from modern plant cuticles (Villena *et al.*, 1996) have been subjected to pyrolysis-coupled gas–liquid chromatography and mass spectrometry. The products included not only those derived from C_{16} and C_{18} fatty acids, but also hydrocarbons in the C_{19}–C_{26} range without any predominance of even numbers of carbon atoms. These probably represent pyrolysis products from cuticular wax-derived materials. It is possible that the resistant polymers found in fossilized cuticles could be derived from diagenetically induced reactions of reactive groups such as epoxides present in the original cuticle. However, the presence of such resistant polymers in modern plants suggests that they are formed on the living plants.

The involvement of lipoxygenase and peroxidase type reactions in the formation of such materials is suggested by the observation that exogenous ^{14}C-labelled polyunsaturated fatty acids were preferentially incorporated into the non-ester bound polymer by developing fruit skin slices; for example, the residues remaining after $LiAlH_4$ hydrogenolysis of apple cutin derived from [1-^{14}C]linoleic acid had 17 times the specific radioactivity as that derived from [1-^{14}C]oleic acid (Kolattukudy *et al.*, 1973b). HI treatment released 40% of this ^{14}C derived from [1-^{14}C]linoleic acid. Even though selective incorporation of the pentadiene-containing fatty acids into the non-ester bound polymer was clear and consistent with the NMR evidence for the presence of methylene chains in the resistant core of the polymer, the nature of the linkages involved in this part of the polymer remains to be elucidated. Since this part of the polymer is most probably covalently attached to the ester-bound part of the polymer and probably derived at least in part from the same type of monomers, it might be premature to coin a new term for such a segment of the polymer as if these are two distinctly different polymers rather than one polymer containing ester bonds and other types of bonds.

3.3.3 *Biosynthesis of cutin*

Since cutin is an insoluble polymer, biosynthetic studies required methods to analyse the polymer derived from exogenous radioactive precursors in small amounts of tissue. $LiAlH_4$ reduction of the insoluble material, obtained from small amounts of rapidly growing tissues incubated with precursors followed by extraction with organic solvents, yielded reduced cutin monomers which could be subjected to chromatographic analyses. Such metabolic studies conducted in the 1970s led to the formulation of metabolic pathways for the major cutin monomers (Figure 3.5; Kolattukudy, 1980). Subsequently, the enzymatic activities postulated to be involved in the biosynthesis of the monomers and enzymatic synthesis of the

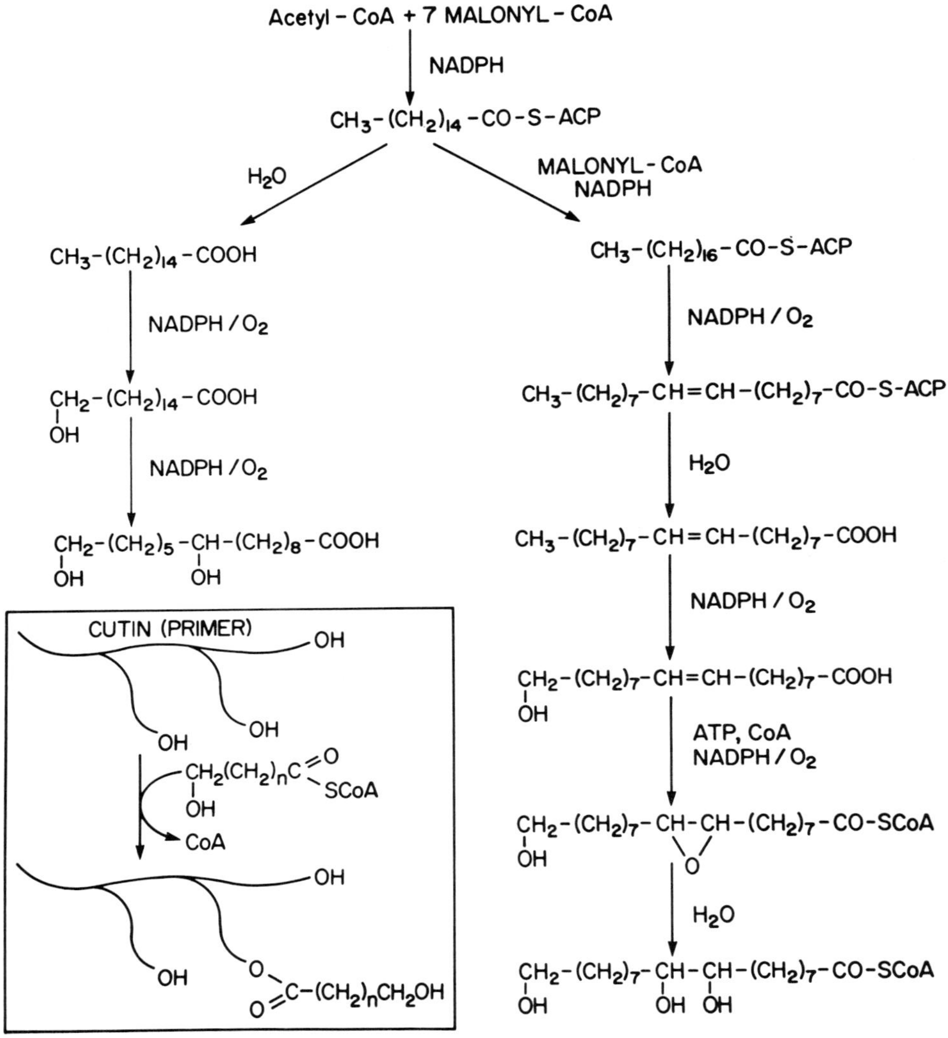

Figure 3.5. *Biosynthetic pathways for the major cutin monomers.*

polymer from the monomer have been demonstrated in cell-free preparations. However, the enzymes have not been purified and characterized and nothing is known about the genes that encode these enzymes.

3.3.4 *Biosynthesis of the C_{16}-family of cutin monomers*

Vicia faba leaf cutin is composed chiefly of C_{16} family of monomers. Rapidly growing leaves of *V. faba* were found to incorporate labelled acetate and palmitate into cutin most rapidly (Kolattukudy, 1970c). Exogenous palmitic acid was incorporated into both 16-hydroxypalmitic acid and 9- or 10,16-dihydroxypalmitic acid and exogenous 16-hydroxypalmitic acid was converted to 9- or 10,16-

dihydroxypalmitic acid (Kolattukudy and Walton, 1972). Midchain hydroxylated C_{16} acid was not converted to the dihydroxy acid. Neither palmitoleic acid nor palmitelaidic acid was converted into dihydroxy C_{16} acid, indicating that desaturation is not involved in this hydroxylation. Thus, it was postulated that palmitic acid, the product of fatty acid synthase, is hydroxylated at C_{16} and then at 9 or 10 position. The positional specificity of the midchain hydroxylase may vary, thus, generating species-specific positional isomer mixtures. Microsomal preparation from the shoots of *V. faba* seedlings catalysed ω-hydroxylation of palmitic acid with NADPH and O_2 as required cofactors (Soliday and Kolattukudy, 1977). This mixed function oxidase was inhibited by CO suggesting the involvement of a *Cyt* P_{450} type enzyme. However, this hydroxylation was unusually sensitive to CO inhibition and light could not reverse the inhibition. Oleic acid was also hydroxylated at comparable rates but stearic was a poor substrate. Recently, ω-hydroxylation catalysed by *Cyt* P_{450} was confirmed in *V. faba* seedlings in which the hydroxylase was induced by clofiltrate treatment and, as expected, antibodies raised against NADPH-*Cyt* P_{450} reductase inhibited the reaction (Pinot *et al.*, 1992). This induced hydroxylase could also ω-hydroxylate epoxides and diol derivatives, possibly indicating the relatively non-specific nature of xenobiotic metabolizing enzymes.

3.3.5 *Biosynthesis of C_{18} family of cutin monomers*

The early metabolic experiments used rapidly expanding fruit tissues that have the C_{18} family of monomers as major components. The purpose was to examine the sequence of reactions involved in the biosynthesis of these monomers. In expanding grape berry skin slices exogenous labelled oleic acid was converted mainly into 18-hydroxyoleic acid and 18-hydroxy-9,10-epoxy C_{18} acid, whereas in skin slices of young apple fruits labelled oleic acid was incorporated into the same two hydroxy and epoxy acids as well as 9,10-18-trihydroxy C_{18} acid reflecting the composition of the C_{18} monomers in the two tissues (Kolattukudy *et al.*, 1973b). Exogenous stearic acid was not incorporated into any midchain modified monomers. Exogenous labelled linoleic acid was incorporated into the corresponding hydroxy and epoxy acids with unmodified double bond at Δ^{12} position, demonstrating the positional specificity of the epoxidation for the double bond at Δ^{9} position. Even though this specificity is the commonly observed pattern in plants, some plants can also epoxidize the Δ^{12} and hydrate the epoxide generating 9,10-12,13,18-pentahydroxy C_{18} acid. In *Rosmarinus officinalis*, for example, exogenous labelled linoleic acid was incorporated into not only the Δ^{9} double bond modified products indicated above but also into 9,10,18-trihydroxy-12,13-epoxy C_{18} acid and the pentahydroxy C_{18} acid (Croteau and Kolattukudy, 1974a). In young apple fruit skin slices exogenous linolenic acid was incorporated into the 18-hydroxy-Δ^{9} modified epoxy and trihydroxy acids with unmodified Δ^{12} and Δ^{15} double bonds. In fact, 18-hydroxy-9,10-epoxy and 9,10,18-trihydroxy C_{18} dienoic acid was found to be a significant component of cutin in young apple fruits, but in mature fruits they were minor components, probably reflecting the decreasing availability of linolenic acid for cutin synthesis during maturation of the fruit (Kolattukudy *et al.*, 1973b).

The sequence of reactions involved in the biosynthesis of the C_{18} family of cutin monomers was demonstrated by the specific incorporation of exogenous 18-hydroxy oleic acid into 18-hydroxy-9,10-epoxy C_{18} acid and 9,10,18-trihydroxy acid of cutin and incorporation of exogenous labelled 18-hydroxy-9,10-epoxy C_{18} acid into 9,10,18-trihydroxy C_{18} acid of cutin in apple fruit skin slices (Croteau and

Kolattukudy, 1974b). Thus, ω-hydroxylation, epoxidation followed by epoxide hydration would generate the most common members of the C_{18} family of cutin monomers.

The unique reactions postulated to be involved in this process, the epoxidation and epoxide hydration, have been demonstrated in cell-free preparations. Epoxidation of 18-hydroxy[18-^{3}H]oleic acid to the corresponding *cis*-epoxy acid was catalysed by a 3000 g particulate preparation from young spinach leaves in which this epoxy acid is a major cutin component (Croteau and Kolattukudy, 1975). This reaction required ATP and CoA, suggesting that the substrate of the epoxidation was the CoA ester. It also required NADPH and O_2 and was photoreversibly inhibited by CO, suggesting that a *Cyt* P_{450} type enzyme was involved in this epoxidation. The biosynthetic relevance of this activity was demonstrated by the substrate specificity. The enzyme required *cis*-Δ^9 and a free ω-hydroxyl group. Recently, two new enzyme activities found in the soybean seedlings were suggested to be involved in the biosynthesis of cutin monomers (Blée and Schuber, 1993). A hydroperoxide-dependent epoxidase found in the microsomes catalysed epoxidation of oleic acid. Even though specificity for *cis* olefin is consistent with its possible involvement in the biosynthesis of cutin monomers, it showed low regioselectivity in that *cis*-double bonds in positions other than Δ^9 in C_{18} monoenoic acid and both double bonds in linoleic acid were epoxidized (Blée and Schuber, 1990). 18-Hydroxyoleic acid was found to be a poor substrate for this epoxidase. A soluble epoxide hydrase was found in the soybean seedlings. This enzyme showed a preference for *cis*-epoxide but 18-hydroxy-9,10-epoxyoleic acid was a poor substrate unlike the particulate epoxide hydrase found in skin slices of young apple fruits. (In fact, the cytosol from the actively cutin-synthesizing apple tissue showed no epoxide hydrase activity.) Whether the well characterized cytosolic epoxide hydrase of soybean seedlings is involved in cutin biosynthesis remains uncertain as no real biological connection to cutin biosynthesis has been established. In developing seeds of *Euphorbia lagascae,* which produce *cis*-12,13-epoxy-9-octadecenoic acid (vernolic acid), both a *Cyt* P_{450} type epoxidase and a hydroperoxide-dependent epoxidase were found, but in germinating seeds, which do not synthesize the epoxy acid only the latter epoxidase was found (Blée *et al.*, 1993) suggesting that the *Cyt* P_{450} type epoxidase may be the biosynthetic enzyme. Probably a similar situation exists in cutin biosynthesis where the *Cyt* P_{450} type epoxidase might also be the biosynthetic enzyme. Whether the hydroperoxide-dependent epoxidase and the cytosolic epoxide hydrase are present in cutin-synthesizing epidermal tissues as previously noted for the O_2-*Cyt* P_{450} dependent epoxidase and particulate epoxide hydrase remains unknown. If the microsomal hydroperoxide-dependent epoxidase and the cytosolic epoxide hydrase were involved in the cutin biosynthesis, their substrate specificities would suggest that oleic acid would be epoxidized first and then the epoxide would be hydrated and, subsequently, the epoxy and dihydroxy acids would undergo ω-hydroxylation. The epoxy acid and dihydroxy acid would have to be excluded from incorporation into the polymer as the naturally occurring cutin monomers have ω-hydroxyl groups. On the other hand, the sequence of events shown in Figure 3.5 is consistent with the observed composition of cutin.

3.3.6 *Synthesis of cutin polymer from monomers*

Synthesis of an insoluble polymer to be deposited outside the epidermal cell walls would have to occur at the site of final deposition. In fact, a cutin-containing 3000 g particulate fraction from the excised epidermal tissue of rapidly growing *V.*

faba leaves incorporated labelled hydroxy C_{16} monomers into insoluble material with ATP and CoA as required cofactors (Croteau and Kolattukudy, 1974c). That the incorporation represents synthesis of cutin was shown by the fact that only chemical treatments that are known to release cutin monomers or treatment with only cutinase but no other hydrolytic enzymes released the incorporated label. Furthermore, only the enzyme preparations from the epidermal tissue of *V. faba* and *S. odorus* but not from mesophyl tissue catalysed this incorporation as expected from its role in cutin biosynthesis. Presumably, the hydroxyacyl moieties from their CoA esters were transferred to the growing polymer and, thus, the cutin contained in the 3000 g particulate enzyme preparation would serve as the receiver of the hydroxyacyl groups. Release of the enzyme from the endogenous cutin by ultrasonic treatment yielded a soluble enzyme preparation that required exogenous primer for incorporation of the hydroxy acids. Cutin from many plant species could serve as primer. Finely divided cutin from the rapidly growing young *V. faba* leaves served as the best primer for this system and the priming efficiency decreased with age of the leaf. Presumably, the relatively open structure of cutin in the young leaves provide maximum free hydroxyl sites for incorporation of the monomer. In fact, chemical modification of cutin to increase the number of primary hydroxyl groups or opening of the cutin structure by brief treatment with cutinase increases the priming capability of the polymer. Thus, hydroxylacyl-CoA:cutin transacylase located in association with the growing polymer catalyses the synthesis of cutin from monomers (Figure 3.5).

3.4 Suberization

Since it is the specialized function of epidermal cells to synthesize cutin and associated waxes, a different type of barrier formation is initiated when a plant organ is wounded. Thus, irrespective of the nature of the natural covering of the organ, wounding induces suberization, which involves essentially the suicide of the outermost one or two layers of cells at the wound area to protect the rest of the organ from desiccation and environmental hazards, including infection. Wound-induced suberization has been demonstrated on fruits, leaves and tubers (Dean and Kolattukudy, 1976). Whenever plants need to erect a barrier during developmental process or to respond to an environmental stress, they seem to resort to suberization of the cell walls. Some examples of suberization induced by developmental conditions include suberization of abscission zone and chalazal region of the seed coat to seal off the seed at maturation (Espelie *et al.*, 1980). Examples of environmental conditions that induce suberization include increased suberization induced by high Mg^{+2} concentration (Pozuelo *et al.*, 1984) and suberization around Ca oxalate crystals in idioblasts (Espelie *et al.*, 1982).

3.5 How are the cuticular components transported to the extracellular site?

Early reports about the occurrence of pits or pores have not been verified and no reliable answer to the question is currently available (see Chapter 2). It is possible that the transport is through vesicles which find their way through anastamosing channels in the wall/cuticle. Rapid cutin synthesis associated with rapid elongation of deep water rice, elicited by gibberellin, showed a 60-fold increase in the incorporation of labelled palmitic acid into cuticle and electron microscopic examination showed what appeared to be vesicles leaving the epidermal plasma membrane (Hoffmann-Benning and Kende, 1994). Electron microscopic examination of suberizing cells of potato tuber slices also showed what appeared to be vesicles outside the plasma membrane (Barckhausen, 1978). Another possibility is diffusion of cuticular waxes and cutin precursors with assistance from carrier-like proteins. Recently, it was found that cuticular wax contains one major protein that was found to be a member of the lipid transfer protein family (Pyee *et al.*,1994). In fact, most of the lipid transfer protein in broccoli leaves was found to be in the cuticular wax. At least four genes coding for this group of lipid transfer proteins were found, cloned and sequenced (Pyee and Kolattukudy, 1995). They all code for secreted proteins. Expression patterns determined by reverse transcriptase-polymerase chain reaction (RT-PCR) with gene-specific primers showed that these genes are expressed in aerial parts such as leaves and flowers expected to be active in cuticle synthesis but not in roots. Similarly, glucuronidase expression driven by the lipid transfer protein promoters also showed specificity for epidermis and trichomes actively involved in cuticle synthesis but not in roots. This pattern was also found in maize, carrot and *Arabidopsis* (Sossountzov *et al.*, 1991; Sterk, *et al.*, 1991; Thoma *et al.*, 1993, 1994). If and how lipid transfer proteins are involved in the synthesis and transport of cuticular components remains to be established.

3.6 Interaction of cuticle with microbes

The cuticle is the first site of contact of plants with microbes in the environment, and therefore is thought to play a significant role in plant–microbe interactions. Fungal interaction with plants can lead to diseases that cause the greatest losses in food and fibre production. I shall briefly mention a few aspects of such interactions that have been recently explored.

The cuticle, being highly hydrophobic, can act as an absorbent for the extremely low level of lipophilic self inhibitors that fungal spores use to help prevent premature germination and differentiation to form infection structures before a plant host is available. A recent observation suggests such a role for the rice blast organism *Magnaportha grisea* (Y. Hegde and P.E. Kolattukudy, unpublished data). The spores of this organism are induced to germinate and differentiate to form the penetration structure called appressorium by rice wax. At higher spore concentrations germination is inhibited and with between 10^5 and 10^6 appressoria ml^{-1}, germination occurred but appressorium formation required rice wax. Below 10^5 spores ml^{-1} both germination and appressorium formation occurred without wax. Direct evidence for a self inhibitor was obtained when a solvent wash of the spores gave material that inhibited germination and appressorium formation. Certain chromatographic fractions from the hydrophobic material obtained from the spore wash inhibited germination and appressorium formation. However, this inhibition could be reversed by rice wax, other plant waxes or even parafilm, thus strongly suggesting that the hydrophobic nature of the plant cuticle could adsorb the self inhibitor and, thus, promote germination and appressorium formation by the fungal spore.

A more host-specific role for cuticular wax was observed in the interaction of *Colletotrichum gloeosporioides* with its host, the avocado fruit. Only the host wax induced germination and appressorium formation of the fungal spore, and spores of other *Colletotrichum* species were not induced to germinate or differentiate by the avocado cuticular wax (Podila *et al.*, 1993). The host wax in this case triggers the expression of a set of genes in the fungal spore allowing germination and differentiation into appressorium (Hwang and Kolattukudy, 1995). Several such genes have been cloned and sequenced. Although the functions of these genes have not been elucidated, disruption of one of them was found to decrease virulence drastically (Hwang *et al.*, 1995). Thus, cuticular waxes can trigger fungal gene expression, which allows the fungi to infect plants.

Many fungi penetrate directly through the cuticle and invade plants (see Chapter 8). Cutin constitutes the major structural barrier that the invading fungus must penetrate and, therefore, extracellular cutinase secreted by the germinating fungus is thought to aid the fungus in the penetration into the host. Such an enzyme was purified in the early 1970s (Purdy and Kolattukudy, 1975) and characterized as a polyesterase that uses a catalytic triad involving active serine for catalysis (Köller and Kolattukudy, 1982). Recent X-ray crystallographic studies on recombinant cutinase suggest that cutinases form a unique class of enzymes that constitutes a bridge between esterases and lipases (Martinez *et al.*, 1992). The following evidence indicates that assistance from cutinase can play a crucial role in pathogenesis.

(a) Plant pathogens produce and secrete cutinase targeted at the penetration point and during actual infection of the host cutinase is produced.
(b) Inhibition of cutinase by chemicals or antibodies including monoclonal antibodies prevents infection (Kolattukudy, 1985).
(c) Cutinase-deficient mutants have significantly reduced virulence but infection can be restored with exogenous cutinase (Dantzig *et al.*, 1986; Dickman and Patil, 1986).
(d) Pathogens that cannot infect a host without a breached cuticle (wound) when genetically engineered to provide cutinase-producing capability, can infect an intact host (Dickman *et al.*, 1989).

How a fungal spore senses the contact with the plant cuticle and triggers cutinase production has been explored. Spores from highly virulent fungi have small amounts of cutinase and upon contact with the cuticle this enzyme releases small amounts of cutin monomers that cause expression of the cutinase gene (Figure 3.6). The presence of cutinase on the spore (Köller *et al.*,

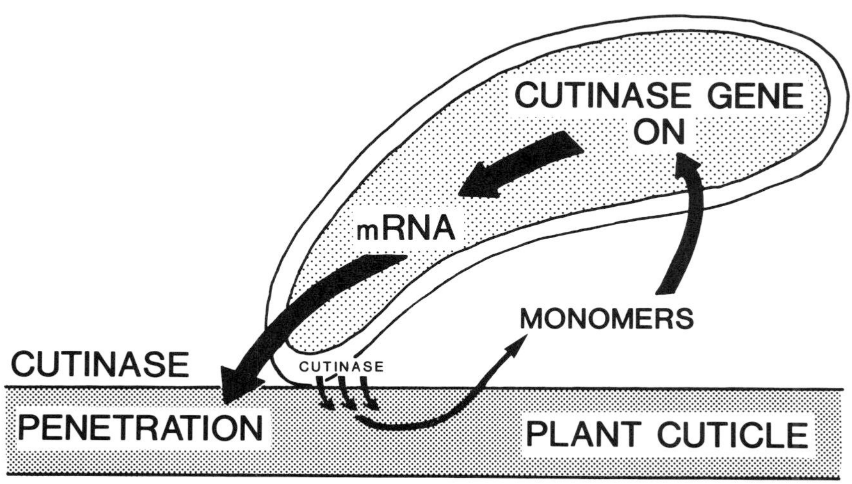

Figure 3.6. *Schematic representation of how the plant cuticle induces cutinase in a fungal spore.*

1982), induction of cutinase gene expression by contact with cutin and activation of transcription of cutinase gene by the unique monomers of cutin (Woloshuk and Kolattukudy, 1986) have been demonstrated. How the hydroxy fatty acids of the plant activate transcription of cutinase gene is not well understood. On the basis of the experimental evidence currently available it is thought that the hydroxy fatty acids are involved in causing phosphorylation of cutinase transcription factor(s) that bind to the cutinase promoter element(s) only when phosphorylated (Bajar *et al.*, 1991).

The cutinase promoter elements required for transcriptional activation by hydroxy fatty acids have been identified (Kämper *et al.*, 1994). Testing of the promoter activity of the 5′-flanking region of cutinase gene fused to a marker gene, chloramphenicol acetyl transferase (CAT), in transformed *Fusarium solani pisi* revealed a G-rich region at 340 bp that is needed for inducible expression, a silencer element that keeps cutinase gene repressed until induced, and a palindromic region at −159 bp that is absolutely essential for transcriptional activation by the cutin monomers.

Some information on transacting factors involved in the transcriptional activation of cutinase gene is also available. Binding of the cutinase promoter to fungal nuclear proteins was demonstrated. Using different promoter fragments for such binding it was shown that nuclear proteins bond to two separate segments of the promoter, one at palindrome 2 and another closer to the transcriptase initiation site (Kämper *et al.*, 1994). The palindrome binding protein was found only in the nuclear extracts from glucose-depleted cultures, whereas the other was found in nuclei from both glucose-containing and glucose-depleted cultures. Analysis of the nuclear protein that bound to the palindrome showed a 50 kDa protein. There is evidence that phosphorylation is involved in the activation of cutinase gene expression by cutin monomer. Gel retardation of cutinase promoter by fungal nuclear extracts showed binding to the promoter and dephosphorylation of the nuclear extract prevented this binding. Phosphorylation of a 50 kDa protein catalysed by the nuclear preparation required the presence of cutin monomer (Bajar *et al.*, 1991). That this phosphorylation is necessary for activation of cutinase transcription in nuclear preparation was shown by the inhibition of cutinase transcription by protein kinase inhibitors and antibodies against phosphotyrosine.

Recently, screening of expression libraries with concatenated promoter elements began to yield clones that encode proteins which bind the cutinase promoter elements (Li and Kolattukudy, 1995). Thus, a 50 kDa protein that autophosphorylates and binds the palindrome region of the cutinase promoter was cloned and expressed. Recently, it was found that this palindrome-binding protein specifically binds the first of the two overlapping palindromes in the promoter region. However, it had been shown that the second palindrome is the one that is essential for transcriptional activation of cutinase gene by hydroxy fatty acids. The cutinase transcription factor (CTF) that binds this palindrome was just cloned and sequenced. Two such proteins called CTF1α and CTF1β have been cloned and expressed in *Escherichia coli* and shown to bind palindrome 2 (D. Li and P.E. Kolattukudy, submitted). Clones that encode proteins that bind the other *cis*-regulatory elements in the cutinase gene are being cloned to seek a molecular level understanding of how plant cutin monomers trigger transcription of the fungal gene.

3.7 Conclusion

The plant cuticle constitutes one of the more chemically complex components of the plant. The diversity of molecules present in the cuticle, as well as the physical and chemical properties of the cuticle, make even chemical analyses and biosynthetic studies extremely difficult. Therefore, even though during the recent decades the basic biochemical pathways involved in the synthesis of the major cuticular components have been elucidated, the nature of the proteins that catalyse the biosynthetic reactions and the genes that encode such enzymes remain obscure. Genetic approaches coupled to the use of molecular biological methods are likely to help solve this problem. Such solutions will be required for exploring how environmental factors affect the biosynthesis of the cuticle at the molecular level. Since the cuticle is the first point of contact between plants and fungal pathogens, understanding of the interactions between the two might help to devise new approaches to control such interaction to help more fully to utilize plants as a renewable resource.

Acknowledgements

The work from the author's laboratory was supported in part by grant IBN9318544 from the National Science Foundation and DE-FG02-93ER20109 from the Department of Energy.

References

Aarts, M.G.M., Keijzer, C.J., Stiekama, W.J. *et al.* (1995) Molecular characterization of the *cer1* gene of *Arabidopsis* involved in epicuticular wax biosynthesis and pollen fertility. *Plant Cell* **7,** 2115–2127.

Agrawal, V.P., Lessire, R. and Stump, P.K. (1984) Biosynthesis of very long chain fatty acids in microsomes from epidermal cells of *Allium porrum* L. *Arch. Biochem. Biophys.* **230,** 580–589.

Bajar, A., Podila, G.K. and Kolattukudy, P.E. (1991) Identification of fungal cutinase promoter that is inducible by plant signal via a phosphorylated transacting factor. *Proc. Natl Acad. Sci. USA* **88,** 8208–8212.

Barckhausen, R. (1978) Ultrastructural changes in wounded plant storage tissue cells. In: *Biochemistry of Wounded Plant Tissues* (ed. G. Kahl). deGruyter, Berlin, pp. 1–42.

Bessoule, J.-J., Lessire, R. and Cassagne, C. (1989) Partial purification of the acyl-CoA elongase of *Allium porrum* leaves. *Arch. Biochem. Biophys.* **268,** 475–484.

Blée, E. and Schuber, F. (1990) Efficient epoxidation of unsaturated fatty acids by a hydroperoxide-dependent oxygenase. *J. Biol. Chem.* **265,** 12887–12894.

Blée, E. and Schuber, F. (1993) Biosynthesis of cutin monomers: involvement of a lipoxygenase/peroxygenase pathway. *Plant J.* **4,** 113–123.

Blée, E., Stahl, U., Schuber, F. *et al.* (1993) Regio- and stereoselectivity of cytochrome P-450 and peroxygenase-dependent formation of cis-12,13-epoxy-9(Z)-octadecenoic acid (vernolic acid) in *Euphorbia lagascae*. *Biochem. Biophys. Res. Commun.* **197,** 778–784.

Blomquist, G.J. and Ries, M.K. (1979) The enzymatic synthesis of wax esters by a microsomal preparation from the honeybee *Apis mellifera* L. *Insect Biochem.* **9,** 183–188.

Blomquist, G.J., Tillman-Wall, J.A., Guo, L. *et al.* (1993) Hydrocarbon and hydrocarbon-derived sex pheromones in insects; biochemistry and endocrine regulation. In: *Insect Lipids: Chemistry, Biochemistry and Biology* (eds D.W. Stanley-Samuelson and D.R. Nelson). University of Nebraska Press, Lincoln, NE., pp. 317–351.

Bognar, A.L., Paliyath, G., Rogers, L. *et al.* (1984) Biosynthesis of alkanes by particulate and solubilized enzyme preparations from pea leaves (*Pisum sativum*). *Arch. Biochem. Biophys.* **235,** 8–17.

Bolton, P. and Harwood, J.L. (1977) Fatty acid biosynthesis by a particulate preparation from germinating pea. *Biochem. J.* **168,** 261–266.

Cassagne, C. and Lessire, R. (1974) Studies on alkane biosynthesis in epidermis of *Allium porrum* L. leaves. *Arch. Biochem. Biophys.* **165,** 274–280.

Cassagne, C. and Lessire, R. (1978) Biosynthesis of saturated very long chain fatty acids by purified membrane fractions from leek epidermal cells. *Arch. Biochem. Biophys.* **191,** 146–152.

Cheesbrough, T.M. and Kolattukudy, P.E. (1984) Alkane biosynthesis by decarbonylation of aldehydes catalyzed by a particulate preparation from *Pisum sativum*. *Proc. Natl Acad. Sci. USA* **81**, 6613–6617.

Cheesbrough, T.M. and Kolattukudy, P.E. (1988) Microsomal preparation from an animal tissue catalyzes release of carbon monoxide from a fatty aldehyde to generate an alkane. *J. Biol. Chem.* **263**, 2738–2743.

Coe, E.H., Hoisington, D.A. and Neuffer, M.G. (1990) Linkage map of corn (maize) *Zea mays* L.) (2N = 20). In: *Genetic Maps* (ed. S.J. O'Brien). Cold Spring Harbor Laboratory Press, Cold Spring Harbor, NY, p. 6.39.

Crisp, C.E. (1965) The Biopolymer Cutin. Ph.D. Thesis, University of California, Davis, California.

Croteau, R. and Kolattukudy, P.E. (1974a) Biosynthesis of pentahydroxystearic acid of cutin from linoleic acid in *Rosmarinus officinalis*. *Arch. Biochem. Biophys.* **162**, 458–470.

Croteau, R. and Kolattukudy, P.E. (1974b) Direct evidence for the involvement of epoxide intermediates in the biosynthesis of the C_{18} family of cutin acids. *Arch. Biochem. Biophys.* **162**, 471–480.

Croteau, R. and Kolattukudy, P.E. (1974c) Biosynthesis of hydroxy fatty acid polymers. Enzymatic synthesis of cutin from monomer acids by cell-free preparations from the epidermis of *Vicia faba* leaves. *Biochemistry* **13**, 3193–3202.

Croteau, R. and Kolattukudy, P.E. (1975) Biosynthesis of hydroxy fatty acid polymers. Enzymatic epoxidation of 18-hydroxyoleic acid to 18-hydroxy-*cis*-9,10-epoxystearic acid by a particulate preparation from spinach (*Spinacia oleracea*). *Arch. Biochem. Biophys.* **170**, 61–72.

Dantzig, A.H., Zuckerman, S.H. and Andonov-Roland, M.M. (1986) Isolation of a *Fusarium solani* mutant reduced in cutinase activity and virulence. *J. Bacteriol.* **168**, 911–916.

Dean, B.B. and Kolattukudy, P.E. (1976) Synthesis of suberin during wound-healing in jade leaves, tomato fruit, and bean pods. *Plant Physiol.* **58**, 411–416.

Deas, A.H.B. and Holloway, P.J. (1977) The intermolecular structure of some plant cutins. In: *Lipids and Lipid Polymers in Higher Plants* (eds M. Tevini and H.K. Lichtenthaler). Springer-Verlag, Berlin, pp. 293–300.

Dennis, M.W. and Kolattukudy, P.E. (1991) Alkane biosynthesis by decarbonylation of aldehyde catalyzed by a microsomal preparation from *Botryococcus braunii*. *Arch. Biochem. Biophys.* **287**, 268–275.

Dennis, M. and Kolattukudy, P.E. (1992) A cobalt-porphyrin enzyme converts a fatty aldehyde to a hydrocarbon and CO. *Proc. Natl Acad. Sci. USA* **89**, 5306–5310.

Dickman, M.B. and Patil, S.S. (1986) Cutinase deficient mutant of *Colletotrichum gloeosporioides* are non-pathogenic to papaya fruit. *Physiol Mol. Plant Pathol.* **28**, 235–242.

Dickman, M.B., Podila, G.K. and Kolattukudy, P.E. (1989) Insertion of cutinase gene into a wound pathogen enables it to infect intact host. *Nature* **342**, 446–448.

Domazetis, G., James, B.R., Tarpey, B. *et al.* (1981) Catalytic decarbonylation of aldehydes using ruthenium (II) porphyrin systems. In: *Catalytic Activation of Carbon Monoxide* (ed. P.C. Ford). Am. Chem. Soc., Washington, DC, pp. 243–252.

Donadio, S., Staver, M.J., McAlpine, J.B. *et al.* (1991) Modular organization of genes required for complex polyketide biosynthesis. *Science* **252**, 675–679.

Espelie, K.E., Dean, B.B. and Kolattukudy, P.E. (1979) Composition of lipid-derived polymers from different anatomical regions of several plant species. *Plant Physiol.* **64**, 1089–1093.

Espelie, K.E., Davis, R.W. and Kolattukudy, P.E. (1980) Composition, ultrastructure and function of the cutin- and suberin-containing layers in the leaf, fruit peel, juice-sac and inner seed coat of grapefruit (*Citrus paradisi* Macfed). *Planta* **149**, 498–511.

Espelie, K.E., Wattendorff, J. and Kolattukudy, P.E. (1982) Composition and ultrastructure of the suberized cell wall of isolated crystal idioblasts from *Agave americana* L. leaves. *Planta* **155**, 166–175.

Friedberg, S.J., Greene, R.C. (1967) The enzymatic synthesis of wax in liver. *J. Biol. Chem.* **242**, 234–237.

Hannoufa, A., McNevin, J. and Lemieux, B. (1993) Epicuticular waxes of *eceriferum* mutants of *Arabidopsis thaliana*. *Phytochemistry* **33**, 851–855.

Harwood, J.L. and Stumpf, P.K. (1971) Fat metabolism in higher plants XLIII. Control of fatty acid synthesis in germinating seeds. *Arch. Biochem. Biophys.* **142**, 281–291.

Hoffmann-Benning, S. and Kende, H. (1994) Cuticle biosynthesis in rapidly growing internodes of deepwater rice. *Plant Physiol.* **104**, 719–723.

Holloway, P.J., Brown, G.A., Baker, E.A. *et al.* (1977a) Chemical composition and ultra structure in the epicuticular wax in three lines of *Brassica napus* (L). *Chem. Phys. Lipids* **19**, 114–127.

Holloway, P.J., Hunt, G.M., Baker, E.A. *et al.* (1977b) Chemical composition and ultrastructure of the epicuticular wax in four mutants of *Pisum sativum* (L.). *Chem. Phys. Lipids*, **20**, 141–155.

Hwang, C.-S. and Kolattukudy, P.E. (1995) Isolation and characterization of genes expressed uniquely during appressorium formation by *Colletotrichum gloeosporioides* conidia induced by the host surface wax. *Mol. Gen. Genet.* **247**, 282–294.

Hwang, C.-S., Flaishman, M.A. and Kolattukudy, P.E. (1995) Cloning of a gene expressed during appressorium formation by *Colletotrichum gloeosporioides* and a marked decrease in virulence by disruption of this gene. *Plant Cell* **7**, 183–193.

Jenks, M.A., Tuttle, H.A., Eigenbrode, S.D. *et al.* (1995) Leaf epicuticular waxes in the *eceriferum* mutants in *Arabidopsis. Plant Physiol.* **108**, 369–377.

Kämper, J., Kämper, U., Rogers, L.M. *et al.* (1994) Identification of regulatory elements in the cutinase promoter from *Fusarium solani* f. sp. *pisi* (*Nectria haematococca*). *J. Biol. Chem.* **269**, 9195–9204.

Khan, A.A. and Kolattukudy, P.E. (1973) Control of synthesis and distribution of acyl moieties in etiolated *Euglena gracilis. Biochemistry* **12**, 1939–1948.

Khan, A.A. and Kolattukudy, P.E. (1974) Decarboxylation of long chain fatty acids to alkanes by cell free preparations of pea leaves (*Pisum sativum*). *Biochem. Biophys. Res. Commun.* **61**, 1379–1386.

Kolattukudy, P.E. (1965) Biosynthesis of wax in *Brassica oleracea. Biochemistry* **4**, 1844–1855.

Kolattukudy, P.E. (1966) Biosynthesis of wax in *Brassica oleracea*. Relation of fatty acid to wax. *Biochemistry* **5**, 2265–2275.

Kolattukudy, P.E. (1967a) Mechanism of synthesis of waxy esters in broccoli (*Brassica oleracea*). *Biochemistry* **6**, 2705–2717.

Kolattukudy, P.E. (1967b) Biosynthesis of paraffins in *Brassica oleracea*: fatty acid elongation-decarboxylation as a plausible pathway. *Phytochemistry* **6**, 963–975.

Kolattukudy, P.E. (1968a) Further evidence for an elongation-decarboxylation mechanism in the biosynthesis of paraffins in leaves. *Plant Physiol.* **43**, 375–383.

Kolattukudy, P.E. (1968b) Species specificity in the biosynthesis of branched paraffins in leaves. *Plant Physiol.* **43**, 1423–1429.

Kolattukudy, P.E. (1970a) Composition of the surface lipids of pea leaves (*Pisum sativum*). *Lipids* **5**, 398–402.

Kolattukudy, P.E. (1970b) Reduction of fatty acids to alcohols by cell-free preparation of *Euglena gracilis. Biochemistry* **9**, 1095–1102.

Kolattukudy, P.E. (1970c) Cutin biosynthesis in *Vicia faba* leaves. *Plant Physiol.* **46**, 759–760.

Kolattukudy, P.E. (1971) Enzymatic synthesis of fatty alcohols in *Brassica oleracea. Arch. Biochem. Biophys.* **142**, 701–709.

Kolattukudy, P.E. (1976) Introduction to natural waxes. In: *Chemistry and Biochemistry of Natural Waxes* (ed. P.E. Kolattukudy). Elsevier, Amsterdam, pp. 1–15.

Kolattukudy, P.E. (1977) Lipid polymers and associated phenols, their chemistry, biosynthesis, and role in pathogenesis. In: *The Structure, Biosynthesis, and Degradation of Wood* (eds F.A. Loewus and V.C. Runeckles). Plenum, New York, pp. 185–246.

Kolattukudy, P.E.. (1980) Cutin, suberin, and waxes. In: *The Biochemistry of Plants*, vol. IV (ed. P.K. Stumpf). Academic Press, London, pp. 571–645.

Kolattukudy, P.E. (1985) Enzymatic penetration of the plant cuticle by fungal pathogens. *Annu. Rev. Phytopathol.* **23**, 223–250.

Kolattukudy, P.E. (1987) Lipid derived defensive polymers and waxes and their role in plant–microbe interaction. In: *The Biochemistry of Plants*, Vol. 9 (ed. P.K. Stumpf). Academic Press, New York, pp. 291–313.

Kolattukudy, P.E. and Buckner, J.S. (1972) Chain elongation of fatty acids by cell-free extract of epidermis from pea leaves (*Pisum sativum*). *Biochem. Biophys. Res. Commun.* **46**, 801–807.

Kolattukudy, P.E. and Espelie, K.E. (1985) Biosynthesis of cutin, suberin and associated waxes. In: *Biosynthesis and Biodegradation of Wood Components* (ed. T. Higuchi). Academic Press, New York, pp. 161–207.

Kolattukudy, P.E. and Espelie, K.E. (1989) Chemistry, biochemistry, and function of suberin and associated waxes. In: *Natural Products of Woody Plants I* (ed. J.W. Rowe). Springer-Verlag, Berlin, pp. 304–367.

Kolattukudy, P.E. and Liu, T.-Y.J. (1970) Direct evidence for biosynthetic relationships among hydrocarbons, secondary alcohols and ketones in *Brassica oleracea. Biochem. Biophys. Res. Commun.* **41**, 1369–1374.

Kolattukudy, P.E. and Rogers, L. (1986) Acyl-CoA reductase and acyl-CoA: fatty alcohol acyl transferase in the microsomal preparation from the bovine meibomian gland. *J. Lipid Res.* **27**, 404–411.

Kolattukudy, P.E. and Walton, T.J. (1972) Structure and biosynthesis of the hydroxy fatty acids of cutin in *Vicia faba* leaves. *Biochemistry* **11**, 1897–1907.

Kolattukudy, P.E. and Walton, T.J. (1973) The biochemistry of plant cuticular lipids. *Progr. Chem. Fats Other Lipids* **8**, 121–175.

Kolattukudy, P.E., Buckner, J.S. and Brown, L. (1972) Direct evidence for a decarboxylation mechanism in the biosynthesis of alkanes in *B. oleracea*. *Biochem. Biophys. Res. Commun.* **47**, 1306–1313.

Kolattukudy, P.E., Buckner, J.S. and Liu, T.-Y.J. (1973a) Biosynthesis of secondary alcohols and ketones from alkanes. *Arch. Biochem. Biophys.* **156**, 613–620.

Kolattukudy, P.E., Croteau, R. and Brown, L. (1974) Structure and biosynthesis of cuticular lipids. Hydroxylation of palmitic acid and decarboxylation of C_{28}, C_{30} and C_{32} acids in *Vicia faba* flowers. *Plant Physiol.* **54**, 670–677.

Kolattukudy, P.E., Rogers, L. and Larson, J.D. (1981) Enzymatic reduction of fatty acids and α-hydroxy fatty acids. *Methods Enzymol.* **71**, 263–275.

Kolattukudy, P.E., Walton, T.J. and Kushwaha, R.P.S. (1973b) Biosynthesis of the C_{18} family of cutin acids: ω-hydroxy oleic acid, ω-hydroxy-9,10-epoxy stearic acid, 9,10,18-trihydroxy stearic acid, and their Δ^{12}-unsaturated analogs. *Biochemistry* **12**, 4488–4498.

Köller, W. and Kolattukudy, P.E. (1982) Mechanism of action of cutinase: chemical modification of the catalytic triad characteristic for serine hydrolases. *Biochemistry* **21**, 3083–3090.

Köller, W., Allan, C.R. and Kolattukudy, P.E. (1982) Role of cutinase and cell wall degrading enzymes in infection of *Pisum sativum* by *Fusarium solani* f sp. *pisi*. *Physiol. Plant Pathol.* **20**, 47–60.

Kunst, L., Taylor, D.C. and Underhill, E.W. (1992) Fatty acid elongation in developing seeds of *Arabidopsis thaliana*. *Plant Physiol. Biochem.* **30**, 425–434

Lemieux, B., Koornneef, M. and Feldman, K.A. (1994) Epicuticular wax and *eceriferum* mutants. In: *Arabidopsis* (eds E.M. Meyerowitz and C.R. Somerville). Cold Spring Harbor Laboratory Press, Cold Spring Harbor, NY, pp. 1031–1047.

Lessire, R., Bessoule, J.-J. and Cassagne, C. (1985) Solubilization of C_{18}-CoA and C_{20}-CoA elongases from *Allium porrum* L. epidermal cell microsomes. *FEBS Lett.* **187**, 314–320.

Lessire, R., Bessoule, J.-J. and Cassagne, C. (1989) Involvement of a β-ketoacyl-CoA intermediate in acyl-CoA elongation by an acyl-CoA elongase purified from leek epidermal cells. *Biochim. Biophys. Acta* **1006**, 35–40.

Li, D. and Kolattukudy, P.E. (1995) Cloning and expression of cDNA encoding a protein that binds a palindromic promoter element essential for induction of fungal cutinase by plant cutin. *J. Biol. Chem.* **270**, 11753–11756.

Macey, M.J.K. and Stumpf, P.K. (1968) Fat metabolism in higher plants XXXVI: long chain fatty acid synthesis in germinating peas. *Plant Physiol.* **43**, 1637–1647.

Martinez, C., de Geus, P., Lauwereys, M. *et al.* (1992) *Fusarium solani* cutinase is a lipolytic enzyme with a catalytic serine accessible to solvent. *Nature* **356**, 615–618.

Mathur, M. and Kolattukudy, P.E. (1992) Molecular cloning and sequencing of the gene for mycocerosic acid synthase, a novel fatty acid elongating multifunctional enzyme, from *Mycobacterium tuberculosis* var. *bovis Bacillus Calmette-Guerin*. *J. Biol. Chem.* **267**, 19388–19395.

McNevin, J.P., Woodward, W., Hannoufa, A. *et al.* (1993) Isolation and characterization of *eceriferum* (*cer*) mutants induced by T-DNA insertions in *Arabidopsis thaliana*. *Genome* **36**, 610–618.

Mikkelsen, J.D. (1979) Structure and biosynthesis of β-diketones in barley spike epicuticular wax. *Carlsberg Res. Commun.* **44**, 133–147.

Mikkelsen, J.D. (1984) Biosynthesis of esterified alkan-2-ols and β-diketones in barley spike epicuticular wax: synthesis of radioactive intermediates. *Carlsberg Res. Commun.* **49**, 391–416.

Mikkelsen, J.D. and von Wettstein-Knowles, P. (1978) Biosynthesis of β-diketones and hydrocarbons in barley spike epicuticular wax. *Arch. Biochem. Biophys.* **188**, 172–181.

Mikkelsen, J.D. and von Wettstein-Knowles, P. (1984) Biosynthesis of esterified alkan-2-ols in barley spike epicuticular wax. In: *Structure, Function and Metabolism of Plant Lipids* (eds P.-A. Siegenthaler and W. Eichenberger). Elsevier, Amsterdam, pp. 517–520.

Mohan, R., Vijayan, P. and Kolattukudy, P.E. (1993) Developmental and tissue-specific expression of a tomato anionic peroxidase (*tap1*) gene by a minimal promoter, with wound and pathogen induction by an additional 5′-flanking region. *Plant Molec. Biol.* **22,** 475–490.

Moore, C. and Snyder, F. (1982) Properties of microsomal acyl coenzyme A reductase in mouse preputial glands. *Arch. Biochem. Biophys.* **214,** 489–499.

Pinot, F., Salaün, J.-P., Bosch, H. *et al.* (1992) ω-Hydroxylation of Z9-octadecenoic, Z9,10-epoxystearic and 9,10-dihydroxystearic acids by microsomal cytochrome P450 systems in *Vicia sativa*. *Biochem. Biophys. Res. Commun.* **184,** 183–193.

Podila, G.K., Rogers, L.M. and Kolattukudy, P.E. (1993) Chemical signals from avocado surface wax trigger germination and appressorium formation in *Colletotrichum gloeosporioides*. *Plant Physiol.* **103,** 267–272.

Pozuelo, J.M., Espelie, K.E. and Kolattukudy, P.E. (1984) Magnesium deficiency results in increased suberization in endodermis and hypodermis of corn roots. *Plant Physiol.* **74,** 256–260.

Purdy, R.E. and Kolattukudy, P.E. (1975) Hydrolysis of plant cuticle by plant pathogens. Purification, amino acid composition, and molecular weight of two isozymes of cutinase and a nonspecific esterase from *Fusarium solani* f. *pisi*. *Biochemistry* **14,** 2824–2831.

Pyee, J. and Kolattukudy, P.E. (1995) The gene for the major cuticular wax-associated protein and three homologous genes from broccoli (*Brassica oleracea*) and their expression patterns. *Plant J.* **7,** 49–59.

Pyee, J., Yu, H. and Kolattukudy, P.E. (1994) Identification of a lipid transfer protein as the major protein in the surface wax of broccoli (*Brassica oleracea*) leaves. *Arch. Biochem. Biophys.* **311,** 460–468.

Rainwater, D.L. and Kolattukudy, P.E. (1985) Fatty acid biosynthesis in *Mycobacterium tuberculosis* var. *bovis Bacillus Calmette-Guérin*. Purification and characterization of a novel fatty acid synthase, mycocerosic acid synthase, which elongates *n*-fatty acyl-CoA with methyl malonyl-CoA. *J. Biol. Chem.* **260,** 616–623.

Reed, J.R., Vanderwal, D., Choi, S. *et al.* (1994) Unusual mechanism of hydrocarbon formation in the housefly: cytochrome P450 converts aldehyde to the sex pheromone component (Z)-9-tricosene and CO_2. *Proc. Natl Acad. Sci. USA* **91,** 10000–10004.

Riley, R.G. and Kolattukudy, P.E. (1975) Evidence for covalently attached *p*-coumaric acid and ferulic acid in cutins and suberins. *Plant Physiol.* **56,** 650–654.

Shanklin, J., White, E. and Fox, B.G. (1994) Eight histidine residues are catalytically essential in a membrane-associated iron enzyme, stearoyl-CoA desaturase, and are conserved in alkane hydrolase and xylene monooxygenase. *Biochemistry* **33,** 12787–12794.

Snyder, F. and Malone, B. (1970) Enzymic interconversion of fatty alcohols and fatty acids. *Biochem. Biophys. Res. Commun.* **41,** 1382–1387.

Soliday, C.L. and Kolattukudy, P.E. (1977) Biosynthesis of cutin. ω-hydroxylation of fatty acids by a microsomal preparation from germinating *Vicia faba*. *Plant Physiol.* **59,** 1116–1121.

Sossountzov, L., Ruiz-Avila, L., Vignols, F. *et al.* (1991) Spatial and temporal expression of a maize lipid transfer protein gene. *Plant Cell* **3,** 923–933.

Sterk, P., Booji, H., Schellekens, G.A. *et al.* (1991) Cell-specific expression of the carrot EP2 lipid transfer protein gene. *Plant Cell* **3,** 907–921.

Tegelaar, E.W., Wattendorff, J. and deLeeuw, J.W. (1993) Possible effects of chemical heterogeneity in higher land plant cuticles on the preservation of its ultrastructure upon fossilization. *Rev. Palaeobot. Palynol.* **77,** 149–170.

Thoma, S., Kaneko, Y. and Somerville, C. (1993) A non-specific lipid transfer protein from *Arabidopsis* is a cell wall protein. *Plant J.* **3,** 427–436.

Thoma, S., Hecht, U., Kippers, A. *et al.* (1994) Tissue-specific expression of a gene encoding a cell wall localized lipid transfer protein from *Arabidopsis*. *Plant Physiol.* **105,** 35–45.

Tulloch, A.P. (1976) Chemistry of waxes of higher plants. In: *Chemistry and Biochemistry of Natural Waxes* (ed. P.E. Kolattukudy). Elsevier, Amsterdam, pp. 235–287.

Villena, J.F., Casado, C.G., Luque, P. *et al.* (1996) Structural characteristics of a resistant biopolymer isolated from plant cuticles. *J. Exptl. Bot.* **47** (suppl.), 56.

von Wettstein-Knowles, P. (1972) Genetic control of β-diketone and hydroxy-β-diketone synthesis in epicuticular waxes of barley. *Planta* **106,** 113–130.

von Wettstein-Knowles, P. (1976) Biosynthetic relationships between β-diketones and esterified alkan-2-ols deduced from epicuticular wax of barley mutants. *Molec. Gen. Genet.* **144,** 43–48.

von Wettstein-Knowles, P. (1985) Effects of inhibitors on synthesis of esterified alkan-2-ols in barley spike epicuticular wax. *Carlsberg Res. Commun.* **50,** 239–262.

von Wettstein-Knowles, P. (1987) Genes, elongases and associated enzyme systems in epicuticular wax synthesis. In: *The Metabolism, Structure and Function of Plant Lipids* (eds P.K. Stumpf,

J.B. Mudd and W.D. Nes). Plenum Press, New York, pp. 489–498.

von Wettstein-Knowles, P.M. (1993) Waxes, cutin and suberin. In: *Lipid Metabolism in Plants* (ed. T.S. Moore, Jr.). CRC Press, Boca Raton, FL, pp. 127–166.

Wang, L., Takayama, K., Goldman, D.S. *et al.* (1972) Synthesis of alcohol and wax ester by a cell-free system in *Mycobacterium tuberculosis. Biochim. Biophys. Acta* **260,** 41–48.

Wang, X. and Kolattukudy, P.E. (1995a) Solubilization and purification of aldehyde-generating fatty acyl-CoA reductase from green alga *Botryococcus braunii. FEBS Letts.* **370,** 15–18.

Wang, X. and Kolattukudy, P.E. (1995b) Solubilization, purification and characterization of fatty acyl-CoA reductase from duck uropygial gland. *Biochem. Biophys. Res. Commun.* **208,** 210–215.

Woloshuk, C.P. and Kolattukudy, P.E. (1986) Mechanism by which contact with plant cuticle triggers cutinase gene expression in the spores of *Fusarium solani* f. sp. *pisi. Proc. Natl Acad. Sci. USA* **83,** 1704–1708.

Yoder, J.A., Denlinger, D.L., Dennis, M.W. *et al.* (1992) Enhancement of diapausing fleshfly puparia with additional hydrocarbons and evidence for alkane biosynthesis by a decarbonylation mechanism. *Insect Biochem. Molec. Biol.* **22,** 237–243.

Zlotnik-Mazori, T. and Stark, R.E. (1988) Nuclear magnetic resonance studies of cutin, an insoluble plant polyester. *Macromolecules* **21,** 2412–2417.

4

Diffusion through cuticles: principles and models

L. Schreiber, T. Kirsch and M. Riederer

4.1 Introduction

The plant cuticle controls the exchange of matter between leaf and atmosphere. Loss of water (Schönherr, 1982) and solutes (Tukey, 1970) from the leaf interior as well as uptake of non-volatile chemicals deposited from the atmosphere to the leaf surface (Schönherr and Riederer, 1989) are strongly dependent on transport properties of the cuticle. Additionally, when stomata are closed or when they are missing, as frequently occurs with fruits, gas exchange is also controlled by the cuticular barrier (Lendzian and Kerstiens, 1991). Thus, in order to evaluate the physiological and ecological significance of the plant cuticle as the interface between the atmosphere and the leaf interior, it is necessary to analyse and describe the function of the cuticle as a transport-limiting barrier. In the following sections, basic equations and parameters describing the steady-state flow of matter across barrier membranes will be introduced. Subsequently, examples from studies on cuticular transport on the level of the intact leaf, the isolated cuticle and recrystallized cuticular wax, will be presented and discussed.

On the basis of these findings a model of cuticular transport will be suggested taking the molecular structure of the wax and the physicochemical properties of the solutes into consideration. Finally, general conclusions concerning the structure and the function of the cuticular transport barrier will be drawn.

4.2 Basic parameters and equations

The steady state flow F (mol s^{-1}) of a substance across the cuticular membrane is given by the following equation (Kerler *et al.*, 1984):

$$F = P \cdot A \cdot \Delta c. \qquad (4.1)$$

A (m^2) is the exposed area across which the flow of substance takes place and Δc (mol m^{-3}) is the driving force given by the concentration difference of the substance in the donor and receiver solution, respectively. The permeance P (m s^{-1}) is the proportionality factor relating F to A and Δc. It is the characteristic quantity describing the permeability of a specific cuticle for a given substance since it is independent of the boundary conditions of the respective experiments (time of

Plant Cuticles, edited by G. Kerstiens.

uptake, donor concentration, area exposed). Using P, barrier properties of cuticles from different species for the same compound (Riederer and Schönherr, 1985) and the barrier properties of the same cuticular membrane (CM) for different chemicals (Kerler and Schönherr, 1988) can be compared on a quantitative basis.

The permeance P itself is a composed quantity:

$$P = \frac{D \cdot K}{\Delta x}. \tag{4.2}$$

D ($m^2\ s^{-1}$) is the diffusion coefficient of the compound within the membrane, K is the dimensionless partition coefficient of the compound between the external donor solution and the membrane and Δx (m) is the path length of diffusion through the membrane. D can be described as the mobility of the respective compounds within the transport-limiting membrane and depends on the size of the permeating molecules. The partition coefficient K, which is the ratio of the equilibrium concentrations of a given solute within the cuticle and in the external aqueous solution, describes the tentative solubility of the substance in the membrane. With a lipophilic membrane, as is the case with the plant cuticle, K will increase with increasing lipophilicity of the molecules. Finally, with a homogeneous membrane, Δx will simply be the thickness of the membrane. However, with a heterogeneous membrane as is the case with the plant cuticle, the path length across the membrane is not simply given by the thickness of the cuticle rather than the tortuosity of the diffusional path within the membrane, as will be shown further below.

A schematic view of the cross-section of the plant cuticle reveals its heterogeneous composition, consisting of cutin and waxes (see Figure 2.13). The cuticular matrix is a lipophilic, amorphous polymer membrane (Holloway, 1982). Cuticular waxes, in most cases dominated by linear, long-chain aliphatics with various functional groups, amount to 1–10 % of the mass of the cuticle (Baker, 1982; Walton, 1990). They are deposited as solid aggregates into the outer regions of the cutin polymer (Sitte and Rennier, 1963), and in some cases in significant amounts onto the outer surface of the cuticle (Barthlott, 1990). While the cutin polymer forms the support into and onto which cuticular waxes are deposited, the main transport-limiting barrier of the plant cuticle is formed by the waxes (Riederer and Schreiber, 1995). This is due to their highly ordered, partially crystalline nature (Reynhardt and Riederer, 1991, 1994). The effectivity of the waxes as a barrier can easily be demonstrated by removing them from the cuticle by an appropriate organic solvent such as chloroform or hexane. As a result, cuticular permeabilities for water or organic chemicals increase by factors ranging between 10 to more than 1000 (Schönherr and Riederer, 1989).

When cutin and wax are treated as transport barriers in series the sum of the reciprocals of the permeances of wax (P_{wax}) and cutin (P_{cutin}) will give the total permeance (P_{total}) of the composed membrane.

$$\frac{1}{P_{total}} = \frac{1}{P_{wax}} + \frac{1}{P_{cutin}}. \tag{4.3}$$

Since P_{cutin} is about a factor of 10 up to 1000 higher than P_{wax}, barrier properties of the intact cuticle (P_{total}) are more or less completely determined by the transport properties of the cuticular wax. From this, it is obvious that it rarely will be the thickness of the whole cuticle which determines the path length of diffusion.

Instead, it is the thickness of the wax layer and, as will be shown below, the tortuosity within the wax layer which essentially determines the path length of diffusion. Thus, besides the investigation of the transport properties of the intact cuticle it is necessary to investigate the transport properties of the cuticular wax.

In the following section, cuticular permeability of non-electrolytes (water and organic compounds) will be dealt with, whereas, cuticular permeability of ions and ionized compounds will not be included. Due to the fact that isolated cuticles carry fixed charges (Schönherr and Bukovac, 1973), rates of cuticular permeability of ionized compounds do not depend only on the chemical but on the electrochemical gradient of the compounds across the cuticle (Tyree *et al.*,1992). However, the electrochemical gradient across the cuticle cannot be measured accurately and thus the driving force needed for the calculation of permeances (see Equation 4.1) is not known. Therefore, a quantitative analysis of cuticular permeability as it can be obtained with non-electrolytes on the basis of the formalism described above cannot be carried out with ionized compounds. Therefore, the following applies solely to the transport of non-electrolytes and investigations on the cuticular permeability of intact leaves and isolated cuticles as well as the transport properties of extracted and subsequently recrystallized wax will be presented.

4.3 Cuticular permeability of the intact leaf

The investigation of foliar uptake of organic chemicals from aqueous solutions into conifer needles revealed biphasic uptake kinetics (Schreiber and Schönherr, 1992). A first, rapid and saturable phase, which lasts only about 30 min, can be separated from a second, slow and non-saturable phase showing a linear rate of uptake over several hours. The first phase of uptake can be interpreted as due to the adsorption of the organic chemicals to the interface between the waxy leaf surface and the aqueous solution. It is rapid and saturable. The second phase, however, represents the diffusion across the waxy transport barrier of the cuticle into the needle interior. This process is significantly slower compared to the surface adsorption and, due to the infinite sink of the needle interior, cannot be saturated under experimental conditions used.

This interpretation of biphasic uptake kinetics, consisting of surface adsorption and cuticular penetration, has been corroborated by experiments measuring desorption of solutes from conifer needles (Schreiber and Schönherr, 1993a). Needles which had been loaded for distinct periods of time with ^{14}C-labelled organic compounds such as pentachlorophenol (PCP) were subsequently desorbed again. Within the first 5 to 10 min more than 90% of the total desorbable material could be desorbed, which corresponds to the first phase of uptake representing the amounts of chemicals adsorbed to the surface of the needle. During the remaining desorption period, lasting up to 24 h, a significant amount of PCP, which increased with increasing times of loading, remained irreversibly associated with the needle. This can be attributed to the second phase of uptake and represents that amount of PCP which penetrated across the waxy transport barrier into the needle interior.

This irreversibility of uptake, clearly shown by desorption experiments, demonstrates that the needle interior acts as an infinite sink. Within the duration of an experiment the substance reaching the needle interior will be hindered from desorption by processes such as translocation, conjugation, metabolism and ionization. Thus, the mechanism of foliar uptake of organics from diluted aqueous solutions can be separated into four consecutive steps: (1) surface adsorption; (2) dissolution in wax and/or cutin forming the transport limiting barrier; (3) diffusion across the transport limiting barrier of wax and across the adjacent polymer matrix; (4) finally the transfer or desorption of the solutes from the CM into the apoplast of the cell wall.

An important conclusion which has to be drawn from biphasic uptake kinetics is the fact that the total amount of a lipophilic solute, whether it is an agrochemical or a xenobiotic compound, associated with needles or leaves does not necessarily represent that amount of chemical which has really been taken up into the mesophyll. After short periods of uptake, the major part of the substance will be located on the leaf surface. Whereas, after longer periods of time significant amounts of the solute will diffuse into the interior of the leaf.

The pronounced tendency of lipophilic solutes to adsorb to the leaf/water interface has been used to determine adsorption isotherms of PCP on the needle surfaces of several conifer species in order to measure their specific leaf surface areas (Schreiber and Schönherr, 1993b). The adsorption isotherms, which are the plot of the amounts of chemicals adsorbed on the needle surface as a function of the increasing external concentrations, exhibited weak but clearly discernible plateaus. The plateau represents that number of PCP molecules which forms a complete monomolecular layer on the needle surface. These isotherms can be linearized using the BET formalism (Brunauer, Emmett and Teller, 1938), assuming a realistic value for the surface area occupied by one single PCP molecule (Bondi, 1964) and can be used for estimating the specific leaf surface area of the respective sample.

The values of specific leaf surface areas for five different conifer species ranged from 30 to 640 mm^2 per mm^2 projected leaf surface areas (Schreiber and Schönherr, 1993b). Due to the microscopic roughness of the leaf surfaces of the different plant species, leaf surfaces areas are enlarged by factors from 10 up to more than 100. Plant communities such as forest ecosystems possess large surface areas on a macroscopic scale (Schulze, 1982), and this surface area is significantly enlarged on a microscopic scale by specific leaf surface areas, due to the pronounced roughness of leaf surfaces. Rates of foliar uptake into conifer needles were linearly correlated to specific leaf surface areas. Needles with higher specific leaf surfaces had higher rates of cuticular uptake.

However, the method suggested for the determination of specific leaf surfaces is not completely unambiguous. It offers the opportunity of measuring the contact area between an aqueous solution and the wetted leaf surface, instead of the complete leaf surface, which will be of significance for the interaction of gaseous compounds with the leaf surface. Since BET constants were relatively low, the plateaus of the isotherms were only weakly pronounced and contact areas calculated from the monolayer capacities contain a relatively large error (Gregg and Sing, 1982). Furthermore, there are some theoretical arguments that the contact area measured does not necessarily represent that area across which foliar uptake occurs. This method of estimating contact areas of leaf surface also includes leaf surface structures like epicuticular wax (EW) crystals, which will significantly enlarge the specific surface area of the leaf. However, it is hardly imaginable that solute diffusion will occur within these highly ordered and densely packed epicuticular crystal structures (Jetter and Riederer, 1994). Nevertheless, with the conifers investigated, rates of foliar uptake were correlated to the measured contact areas, which underlines the possible significance of the size of the specific leaf surface area in studies on cuticular permeability.

4.4 Diffusion across the isolated cuticular membrane

In the past, most investigations concerning cuticular permeability have been carried out investigating transport properties of isolated CM (Bukovac and Petracek, 1993; Chamel, 1986; Kerler *et al.*, 1984). The common objective was a general description of cuticular permeability on the basis of independent, physicochemical parameters of the solutes, which would allow the prediction of cuticular permeability from certain independent descriptors. From Equation 4.2 it is clear that P is directly proportional to D and K and indirectly proportional to Δx. A good and simple model, which works fairly well, correlated cuticular permeabilities of different chemicals with the lipophilicity of the respective compounds (Kerler and Schönherr, 1988) usually expressed by the octanol/water (K_{ow}) or cuticle/water partition coefficient (K_{cw}). The reason for the good correlation of the permeance with lipophilicity, without having to consider the parameter D, probably is due to the fact that a spectrum of substances varying by several orders of magnitude in lipophilicity has been investigated. In contrast, the size of the compounds investigated varied only by a factor of two, and thus the dependence of P on D obviously disappeared behind the predominating dependence on K.

Permeances calculated according to Equation 4.1 from the rates of cuticular diffusion of organic chemicals into intact conifer needles were linearly correlated to the lipophilicity of the compounds (Schreiber and Schönherr, 1992), expressed as the cuticle/water partition coefficient (K_{cw}). Furthermore, permeances measured with conifers were highly correlated to permeances calculated from the model established investigating solute permeability of isolated CM of *Citrus aurantium* L. (Kerler and Schönherr, 1988). Thus, if molecular size does not vary over more than a factor of two, cuticular permeabilities can be predicted from the lipophilicity of the compounds.

Nevertheless, dependence of P on the solute mobility D within the transport-limiting barrier of the plant cuticle can be shown by either correcting measured P values for lipophilicity or by measuring mobilities of solutes within the transport-limiting barrier of the cuticle applying an experimental set-up which measures cuticular permeability independently from solute lipophilicity. When permeances of the conifers (Schreiber and Schönherr, 1992) are divided by K_{cw}, a significant linear correlation between P/K and the molar volume is obtained (Figure 4.1), demonstrating the importance of the solute mobility within the transport-limiting barrier of the cuticle. Alternatively, solutes which have been sorbed to the inner amorphous compartment of the cuticle can be desorbed from the cuticle across the outer transport-limiting wax barrier (Bauer and Schönherr, 1992; and see Chapter 6). This yields rate constants k (s^{-1}) of desorption, which again correspond to P/K and thus are independent of solute lipophilicity. Rate constants have also been shown to be strongly dependent on molar volume of the solutes, demonstrating again the important influence of solute mobility within the transport-limiting barrier of the cuticle.

4.5 Transport properties of reconstituted cuticular wax

In order to obtain a better understanding of the waxy barrier of plant cuticles, it is necessary to investigate the transport properties of cuticular waxes alone. This became possible by applying a new method, which allows the determination of

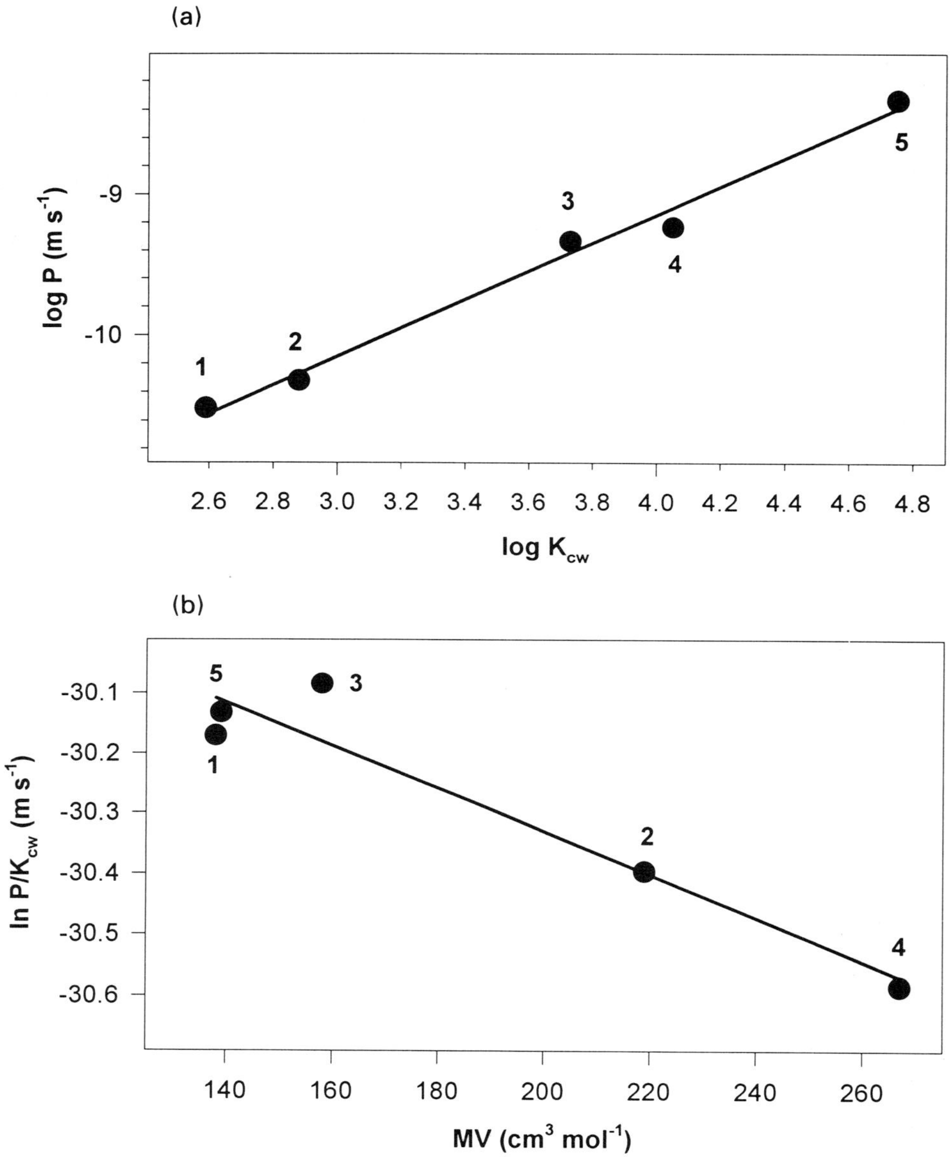

Figure 4.1. *(a) Correlation of cuticular permeability (P) of conifer needles for five compounds with cuticle/water partition coefficients (K_{cw}) of the compounds (r^2 = 0.98). (b) Correlation of cuticular permeability (P) of conifer needles divided by cuticle/water partition coefficients (K_{cw}) with the molar volumes of the compounds (r^2 = 0.93). (1 = 2,4-dichlorophenoxyacetic acid; 2 = triadimenol; 3 = lindane; 4 = bitertanol; 5 = pentachlorophenol (PCP).) (Data from Schreiber and Schönherr, 1992.)*

diffusion coefficients of radiolabelled chemicals in extracted and subsequently recrystallized waxes (Schreiber and Schönherr, 1993c).

Prior to the recrystallization, a small amount of a radiolabelled substance (0.01–0.1% by mass) is added to the wax. Subsequently, the radioactivity is desorbed again from the wax sample. The desorption kinetics measured were non-linear and asymptotically approached 100% desorption. The lower the rate of desorption the higher is the resistance of the wax towards the mobility of the radiolabelled probe. The application of the appropriate form of Fick´s second law of diffusion (Felder and Huvard, 1980) allows the estimation of molecular diffusion coefficients:

$$\frac{M_t}{M_0} = \frac{4}{\Delta x} \cdot \sqrt{\frac{D}{\pi}} \cdot \sqrt{t}. \tag{4.4}$$

M_t/M_0 represents the fractional amount of substance desorbed at time t (s), Δx (m) represents the path length of diffusion in the wax, which is taken in a first approximation as the thickness of the wax layer, D ($m^2\ s^{-1}$) is the molecular diffusion coefficient and t (s) is the time. By plotting M_t/M_0 vs. $t^{1/2}$, a linear graph is obtained up to 50% desorption. The slope of this graph is equal to $4/\Delta x \cdot (D/\pi)^{1/2}$. Thus, by rearranging Equation 4.4 the molecular diffusion coefficient D, a quantitative measure of the transport properties of the respective wax sample, can be estimated.

Using ^{14}C-labelled octadecanoic acid as a molecular probe the transport properties of cuticular waxes of 24 different plant species have been characterized so far (Schreiber and Riederer, 1996a). Values of D ranged from 10^{-19} to $10^{-17}\ m^2\ s^{-1}$. Lowest diffusion coefficients were observed in the wax of plants with perennial leaves, intermediate D were found in the wax of plants with deciduous leaves and highest D occurred in fruit waxes. In order to put these findings into perspective, diffusion coefficients in cuticular waxes can be compared to D of biomembranes which are in the range of $10^{-12}\ m^2\ s^{-1}$ (Stein, 1986), and thus five to seven orders of magnitude higher than D in cuticular waxes. This comparison clearly shows that diffusion coefficients in waxes are extremely low. This is a direct consequence of the highly ordered, solid-crystalline arrangement of the waxes and thus demonstrates the high efficiency of the cuticular transport barrier.

During evolution, the occurrence of the water-resistant plant cuticle certainly was one of several important prerequisites for the colonization of the land by plants (Takhtajan, 1969; and see Chapter 1), since one of the most important functions of the plant cuticle is the protection against uncontrolled water loss. Under moderate environmental conditions, transpiration from leaf surfaces of higher plants is controlled by stomata. In situations of severe water stress, however, stomata have to be kept closed for longer periods of time and rates of water loss from leaf surfaces of higher plants are essentially determined by the water permeability of the plant cuticles (cuticular transpiration; see also Chapter 5). Therefore, water permeabilities of enzymatically isolated plant cuticles of the 24 plant species whose waxes had been used for determining diffusion coefficients were also measured according to a method of Schönherr and Lendzian (1981). This study was undertaken with the aim of evaluating whether mobilities of the ^{14}C-labelled octadecanoic acid obtained with isolated and subsequently reconstituted cuticular waxes indeed reflected transport properties of intact plant CM containing cuticular waxes in their native conditions.

Water permeabilities of isolated CM were neither correlated to wax coverage (μg cm^{-2}) of the cuticles nor to their thickness (Schreiber and Riederer, 1996b).

However, it was discovered that water permeabilities measured with intact CM were highly correlated to the mobilities of radiolabelled octadecanoic acid in isolated cuticular wax of the same plant species (Schreiber and Riederer, 1996a). This gives new insights into the molecular organization of the waxy transport barrier and allows us to draw several important conclusions.

(a) The model system, simulation of transport properties of plant cuticles using isolated cuticular waxes, indeed reflects transport properties of the intact CM.
(b) Applying this method, it is possible to show directly that barrier properties of intact cuticles are almost solely a function of cuticular waxes. This underlines the importance of Equation 4.3 as a practicable working hypothesis, treating the transport property of the intact cuticle as a membrane composed of barriers in series, one with a high and another with low contribution to overall permeability.
(c) The fact that isolated cuticular waxes reflect the properties of intact CM leads to the conclusion that the cuticular wax barrier must be a self-organizing system. Obviously, cuticular waxes spontaneously arrange themselves at the interface leaf surface/atmosphere forming the effective waxy transport barrier. Thus, the molecular structure and consequently the transport properties of cuticular waxes are in good agreement between intact plant cuticles and recrystallized wax samples.
(d) These results provide evidence that irrespective of the size and the polarity of the compounds (here water and octadecanoic acid) the waxy transport barrier of the cuticle behaves like a sorption/diffusion membrane. There is no need for the assumption of pores or ectodesmata (Franke, 1967) in order to describe and understand the permeability of the CM for water and solutes.
(e) Finally, this experimental system will be helpful to estimate cuticular permeability from leaves where an isolation of the cuticle is not possible or where the cuticles have stomata, for example on the abaxial side of hypostomatic leaves.

By means of this experimental system it became possible to analyse the mechanisms of cuticular transport on a molecular level. In recent investigations dealing with molecular diffusion in human skin (Potts and Guy, 1992) and isolated CM (Schönherr and Baur, 1994) the following model was suggested for analysing lipid transport barriers:

$$D = D_0 \cdot e^{-\beta \cdot MV}. \tag{4.5}$$

D ($m^2\ s^{-1}$) is the diffusion coefficient of the respective compound, D_0 ($m^2\ s^{-1}$) is the diffusion coefficient of a molecule with a molar volume of zero, β (mol cm^{-3}) is the size selectivity describing the dependence of D on the size of the molecules, and MV ($cm^3\ mol^{-1}$) is the molar volume of the investigated molecules. The great advantage of Equation 4.5 lies in the two parameters β and D_0, both of which can directly be related to distinct physical structures of the wax. On the basis of recent spectroscopic investigations of synthetic as well as natural waxes the molecular structure of wax is composed in principle of two distinct phases, an amorphous and a crystalline phase (Basson and Reynhardt, 1992; Reynhardt and Riederer, 1991, 1994). The crystalline wax phase is thought to consist of the linear long-chain aliphatic molecules forming more or less perfect crystals. The amorphous wax phase should be composed of the chain ends of the aliphatic molecules. Diffusion across the waxy transport barrier is thought to occur solely in the amorphous wax phase. On the basis of these investigations, a molecular model for the diffusion across the waxy transport barrier has been suggested (Riederer and Schreiber, 1995). The amorphous phase provides a tortuous diffusional path across the wax, whereas the crystalline wax phases can be regarded as excluded volumes with respect to diffusion.

Size selectivity, β, is a measure of the properties of the amorphous wax phase, whereas D_0 characterizes the crystalline wax phase. Measuring diffusion coefficients of various molecules in reconstituted cuticular waxes of *Fagus sylvatica* L. (beech) and *Picea abies* (Karst.) L. (spruce) revealed that diffusion coefficients were, in most cases, significantly lower in spruce than in beech (Schreiber *et al.*, 1996). The size selectivity, β, of both waxes, however, was equal, indicating that the molecular structure of the amorphous phase of both plant waxes was similar. However, D_0 was significantly lower in spruce compared to beech wax indicating that the degree of crystallinity and/or the tortuosity within the wax was larger in spruce than in beech. Thus, comparing these two species it can be concluded that the barrier properties of spruce wax were enhanced in comparison to that of beech by an increased crystallinity of the wax and the resulting higher tortuosity for the permeating solutes within the wax. Since the true path length of diffusion is not known, diffusion coefficients measured by this method, taking the thickness of the wax layer for the calculation of D, are 'apparent diffusion coefficients', instead of the true diffusion coefficients relevant for the amorphous wax phase.

Recently, a very promising approach estimating the path length of diffusion across the transport-limiting barrier in the plant cuticle has been suggested (Baur *et al.*, 1996; Schönherr and Baur, 1994). Measuring rate constants of desorption from intact CM and polymer matrix membranes (MX), where cuticular wax had been removed, revealed that size selectivities in both types of membranes (CM and MX) were equal, while D_0 was by more than three orders of magnitude higher in the MX than in the CM. This clearly shows that the path of diffusion of solutes within the transport-limiting barrier of the cuticle has a large tortuosity which is significantly reduced, if not completely removed, when waxes are extracted. A similar type of composed membrane can also be found for technical membranes (Cussler *et al.*, 1988), where an amorphous polymer membrane contains impermeable flakes of mica leading to a substantial increase of the path length of diffusion and thus to a significant reduction of the apparent diffusion coefficients, calculated by taking the thickness of the membrane as Δx.

4.6 General conclusions

The diffusion across the plant cuticle follows basic laws of passive diffusion across lipophilic membranes (Cussler, 1984; Stein, 1986). Models of cuticular permeability obtained from investigations of isolated CM are also valid for intact leaves. Working with isolated, recrystallized cuticular wax is a suitable experimental system for mechanistic investigations of cuticular transport properties on a molecular level. Thus, by the consequent application of different experimental techniques (intact leaves, isolated cuticles and reconstituted cuticular wax) it becomes possible to analyse and understand the cuticular transport barrier, from the organ level represented by the intact leaf over isolated cuticles, down to the molecular level represented by the transport properties of reconstituted cuticular wax.

References

Baker, E.A. (1982) Chemistry and morphology of plant epicuticular waxes. In: *The Plant Cuticle* (eds D.F. Cutler, K.L. Alvin and C.E. Price). Academic Press, London, pp. 139-165.

Barthlott, W. (1990). Scanning electron microscopy of the epidermal surface in plants. In: *Scanning Electron Microscopy in Taxonomy and Functional Morphology* (ed. D. Claugher). Clarendon Press, Oxford, pp. 69-94.

Basson, I. and Reynhardt, E.C. (1992) The structure and melting of paraffinic Fischer-Tropsch waxes. *Chem. Phys. Lett.* **198**, 367-372.

Bauer, H. and Schönherr, J. (1992) Determination of mobilities of organic compounds in plant cuticles and correlation with molar volumes. *Pestic. Sci.* **35**, 1-11.

Baur, P., Marzouk, H., Schönherr, J. *et al.* (1996) Mobilities of organic compounds in plant cuticles as affected by structure and molar volumes of chemicals and plant species. *Planta* **199**, 404-412.

Bondi, A. (1964) Van der Waals volumes and radii. *J. Phys. Chem.* **68**, 441-451.

Brunauer, S., Emmett, P.H. and Teller, E. (1938) Adsorption of gases in multimolecular layers. *J. Am. Chem. Soc.* **60**, 309-319.

Bukovac, M.J. and Petracek, P.D. (1993) Characterizing pesticide and surfactant penetration with isolated plant cuticles. *Pestic. Sci.* **37**, 179-194.

Chamel, A. (1986) Foliar absorption of herbicides: study of the cuticular penetration using isolated cuticles. *Physiol. Végétale* **24**, 491-508.

Cussler, E.L. (1984) *Diffusion. Mass Transfer in Fluid Systems.* Cambridge University Press, Cambridge.

Cussler, E.L., Hughes, S.E., Ward, W.J. *et al.* (1988) Barrier membranes. *J. Membr. Sci.* **38**, 161-174.

Felder, R.M. and Huvard, G.S. (1980) Permeation, diffusion, and sorption of gases and vapors. In: *Methods of Experimental Physics,* Vol. 16 (eds L. Marton and C. Marton). Academic Press, New York, pp. 315-377.

Franke W. (1967) Mechanisms of foliar penetration of solutions. *Annu. Rev. Plant Physiol.* **18**, 281-300.

Gregg, S.J. and Sing, K.S.W. (1982) *Adsorption, Surface Area and Porosity.* Academic Press, London.

Holloway, P.J. (1982) Structure and histochemistry of plant cuticular membranes:an overview. In: *The Plant Cuticle* (eds D.F. Cutler, K.L. Alvin and C.E. Price). Academic Press, London, pp. 1-32.

Jetter, R. and Riederer, M. (1994) Epicuticular crystals of nonacosan-10-ol: in-vitro reconstruction and factors influencing crystal habits. *Planta* **195**, 257-270.

Kerler, F. and Schönherr, J. (1988) Permeation of lipophilic chemicals across plant cuticles: prediction from partition coefficients and molar volumes. *Arch. Environm. Contam. Toxicol.* **17**, 7-12.

Kerler, F., Riederer, M. and Schönherr, J. (1984) Non-electrolyte permeability of plant cuticles: A critical evaluation of experimental methods. *Physiol. Plant.* **62**, 599-602.

Lendzian, K.J. and Kerstiens, G. (1991) Sorption and transport of gases and vapors in plant cuticles. *Rev. Environm. Contam. Toxicol.* **121**, 65-128.

Potts, R.O. and Guy, R.H. (1992) Predicting skin permeability. *Pharmacol. Res.* **9**, 663-669.

Reynhardt, E.C. and Riederer, M. (1991) Structure and molecular dynamics of the cuticular wax from leaves of *Citrus aurantium* L. *J. Phys. D: Appl. Phys.* **24**, 478-486.

Reynhardt, E.C. and Riederer, M. (1994) Structures and molecular dynamics of plant waxes. II Cuticular waxes from leaves of *Fagus sylvatica* L. and *Hordeum vulgare* L. *Eur. Biophys. J.* **23**, 59-70.

Riederer, M. and Schönherr, J. (1985) Accumulation and transport of (2,4-Dichlorophenoxy) acetic acid in plant cuticles: 2. Permeability of the cuticular membrane. *Ecotox. Environ. Safety* **9**, 196-208.

Riederer, M. and Schreiber, L. (1995) Waxes: the transport barriers of plant cuticles. In: *Waxes: Chemistry, Molecular Biology and Functions* (ed. R.J. Hamilton). The Oily Press, Dundee, pp. 131-156.

Schönherr, J. (1982) Resistance of plant surfaces to water loss: transport properties of cutin, suberin and associated lipids. In: *Encyclopedia of Plant Physiology,* New Series, Vol. 12B (eds O.L. Lange, P.S. Nobel, C.B. Osmond and H. Ziegler). Springer, Berlin, pp. 153-179.

Schönherr, J. and Baur, P. (1994) Modelling penetration of plant cuticles by crop protecting agents (CPA) and effects of adjuvants on rates of penetration. *Pestic. Sci.* **42**, 185-208.

Schönherr, J. and Bukovac, M.J. (1973) Ion exchange properties of isolated tomato fruit cuticular membrane: exchange capacity, nature of fixed charges and cation selectivity. *Planta* **109**, 73–93.

Schönherr, J. and Lendzian, K.J. (1981) A simple and inexpensive method of measuring water permeability of isolated plant cuticular membranes. *Zeitsch. Pflanzenphys.* **102**, 321–327.

Schönherr, J. and Riederer, M. (1989) Foliar penetration and accumulation of organic chemicals in plant cuticles. *Rev. Environm. Cont. Toxicol.* **108**, 1–70.

Schreiber, L. and Riederer, M. (1996a) Determination of diffusion coefficients of octadecanoic acid in isolated cuticular waxes and their relationship to cuticular water permeabilities. *Plant, Cell Environ* (in press).

Schreiber, L. and Riederer, M. (1996b) Ecophysiology of cuticular transpiration: comparative investigation of cuticular water permeability of plant species from different habitats. *Oecologia* **107**, 426–432.

Schreiber, L. and Schönherr, J. (1992) Uptake of organic chemicals in conifer needles: surface adsorption and permeability of cuticles. *Environ. Sci. Technol.* **26**, 153–159.

Schreiber, L. and Schönherr, J. (1993a) Uptake of two chlorinated chemicals in conifer needles: reversibility and compartmental analysis. *New Phytol.* **123**, 547–554.

Schreiber, L. and Schönherr, J. (1993b) Contact areas between leaf surfaces and aqueous solutions: quantitative determination of specific leaf surface contact areas. *J. Exp. Bot.* **44**, 1653–1661.

Schreiber, L. and Schönherr, J.(1993c) Mobilities of organic compounds in reconstituted cuticular wax of barley leaves: determination of diffusion coefficients. *Pestic. Sci.* **38**, 353–361.

Schreiber, L., Kirsch, T. and Riederer, M. (1996) Transport properties of cuticular waxes of *Fagus sylvatica* L. and *Picea abies* (L.) Karst.: Estimation of size selectivity and tortuosity from diffusion coefficients of aliphatic molecules. *Planta* **198**, 104–109.

Schulze, E.D. (1982) Plant life forms and their carbon, water and nutrient relations. In: *Encyclopedia of Plant Physiology*, New Series, Vol. 12B (eds O.L. Lange, P.S. Nobel, C.B. Osmond and H. Ziegler). Springer, Berlin, pp. 615–676.

Sitte P. and Rennier R. (1963) Untersuchungen an cuticularen Zellwandschichten. *Planta* **60**, 19–40.

Stein W.D. (1986) *Transport and Diffusion Across Cell Membranes*. Academic Press, London.

Takhtajan, A. (1969) *Flowering Plants. Origin and Dispersal.* Oliver and Boyd, Edinburgh.

Tukey, H.B. (1970) The leaching of substances from plants. *Annu. Rev.Plant Physiol.* **21**, 305–324.

Tyree, M.T., Wescott, C.R., Tabor, C.A. *et al.* (1992) Diffusion and electric mobility of KCl within isolated cuticles of *Citrus aurantium*. *Plant Physiol.* **99**, 1057–1061.

Walton, T.J. (1990) Waxes, cutin and suberin. *Methods Plant Biochem.* **4**, 105–158.

5

Diffusion of water vapour and gases across cuticles and through stomatal pores presumed closed

G. Kerstiens

5.1 Introduction

The cuticular envelope enclosing most above-ground, non-woody plant organs provides essential protection against desiccation. Stomatal closure can lower leaf conductance by between 94% (herbaceous shade plants and wild graminoids) and 99.7% (succulents) of its maximum value (Körner, 1994). Whereas averages of maximum conductance within functional groups vary by only a factor of three between groups, those of minimum conductance (g_{min}) vary almost 50-fold (Körner, 1994). This raises the question how much water loss from incompletely closed stomata contributes to g_{min} in addition to transpiration across the cuticle. Specific problems of interpretation may be associated with the various ways of measuring g_{min} or cuticular permeance to water (P_{H2O}). Similar uncertainties exist with respect to minimum deposition velocities of and cuticular permeance to various gases. These are the topics that will be discussed in this chapter.

5.2 Cuticular water permeance and minimum conductance

Cuticular permeance, stomatal or overall conductance (the latter including cuticular permeance) and deposition velocity are all defined as the quotient of flow rate density (g m^{-2} s^{-1}) and the difference in concentration (g m^{-3}) of water vapour or any other gas between leaf interior and leaf boundary layer. The resulting unit is m s^{-1}. All data refer to total leaf surface (twice the projected leaf area). Stomatal conductances are usually expressed in terms of molar units today (using the difference in mole fraction of water vapour rather than concentration as the driving force), but the usage of these units should be avoided to describe permeances (Kerstiens, 1996b). The driving force for diffusion through a solid membrane like the cuticle is the concentration gradient, not the gradient in mole fraction. A concentration-based value of 1.0×10^{-5} m s^{-1} corresponds to a mole fraction-based value of 0.41 mmol m^{-2} s^{-1} at 25°C, provided the latter refers to total leaf surface.

Plant Cuticles, edited by G. Kerstiens.

5.2.1 *Experimental techniques*

Category A: Water vapour flux from stomata-free systems. This group unites measurements with isolated astomatous cuticles and direct measurements of water vapour flux from astomatous leaf surfaces. Any contribution from stomata to water flux was physically impossible. The two most frequently used techniques are: (i) mounting of enzymatically isolated cuticles on small vessels filled with water and measurement of water loss from the vessels sealed by the cuticle over periods of several weeks under controlled conditions (Schönherr and Lendzian, 1981; Schreiber and Riederer, 1996); (ii) airtight attachment of a cup to the leaf surface and measurement of the water flux into the cup (Kerstiens, 1995; Kerstiens and Lendzian, 1989b). Two problems may contribute to the underestimation of 'true' values of P_{H2O} (i.e. *in vivo* in the field) by these techniques; these are the decrease of permeance of isolated cuticles during storage (Geyer and Schönherr, 1990) and the effect of air humidity on permeance (Gäumann and Jaag, 1936; Schönherr, 1982; van Gardingen and Grace, 1992). For experimental reasons, virtually all experiments in this group were carried out with 0% relative humidity (r.h.) at the outer surface of the cuticle. Both effects are discussed in detail by Kerstiens (1996b). In combination they may lead to an underestimate of the 'true' P_{H2O} by a factor of 2–3.

Category B: Transpiration decline curves of hypostomatous leaves with sealed abaxial side. Hypostomatous leaves are detached and allowed to desiccate over a period of time under monitored and more or less constant conditions of temperature and air humidity. The abaxial side is covered with low-melting paraffin, petroleum jelly (petrolatum) or vaseline. Following an initial steep decrease in leaf mass while stomata close, the rate of mass loss usually reaches a constant level which is maintained for several hours. Hygen (1951) suggested that the rate of water loss following stomatal closure is not constant, but decreases with time because of the decrease in internal water concentration of the leaf (which is a strong simplification of the true relationships). Therefore he recommended that g_{min} be calculated from a plot of log (mass) vs. time rather than mass vs. time. However, the effect of declining leaf water content on driving force is usually minimal (Kerstiens, 1996b) and makes it impossible to distinguish between linear and logarithmic decline of leaf mass within the experimental error. Although the Hygen method may be useful if transpiration rates are very low and experiments are conducted over a period of several days, it is my feeling that if usage of the Hygen technique is *necessary* in a typical short-term experiment (up to several hours) because the decline curve cannot otherwise be linearized, something is wrong with the experiment anyway. The most likely reason is that the relative water content is falling so fast that the decline in water content does indeed affect the driving force for water loss. This should not normally occur unless the water loss during the initial phase of stomatal closure was very high or stomata close so imperfectly that the rate remains relatively high all the time almost until complete desiccation. In both cases, leaf water content will be very far away from physiological values during the period the 'cuticular' transpiration is determined.

Often the effectiveness of the sealant on the abaxial leaf surface is not evaluated. This could be achieved, for instance, by covering both leaf surfaces and measuring mass loss. With pubescent, undulated and/or highly dissected leaves it may be quite difficult to accomplish a perfect cover, that is one *completely* devoid of gaps that would allow some stomatal transpiration. The contribution of stomatal transpiration may otherwise be quite high in relative terms, as will be demonstrated below. I am aware of only one measurement of the water permeance of a vaseline cover itself, by Kamp (1930). He found a value of approximately 3×10^{-6} m s^{-1}, which may not be representative for any other experiment.

Recently, Hoad *et al.* (1996) introduced a very promising new technique. The abaxial side is not covered by the application of a sealant, but leaf discs are sealed within a very impermeable envelope which exposes part of the adaxial surface. The envelope also contains a small water reservoir to delay the decline in leaf water content.

Category C: Water vapour flux from hypostomatous leaves with sealed abaxial side. This category comprises only a single paper, that by Holmgren *et al.* (1965). These authors sealed the underside of hypostomatous leaves with vaseline and measured gas exchange rates of leaves enclosed in a cuvette.

Category D: Transpiration decline curves of unsealed leaves. These experiments are carried out as described for category B, with the exception that water can transpire freely from both leaf surfaces.

Category E: Water vapour flux from unsealed leaves, stomata presumed closed. This category unites all experiments with intact leaves or detached, but not desiccating leaves mounted in a porometer. Water vapour flux was measured with whole-leaf cuvettes, by applying a porometer head to one stomata-bearing surface, or by measuring water loss from the potometer. In order to maximize stomatal closure, measurements were taken in darkness, under high CO_2 concentration, with severely water-stressed plants or after feeding a high concentration of abscisic acid in potometer experiments. The results of Moreshet (1970) and Šantrůček (1991), who attempted to measure cuticular transpiration of a leaf with open stomata, have also been included in this group. Moreshet (1970) measured conductances (including cuticular permeance) of adaxial and abaxial leaf surfaces and the combined resistance to viscous flow of argon across amphistomatous leaves in the light (*Helianthus annuus*: $g_{min} = 8 \times 10^{-5}$ at 30% relative air humidity). The argon flow depended on stomatal conductance only, which was deducted from the overall conductance to obtain cuticular permeance. Šantrůček (1991) forced a mass flow of air across amphistomatous leaves to suppress stomatal transpiration while measuring transpiration rate of the surface exposed to high pressure. As Šantrůček (1991) points out, however, it is unclear how much stomata contributed to the observed minimum conductances (four species; $g_{min} = 4.5 \times 10^{-5}$ - 6.7×10^{-4} m s^{-1} at low r.h.).

5.2.2 *Comparison of results obtained with different techniques*

It was demonstrated recently that there is a large difference between the bulk of P_{H2O} values determined with astomatous leaf cuticles in systems where any contribution from stomatal transpiration could be completely excluded (category A) and field measurements of g_{min} when environmental conditions favoured maximum stomatal closure (category E) (Kerstiens, 1996b). Transpiration decline curves of detached, desiccating leaves (categories B and D) yielded results that lay between the other two groups. There was no difference between categories B and D. Apparently, it does not matter whether stomata-bearing surfaces of hypostomatous leaves are sealed or not. That review pooled results for 200 species, without taking account of the type of leaf or plant (e.g. evergreen, xeromorphic leaves; conifer needles; herbaceous, broad leaves; grass blades) or its relative abundance in each category. Figures 5.1a and b, respectively, present the same type of analysis for subsets of 121 more or less mesophytic species (including all graminoid species) and 61 evergreen and xeromorphic or succulent species. Both analyses contain multiple entries from different studies for some species. Conifers have been excluded from the analysis because there are no data for categories A–C. The distribution pattern of values in the experimental categories of each subset is similar to that of the complete data set. The persistent lack of congruence between categories A and E indicates these techniques are likely to have measured different aspects of minimum leaf conductance, namely cuticular permeance on the one hand and maximum (but usually incomplete) stomatal closure on the other.

Although there is a possibility that ageing of isolated cuticles and their relatively low moisture content in most experiments of category A may have led to an underestimate of P_{H2O} under field conditions by a factor of 2–3 (Kerstiens, 1996b), this cannot explain the large difference between categories A and E in Figure 5.1. Instead, stomata may not always, or perhaps hardly ever, shut completely. Figure 5.2 illustrates the contribution of cuticular permeance to g_{min} in the presence of very small residual stomatal apertures (assuming no differences between individual stomata). The values of P_{H2O} used were 5 and 15×10^{-5} m s^{-1}. They represent approximately the 20th and 80th percentile of values in category A of Figure 5.1a, corrected for possible underestimation in the experiment (i.e. original values multiplied by 3). These values also correspond to g_{min} values of *Avena sativa* (cultivated oat) determined by Bengtson *et al.* (1978). Stomatal conductance was calculated using the formula of Parlange and Waggoner (1970), assuming a slit-like pore, and data for stomata of *Avena fatua* (wild oat) from van Gardingen *et al.* (1989). For aperture widths up to and including 0.1 μm, the Knudsen diffusion coefficient was used (Cussler, 1984). The transition between Knudsen diffusion (when water molecules collide with the aperture walls more often than with other gas molecules) and unhindered diffusion occurs at a pore width of 0.05–0.1 μm. Using the Knudsen diffusion coefficient up to 0.1 μm gives a slightly more conservative estimate for the stomatal contribution to overall conductance. Figure 5.2 clearly shows that stomata may contribute the major share of g_{min} when pores appear completely closed. Even with the most powerful observation technique, the scanning electron microscope, it is unlikely that apertures less than 0.1–0.2 μm wide would be detected.

The two species for which estimates of P_{H2O} and g_{min} exist from the widest range of techniques and number of studies are *Fagus sylvatica* and *Hedera helix*. As Table 5.1 demonstrates, agreement between different investigations is quite good for both species. However, these cases seem to be the exception rather than the rule. Values for *Coffea arabica*, for instance, vary from 2×10^{-6} m s^{-1} (Garrec and Plebin, 1986) to 1.5×10^{-4} m s^{-1} (Kamp, 1930). Estimates of g_{min} for *Sorghum bicolor* and *Zea mays*, to take two examples of amphistomatous plants for which there are no category A–C data, range from 1×10^{-5} m s^{-1} (Premachandra *et al.*, 1992) to 3.4×10^{-4} m s^{-1} (Saneoka and Ogata, 1987), and from 1×10^{-5} m s^{-1} (Dubé *et al.*, 1975) to 4.0×10^{-4} m s^{-1} (Knittel and Pell, 1991), respectively.

The fact that many evergreen, xeromorphic leaves have thick cuticles and low water permeances lies probably at the heart of the still widespread misconception that thicker cuticles are better water barriers. In the data set shown in Figure 5.1, evergreen leaves had significantly ($p < 0.001$) lower values of P_{H2O} or g_{min} than deciduous ones if determined in stomata-free systems (category A) or with transpiration decline curves (results in categories B and D pooled for the analysis; Mann-Whitney rank sum test). However, the overall thickness of the cuticle is *not*

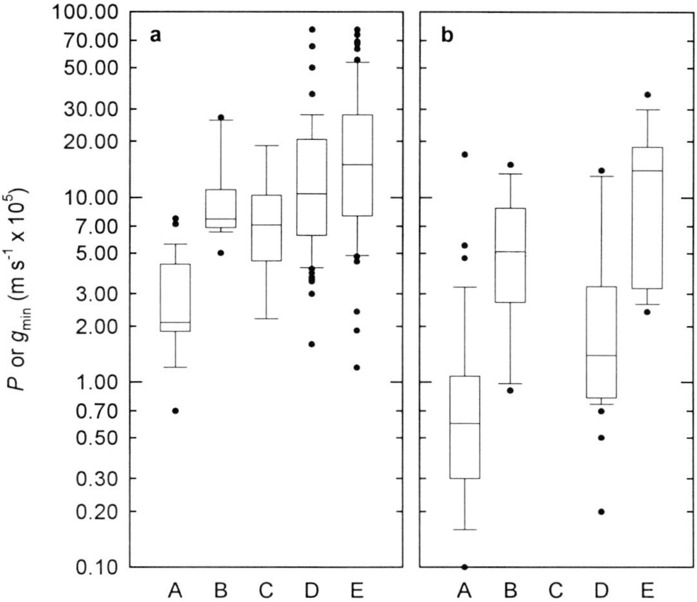

Figure 5.1. *Distribution of* P_{H2O} *and* g_{min} *values of non-xeromorphic deciduous leaves (a) and non-coniferous evergreen leaves (b) obtained with different experimental techniques. See text for description of categories A-E. Boxes represent 25th-75th percentiles and capped bars the 10th and 90th percentiles. The line inside boxes represents the median. Dots represent data outside the 10th-90th percentiles. Number of data in each category (deciduous/evergreen species): A, 25/31; B, 14/13; C, 5/0; D, 87/31; E, 58/9. For both types of leaves, the distribution of data in category A is significantly different from category E and pooled categories B and D, but not between B+D and E (two-sided Kolmogorov-Smirnov test). A detailed list of data and techniques can be found in Kerstiens (1996b). (Data from Araus* et al., *1991; Bakker, 1991; Bataillé* et al., *1990; Becker* et al., *1986; Bengtson* et al., *1978; Benzing and Burt, 1970; Blom-Zandstra* et al., *1995; Clarke and Richards, 1988; Clarke* et al., *1991; Cochrane and Slatyer, 1988; Darwin, 1916; Davies and Kozlowski, 1974; Dubé* et al., *1975; Ellsworth* et al., *1995; Febrero and Araus, 1994; Garrec and Kerfourn, 1989; Garrec and Plebin, 1986; Gäumann and Jaag, 1936; Gauslaa, 1990; Grace, 1974; Greger and Johansson, 1992; Haas and Schönherr, 1979; Hajibagheri* et al., *1983; Häne* et al., *1982; Hoad* et al., *1994, 1996; Holbrook and Sinclair, 1992; Holmgren* et al., *1965; Jefferson* et al., *1989; Jenks* et al., *1994; Jordan and Ritchie, 1971; Jordan* et al., *1984; Kamp, 1930; Kerstiens, 1994, 1995, 1996b; Kerstiens and Lendzian, 1989b; Kirschbaum and Pearcy, 1988; Knittel and Pell, 1991; C. Körner (personal communication); Kuiper, 1961; Larsson and Svenningsson, 1986; Lendzian and Kerstiens, 1991; Meinzer* et al., *1990; Mooney and Miller, 1985; Moreshet, 1970; Morrissey* et al., *1993; Muchow and Sinclair, 1989; Nagarajah, 1979; Nazrul-Islam, 1983; Oosterhuis* et al., *1988; O'Toole* et al., *1979; Paje* et al., *1988; Pallardy and Rhoads, 1993; Pisek and Berger, 1938; Premachandra* et al., *1992; Quisenberry* et al. *(1982; Rawson and Clarke, 1988; Reiling and Davison, 1995; Saneoka and Ogata, 1987; Šantrůček, 1991; Schmidt* et al., *1981; Schönherr and Lendzian, 1981; Schreiber and Riederer, 1996; Shepherd, 1964; Sinclair and Ludlow, 1986; Slatyer and Bierhuizen, 1964; Stickan and Zhang, 1992; van Gardingen and Grace, 1992; von Stamm, 1994; Wang* et al., *1988; Wedler* et al., *1996; West and Gaff, 1976; Williams and Amer, 1957; Yoshida and de los Reyes, 1976.)*

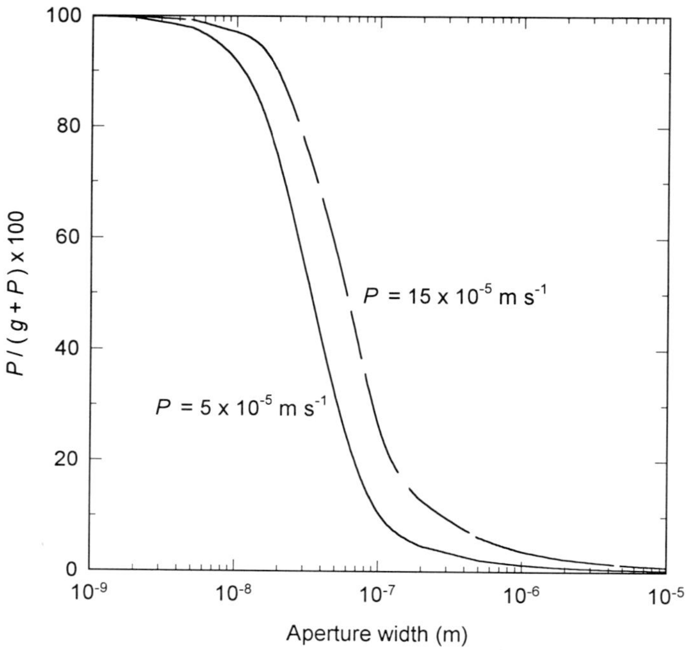

Figure 5.2. *Relative contribution of cuticular water permeance* P *to overall conductance (stomatal conductance* g *plus permeance* P*) as a function of stomatal aperture. See text for details of calculations.*

related to its permeance (Becker *et al.,* 1986; Kamp, 1930; see Chapter 4). Very thin cuticles can be very good barriers, and vice versa. Long-living leaves tend to direct relatively more resources into initially expensive, long-lasting defence structures, and a thick cuticle may be one of them (Coley, 1988; Kerstiens, 1996a). The development of low cuticular water permeance, possibly as an adaptation to the water-limited habitats of many xerophytes, should be regarded as a completely independent process.

5.3 Cuticular permeance to pollutant gases and minimum deposition velocities

5.3.1 *Dependence of cuticular permeance on gas and cuticle properties*

Permeance of isolated *Citrus aurantium* leaf cuticles and *Lycopersicon esculentum* (tomato) fruit cuticles to different gases and vapours is closely correlated with the boiling temperature of the latter (Kerstiens *et al.,* 1992). Permeance varies by 3-4 orders of magnitude between oxygen and water vapour, with CO_2, H_2S, SO_2, CH_3SH and NO_2 having intermediate values. Permeance of both cuticle types rises by just over an order of magnitude for each increase in boiling point of 100 K. This indicates that differences in permeance between gases are mainly determined by their solubility in the cuticle, a finding similar to the pattern observed for organic compounds (see Section 4.4). The boiling temperature is a reasonably good measure

Table 5.1. *Cuticular permeance (P_{H2O}) and minimum conductance (g_{min}) of* Fagus sylvatica *and* Hedera helix

Technique (category)	*Fagus sylvatica*		*Hedera helix*	
	P, g_{min}	(reference)	P, g_{min}	(reference)
Isolated cuticles (A)			0.3	(1)
Isolated cuticles (A)			0.2–0.6	(7)
WVF from adaxial surface (A)	3.7	(7)	0.3	(7)
WVF from adaxial surface (A)	1.4–2.9	(5)	0.6	(6)
MLD, abaxial surface sealed (B)	4–7	(11)	~1	(2)
MLD, abaxial surface sealed (B)	6	(3)	0.6–1.3	(4)
MLD, disc in envelope (B)	8	(3)		
MLD (D)	4.6–7.3	(9)	1.0	(9)
MLD (D)	4.2	(7)		
MLD (D)	5.6	(5)		
MLD (D)	3.4–5.6	(8)		
Nocturnal WVF in the field (E)	~1–10	(10)		

Units are 10^{-5} m s^{-1}.
WVF , water vapour flux (determined directly); MLD, mass loss of detached leaves.
Ranges denote means from different experiments or different leaf populations.
References: (1) Becker *et al.* (1986); (2) Darwin (1916); (3) Hoad *et al.* (1996); (4) Kamp (1930); (5) Kerstiens (1995); (6) Kerstiens (1996b); (7) Kerstiens and Lendzian (1989b); (8) C. Körner (personal communication); (9) Pisek and Berger (1938); (10) Stickan and Zhang (1992); (11) van Gardingen and Grace (1992).

of gas condensability, that is, of the forces acting between identical gas molecules, which in turn to some degree predict the strength of interactions between the cuticle and gas molecules dissolved in its polymer network (Bixler and Sweeting, 1971); the stronger the interaction, the higher the ratio of gas concentration in the cuticle itself and in the medium surrounding the cuticle.

Ozone has a lower permeance than predicted from this relationship because it decomposes while crossing the cuticle (Kerstiens and Lendzian, 1989a). In fact, ozone permeance can only be measured if extremely high concentration differences are applied across the cuticle. For environmentally relevant ozone concentrations it can safely be said that cuticles are completely impermeable to this pollutant.

Cuticular permeance to a particular gas may depend on its own concentration in the atmosphere and on the concentrations of other gases and vapours, in particular those which are more easily condensable (e.g. water vapour). These interactions have not been investigated systematically. The reader is referred to Lendzian and Kerstiens (1991) and Kerstiens *et al.* (1992) for a theoretical discussion.

5.3.2 *Comparison between cuticular permeance and minimum deposition velocity*

The deposition velocity (v_d) describes the rate of transport of a gas from the atmosphere towards the ground. It is usually based on ground area; for instance, a v_d of 2 mm s^{-1} is obtained if at an atmospheric concentration of 40 μg m^{-3} a downward flux of 80 ng s^{-1} m^{-2} land surface is observed and the concentration is 0 at ground level. In order to compare v_d with conductances or cuticular permeances, it has to be based on total leaf surface. In the case of deposition to a canopy in the field, this parameter would be obtained by dividing v_d by twice the leaf area index (the ratio of projected leaf area to ground area), assuming that deposition to other sinks (soil, trunks, branches) was negligible. This is usually not the case in the field. Unless the contribution of other sinks to overall deposition rate can be estimated and deducted, laboratory studies where (most) other sinks are excluded are the only way to assess uptake by leaves. The deposition velocity based on total leaf area is called surface flux velocity (SFV) if the following conditions are fulfilled: any other sinks are absent, negligible or being deducted; environmental conditions favour stomatal closure; the uptake by the leaf (including adsorption to the surface) is not notably limited by gas diffusion in the atmosphere (Kerstiens *et al.*, 1992).

Table 5.2 summarizes SFV values obtained with a range of different experimental approaches, both field studies with enclosed branches (Grennfelt *et al.*, 1983; Hällgren *et al.*, 1982; Johansson *et al.*, 1983) and various laboratory experiments. In the case of field studies, data represent nocturnal values, and other sinks were considered negligible. The important point that comes out of this survey is that no value of SFV was lower than $1\text{–}2 \times 10^{-5}$ m s^{-1} for either SO_2 or NO_2. Similarly, the lowest SFV for NO_2 ever observed under daytime conditions was 1×10^{-5} m s^{-1} (Table 1 in Hanson and Lindberg, 1991). The cuticular permeances for these gases are much lower (Table 5.2). What may have caused this apparent contradiction? Obviously, failure of stomata to close completely at night is one. However, several studies have plotted v_d vs. conductance (obtained from transpiration measurements) and found a positive intercept on the y-axis, indicating a certain flux when transpiration rate was zero. The uncertainty in the regression analysis suggests that this intercept may not always have been significantly different from zero. If it was, however, it would indicate a non-stomatal flux. This could either be a sink in the system that has wrongly been assumed to be negligible, or it could indicate accumulation on the leaf surface. It cannot, however, indicate uptake through the cuticle unless the permeance to the gas in question is greater than to water. This is because cuticular transpiration contributes to conductance which was used in the regression analysis. For physical reasons, it is extremely unlikely that permeance to SO_2 or NO_2 is higher than water permeance: as solubility of SO_2 and NO_2 in cuticles (and in any other polymer) is always lower than that of water because of their lower condensability, their diffusion coefficient would have to be much larger than that of water (see Section 4.2). All measurements with isolated cuticles show lower permeances to SO_2 and NO_2 than to water vapour. The covalent binding of NO_2 to the cuticle reported by Kisser–Priesack *et al.* (1987) and Lendzian and Kerstiens (1991) occurs at very low rates and cannot explain a large difference between uptake into the cuticle and transport across it, that is, between SFV and permeance (Kerstiens *et al.*, 1992).

One other possibility that could lead to trans-cuticular uptake rates approaching SFV is that the gases do not diffuse across the cuticle as SO_2 and NO_2, but in a chemically different form which is more soluble in the cuticle. This could not be an anion, because the cutin polymer contains carboxyl groups which are dissociated at physiological pH. Therefore, one would expect the cuticular matrix virtually to exclude anions, resulting in a very low permeance (Tyree, 1994). As far as NO_2 is concerned, one can speculate that oxidation to HNO_3 might be involved. I am not aware of any measurements related to cuticular permeance to HNO_3 vapour except those of Marshall and Cadle (1989) who found steady-state flux constants of 3–6 $\times 10^{-4}$ m s^{-1} with dormant branches of *Pinus strobus*. It is, however, not clear if

Table 5.2. *Cuticular permeance (P) and surface flux velocity (SFV) for SO_2 and NO_2*

Species	SO_2	NO_2	Reference
	P or SFV (10^{-4} m s^{-1})		
Permeances			
Citrus aurantium	0.002	0.018	Lendzian and Kerstiens (1988)
Citrus aurantium	0.006		Lendzian and Kerstiens (1991)
Capsicum annuum fruit	0.004	0.009	Lendzian and Kerstiens (1988)
Abies alba	0.008	0.056	Lendzian and Kerstiens (1988)
Lycopersicon esculentum fruit	0.012	0.015	Lendzian and Kerstiens (1988)
Lycopersicon esculentum fruit	0.09		Lendzian and Kerstiens (1991)
Surface flux velocities			
Seven ornamental species		0.10–0.60	Saxe (1986)
Hordeum vulgare	0.19–1.2		Spedding (1969)
Sorbus aria	0.67	0.29	Elkiey *et al.* (1982)
Acer platanoides	0.75	0.33	Elkiey *et al.* (1982)
Ficus benjamina	1.0	0.58	Elkiey *et al.* (1982)
Nicotiana tabacum		0.85	Neubert *et al.* (1993)
Poa pratensis (nine cultivars)	1.4–4.5	1.7–5.2	Elkiey and Ormrod (1981)
Prunus sargentii	1.5	0.83	Elkiey *et al.* (1982)
Betula pendula	1.7	1.3	Elkiey *et al.* (1982)
Pinus sylvestris	~2		Johansson *et al.* (1983)
Phaseolus vulgaris	2.5		van Hove *et al.* (1990)
Vicia faba	2.8		Black and Unsworth (1979)
Phaseolus vulgaris	3.2		Taylor *et al.* (1983)
Quercus palustris	3.8		Dasch (1989)
Pisum sativum	3.8		Olszyk and Tingey (1985)
Ulmus americana	4.7		Dasch (1989)
Helianthus annuus		4.1	Neubert *et al.* (1993)
Pinus nigra	5.0		Dasch (1989)
Pinus sylvestris	5–11		Hällgren *et al.* (1982)
Pinus sylvestris		0.6–78	Grennfelt *et al.* (1983)
Glycine max	6.5		Taylor *et al.* (1983)
Geranium carolinianum	8–11		Taylor and Tingey (1983)
Diplacus aurantiacus	12		Olszyk and Tingey (1985)
Lycopersicon esculentum	20		Olszyk and Tingey (1985)

transport *across* the cuticle occurred in this experiment at all (Lendzian and Kerstiens, 1991). As the boiling point of nitric acid (86°C) is close to that of water, its permeance may be 1–2 orders of magnitude higher than that for NO_2. However, *if* a real non-stomatal flux to the leaf surface did occur in the experiments quoted in Table 5.2, by far the most likely explanation would be accumulation on the leaf surface. Chapter 15 reviews the processes involved.

5.4 Conclusions

It was shown that none of the various techniques used to assess 'cuticular' permeances is free from possible pitfalls. Incompletely closed stomata present the most significant problems. Possibly the best technique for the determination of permeances is to use isolated astomatous cuticles and measure the fluxes across them at a constant air humidity representing average field conditions. Where cuticles cannot be isolated intact, the envelope technique recently developed by Hoad *et al.* (1996) may be the best way forward for flat and astomatous leaf surfaces. Where there is a theoretical possibility of stomata contributing to minimum leaf conductance, it is most likely that they will do so. Diffusion rates through stomatal pores less than 0.1 μm wide may exceed cuticular rates several-fold. Even under conditions promoting maximum stomatal closure, the occurrence of purely cuticular transpiration or pollutant uptake appears to be the exception rather than the rule.

Acknowledgements

Part of the work reported here was funded by various grants and fellowships from the Commission of the European Communities and the European Environmental Research Organisation. I thank Dr Paul van Gardingen for his comments on a draft of this paper.

References

Araus, J.L., Febrero, A. and Vendrell, P. (1991) Epidermal conductance in different parts of durum wheat grown under Mediterranean conditions — the role of epicuticular waxes and stomata. *Plant Cell Environ.* **14,** 545-558.

Bakker, J.C. (1991) Leaf conductance of four glasshouse vegetable crops as affected by air humidity. *Agric. For. Meteorol.* **55,** 23-36.

Bataillé, A., Eveling, D.W., McCall, R.D. *et al.* (1990) The effects of fungicides on water loss of *Coleus blumei* leaves and *Pelargonium* hybrid petals. *Ann. Bot.* **66,** 579-586.

Becker, M., Kerstiens, G. and Schönherr, J. (1986) Water permeability of plant cuticles: permeance, diffusion and partition coefficients. *Trees* **1,** 54-60.

Bengtson, C., Larsson, S. and Liljenberg, C. (1978) Effects of water stress on cuticular transpiration rate and amount and composition of epicuticular wax in seedlings of six oat varieties. *Physiol. Plant.* **44,** 319-324.

Benzing, D.H. and Burt, K.M. (1970) Foliar permeability among twenty species of the Bromeliaceae. *Bull. Torrey Bot. Club* **97,** 269-279.

Bixler, H.J. and Sweeting, O.J. (1971) Barrier properties of films. In: *The Science and Technology of Polymer Films,* Vol. II (ed. O.J. Sweeting). Wiley-Interscience, New York, pp. 8-130.

Black, V.J. and Unsworth, M.H. (1979) Resistance analysis of sulphur dioxide fluxes to *Vicia faba*. *Nature* **282,** 68-69.

Blom-Zandstra, M., Pot, C.S., Maas, F.M. *et al.* (1995) Effects of different light treatments on the nocturnal transpiration and dynamics of stomatal closure of two Rose cultivars. *Sci. Hort.* **61,** 251-262.

Clarke, J.M. and Richards, R.A. (1988) The effects of glaucousness, epicuticular wax, leaf age, plant height, and growth environment on water loss rates of excised wheat leaves. *Can. J. Plant Sci.* **68,** 975-982.

Clarke, J.M., Richards, R.A. and Condon, A.G. (1991) Effect of drought stress on residual transpiration and its relationship with water-use of wheat. *Can. J. Plant Sci.* **71,** 695-702.

Cochrane, P.M. and Slatyer, R.O. (1988) Water relations of *Eucalyptus pauciflora* near the alpine tree line in winter. *Tree Physiol.* **4,** 45-52.

Coley, P.D. (1988) Effects of plant growth rate and leaf lifetime on the amount and type of anti-herbivore defense. *Oecologia* **74,** 531-536.

Cussler, E.L. (1984) *Diffusion. Mass Transfer in Fluid Systems.* Cambridge University Press, Cambridge.

Darwin, F. (1916) On the relation between transpiration and stomatal aperture. *Phil. Trans. R. Soc.* B **207,** 413-437.

Dasch, J.M. (1989) Dry deposition of sulfur dioxide or nitric acid to oak, elm and pine leaves. *Environ. Pollut.* **59,** 1-16.

Davies, W.J. and Kozlowski, T.T. (1974) Stomatal responses of five woody angiosperms to light intensity and humidity. *Can. J. Bot.* **52,** 1525-1534.

Dubé, P.A., Stevenson, K.R., Thurtell, G.W. *et al.* (1975) Effects of water stress on leaf respiration, transpiration rates in the dark and cuticular resistance to water vapor diffusion of two corn inbreds. *Can. J. Plant Sci.* **55,** 565-572.

Elkiey, T. and Ormrod, D.P. (1981) Sorption of O_3, SO_2, NO_2 or their mixture by nine *Poa pratensis* cultivars of differing pollutant sensitivity. *Atmos. Environ.* **9,** 1739-1743.

Elkiey, T., Ormrod, D.P. and Marie, B. (1982) Foliar sorption of sulphur dioxide, nitrogen dioxide, and ozone by ornamental woody plants. *HortScience* **17,** 358-360.

Ellsworth, D.S., Tyree, M.T., Parker, B.L. *et al.* (1995) Impact of pear thrips damage on sugar maple physiology: a whole-tree experiment. In: *Towards Understanding Thrips Management,* NATO ASI Series A, Vol. 276, (eds B.L. Parker, M. Skinner and T. Lewis). Plenum Publishing, New York, pp. 53-58.

Febrero, A. and Araus, J.L. (1994) Epicuticular wax load of near-isogenic barley lines differing in glaucousness. *Scanning Microsc.* **8,** 735-748.

Garrec, J.P. and Kerfourn, C. (1989) Effets de pluies acides et de l'ozone sur la perméabilité à l'eau et aux ions de cuticules isolées. *Environ. Exp. Bot.* **29,** 215-228.

Garrec, J.P. and Plebin, R. (1986) Perméabilité au fluorure d'hydrogène (HF) des cuticules avec ou sans stomates de feuilles: influence de la présence des stomates et comparaisons avec la perméabilité à l'eau. *Environ. Exp. Bot.* **26,** 299-308.

Gäumann, E. and Jaag, O. (1936) Untersuchungen über die pflanzliche Transpiration. *Ber. Schweiz. Bot. Ges.* **45,** 411-518.

Gauslaa, Y. (1990) Water relations and mineral nutrients in *Melampyrum pratense* (Scrophulariaceae) in oligo- and mesotrophic boreal forests. *Acta Oecol.* **11,** 525-537.

Geyer, U. and Schönherr, J. (1990) The effect of the environment on the permeability and composition of *Citrus* leaf cuticles. I. Water permeability of isolated cuticular membranes. *Planta* **180,** 147-153.

Grace, J. (1974) The effect of wind on grasses. I. Cuticular and stomatal transpiration. *J. Exp. Bot.* **25,** 542-551.

Greger, M. and Johansson, M. (1992) Cadmium effects on leaf transpiration of sugar beet (*Beta vulgaris*). *Physiol. Plant.* **86,** 465-473.

Grennfelt, P., Bengtson, C. and Skärby, L. (1983) Dry deposition of nitrogen dioxide to Scots pine needles. In: *Precipitation Scavenging, Dry Deposition, and Resuspension* (eds H.R. Pruppacher, R.G. Semonin and W.G.N. Slinn). Elsevier, New York, pp. 753-761.

Haas, K. and Schönherr, J. (1979) Composition of soluble cuticular lipids and water permeability of cuticular membranes from *Citrus* leaves. *Planta* **146,** 399-403.

Hajibagheri, M.A., Hall, J.L. and Flowers, T.J. (1983) The structure of the cuticle in relation to cuticular transpiration in leaves of the halophyte *Suaeda maritima* (L.) Dum. *New Phytol.* **94,** 125-131.

Hällgren, J.E., Linder, S., Richter, A. *et al.* (1982) Uptake of SO_2 in shoots of Scots pine: field measurements of net flux of sulphur in relation to stomatal conductance. *Plant Cell Environ.* **5,** 75-83.

Häne, B., Eller, B.M. and Wanner, H. (1982) Transpiration von 3 *Coffea*-Arten, gemessen mit verschiedenen Methoden. *Flora* **172,** 180-192.

Hanson, P.J. and Lindberg, S.E. (1991) Dry deposition of reactive nitrogen compounds: a review of leaf, canopy and non-foliar measurements. *Atmos. Environ.* **25A,** 1615-1634.

Hoad, S.P., Jeffree, C.E. and Grace, J. (1994) Effects of wind and simulated acid mist on leaf cuticles. In: *Air Pollutants and the Leaf Cuticle,* NATO ASI Series G, Vol. 36 (eds K.E. Percy, J.N. Cape, R. Jagels and C.J. Simpson). Springer-Verlag, Berlin, pp. 225-237.

Hoad, S.P., Grace, J. and Jeffree, C.E. (1996) A leaf disc method for measuring cuticular conductance. *J. Exp. Bot.* **47,** 431-437.

Holbrook, N.M. and Sinclair, T.R. (1992) Water balance in the arborescent palm, *Sabal palmetto.* I. Stem structure, tissue water release properties and leaf epidermal conductance. *Plant Cell Environ.* **15,** 393-399.

Holmgren, P., Jarvis, P.G. and Jarvis, M.S. (1965) Resistances to carbon dioxide and water vapour transfer in leaves of different plant species. *Physiol. Plant.* **18,** 557-573.

Hygen, G. (1951) Studies in plant transpiration I. *Physiol. Plant.* **4,** 57-183.

Jefferson, P.G., Johnson, D.A. and Asay, K.H. (1989) Epicuticular wax production, water status and leaf temperature in Triticeae range grasses of contrasting visible glaucousness. *Can. J. Plant Sci.* **69,** 513-519.

Jenks, M.A., Joly, R.J., Peters, P.J. *et al.* (1994) Chemically-induced cuticle mutation affecting epidermal conductance to water vapor and disease susceptibility in *Sorghum bicolor* (L.) Moench. *Plant Physiol.* **105,** 1239–1245.

Johansson, C., Richter, A. and Granat, L. (1983) Dry deposition on coniferous forest of SO_2 at ppb level. In: *Precipitation Scavenging, Dry Deposition, and Resuspension* (eds H.R. Pruppacher, R.G. Semonin and W.G.N. Slinn). Elsevier, New York, pp. 775–783.

Jordan, W.R. and Ritchie, J.T. (1971) Influence of soil water stress on evaporation, root absorption, and internal water status of cotton. *Plant Physiol.* **48,** 783–788.

Jordan, W.R., Shouse, P.J., Blum, A. *et al.* (1984) Environmental physiology of sorghum. II. Epicuticular wax load and cuticular transpiration. *Crop Sci.* **24,** 1168–1173.

Kamp, H. (1930) Untersuchungen über Kutikularbau und kutikuläre Transpiration von Blättern. *Jb. wissenschaft. Bot.* **72,** 403–465.

Kerstiens, G. (1994) Effects of low light intensity and high air humidity on morphology and permeability of plant cuticles, with special respect to plants cultured *in vitro*. In: *Physiology, Growth and Development of Plants in Culture* (eds P.J. Lumsden, J.R. Nicholas and W.J. Davies). Kluwer Academic Publishers, Dordrecht, pp. 132–142.

Kerstiens, G. (1995) Cuticular water permeance of European trees and shrubs grown in polluted and unpolluted atmospheres, and its relation to stomatal response to humidity in beech (*Fagus sylvatica* L.). *New Phytol.* **129,** 495–503.

Kerstiens, G. (1996a) Signalling across the divide: a wider perspective of cuticular structure-function relationships. *Trends Plant Sci.* **1,** 125–129.

Kerstiens, G. (1996b) Cuticular water permeability and its physiological significance. *J. Exp. Bot.* (in press).

Kerstiens, G. and Lendzian, K.J. (1989a) Interactions between ozone and plant cuticles. I. Ozone deposition and permeability. *New Phytol.* **112,** 13–19.

Kerstiens, G. and Lendzian, K.J. (1989b) Interactions between ozone and plant cuticles. II. Water permeability. *New Phytol.* **112,** 21–27.

Kerstiens, G., Federholzner, R. and Lendzian, K.J. (1992) Dry deposition and cuticular uptake of pollutant gases. *Agric. Ecosyst. Environ.* **42,** 239–253.

Kirschbaum, M.U.F. and Pearcy, R.W. (1988) Gas exchange analysis of the relative importance of stomatal and biochemical factors in photosynthetic induction in *Alocasia macrorrhiza*. *Plant Physiol.* **86,** 782–785.

Kisser-Priesack, G.M., Scheunert, I., Gnatz, G. *et al.* (1987) Uptake of $^{15}NO_2$ and ^{15}NO by plant cuticles. *Naturwissenschaften* **74,** 550–551.

Knittel, R. and Pell, E.J. (1991) Effects of drought stress and simulated acidic rain on foliar conductance of *Zea mays* L. *Environ. Exp. Bot.* **31,** 79–90.

Körner, C. (1994) Scaling from species to vegetation: the usefulness of functional groups. In: *Biodiversity and Ecosystem Function* (eds E.D. Schulze and H.A. Mooney). Springer-Verlag, Berlin, pp. 117–140.

Kuiper, P.J.C. (1961) The effects of environmental factors on transpiration of leaves, with special reference to stomatal light response. *Mededel. Landbouwhogeschool Wageningen* **61,** 1–49.

Larsson, S. and Svenningsson, M. (1986) Cuticular transpiration and epicuticular lipids of primary leaves of barley (*Hordeum vulgare*). *Physiol. Plant.* **68,** 13–19.

Lendzian, K.J. and Kerstiens, G. (1988) Interactions between plant cuticles and gaseous air and pollutants. *Aspects Appl. Biol.* **17,** 97–104.

Lendzian, K.J. and Kerstiens, G. (1991) Sorption and transport of gases and vapors in plant cuticles. *Rev. Environ. Contam. Toxicol.* **121,** 65–128.

Marshall, J.D. and Cadle, S.H. (1989) Evidence for trans-cuticular uptake of HNO_3 vapor by foliage of eastern white pine (*Pinus strobus* L.). *Environ. Pollut.* **60,** 15–28.

Meinzer, F.C., Wisdom, C.S., Gonzales-Coloma, A. *et al.* (1990) Effects of leaf resin on stomatal behaviour and gas exchange of *Larrea tridentata* (DC.) Cov. *Funct. Ecol.* **4,** 579–584.

Mooney, H.A. and Miller, P.C. (1985) Chaparral. In: *Physiological Ecology of North American Plant Communities* (eds B.F. Chabot and H.A. Mooney). Chapman and Hall, New York, pp. 213–231.

Moreshet, S. (1970) Effect of environmental factors on cuticular transpiration resistance. *Plant Physiol.* **46,** 815–818.

Morrissey, L.A., Zobel, D.B. and Livingston, G.P. (1993) Significance of stomatal control on methane release from *Carex*-dominated wetlands. *Chemosphere* **26,** 339–355.

Muchow, R.C. and Sinclair, T.R. (1989) Epidermal conductance, stomatal density and stomatal size

among genotypes of *Sorghum bicolor* (L.) Moench. *Plant Cell Environ.* **12,** 425-431.
Nagarajah, S. (1979) Differences in cuticular resistance in relation to transpiration in tea (*Camellia sinensis*). *Physiol. Plant.* **46,** 89-92.
Nazrul-Islam, A.K.M. (1983) Tissue water relations in a range of plant species and its ecological significance. *Flora* **174,** 329-337.
Neubert, A., Kley, D., Wildt, J. *et al.* (1993) Uptake of NO, and NO_2 and O_3 by sunflower (*Helianthus annuus* L.) and tobacco (*Nicotiana tabacum* L.): dependence on stomatal conductivity. *Atmos. Environ.* **27A,** 2137-2145.
Olszyk, D.M. and Tingey, D.T. (1985) Interspecific variation in SO_2 flux. *Plant Physiol.* **79,** 949-956.
Oosterhuis, D.M., Parker, M.L., Wullschleger, S.D. *et al.* (1988) The citrus leaf cuticle in relation to measurement of leaf water potential using thermocouple psychrometers. *Plant Cell Environ.* **11,** 129-135.
O'Toole, J.C., Cruz, R.T. and Seiber, J.N. (1979) Epicuticular wax and cuticular resistance in rice. *Physiol. Plant.* **47,** 239-244.
Paje, M.C.M., Ludlow, M.M. and Lawn, R.J. (1988) Variation among soybean (*Glycine max* (L.) Merr) accessions in epidermal conductance of leaves. *Austr. J. Agric. Res.* **39,** 363-373.
Pallardy, S.G. and Rhoads, J.L. (1993) Morphological adaptations to drought in seedlings of deciduous angiosperms. *Can. J. For. Res.* **23,** 1766-1774.
Parlange, J.Y. and Waggoner, P.E. (1970) Stomatal dimensions and resistance to diffusion. *Plant Physiol.* **46,** 337-342.
Pisek, A. and Berger, E. (1938) Kutikuläre Transpiration und Trockenresistenz isolierter Blätter und Sprosse. *Planta* **28,** 124-155.
Premachandra, G.S., Saneoka, H., Fujita, K. *et al.* (1992) Leaf water relations, osmotic adjustment, cell membrane stability, epicuticular wax load and growth as affected by increasing water deficits in sorghum. *J. Exp. Bot.* **43,** 1569-1576.
Quisenberry, J.E., Roark, B. and McMichael, B.L. (1982) Use of transpiration decline curves to identify drought-tolerant cotton germplasm. *Crop Sci.* **22,** 918-922.
Rawson, H.M. and Clarke, J.M. (1988) Nocturnal transpiration in wheat. *Austr. J. Plant Physiol.* **15,** 397-406.
Reiling, K. and Davison, A.W. (1995) Effects of ozone on stomatal conductance and photosynthesis in populations of *Plantago major* L. *New Phytol.* **129,** 587-594.
Saneoka, H. and Ogata, S. (1987) Relationship between water-use efficiency and cuticular wax deposition in warm season forage crops grown under water deficit conditions. *Soil Sci. Plant Nutr.* **33,** 439-448.
Šantrůček, J. (1991) Transpiration efficiency and apparent cuticular transpiration in some C_3 and C_4 plants. *Biol. Plant.* **33,** 192-199.
Saxe, H. (1986) Stomatal-dependent and stomatal-independent uptake of NO_x. *New Phytol.* **103,** 199-205.
Schmidt, H.W., Mérida, T. and Schönherr, J. (1981) Water permeability and fine structure of cuticular membranes isolated enzymatically from leaves of *Clivia miniata* Reg. *Z. Pflanzenphysiol.* **105,** 41-51.
Schönherr, J. (1982) Resistance of plant surfaces to water loss: transport properties of cutin, suberin and associated lipids. In: *Encyclopedia of Plant Physiology,* Vol. 12B, *Physiological Plant Ecology 2* (eds. O.L. Lange, P.S. Nobel and C.B. Osmond). Springer-Verlag, Berlin, pp. 153-179.
Schönherr, J. and Lendzian, K.J. (1981) A simple and inexpensive method of measuring water permeability of isolated plant cuticular membranes. *Z. Pflanzenphysiol.* **102,** 321-327.
Schreiber, L. and Riederer, M. (1996) Determination of diffusion coefficients of octadecanoic acid in isolated cuticular waxes and their relationship to cuticular water permeabilities. *Plant Cell Environ.* (in press).
Shepherd, W. (1964) Paths and mechanisms of moisture movements in detached leaves of white clover (*Trifolium repens* L.). I. Losses of petiole moisture direct from petioles and via laminae. *Ann. Bot.* **28,** 207-220.
Sinclair, T.R. and Ludlow, M.M. (1986) Influence of soil-water supply on the plant water balance of 4 tropical grain legumes. *Austr. J. Plant Physiol.* **13,** 329-341.
Slatyer, R.O. and Bierhuizen, J.F. (1964) Transpiration from cotton leaves under a range of environmental conditions in relation to internal and external diffusive resistances. *Austr. J. Biol. Sci.* **17,** 115-130.

Spedding, D.J. (1969) Uptake of sulphur dioxide by barley leaves at low sulphur dioxide concentrations. *Nature* **224,** 1229–1230.

Stickan, W. and Zhang, X. (1992) Seasonal changes in CO_2 and H_2O gas exchange of young European beech (*Fagus sylvatica* L.). *Trees* **6,** 96–102.

Taylor, G.E., Jr. and Tingey, D.T. (1983) Sulfur dioxide flux into leaves of *Geranium carolinianum* L. *Plant Physiol.* **72,** 237–244.

Taylor, G.E., Jr., McLaughlin, S.B., Shriner, D.S.*et al.* (1983) The flux of sulfur-containing gases to vegetation. *Atmos. Environ.* **17,** 789–796.

Tyree, M.T. (1994) Ion transport across leaf cuticles: concepts and mechanisms. In: *Air Pollutants and the Leaf Cuticle,* NATO ASI Series G, Vol. 36, (eds K.E. Percy, J.N. Cape, R. Jagels and C.J. Simpson). Springer-Verlag, Berlin, pp. 55–65.

van Gardingen, P.R. and Grace, J. (1992) Vapour pressure deficit response of cuticular conductance in intact leaves of *Fagus sylvatica* L. *J. Exp. Bot.* **43,** 1293–1299.

van Gardingen, P.R., Jeffree, C.E. and Grace, J. (1989) Variation in stomatal aperture in leaves of *Avena fatua* L. observed by low-temperature scanning electron-microscopy. *Plant Cell Environ.* **12,** 887–898.

van Hove, L.W.A., Vredenberg, W.J. and Adema, E.H. (1990) The effect of wind velocity, air temperature and humidity on NH_3 and SO_2 transfer into bean leaves (*Phaseolus vulgaris* L.). *Atmos. Environ.* **24A,** 1263–1270.

von Stamm, S. (1994) Linked stomata and photosynthesis model for *Corylus avellana* (hazel). *Ecol. Model.* **75/76,** 345–357.

Wang, W.L., Lin, Z.P. and Zhang, X.Y. (1988) Some aspects of plant cuticular transpiration. *Acta Phytophysiol. Sinica* **14,** 123–129.

Wedler, M., Geyer, R., Heindl, B. *et al.* (1996) Leaf-level gas exchange and scaling-up of forest understory carbon fixation rates with a 'patch-scale' canopy model. *Theor. Appl. Climatol.* **53,** 145–156.

West, D.W. and Gaff, D.F. (1976) The effect of leaf water potential, leaf temperature and light intensity on leaf diffusion resistance and the transpiration of leaves of *Malus sylvestris*. *Physiol. Plant.* **38,** 98–104.

Williams, W.T. and Amer, F.A. (1957) Transpiration from wilting leaves. *J. Exp. Bot.* **8,** 1–19.

Yoshida, S. and de los Reyes, E. (1976) Leaf cuticular resistance of rice varieties. *Soil Sci. Plant Nutr.* **22,** 95–98.

6

Effects of temperature, surfactants and other adjuvants on rates of uptake of organic compounds*

J. Schönherr and P. Baur

6.1 Introduction

Foliar uptake can occur from vapour, liquid or solid donor phases. Most environmental pollutants are taken up as vapours or dissolved in water (Riederer, 1995). The situation is more complex with agricultural sprays as active ingredients are usually formulated using a large variety of additives, called adjuvants. Adjuvants affect the physical properties of the spray liquid and they may serve as emulsifiers, wetting agents, spreader, sticker, antifoaming agents, or buffers (McWhorter, 1982). Adjuvants which improve biological effectiveness of active ingredients have been termed activator adjuvants, but their modes of action are still not fully understood. Cuticular penetration, diffusion in the apoplast, uptake into leaf cells and organelles, metabolism and long-distance translocation are involved before an active ingredient reaches its biochemical site of action inside the plant. Adjuvants can influence all of these processes.

Cuticles are the first and, in most cases, the rate-limiting barriers in foliar uptake. Foliar uptake as affected by adjuvants has been studied using radiolabelled organics and intact or detached leaves. With this approach rates of disappearance of solutes from the leaf surface are measured and effects of adjuvants on permeability of cuticles and on driving forces of uptake cannot be separated (Stock and Holloway, 1993; Stock *et al.*, 1993). This limits the usefulness of this approach as an analytical tool. In a different approach isolated cuticles were used and effects of solute properties (molar volume, lipophilicity) and adjuvant properties on solute mobility in and permeability of cuticles were studied (Bauer and Schönherr, 1992; Baur *et al.*, 1996a,b; Kerler and Schönherr, 1988a; Riederer and Schönherr, 1985; Riederer and Schreiber, 1995; Schönherr, 1993a,b; Schönherr and Baur, 1994; Schönherr and

Plant Cuticles, edited by G. Kerstiens.

* This chapter is dedicated to Prof. Dr Drs h.c. G. Bünemann on the occasion of his 70th birthday.

Riederer, 1989). However, permeabilities of cuticles alone are insufficient for predicting rates of foliar uptake, because they are proportional to permeability *and* driving forces (Schönherr and Baur, 1994). Adjuvants can affect both solute mobilities in cuticles and driving forces of foliar uptake and this will be the main topic of this chapter.

6.2 Theoretical considerations

Plant cuticles are lipid membranes and theories describing mass transfer across synthetic membranes (Crank and Park, 1968; Cussler, 1984) can also be used with cuticles. Rates of penetration (J) are proportional to the driving force, the proportionality coefficients being the permeance (P) or permeability coefficient:

$$\frac{M}{At} = J = P\left(C_{\text{don}} - C_{\text{rec}}\right). \qquad (6.1)$$

In Equation 6.1 rate of penetration (J) is defined as amount (M) which diffuses per area (A) and time (t) and the difference in concentration between donor and receiver is used as the driving force. Identical solvents (e.g. water or aqueous solutions of adjuvants) must be used on both sides of the cuticle. This, however, is rarely the case in foliar penetration of formulated active ingredients. Usefulness of Equation 6.1 is further limited because P is a mixed quantity as seen in Equation 6.2:

$$P = \frac{DK}{l_{\text{cm}}} \qquad (6.2)$$

where D is the diffusion coefficient of the solutes in the membrane, K is the membrane/solution partition coefficient and l_{cm} is the thickness of the cuticular membrane (CM). D is a measure of mobility of a solute in the membrane and K accounts for differential solubility of solutes in the membrane and adjacent solutions (Schönherr and Riederer, 1989).

Effects of temperature (T) on D and K differ. D increases substantially with rising T (Baur and Schönherr, 1995), while K is affected much less and may either increase or decrease (Riederer and Schönherr, 1986). Certain adjuvants called accelerators can increase D substantially (Schönherr and Baur, 1994; Schreiber, 1995; Schreiber *et al.* 1996b) and this effect is temperature dependent (Baur, 1993). These facts demonstrate that effects of temperature and adjuvants on D and K must be studied separately.

Plant cuticles are heterogeneous membranes made up of a very thin layer composed of cutin and cuticular waxes (the limiting skin) located at the morphological outer surface of cuticles, while the bulk of the cuticle between the limiting skin and the epidermal cell wall adds little to the total transport resistance of the cuticle (Schönherr and Baur, 1994; Schönherr and Riederer, 1988). This lamellar nature of the cuticle is illustrated schematically in Figure 6.1a, where the limiting skin amounts to only about 10% of the total thickness of the cuticle. Rates of cuticular penetration are determined by the limiting skin, because here both mobility and solubility are much lower than in the sorption compartment underneath. These arguments have been substantiated elsewhere (Bauer and Schönherr, 1992; Schönherr and Bauer, 1992; Schönherr and Baur, 1994).

Cuticles are heterogeneous membranes and D and K in the limiting skin and the sorption compartment differ. For this reason the hold-up time method for estimating diffusion coefficients yields erroneous values (Schönherr and Riederer, 1989),

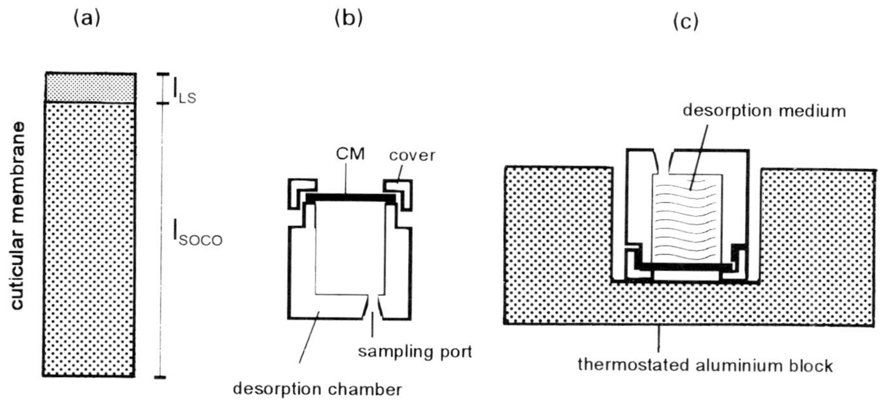

Figure 6.1. *Schematic drawings of: (a) a cross-section of an isolated cuticular membrane composed of a limiting skin (LS) and a sorption compartment (SOCO); (b) transport apparatus used for UDOS and SOFU experiment; and (c) a transport apparatus positioned in the thermostated aluminium block.*

which are much higher than those measured in reconstituted cuticular waxes (Schreiber and Schönherr, 1993; Schreiber *et al.*, 1996b). Since solute mobility in the limiting skin cannot be determined using the hold-up time method a new method called unilateral desorption from the outer surface (UDOS; Bauer and Schönherr, 1992; Schönherr and Bauer, 1992) was developed. With this method a mobility parameter k^* is determined which is directly proportional to D in the limiting skin:

$$D = k^* \cdot l_{ls} \cdot l_{soco}. \tag{6.3}$$

The magnitudes of the thickness terms l_{ls} (thickness of the limiting skin) and l_{soco} (thickness of sorption compartment, c.f. Figure 6.1a) are not known precisely, but this introduces no error when studying effects of adjuvants and temperature on solute mobility for a given type of cuticle. Errors arise only when D values for cuticles from different plant species are compared for which the ratio (l_{ls}/l_{soco}) is not constant. To avoid these complications we shall use k^* as a relative mobility parameter and not convert it to D.

The meaning of the partition coefficient in Equation 6.2 deserves some further consideration. In the past octanol/water (K_{ow}) and cuticle/water partition coefficients (K_{cw}) have been used for modelling permeance of cuticles (Kerler and Schönherr, 1988a,b; Riederer and Schönherr, 1984; Sabljic *et al.*, 1990; Schönherr and Riederer, 1989). The validity of this approach must be questioned, since wax/water partition coefficients (K_{wxw}) are much lower and account probably much better for differential solute solubility in the limiting skin (Schreiber and Schönherr 1993; Schreiber *et al.*, 1996b).

Uptake of agricultural sprays proceeds from hydrated formulation (adjuvant) residues (Schönherr and Baur, 1994), not from water. The appropriate partition coefficient characterizing differential solubility in the waxy limiting skin and in the formulation residue is K_{wxfr}, while the cuticle/water (K_{cw}), the polymer matrix/water partition coefficient (K_{mxw}) or the octanol/water partition coefficient

(K_{ow}) can account for differential solubility between sorption compartment and water in the cell wall (the polymer matrix (MX) is the residue obtained by extracting cuticular waxes from isolated CM). The latter three partition coefficients do not differ much (Kerler and Schönherr, 1988b; Riederer and Schönherr, 1984; Sabljic *et al.*, 1990).

Taking these considerations into account an equation can be derived, which relates rates of solute flow across CM to properties of solutes, adjuvants, limiting skins and cuticles. Substituting Equation 6.3 into Equation 6.2 and combining with Equation 6.1 we obtain

$$J = \frac{k^* \cdot l_{ls} \cdot l_{soco}}{l_{cm}} K(C_{don} - C_{rec}), \tag{6.4}$$

which can be simplified, since $l_{cm} \approx l_{soco}$ and we obtain:

$$J = k^* \cdot l_{ls} \cdot \left(K_{wxfr} C_{fr} - K_{mxw} C_{apo}\right). \tag{6.5}$$

The term in parenthesis represents the driving force, composed of differential solubilities in wax and formulation residue (K_{wxfr}), polymer matrix and water (K_{mxw}) and the concentrations in the formulation residue (C_{fr}) and in the apoplast (C_{apo}), respectively. If the solute concentration in the apoplast remains negligible, because uptake into cells and translocation are rapid, the second term in parenthesis vanishes and the driving force depends only on the product $K_{wxfr}C_{fr}$. This simplified equation will be used for a semiquantitative discussion about effects of temperature, type of adjuvant and adjuvant volume on rates of cuticular penetration. This treatment separates effects of adjuvants on solute mobility (***k****) from effects on differential solubility and driving forces ($K_{wxfr}C_{fr}$).

6.3 Experimental methods

A considerable amount of data on solute mobilities in cuticles (Bauer and Schönherr, 1992; Baur *et al.*, 1996a; Schönherr and Baur, 1994) and on effects of adjuvants on solute mobilities (Baur, 1993; Schönherr, 1993a,b; Schönherr and Bauer, 1992; Schönherr and Baur, 1994) have been published. In these publications methods are described in detail, and here we give only a brief introduction to UDOS to help the reader to better understand the data as well as the potentials and limitations of the method.

6.3.1 *Unilateral desorption from the outer surface (UDOS)*

Astomatous isolated CM are inserted into the desorption apparatus as shown in Figure 6.1b. The sorption compartments of CM are loaded with radiolabelled solutes by placing a small drop (50 µl) of aqueous solutions on the morphological inner surface of the CM exposed in the centre of the cover. As the water evaporates lipophilic solutes diffuse in the sorption compartment of the CM. If polar solutes are used which are not quantitatively sorbed in the CM, a solid residue of solutes will form on the inner surface of the CM leading to erroneous mobilities. Thus, the method is limited to relatively lipophilic solutes. Once sorption equilibrium within the CM has been obtained, solutes are desorbed from the outer surface of the CM using a 1% phospholipid suspension (PLS) which is pipetted into the desorption chamber. Chambers are rocked and thermostated (25°C) during desorption and the radiolabelled solutes can now diffuse through the limiting skin into the desorption medium. Lipophilic solutes are sorbed in the phospholipid vesicles (soybean lecithin) keeping the solute concentration in the water practically zero. These phospholipid vesicles serve as scavengers for solutes but they do not penetrate the limiting skin, nor do their monomers (Zellmer, 1992). These inert PLS are used to establish solute mobility in native, unaltered CM. Desorption media are withdrawn periodically and replaced by fresh ones. After two to five consecutive samples desorption is continued using micellar surfactant solutions, or aqueous solutions of other adjuvants. Three to five additional samples are withdrawn and as the last step, the CM is cut out, extracted with scintillation cocktail and counted together with the samples of desorption media. If the amount desorbed at the time t is called M_t and the amount initially applied is M_0 the data can be plotted as ln $(1-M_t/M_0)$ vs. t, which always results in straight lines. The term in parenthesis is equivalent to the fraction of radioactivity remaining in the CM at time t and the slope of the line is the rate constant of desorption k^*_{pls}. The subscript denotes that desorption was carried out using inert PLS. Rate constants measured using aqueous solutions of adjuvants will be marked k^*_{adj} . If $k^*_{adj} > k^*_{pls}$, the adjuvant increases solute mobility in the limiting skin and these adjuvants are called accelerators or accelerator adjuvants (Schönherr, 1993a). The ratio k^*_{adj}/k^*_{pls} is called effect and is calculated for each CM separately.

In UDOS experiments radiolabelled solutes are contained in the sorption compartment of the CM from the beginning and they diffuse through the limiting skin into the desorption medium. This is the only flux if PLS is used for desorption, since it cannot penetrate into the cuticle. If adjuvants which can penetrate the limiting skin are used for desorption, they will accumulate in the cuticle until equilibrium between desorption medium and CM has been established. Depending on permeability this can take minutes, hours or days and it is perceptible from slopes of desorption graphs which increase with time.

6.3.2 *Simulation of foliar uptake (SOFU)*

This method was developed to study the time course of cuticular penetration for each CM separately. The same apparatus as in UDOS is used, but the CM are oriented such that the morphological outer surface is exposed in the centre hole of the cover and the inner surface is in contact with the desorption medium. The chambers are filled with desorption medium (PLS), the sample port is closed with Tesa film (Beiersdorf, Hamburg, Germany) and the chambers are positioned upright into the wells of a thermostated aluminium block which is rocked for better mixing of desorption media (Figure 6.1c). When the desired temperature (25°C) has been established, a small drop (5 or 10 µl) of aqueous test solution containing radiolabelled solutes and non-radioactive adjuvants is applied to the centre of the CM. The water evaporates within 20 to 30 min. At this time the chambers are inverted such that the cuticles face the bottom of the well of the aluminium block (Figure 6.1c). This type of arrangement will be referred to as open donor compartment (d.c. open). In experiments marked 'closed with steel disc' a thin (50 µm) stainless steel disc was positioned over the CM on top of the air-dry residue before the cover was added.

Desorption media were periodically withdrawn for scintillation counting and were replaced by fresh PLS. After the last PLS sample the CM was cut out, extracted in scintillation cocktail and counted, to determine the amounts of solute left either on the surface or in the CM. M_0 was obtained by summing up amounts desorbed plus the amount left on/in the CM for each CM individually. Recoveries ranged from

90 to 110% and data were discarded if recoveries were lower than 90%. It should be pointed out that in these SOFU experiments penetration is defined as amounts of solutes appearing in the desorption medium and the balance to M_0 will be either on the surface or sorbed in the CM.

Data are again plotted as $\ln(1-M_t/M_0)$ vs. t, but slopes (k) were in most instances not linear. These two slopes k and k^* should not be confused since the driving forces differ in the two sets of experiments. In UDOS, the driving force is the solute concentration in the sorption compartment of the CM and the rate constant is marked k^*, while in SOFU the driving force is $K_{wxfr}C_{fr}$. and the rate constant is k.

6.3.3 *Chemicals*

^{14}C-labelled solutes were used as model compounds and the amounts applied ranged from 0.1 to 2 µg/CM. Radiolabelled urea, 3-*O*-methyl-α-D glucose and dichlorophenoxyacetic acid (2,4-D) were obtained from Amersham Buchler (Braunschweig, Germany), paraquat dichloride from Sigma Chemicals (Deisenhofen, Germany) and tebuconazol was a gift from Bayer (Leverkusen, Germany). They had radiochemical purities > 98%. Other relevant data are given in Table 6.1.

All other chemicals were non-radioactive and had purities better than 98%. Monodisperse ethoxylated alcohols were purchased from Fluka (Neu-Ulm, Germany). Tris-(2-ethylhexyl)-phosphate, tripropylphosphate and tributylphosphate were obtained from Aldrich Chemicals (Steinheim, Germany), lactic acid and ammonium sulphate from Merck (Darmstadt, Germany), Tween 80 from Serva (Heidelberg, Germany) and soybean lecithin from Roth (Karlsruhe, Germany). AgsolEx 12 (*N*-dodecyl-2-pyrrolidone) was a gift from GAF Europe (Guildford, UK).

6.3.4 *Cuticular membranes (CM)*

Cuticular membranes (CM) were isolated enzymatically as described by Schönherr and Riederer (1986). Only the astomatous cuticles from the upper surfaces were

Table 6.1. *Properties of ^{14}C-labelled model compounds used*

Compound	Chemical name	Specific activity (MBq mmol^{-1})	log K_{cw}	V_x (cm^3 mol^{-1})
Urea	Carbamide	1987	-1.09	47
Me-Glucoside	3-*O*-Methyl-α-D glucose	318	-0.87	134
2,4-D	2,4-Dichlorophenoxy acetic acid	329	2.89	138
Paraquat	1,1′-Dimethyl-4,4′-dipyridilium dichloride	541	?	199
Tebuconazol	(*R,S*)-1-*p*-Chlorophenyl-4,4-dimethyl-3-(1*H*,-1,2,4,-triazol-1ylmethyl)pentan-3-ol	1730	3.37	241

Molar volumes (V_x) were calculated according to Abraham and McGowan (1987).

utilized. *Citrus aurantium* L. trees were grown in growth chambers (Geyer and Schönherr, 1990), *Camellia sinensis* (L.) O. Kuntze, and *Stephanotis floribunda* Brogn. were raised in a greenhouse, *Parrotia persica* (DC.) C.A. Mey. leaves came from a tree grown in New Zealand. Young, but fully expanded leaves of *Prunus laurocerasus* L. were collected in Northern Italy. Small samples of *Citrus aurantium* and *Citrus limon* (L.) Burm. leaves were taken in an abandoned orchard near Malaga, Spain.

6.3.5 *Statistics*

Rate constants were determined for each CM separately. Effects of adjuvants on rate constants (UDOS) were determined using the method of paired observations by first desorbing with PLS followed by a second desorption period with adjuvant. For each treatment 10–20 CM were used and vertical bars in figures represent 95% confidence intervals.

6.4 Results and discussion

6.4.1 *Effects of adjuvants and temperature on solute mobilities*

Typical time courses of desorption of 2,4-D with PLS (up to the arrow) and adjuvants are depicted in Figure 6.2 for seven to nine different *Citrus* CM. The first three points were obtained using PLS as desorption medium, followed by desorption with aqueous adjuvant solutions. Initial slopes (k^*_{pls}) are linear and slopes increase after changing desorption media. With AgsolEx 12 and pentafluorophenol this change occurred almost instantaneously. However, with some CM having low k^*_{pls} it took about 24 h until the steepest slopes were realized (Figure 6.1a). Slopes increased throughout the experiments when desorption was carried out with tripropylphosphate or tris-ethylhexylphosphate and from desorption graphs it cannot be decided if the slopes obtained during the last interval are maximum slopes or not. Response graphs shown in Figure 6.3 help to decide this question. Here the effects of adjuvants on solute mobility (k^*_{adj}/k^*_{pls}) of individual CM are plotted against the reciprocal value of the initial mobility ($1/k^*_{pls}$) and in the calculation the largest slope obtained with adjuvant was used. Good or very good linearity was observed with pentafluorophenol, AgsolEx 12 and tripropylphosphate, while considerable scatter was obtained with tris-ethylhexylphosphate. The slopes of the graphs are measures of effectiveness of the accelerators and they indicate that the effect depends on initial mobility.

Pentafluorophenol increased mobility by up to 220-fold and was most effective, but this might be due to highest amounts sorbed in cuticular waxes. The magnitude of the effect depends on the amounts of adjuvants sorbed in the CM (Riederer *et al.*, 1995) and this was not measured in the present experiments. Amounts of adjuvant sorbed are likely to differ among the adjuvants and for this reason we cannot decide which adjuvant had the highest intrinsic activity. The scatter seen in the response graph for tris-ethylhexylphosphate indicates that this adjuvant penetrated very slowly and sorption equilibrium between desorption medium and CM was not obtained in this experiment. On the other hand, steep response graphs having good linearity are indicative of sorption equilibrium, hence slopes observed with tripropylphosphate during the interval 120–144 h appear to be maximum possible slopes.

Linearity of response graphs demonstrates that effects of adjuvants on solute mobility are larger with CM having low initial mobilities (k^*_{pls}). This inverse relation

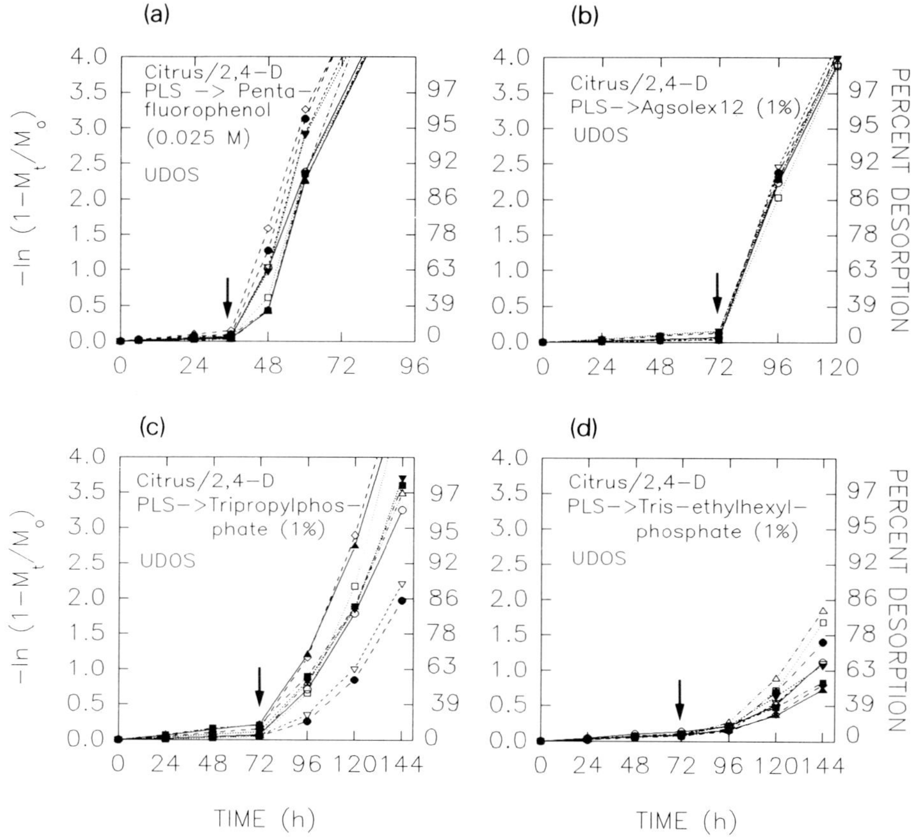

Figure 6.2. *Desorption graphs showing time course of desorption (UDOS) of 2,4-D from* Citrus *CM using first 1% aqueous phospholipid suspension (PLS) followed by aqueous solutions of: (a) pentafluorophenol; (b) AgsolEx 12; (c) tripropylphosphate; or (d) tris-ethylhexylphosphate. Desorption media were changed at the time marked by the arrow. Concentration of desorption media is given in the graphs. Each curve represents data from one CM.*

appears to be a general one and will be commented on later. Here we want to point out another important feature generally observed with accelerator adjuvants (Schönherr, 1993a,b). Reference to Figure 6.3 demonstrates that initial mobilities of 2,4-D in *Citrus* CM vary widely over about an order of magnitude. The set of CM used with Agsolex 12 had an initial mean k^*_{pls} of 4.76×10^{-7} s^{-1} and a standard deviation of 7.40×10^{-7} s^{-1}, that is the coefficient of variation amounted to 155%. In contrast, the rate constant measured with AgsolEx 12 was 1.70×10^{-6} s^{-1}, the standard deviation was 4.00×10^{-7} s^{-1} and the coefficient of variation was only 19%. Thus, the large natural variability in 2,4-D mobility of *Citrus* CM is tremendously reduced by accelerator adjuvants. Mobilities are not only much higher, they are also much more uniform.

Environmental conditions are known to affect wax composition (Baker, 1982; Riederer and Schneider, 1990) and rates of foliar uptake. Furthermore, large

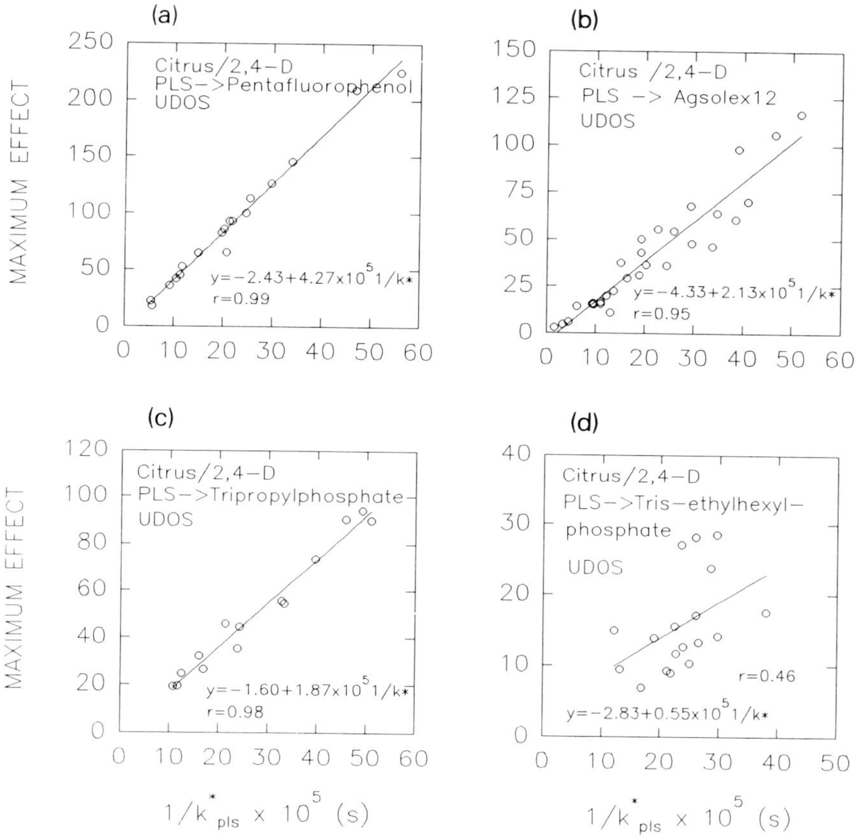

Figure 6.3. *Response graphs showing the effect of desorption media on solute mobility in CM. The effect is the ratio of the maximum slopes observed with the desorption media divided by the slope obtained during desorption with PLS (k^*_{pls}). The effect was calculated for each CM separately and was plotted against the reciprocal value of the initial slope measured with PLS ($1/k^*_{pls}$).*

differences in permeability of cuticles (Baur *et al.*, 1996a; Riederer and Schönherr, 1985; Schönherr and Riederer, 1989) exist among plant species. Using tributylphosphate we have tested accelerator effects of CM from different species and origins. 2,4-D mobilities in *Citrus aurantium* CM collected either in Spain or in growth chambers (Geyer and Schönherr, 1990) and *Citrus limon* CM from Spain had different initial mobilities (k^*_{pls}) but all data fall on a single response graph (Figure 6.4a) showing that responses to this accelerator depended only on $1/k^*_{pls}$ and when treated with tributylphosphate mobilities became very similar. Response graphs obtained with CM from different species differed somewhat in slope, but good linearity was obtained with all species over the entire range of $1/k^*_{pls}$ of about a factor of 100 (Figure 6.4b). Clearly, tributylphosphate acts as accelerator with cuticles of all of these (and probably all other) plant species and its effect is greater the smaller k^*_{pls} This inverse relation appears to be general, provided accelerators penetrate rapidly. Comparable results have been obtained with other accelerator

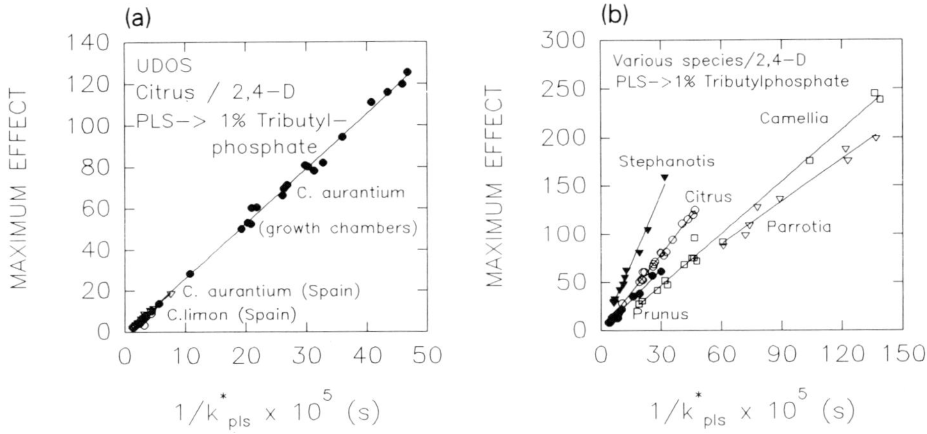

Figure 6.4. *Response graphs showing the effect of tributylphosphate on 2,4-D mobility in CM from different plants and different origins.*

adjuvants such as monodisperse and polydisperse ethoxylated fatty alcohols (Bauer and Schönherr, 1992; Schönherr, 1993a,b), alkoxylated tallow amines (Baur and Schönherr, 1996), *n*-alkylester of dicarboxylic acids (Baur, 1993), the insecticide chlorfenvinphos (Baur *et al.*, 1996b) and such simple compounds as pentafluorophenol (Figure 6.3a). The inverse relation appears to be even more general, for if the effect of the accelerator tetraethyleneglycol monooctylether (C_8E_4) was plotted against log $1/k^*_{pls}$ of different solutes in *Citrus* CM a linear graph was again obtained (J. Schönherr, unpublished data) showing that mobilities of solutes having low k^*_{pls} by virtue of the large molar volume (Bauer and Schönherr, 1992; Baur, 1993) are much more increased than those of relatively small solutes. As a consequence, differences in solute mobilities caused by differences in their molar volumes are largely eliminated by accelerators. This was shown for *Citrus* and pear leaf CM and the polydisperse ethoxylated fatty alcohol surfactant Genapol C-050 (Baur, 1993).

Which properties of organic compounds are crucial for accelerator action remains obscure so far, but it is clear that they must be lipophilic to be sorbed in cuticular waxes (Baur, 1993; Baur *et al.*, 1996b; Riederer *et al.*, 1995; Schönherr 1993a,b; Schreiber *et al.*, 1996b) and decrease viscosity of the amorphous fraction of cuticular waxes (Schreiber *et al.*, 1996b).

Increasing temperatures also leads to an increase in fluidity (a decrease in viscosity) of liquid and solid phases and a concomitant increase in solute mobility (Crank and Park, 1968; Cussler, 1984). Such an effect has also been observed with plant cuticles (Baur and Schönherr, 1995; Norris and Bukovac, 1969). In the temperature range 17–34°C linear Arrhenius graphs for k^*_{pls} were obtained with *Citrus* and *Pyrus* leaf CM and a number of solutes differing in molar volumes indicating the absence of phase transitions in cuticles. The slopes of these graphs represent the energies of activation for solute diffusion (E_a) and they ranged from 120 to 170 kJ mol^{-1}. When these activation energies for cuticles from two different species and four different solutes were plotted against $1/k^*_{pls}$ (at 25°C) a perfectly linear graph (r^2=0.998) was obtained (Baur and Schönherr, 1995). Thus activation energies increased with increasing $1/k^*_{pls}$ and it made no difference if this was caused by increasing molar volumes of solutes or endogenous properties of CM of different plant species (mobility of a given solute is considerably higher in *Pyrus* leaf CM than in *Citrus* CM).

Extracting cuticular waxes from CM increases permeability to water and solutes by up to three orders of magnitude (Schönherr and Riederer, 1989). This is convincing evidence that low permeability and mobility in cuticles are caused by cuticular waxes. It is not entirely clear how this is accomplished, but Baur *et al.* (1996a), Riederer and Schreiber (1995), Schönherr and Baur (1994), Schreiber (1995), and Schreiber *et al.* (1996b) have argued that tortuosity of the diffusion path in the amorphous wax fraction is greatly increased by platelets of crystalline wax. This model implies that viscosity of cutin and amorphous waxes are similar (Baur *et al.*, 1996a), and both permeability of cuticles and apparent diffusion coefficients in waxes are very low because the tortuous diffusion path in the waxes is very much longer than the thickness of the waxy layer. Accelerators and temperature may increase permeability of cuticles by decreasing viscosity of the amorphous wax and (possibly) by decreasing tortuosity. Whatever the causes, the presence of accelerator adjuvants in cuticular waxes has three main effects: (i) size selectivity (effects of molar volumes on mobility); (ii) interspecific differences in solute mobility; and (iii) effects of decreasing temperatures on solute mobility are greatly reduced. Thus, differences in permeabilities and mobilities in cuticles that might be caused by plant species, solute size or temperature are greatly reduced by accelerators. This is a very remarkable feature that could be utilized in formulating herbicides, growth regulators and plant protection agents.

6.4.2 *Effects of adjuvants on rates of cuticular penetration (SOFU)*

Penetration in absence of adjuvants. If cuticular penetration is plotted as the logarithm of the relative amount of active ingredient present on the surface of the CM vs. time, the resulting graphs are initially linear, but level off at later times. In Figure 6.5 this is shown for penetration of the large lipophilic solute tebuconazol across *Citrus* and pear leaf CM. Results obtained for individual CM are plotted to illustrate variability among CM and plant species. Linearity implies that rates of penetration follow a first order kinetics, that is, the amounts of solute on the surface decrease exponentially with time. This can be characterized by the rate constant k, which is equivalent to the slope. In the graph for *Citrus* it can be seen that straight lines do not go through the origin and about 2 h elapse before solutes have crossed the CM and appear in the receiver.

These experiments were conducted with open donor compartments (d.c. open) and adjuvants were not added. During evaporation of the water of the donor droplet solutes diffuse into the surface wax, but it is not known if all solutes are sorbed in the epicuticular waxes during the initial 20 to 30 min. Solutes not sorbed in the surface wax during droplet drying may form a solid residue or a concentrated aqueous solution. To exclude the possibility of formation of a solid residue on the cuticles experiments were conducted using solutes differing in polarity and 100% humidity. As soon as the donor droplets were apparently dry, a stainless steel disc was placed directly on the outer surface of each CM before applying the cover (Figure 6.1b). This assured 100% humidity, since water penetrating the CM during the experiment could not escape. At the end of the experiments when the chambers were dissembled it was noted that visible water films had not formed between CM and steel disc, even though the atmosphere must have been saturated

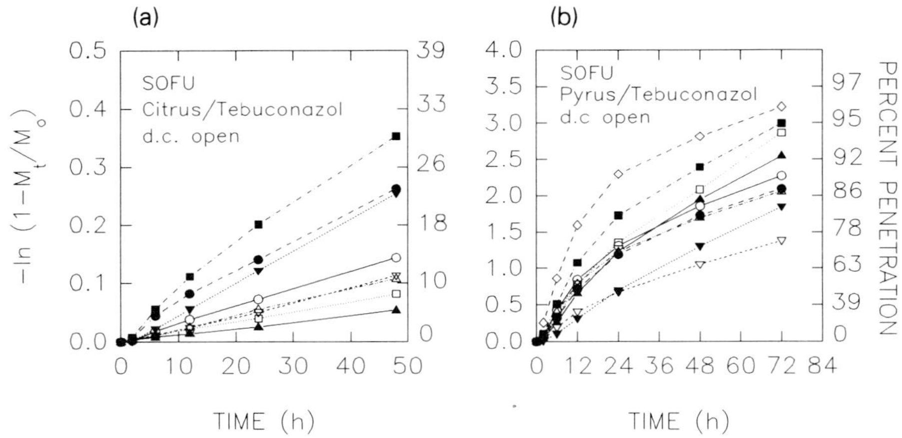

Figure 6.5. *Time course of penetration of tebuconazol across* Citrus *and pear leaf CM. Aqueous solutions of tebuconazol were applied as 10 μl droplets to the outer surface of each CM and after evaporation of the water rates of appearance of tebuconazol in the receiver were measured. Data for individual CM are shown to demonstrate variability among CM in rates of penetration.*

with moisture. Urea is the smallest and most polar one and it penetrated most rapidly. After 8 h 95% of the dose applied had reached the receiver (Figure 6.6a). Rather high penetration rates were also observed with methyl glucose and paraquat dichloride. Rates were lowest (but similar to those shown in Figure 6.5b) with the large lipophilic molecule tebuconazol. All plots have an initial linear portion and hold-up times are visible with methyl glucose, paraquat and tebuconazol (Figure 6.6b–d).

Under the conditions chosen, water diffusing from the receiver through the CM could not escape and it appears likely that after some time saturated solutions served as donor. From this scenario one would expect penetration to be a first order process throughout. However, penetration graphs levelled off after some time. With urea first order penetration prevailed until about 95% had penetrated, while with methyl glucose curves levelled off at about 86%. With paraquat and tebuconazol deviation from linearity occurred even earlier. As mentioned, a visible water film between cuticle and steel disc did not form and this renders it unlikely that levelling off was caused by a dilution of the donor solutions. A more probable cause can be offered. We measured the rates at which solutes appeared in the receiver, not the rates of disappearance from the surface. Initially, all solutes are dissolved in the surface wax or in an aqueous film on the surface of the cuticle. As solutes diffuse through the cuticles towards the receiver some solutes are temporarily retained in the CM. The proportions of solutes on the surface and in the CM change during the experiment. This distribution depends on lipophilicity of solutes and the CM will contain more lipophilic tebuconazol than polar urea. It follows from this argument that levelling off is due to a violation of our implicit assumption that solutes are either on the surface or in the receiver. Penetration curves level off as soon as a significant amount of solute is contained in the CM relative to the amount in the aqueous film on the surface.

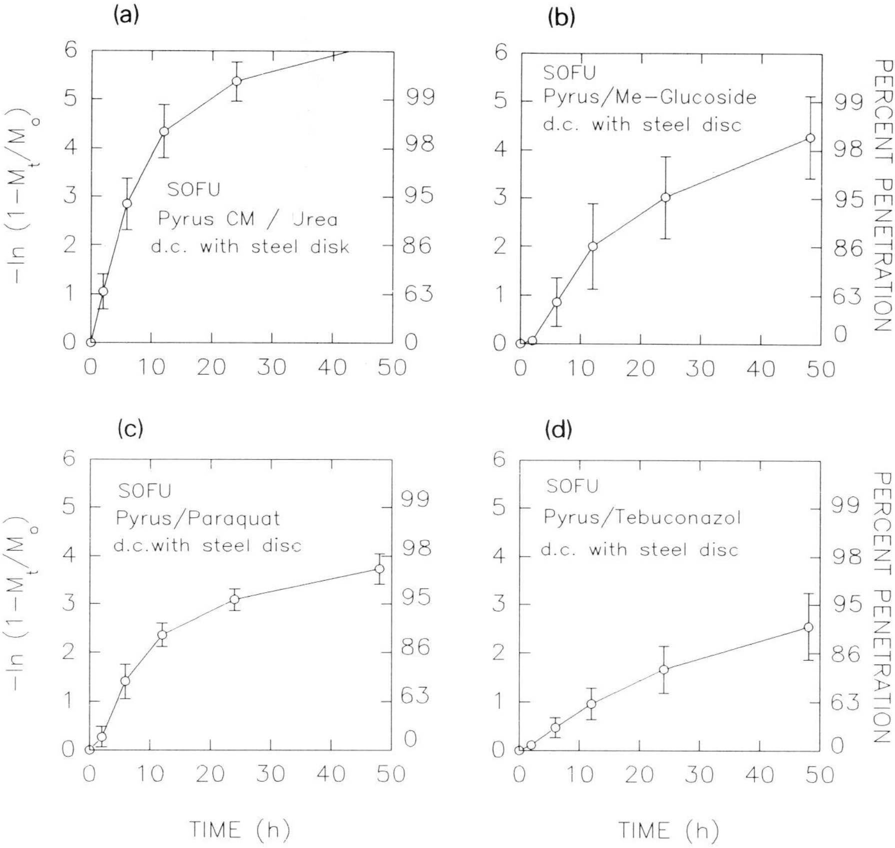

Figure 6.6. *Time course of penetration of various solutes across pear leaf CM. Means and 95% confidence intervals obtained from 10 to 20 CM are plotted.*

Effect of ionization of weak electrolytes on rates of penetration. Dissociation of weak acids makes them more polar by up to four orders of magnitude (Schönherr and Riederer, 1989) and the partition coefficient decreases in proportion. According to Equation 6.5 the partition coefficients are considered a part of the driving force. Ionization of weak acids decreases the driving force and if all molecules are ionized, the driving force and rates of penetration decrease by four orders of magnitude. It follows that reducing ionization is most important when trying to maximize rates of penetration of weak acids, such as 2,4-D, which has a pK_a of 2.76. This can be done by buffering at pH values at or below the pK_a. When 2,4-D is dissolved in lactic acid buffer (LAB) having a pH of 3.0 about 35% of the molecules are non-ionized. They are lipophilic (Table 6.1) and penetrate the cuticles. The ionized and non-ionized fractions are in equilibrium and penetration can go to completion even at this relatively high pH since more and more ionized molecules become non-ionized as non-ionized molecules enter the cuticle. When 0.2 µg 2,4-D is dissolved in 10 µl of 0.25% LAB, the 2,4-D will eventually be dissolved in 2.50 μg lactic acid once the water has evaporated. There will be some hydration water associated with the residue, the amounts of which depend on relative humidity. When humidity is 100% as in Figure 6.7a penetration is rapid and nearly linear up to 90%. Conducting the same experiment with d.c. open and at lower humidity (Figure 6.7b) resulted in a drastic decrease in rates of penetration. The quick levelling off suggests that hydration and buffering were insufficient due to loss of water and lactic acid by evaporation. Doubling the LAB concentration improved the situation and near linearity was re-established (Figure 6.7c), still rates remained much lower than in Figure 6.7a. A further increase in LAB concentration to 2% lead to a very large surface residue as 0.2 μg 2,4-D are now initially dissolved in 20 μl lactic acid (plus hydration water). Under these conditions the penetration graph was perfectly linear (Figure 6.7d), but rates were very low, as the 2,4-D on the surface is now greatly diluted which reduced the driving force in proportion (Equation 6.5). These data show that buffering with lactic acid is of limited

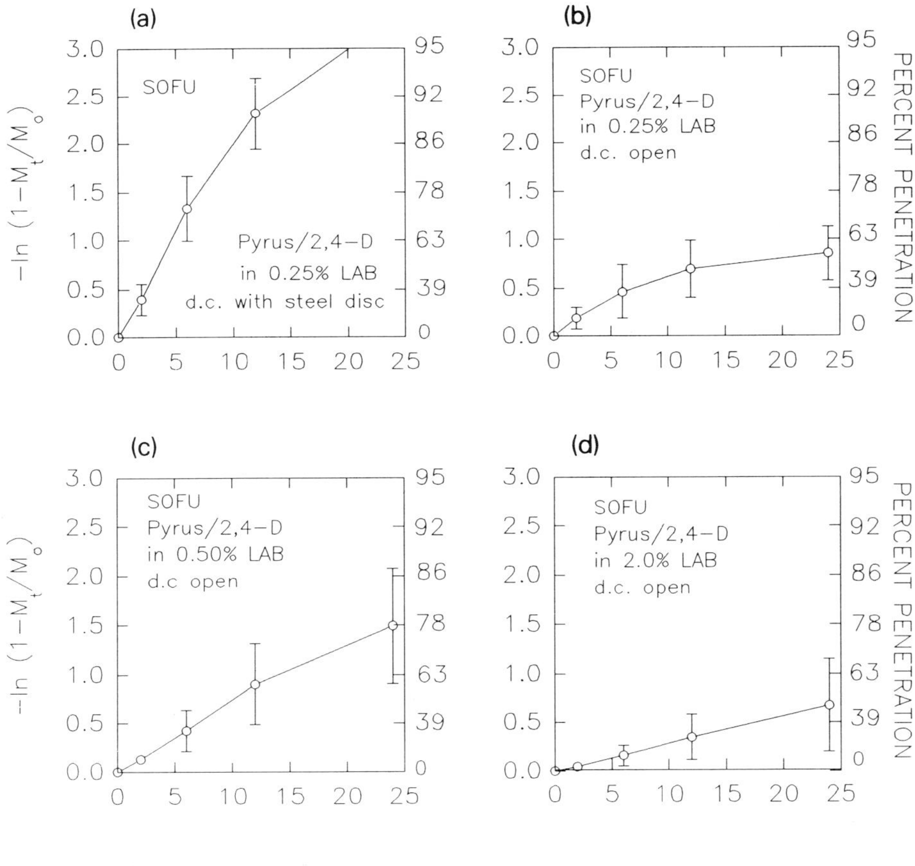

Figure 6.7. *Time course of penetration of 2,4-D across pear leaf CM as affected by concentration of LAB. 10 μl droplets of aqueous LAB having pH 3.0 were applied to the centre of each CM. Means and 95% confidence intervals obtained from 10 to 20 CM are plotted.*

usefulness. If low concentrations are used, volatility is a problem, and at higher concentrations, dilution reduces rates of penetration. Furthermore, if sprayed on leaves, high LAB concentrations having pH 3.0 are likely to cause leaf damage. We have tried to overcome these problems by depressing ionization by adding ammonium sulphate (AS), which increases ionic strengths in the surface residue.

Effects of adjuvants on rates of penetration. With d.c. closed very high rates of penetration were observed (data not shown) but with d.c. open (80% humidity) rates of penetration were very low (Figure 6.8a line marked zero). We attributed this to an insufficient amount of hydration water associated with the residue which appeared white and we tried to overcome this problem by adding some Tween 80 as a solvent. This polydisperse surfactant (polyoxyethylene sorbitan monooleate) is quite polar and it is not an accelerator adjuvant (Schönherr and Baur, 1994). By virtue of its high molar volume and the large numbers of polar ethoxy groups (20) most oligomers cannot penetrate cuticles and stay on the surface. Addition of Tween 80 (which is a liquid at ambient temperatures) greatly increased rates of penetration, even at only 0.05% (Figure 6.8a). At 0.15 % rates were maximum and after a hold-up time of about 2 h, penetration graphs were perfectly linear. A further increase in Tween 80 concentration to 0.50 or 1.50% decreased rates, but linearity was maintained. These data suggest that Tween 80 served indeed as a solvent for both 2,4-D and AS. Hydration water may have added to the effect. Linearity of the penetration graphs demonstrates that most of the 2,4-D which had not yet penetrated remained in the fairly large surface residues. Relatively little 2,4-D was sorbed in the CM because the mass of the residue was large compared to the volume of the CM and only a small fraction of the 2,4-D was probably non-ionized.

In the previous figures of this section rates of penetration were manipulated using passive media which do not affect solute mobility in the limiting skin. With hydration water, LAB, Tween 80 as passive solvent and AS we only manipulated

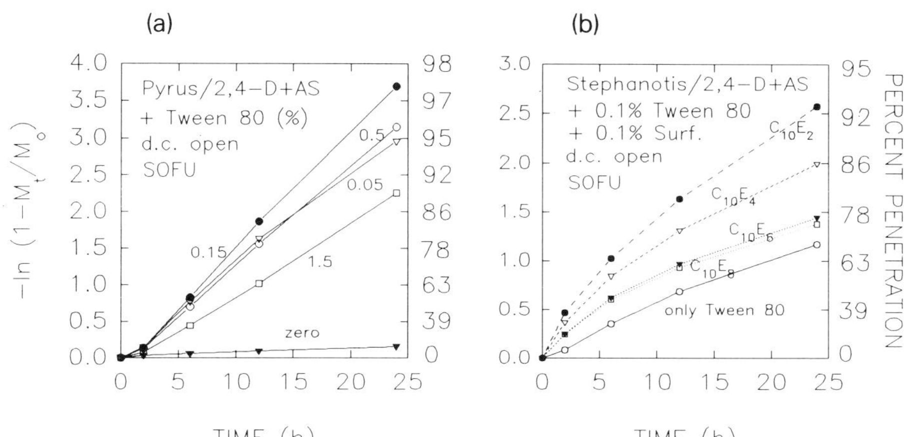

Figure 6.8. *Time course of penetration of 2,4-D across pear leaf and* Stephanotis *CM as affected by 0.1 M AS, Tween 80, and selected monodisperse ethoxylated decanols. Volume of droplets was 5 µl. Means of 15 to 20 CM are plotted and confidence intervals were omitted for clarity. (Modified from Schönherr and Baur, 1994.)*

concentrations of solutes in the surface residue, the degree of ionization of the weak acid 2,4-D, and the partition coefficient (K_{wxfr}). Thus, all the effects observed can be attributed to effects of these compounds on driving forces (Equation 6.5). In the following experiments we have included accelerator adjuvants to demonstrate their effects on rates of penetration.

Pear leaf CM are relatively permeable and nearly complete penetration can be obtained even in the absence of accelerator adjuvant as seen in Figure 6.8a. Effects of accelerator adjuvants were therefore tested using CM from *Stephanotis* leaves, in which solute mobilities and permeabilities are lower as seen in Figure 6.8b (curve marked 'only Tween 80'). This treatment was taken as base formulation to ensure that 2,4-D stays dissolved in hydrated Tween 80 and ionization is sufficiently reduced by AS. To this base formulation 0.1% of monodisperse ethoxylated decanols varying in extent of ethoxylation were added. These accelerators increased rates of penetration and eliminated the hold-up time (Figure 6.8b). $C_{10}E_2$ was most effective and rates decreased with increasing ethoxylation. The same sequence of effectiveness had been observed in UDOS experiments (Schönherr, 1993a), but this is fortuitous. In UDOS experiments amounts of surfactants sorbed in CM increase with time until equilibrium between CM and desorption medium is obtained, while in SOFU experiments a small and finite amount of surfactant (in this experiment 5 μg Tween 80 and 5 μg ethoxylated alcohol (EA) dissolved in an aqueous 5 μl drop) is added on top of the CM. While Tween 80 remains on the surface the EA diffuse into and through the limiting skin and the sorption compartment. Eventually they reach the desorption media where they are first scavenged by the phospholipid vesicles and finally removed from the system when the receiver is withdrawn. Diffusion proceeds from a solution of EA in hydrated Tween 80. As EA diffuse into the CM their concentration in the Tween 80 residue decreases and this reduces the driving force for penetration of EA. At the same time increasing amounts of EA disappear from the limiting skin by diffusing into the large sorption compartment and the receiver. Thus, the amounts of EA in the limiting skin increases initially, but after reaching a transient maximum the concentration of EA in the limiting skin decreases and approaches zero eventually. The same applies for 2,4-D and the greatest acceleration of 2,4-D diffusion can be expected when 2,4-D molecules diffuse at the same speed as the EA, for only in this case can 2,4-D molecules diffuse in cuticular waxes which are plasticized by EA. We have no data on partition coefficients of EA in this system nor do we have information concerning their mobilities in *Stephanotis* wax. However, cuticle/water and wax/water partition coefficients of ethoxylated alcohols decrease with increasing ethoxylation (Riederer *et al.*, 1995; Schreiber *et al.*, 1996b). Furthermore, solute mobilities in limiting skins decrease exponentially with increasing molar volumes (Baur *et al.*, 1996a; Schönherr and Baur, 1994). This means that mobilities of EA can be expected to decrease with increasing ethoxylation. These facts and the above scenario suggest that $C_{10}E_2$ penetrated most rapidly and had higher transient concentrations in the limiting skin and rates of diffusion of this accelerator and 2,4-D were matched best.

The situation is simpler when Tween 80 can be omitted as solvent and AS is not needed for depressing ionization. In this case the hydrated accelerator occurs in pure form on the surface of the CM and serves both as solvent and accelerator. This condition was investigated using urea penetration across *Stephanotis* CM and the same EA as in Figure 6.8b. If urea was dissolved in aqueous 0.1% EA, penetration graphs were nearly linear if the degree of ethoxylation was four or greater (Figure 6.9a). Rates observed with $C_{10}E_8$ were identical to the water control. Urea is hygroscopic and in the water control penetration apparently occurred from a concentrated aqueous urea solution. If applied together with $C_{10}E_8$, urea penetration proceeded from a hydrated neat surfactant solution having a larger volume and a lower urea concentration as is the case with the water control. As seen in Figure 6.8b this EA is an accelerator, but at high concentrations, dilution and a possible effect on the partition coefficient (Equation 6.5) eliminated the accelerator effect. EA having lower degrees of ethoxylation are smaller and more lipophilic. They penetrate more rapidly and this leads to a decrease in accelerator volume in the residue and to a concomitant increase in urea concentration. Accelerating properties of these compounds together with the reduction in residue volume resulted in higher rates of penetration, which were highest with $C_{10}E_2$. If the initial concentrations of the surfactants in the treating solution were reduced by a factor of 100, the effect of ethoxylation on penetration was reversed (Figure 6.9b). The dilution effect of these minute amounts of accelerators was negligible and effects are mainly due to their accelerator properties. They were best with $C_{10}E_8$ because this relatively large and more polar accelerator penetrated more slowly into the CM and accelerator action lasted longer than with the smaller more lipophilic accelerators.

Tributylphosphate was shown to be a powerful accelerator (see Figure 6.4) of 2,4-D diffusion and Figure 6.10 presents evidence that this effect is not limited to 2,4-D. Penetration rates of urea across *Stephanotis* CM were greatly increased by tributylphosphate. A very characteristic feature of accelerator action is the elimination of the hold-up time, which is clearly discernible in Figure 6.10 at the 2 h interval. Even the very low amount of 1.25 μg tributylphosphate (0.025% in 5 μl) was very effective, but increasing the amount to 0.75% gave a longer lasting acceleration, while a 100-fold increase (2.5%) resulted in urea dilution and a somewhat reduced effect, if compared to the 0.75% treatment (Figure 6.10a). The reduction of the hold-up time is direct evidence showing that accelerators increase

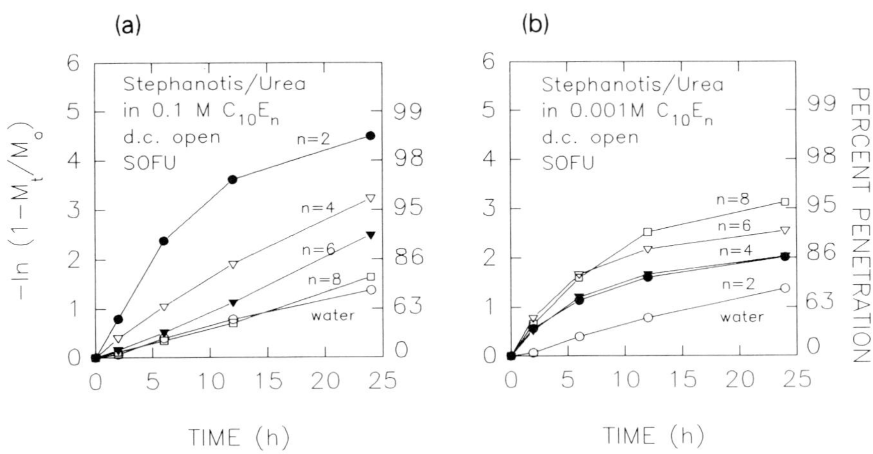

Figure 6.9. *Rates of penetration of urea across* Stephanotis *CM as affected by concentration (in % (w/w) given next to each curve) of monodisperse ethoxylated decanols. Droplet volume was 5 μl. In the control treatment (water) surfactants were omitted. Means of 15 to 20 CM are plotted and confidence intervals were omitted for clarity.*

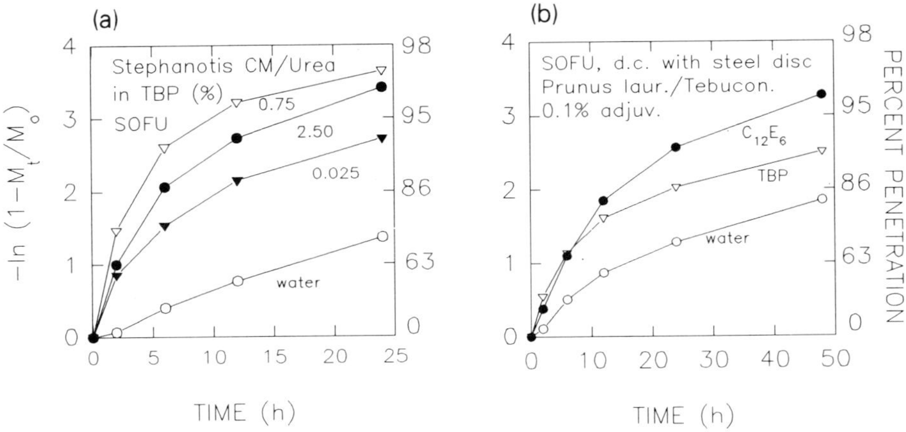

Figure 6.10. *Rates of penetration of urea and tebuconazol as affected by: (a) species and various concentrations of tributylphosphate; or (b) by monodisperse hexaethylene monododecyl ether ($C_{12}E_6$). Droplet volume was 5 µl. Means of 15 to 20 CM are plotted and confidence intervals were omitted for clarity.*

diffusion coefficients in cuticles, as hold-up times depend only on diffusion coefficients and membrane thickness (Cussler, 1984), but not on lipophilicity or partition coefficients.

Figure 6.10b was included to demonstrate that an accelerator may work well with a given plant species and compound but may not be optimum with other plant species or solutes. With tebuconazol and *Prunus laurocerasus* $C_{12}E_6$ had a longer lasting effect than tributylphosphate, probably because it penetrated more slowly. For this reason tributylphosphate was slightly better at the 2 h interval, while at 6 h both accelerators performed equally well.

6.5 Summary and conclusions

We have demonstrated that maximizing driving forces and adding accelerator adjuvants are two independent tools for achieving maximum rates of cuticular penetration. Driving forces are maximum if a saturated solution of active ingredient is formed on the cuticle as soon as the carrier (usually water) has disappeared. In the laboratory this goal can often be obtained by using high humidity which leads to a thin aqueous film on the surface of the cuticle. In the field it is desirable to have rates of penetration which are independent of humidity, and it is necessary to include into the formulation a non-volatile solvent in amounts just sufficient to prevent solidification of active ingredients. The component of the driving force which is accessible to manipulation is the product of concentration of solute times partition coefficient wax/formulation residue. The product must be maximized, not each quantity separately. If a solute is highly soluble in an adjuvant, K_{wxfr} will be small, and a smaller volume is needed to obtain a saturated solution.

Many active ingredients such as growth regulators or herbicides are weak acids. With these compounds care must be taken to obtain conditions in the formulation residue which lead to a high fraction of dissolved non-ionized species. Water solubility of the non-ionized species is much lower than solubility of the sodium or potassium or ammonium salts, and the adjuvant must dissolve all the non-ionized species. With compounds having a pK_a below pH 3.0 buffering is likely to cause phytotoxicity. We successfully reduced ionization of 2,4-D by adding AS. The salt is likely to work with other weak acids just as well and better performance of many agricultural sprays in the presence of AS can be attributed to reduction of ionization, prevention of formation of insoluble calcium salts of active ingredients or possibly to the hygroscopicity of the salt.

If maximum driving forces are realized, rates of penetration can be increased further by adding accelerator adjuvants. Performance of these accelerators depends on plant species, type of solute, amounts of accelerator and other adjuvants. Accelerators penetrate into the cuticle, and type, as well as amounts, of accelerator must be judiciously selected to ensure that accelerator and active ingredient penetrate at similar rates. If the accelerator penetrates too rapidly, large amounts of active ingredient remain on the cuticle and may even form a solid precipitate. If a slow accelerator is selected and amounts are high, dilution of active ingredients will result in lower than maximum driving forces and many hours may have to pass before sufficient accelerator has penetrated into the cuticle and increased solute mobility in the limiting skin. If accelerators and solutes penetrate at equal speed, a concentrated saturated solution of active ingredient can be maintained on the cuticle, while with passive non-penetrating adjuvants the solute concentration and the driving force decrease with time.

As shown here and in previous communications (Baur, 1993; Baur and Schönherr, 1996; Baur *et al.,* 1996b; Schönherr, 1993a,b; Schönherr and Bauer, 1992; Schönherr and Baur, 1994) a great number of organic compounds have accelerator properties. Some of the accelerators are surface active, but this property is not essential as can be seen from the fact that pentafluorophenol and the alkylesters of phosphoric acid, dicarboxylic acid esters and chlorphenvinfos are powerful accelerators. We do not know which structural properties make a good accelerator, but it is clear that accelerators must be sufficiently lipophilic to be sorbed in cuticular waxes in significant (2–10%) amounts (Baur *et al.*, 1996b; Riederer *et al.,* 1995). Thus, accelerators must be lipophilic and their rates of diffusion into and across the limiting skin must be similar to that of the active ingredient. Their rates of penetration into the cuticles depend on properties of waxes of plant species, on molar volumes of accelerators and on intrinsic accelerating properties. Rates of diffusion of active ingredients in waxes increase with increasing concentrations of accelerator sorbed in waxes (Baur *et al.*, 1996b; Riederer *et al.,* 1995), which implies that the optimum concentration of accelerator varies depending on the amount of wax per unit leaf area. Since accelerators only change viscosity of waxes, the type (polarity) of accelerator which results in highest solute mobility does not depend on polarity of solute. Furthermore, accelerator effects on solute mobility in cuticular waxes increase with decreasing temperatures (Baur, 1993; Baur and Schönherr, 1996). This shows that selecting the proper accelerator and finding the optimum concentration for applications with a given active ingredient and plant species is very complex, even though we have only considered adjuvant effects on rates of diffusion in cuticles. Biological effectiveness also depends on applied dose and retention, on driving forces for uptake and on rates of translocation. It remains to be seen which molecular descriptors can account for all these factors, but it is unlikely that hydrophile–lipophile balance (HLB) (van Valkenburg, 1982), degree of ethoxylation and lipophilicity of adjuvants alone (Kirkwood, 1993; Stock and Holloway, 1993; Stock *et al.*, 1993) are sufficient for modelling rates of cuticular penetration or biological effectiveness.

Accelerators work by temporarily decreasing viscosity of cuticular waxes. As a consequence, solute mobility increases, while size selectivity of cuticles and temperature effects on solute mobility are greatly reduced. Energies of activation for diffusion in cuticles are very high, this leads to rather low permeabilities and rates of penetration especially at low temperatures (Baur and Schönherr, 1995). This problem can be overcome by using accelerators in low-temperature formulations or low-temperature additives, which will restore rates of penetration to levels nearly as high as at 20°C.

We have sketched a rational approach towards realizing maximum rates of penetration whenever wanted. Unfortunately, at this time, optimum solutions can be found only empirically using SOFU, because we do not have all important information such as solubilities of solutes in cuticular waxes, adjuvants and accelerators or rates of penetration of accelerator adjuvants. These data must be generated before good solute/accelerator combinations can be predicted and empirical testing can be reduced to a minimum.

References

Abraham, M.H. and McGowan, J.C. (1987) The use of characteristic volumes to measure cavity terms in reversed phase liquid chromatography. *Chromatographia* **23**, 243-246.

Baker, E.A. (1982) Chemistry and morphology of plant epicuticular waxes. In: *The Plant Cuticle* (eds D.F. Cutler, K.L. Alvin and C.E. Price). Academic Press, London, pp. 139-165.

Bauer, H. and Schönherr, J. (1992) Determination of mobilities of organic compounds in plant cuticles and correlation with molar volumes. *Pestic. Sci.* **35**, 1-11.

Baur, P. (1993) Barriereeigenschaften pflanzlicher Kutikularmembranen und ihre Beeinflussung durch organische Xenobiotika. Doctoral Dissertation, Technische Universität München, Germany.

Baur, P. and Schönherr, J. (1995) Temperature dependence of the diffusion of organic compounds across plant cuticles. *Chemosphere* **30**, 1331-1340.

Baur, P. and Schönherr, J. (1996) Die Aufnahme systemischer Wirkstoffe über Blätter: Grundlagen und Optimierung. *Gartenbauwissenschaften* **61**, 105-115.

Baur, P., Hend Marzouk, Schönherr J. *et al.* (1996a) Mobilities of organic compounds in plant cuticles as affected by structure and molar volumes of chemicals and plant species. *Planta* **199**, 404-412.

Baur, P., Grayson, B.T. and Schönherr, J. (1996b) Concentration dependent mobility of chlorfenvinphos in isolated plant cuticles. *Pestic. Sci.* **47**, 171-180.

Crank, J. and Park, G.S. (1968) *Diffusion in Polymers*. Academic Press, London.

Cussler, E.L. (1984) *Diffusion. Mass Transfer in Fluid Systems.* Cambridge University Press, Cambridge

Geyer, U. and Schönherr, J. (1990) The effect of environment on water permeability and composition of *Citrus* leaf cuticles: I. Water permeability of isolated cuticular membranes. *Planta* **180**, 147-153.

Kerler, F. and Schönherr, J. (1988a) Permeation of lipophilic chemicals across plant cuticles: Prediction from partition coefficients and molar volumes. *Arch. Environ. Contam. Toxicol.* **17**, 7-12.

Kerler, F. and Schönherr, J. (1988b) Accumulation of lipophilic chemicals in plant cuticles: Prediction from octanol/water partition coefficients. *Arch. Environ. Contam. Toxicol.* **17**, 1-6.

Kirkwood, R.C. (1993) Use and mode of action of adjuvants for herbicides: A review of some current work. *Pestic. Sci.* **38**, 93-102.

McWhorter, C.G. (1982) The use of adjuvants. In: *Adjuvants for Herbicides* (ed. R.H. Hodgson). Weed Sci. Soc. of America, Champaign, IL, pp. 10-25

Norris, R.F. and Bukovac, M.J. (1969) Some physical-kinetic considerations in penetration of naphthaline acetic acid through isolated pear leaf cuticle. *Physiol. Plant.* **22**, 701-712.

Riederer, M. (1995) Partitioning and transport of organic chemicals between the atmospheric environment and leaves. In: *Plant Contamination: Modelling and Simulation of Organic Chemical Processes* (eds S. Trapp and C. McFarlane). Lewis Publishers, Boca Raton, FL, pp. 153-190.

Riederer, M. and Schneider, G. (1990) The effect of environment on the permeability and composition of *Citrus* leaf cuticles. II. Composition of soluble cuticular lipids and correlation with transport properties. *Planta* **180,** 154–165.

Riederer, M. and Schönherr, J. (1984) Accumulation and transport of (2,4-dichlorophenoxy) acetic acid in plant cuticles: I. Sorption in the cuticular membranes and its components. *Ecotoxicol. Environ. Safety* **8,** 236–247.

Riederer, M. and Schönherr, J. (1985) Accumulation and transport of (2,4-dichlorophenoxy) acetic acid in plant cuticles: II. Permeability of the cuticular membrane. *Ecotoxicol. Environ. Safety* **9,** 196–208.

Riederer, M. and Schönherr, J. (1986) Thermodynamic analysis of nonelectrolyte sorption in plant cuticles:The effect of concentration and temperature on sorption of 4-nitrophenol. *Planta* **169,** 69–80.

Riederer, M. and Schreiber, L. (1995) Waxes: The transport barriers of plant cuticles. In: *Waxes: Chemistry, Molecular Biology and Functions* (ed. R.D. Hamilton). The Oily Press, Dundee, pp. 131–156.

Riederer, M., Burghardt, M., Mayer, S. *et al.* (1995) Sorption of monodisperse alcohol ethoxylates and their effects on the mobility of 2,4-D in isolated plant cuticles. *J. Agric. Food Chem.* **43,** 1067–1075.

Sabljic, A., Güsten, H., Schönherr, J. *et al.* (1990) Modelling plant uptake of airborne organic chemicals. 1. Plant cuticle/water partitioning and molecular connectivity. *Environ. Sci. Technol.* **24:** 1321–1326.

Schönherr, J. (1993a) Effects of monodisperse alcoholethoxylates on mobility of 2,4-D in isolated cuticles. *Pestic. Sci.* **38,** 155–164.

Schönherr, J. (1993b) Effects of alcohols, glycols and monodisperse ethoxylated alcohols on mobility of 2,4-D in isolated plant cuticles. *Pestic. Sci.* **39,** 213–223.

Schönherr, J. and Bauer, H. (1992) Analysis of effects of surfactants on permeability of plant cuticles. In: *Adjuvants and Agrichemicals,* Vol. 2 (ed. C.L. Foy). CRC Press, Boca Raton, FL, pp. 17–35.

Schönherr, J. and Baur, P. (1994) Modelling penetration of plant cuticles by crop protection agents and effects of adjuvants on their rates of penetration. *Pestic. Sci.* **42,** 185–208.

Schönherr, J. and Riederer, M. (1986) Plant cuticles sorb lipophilic compounds during enzymatic isolation. *Plant Cell Environ.* **9,** 459–466.

Schönherr, J. and Riederer, M. (1988) Desorption of chemicals from plant cuticles: evidence for asymmetry. *Arch. Environ. Contam. Toxicol.* **17,** 13–19.

Schönherr, J. and Riederer, M. (1989) Foliar penetration and accumulation of organic chemicals in plant cuticles. *Rev. Environ. Contamin. Toxicol.* **108,** 1–70.

Schreiber, L. (1995) A mechanistic approach towards surfactant/wax interactions: Effects of octaethyleneglycolmonododecylether on sorption and diffusion of organic chemicals in reconstituted cuticular wax of barley leaves. *Pestic. Sci.* **45,** 1–11.

Schreiber, L. and Schönherr, J. (1993) Mobilities of organic compounds in recrystallized cuticular wax of barley leaves: Determination of diffusion coefficients. *Pestic. Sci.* **38,** 353–361.

Schreiber, L., Kirsch, T. and Riederer, M. (1996a) Transport properties of cuticular waxes of *Fagus sylvatica* L. and *Picea abies* (L.) Karst.: Estimation of size selectivity and tortuosity from diffusion coefficients of aliphatic molecules. *Planta* **198,** 104–109.

Schreiber, L., Riederer, M. and Schorn, K. (1996b) Mobilities of organic compounds in reconstituted cuticular wax of barley leaves: Effects of monodisperse alcoholethoxylates on diffusion of pentachlorophenol and tetracosanoic acid. *Pestic. Sci.* (in press).

Stock, D. and Holloway, P.J. (1993) Possible mechanisms of surfactant induced foliar uptake of agrochemicals. *Pestic. Sci.* **38,** 165–177.

Stock, D., Holloway, P.J., Grayson, B.T. *et al.* (1993) Development of an uptake model to rationalise selection of polyoxyethylene surfactant adjuvants for foliage-applied agrochemicals. *Pestic. Sci.* **37,** 233–245.

Van Valkenburg, J.W. (1982) Terminology, classification and chemistry.In: *Adjuvants for Herbicides* (ed. R.H. Hodgson). Monograph, Weed Sci. Soc. America, Champaign, IL.

Zellmer, S. (1992) Die Entwicklung thermolabilier, fusionsfähiger Liposomen für die Tumortherapie und von Transferosomen für den Einsatz an Pflanzen. Doctoral Dissertation, Technische Universität München, Germany.

7

Interactions between electromagnetic radiation and the plant cuticle

J.D. Barnes and J. Cardoso-Vilhena

7.1 Introduction

The optical properties of leaves, and other plant components, have received much interest in the past and a number of comprehensive reviews are available which discuss different aspects (Brakke *et al.,* 1993; Ehleringer, 1981; Fukshansky, 1981; Gausman, 1974, 1985; Grant, 1987; Knipling, 1970; Myers, 1983; Osborne and Raven, 1986; Vogelmann, 1993). The purpose of this chapter is not to repeat the efforts of previous authors, but to review those works pertinent to developing a better understanding of the significance and way in which the cuticle, and epicuticular waxes (EW), influence the interception, scattering and redirection of incident radiation. Particular emphasis is placed on the ecophysiological significance of light attenuation at the air-cuticle interface, as well as the subtle effects of light quantity and quality on the cuticle and EW *per se*.

7.2 Mechanisms underlying leaf surface reflectance

Leaf reflectance spectra can be divided into three rather distinct regions: the ultraviolet (280–400 nm) and visible wavelengths (400–700 nm), where reflection is relatively low because of the high absorptivity of the photosynthetic pigments at these wavelengths; the near infrared (700–1300 nm) where reflectance (and transmittance) is characteristically very high; and the middle-far infrared where there is moderate reflectance because of the strong absorptivity of water at these wavelengths, with bands of decreasing strength located at approximately 2600, 1950, 1430, 1175 and 970 nm. It is often implied that this reflectance is solely determined by the proportion of incident radiation that enters leaves, is randomly scattered within the leaf tissue and is eventually directed back to the leaf surface to be measured as diffuse reflectance (see Gausman, 1985; Grant, 1987). Yet, when leaves are vacuum infiltrated with water, or oils of different refractive indices, to eliminate the cell wall-air refractive discontinuities chiefly responsible for the diffuse reflectance measured at the leaf surface (Kumar and Silva, 1973; Sinclair *et al.*,

Plant Cuticles, edited by G. Kerstiens.

1973), a significant proportion of the incident visible and near-infrared radiation continues to be reflected (Woolley, 1971). There is also considerable difference in leaf reflectance characteristics between genotypes (Knipling, 1970), between leaves in the same canopy (Gates, 1980) and even between upper and lower surfaces of the same leaf (Walter-Shea *et al.*, 1991) which are considered to be primarily determined by differences in internal leaf structure, pigmentation and water content (Gausman, 1985; Grant, 1987). However, other factors, such as the presence and condition of crystalline EW deposits, may also play an important role; for example, Eller's (1979) data for *Kalanchoë pumila* show that the presence of an intact EW layer can substantially increase reflectance of visible, and, to a lesser extent, near-infrared, radiation (Figure 7.1). More recent studies provide evidence of a linear relationship between the radiation reflected/absorbed by leaves and the amount and/or configuration of the EW on aerial plant surfaces (Grant *et al.*, 1995; Johnson *et al.*, 1983). Two optical phenomena are considered responsible for this surface reflectance: specular (or non-Lambertian) reflectance and light scattering.

7.2.1 *Specular reflectance*

Thin, amorphous wax coatings, and featureless regions of the cuticle bearing a superficial crystalline wax layer, act like any other optically smooth surface (or, to be more precise, any surface where the irregularities in it are large compared to the incident wavelength) and specularly reflect radiation (Gates *et al.*, 1965; Mestre, 1935). This type of surface reflection is responsible for the everyday observation that certain field crops may appear white instead of green when viewed at oblique angles (Vanderbilt *et al.*, 1985), the white light resulting from reflectance at the leaf surface visually overwhelming the much smaller amounts of green light scattered from the leaf interior at such off-normal angles (Grant *et al.*, 1993). Specular reflectance can be described by the Fresnel equations and obeys Snell's law of reflection (see Hecht and Zajac, 1974). Basically, this means that the magnitude, directionality and degree of polarization of the beam reflected from the leaf surface depends upon the angle of incidence, surface roughness features, and the refractive index (RI) of the waxes exposed at the surface (Brakke *et al.*, 1993). At near-normal angles of incidence specular reflectance is minimal and makes little contribution to total leaf hemispherical reflectance, but its relative importance increases with the angle of incidence (Walter-Shea *et al.*, 1989). Indeed, some authors indicate that the angular dependency of leaf reflection is dependent upon the extent of specular reflection from the leaf surface (Gates *et al.*, 1965). Recent data shown in Table 7.1 (from Grant *et al.*, 1993; see also McClendon, 1984) indicates that the contribution of specular (polarized) reflectance to total reflectance may be considerable at large off-normal angles of incidence, and cannot be ignored even for pubescent or waxy leaves.

7.2.2 *Light scattering*

The size, distribution and orientation of surface features (e.g. crystalline EW deposits) determine the extent to which light is scattered at the leaf surface and also the degree of polarization. When surface features are of the same order of magnitude, or larger, than the wavelength of incident radiation, the scattered radiation will be partially polarized and randomly scattered in all directions (Hecht and Zajac, 1974). When surface features have dimensions much smaller than the wavelength of the incident radiation the criteria for Raleigh scattering may be met if the particles are sparsely distributed over the leaf surface. In this case, the scattered radiation will be polarized at 90° to the incident beam, partially polarized

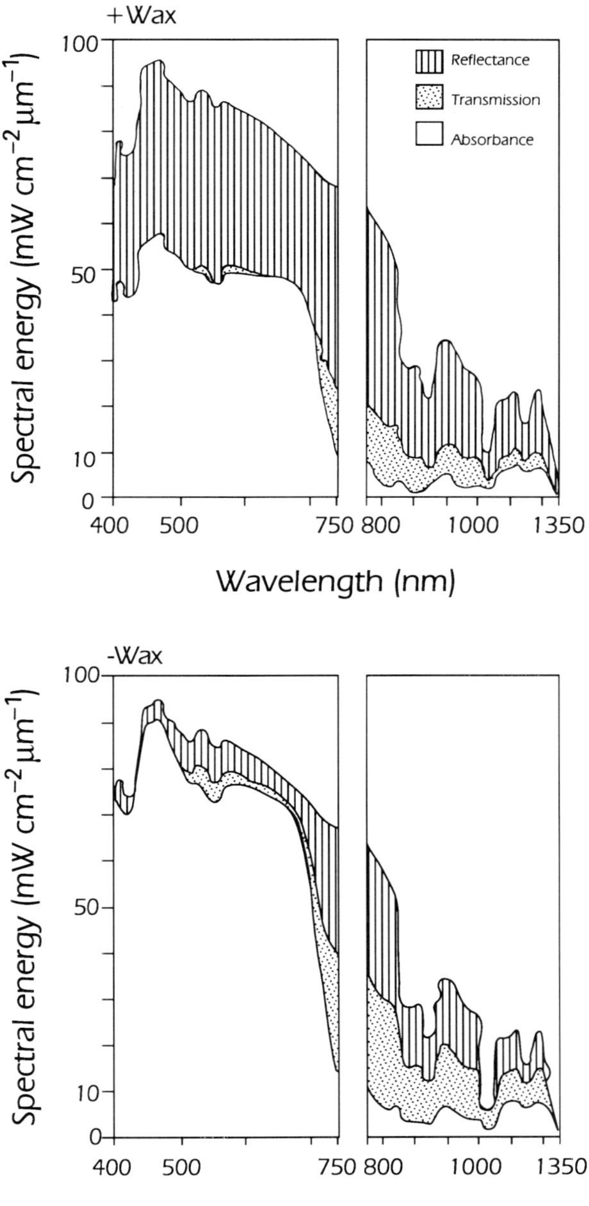

Figure 7.1. *Optical characteristics of leaves of* Kalanchoë pumila *Baker before and after removal of the surface waxes. (Reproduced from Eller (1979) Die strahlungsökologische Bedeutung von Epidermisauflagen.* Flora ***168:*** *146–192, with permission from Gustav Fischer Verlag.)*

Table 7.1. *Polarized reflectance, R_Q, non-polarized diffuse reflectance, R_N, and degree of polarization, P, from the adaxial and abaxial surfaces of leaves of 18 species measured at the Brewster angle at a wavelength of 650 nm*

	Adaxial						Abaxial					
	R_Q		R_N		*P*		R_Q		R_N		*P*	
Species	x	sd	x	sd	x	sd	x	sd	x	sd	x	sd
Abutilon theophrasti [1,1]	14	5	19	5	41	7	16	5	29	4	35	7
Acer saccharum, common [1,4]	21	6	9	6	69	7	14	2	25	4	36	4
Acer saccharum, Indiana [1,4]	27	1	9	1	74	6	10	2	26	4	28	3
Asclepias syriaca [1,1]	20	2	15	3	58	5	13	4	36	5	27	5
Caladium bicolor [1,1]	13	3	38	6	25	3	11	5	38	7	21	5
Chenopodium album [1,1]	9	4	14	4	58	5	7	2	18	1	28	6
Cucurbita pepo [1,1]	12	2	13	3	52	6	4	1	22	3	16	4
Glycine max [1,1]	23	6	11	3	67	7	9	2	25	4	26	5
Gynura aurantiaca [1,3]	19	4	9	1	67	5	30	6	27	2	52	5
Helianthus annuus [1,4]	26	9	19	4	56	7	17	6	40	9	29	9
Lycopersicon esculentum [1,1]	16	3	10	3	61	6	13	5	28	7	32	3
Quercus borealis [3,4]	36	13	7	1	83	6	25	10	22	2	51	11
Quercus prinus [3,4]	47	18	7	1	86	5	15	4	22	2	43	5
Quercus velutina [3,4]	56	22	7	1	88	6	49	24	23	6	64	16
Solanum pycanthum [1,1]	19	4	19	3	56	5	15	4	26	6	37	4
Sorghum bicolor [2-3, 2-3]	70	20	18	9	79	7	60	23	18	5	75	8
Ulmus americana [3,1]	48	14	6	1	88	3	19	5	31	3	38	4
Zea mays [1-3, 2-3]	48	16	15	4	76	4	101	14	19	5	84	4

Modified from Grant *et al.* (1993) Polarized and specular reflectance: variation with leaf surface features. *Physiol. Plant.* **88:** 1–9, with permission from *Physiologia Plantarum.*

Data presented as mean percentages, x, ± standard deviations, sd, of two measurements of six leaves per species.

Leaf surface: [1] hairy, pubescent, tomentose or downy, [2] glaucous or waxy, [3] shiny, [4] matt or mealy. First number refers to adaxial surface, second to abaxial surface.

at other angles and enhanced in the blue region of the spectrum (Grant *et al.*, 1993). In circumstances where the wax deposits are of the correct dimensions to act as Raleigh scatterers, but the waxes form a dense and intricate network of fine crystals projecting from a surface covered by an amorphous layer of wax, the probability that photons will be multiply scattered is increased many-fold. In such circumstances the reflected beam is depolarized, yet is still enhanced in the blue region of the spectrum. This form of light scattering is responsible for the blueish appearance of certain leaves and fruits (Martin and Juniper, 1970), as well as that of certain insects (Hadley, 1989), and explains why visual selections for the degree of glaucousness are commonly a poor indication of the amount of wax deposited on the aerial surfaces of plants (Gonzalez *et al.*, 1996).

The configuration of wax deposits on the leaf surface therefore influences the manner in which radiation is intercepted, scattered and redirected. Since the physicochemical nature of the surface is influenced by a great many factors (including genotype, leaf age, light environment, temperature, humidity and pollution-climate, to name but a few), no generalized statement can be made on the way that the EW influence leaf reflectance, other than the fact that leaves are generally neither purely diffuse (i.e. Lambertian) nor purely specular (non-Lambertian) reflectors of incident radiation (Brakke *et al.*, 1993; Breece and Holmes, 1971; Walter-Shea *et al.*, 1989). Leaf reflectance represents the sum of diffuse and specular components. The diffuse component is non-polarized, varies little with changing angles and emanates predominantly from the leaf interior (Kumar and Silva, 1973; Sinclair *et al.*, 1973), while the non-diffuse component is polarized, emanates solely from the leaf surface, is spread about the specular angle and becomes more important as the angle of incidence increases (Grant, 1987). Visibly shiny leaves tend to have higher specular reflection than matt or pubescent leaves, but leaves that are glaucous or pubescent can still specularly reflect a considerable amount of radiation (Table 7.1). In the typical circumstance where wax deposits on the surface range from small to large in comparison with the wavelength of incident radiation, small particles will predominantly scatter light away from the specular direction, while large features will control the amount of light scattered into the specular direction (Fung, 1984). Changes in reflection as a function of the angle of incidence can therefore be considered a function of the distribution, size and orientation of the reflecting surfaces which comprise the leaf surface (Torrance and Sparrow, 1967).

7.3 Ecophysiological significance of reflectance at the air–cuticle interface

7.3.1 *Remote sensing*

In recent years, remote sensing has been widely employed to map communities, identify vegetation types and study normal successional changes in vegetation, at varying scales of resolution. Conventional methods (excluding thermal infrared imagery and passive microwave sensors) rely on the detection of electromagnetic radiation, in the visible and infrared wavebands, that is reflected or emitted from the complex plant assemblages at the earth's surface. Consequently, the interpretation and meaningfulness of data recorded from aircraft and space-borne sensors, as well as the development of new-improved systems, is reliant on a fundamental understanding of the factors that may contribute to variations in the intensity and quality of leaf spectral reflectance (Knipling, 1970).

Remote sensing technology is being increasingly applied to estimate optical and geometric properties of plant canopies from airborne and satellite measurements of reflectance through the application of complex algorithms to solve inverse problems (Antyufeev and Marshak, 1990). Canopy reflectance and radiative transfer (CRT) models which simulate multispectral reflectance as a function of leaf structural and optical properties play a key role in solving these problems (Baret, 1995). Models are based on the reflectance properties of single leaves, but cannot be applied to canopies without direct modifications because of qualitative and quantitative differences in the two types of reflectance spectra. On a percentage basis the reflectance from a canopy is considerably less than that from a single leaf because of a general attenuation of radiation by variations in illumination angle, leaf orientation, specular reflection, shadows and non-foliage background surfaces such as soil. CRT models are important, therefore, in scaling between the level of the individual leaf and the canopy. As our knowledge of the factors influencing leaf optical properties has improved, there has been an evolutionary refinement of CRT models (Jacquemoud and Baret, 1990) and in recent years major advances have been made in incorporating factors such as individual leaf bidirectional scattering properties (Brakke *et al.*, 1993; Ross and Marshak, 1989). This evolution has considerably improved the connection between large-scale, remote sensing observations and leaf optical properties and in so doing enabled the refinement of airborne and satellite imagery (Danson and Plummer, 1995).

It is perhaps fortunate that the majority of the radiation reflected at the air–cuticle interface is linearly polarized (Grant *et al.*, 1993). This means that measurements of the degree of polarization of reflected radiation provide a readily available means to differentiate between diffuse (predominantly originating from the leaf interior) and specular reflectance (predominantly originating from the leaf surface). Measurements of the degree of polarization are further simplified by the fact that specularly reflected radiation shows no wavelength specificity because the incident radiation never enters the leaf to interact with photosynthetic pigments, cell walls, water or other cellular constituents (Grant *et al.,* 1987, 1993). Consequently, measurements of the degree of polarization of reflected radiation contain specific information about the leaf surface which is independent of that provided by measurements of diffuse spectral reflectance, providing an 'optical fingerprint' that contains potentially useful information to aid the classification of species by conventional remote sensing methods (Danson and Plummer, 1995; Grant, 1987; Grant *et al.*, 1995). In this respect, measurements of the degree of polarization of the reflected beam are known to be influenced by a host of factors including plant species (McClendon, 1984), leaf orientation, that is adaxial vs. abaxial surface (Grant *et al.*, 1993), leaf development stage (Vanderbilt and Grant, 1985) and length of exposure to field conditions (Grant *et al.*, 1987), while measurements at the canopy level are influenced by leaf angle distributions (Vanderbilt *et al.*, 1985), leaf area index (Vanderbilt and Grant, 1985), canopy development stage (Grant *et al.*, 1995), leaf water content (Egan and Hallock, 1969) and air pollutant injury (Eguchi *et al.*, 1982).

7.3.2 *Photoprotection and leaf water economy*

Ultraviolet radiation (280–400 nm). Ultraviolet (UV) radiation is generally considered detrimental to plant growth (Caldwell, 1971, 1981; Teramura, 1983; Ziska *et al.*, 1992, 1993). However, like other organisms, plants sense and respond to potentially detrimental levels of UV radiation, so damage is generally restricted to occasions when the level of UV radiation exceeds the plant's natural capacity to avoid exposure and defend itself. As concern over stratospheric ozone depletion has grown, greater attention has been paid to determining the mechanisms underlying interspecific and intraspecific differences in the sensitivity of plants to both UV-A (320–400 nm) (Caldwell, 1971) and UV-B (280–320 nm) radiation (Day, 1993; Day *et al.*, 1992, 1994; Vogelmann, 1993). Probably the most easily envisaged protective mechanism is the exclusion of UV radiation at the air–cuticle interface, the first refractive discontinuity experienced by incident radiation. Yet, measurements have revealed that the majority of leaves reflect little UV radiation, commonly less than 10% of the incident flux (Caldwell, 1971; Robberecht *et al.*, 1980). There are known to be exceptions and some species exhibit higher reflectance in the UV (Barnes *et al.*, 1996; Caldwell, 1971; Clark and Lister, 1975). However, little is known about the relative importance of reflectance of UV at the leaf surface. Interestingly, recent research conducted by Grant and colleagues (1995), on two mutants of sorghum (*Sorghum bicolor*) exhibiting contrasting patterns of wax deposition, links the intensity of backward and forward reflectance of UV from plant canopies with the presence or absence of crystalline EW deposits, suggesting a need for more detailed research.

In addition to a possible role for the EW in the reflectance of UV radiation, there is growing evidence that both the waxes and the cuticular membrane may be of direct significance in attenuating the level of UV radiation reaching sensitive targets in the leaf mesophyll. Fibre-optic microprobe measurements show that UV radiation is generally attenuated much more readily than visible wavelengths and that the majority of UV screening occurs in the outer layers of tissue (Cen and Bornman, 1993; Vogelmann *et al.*, 1988). To date, interest has remained firmly focused on the concentration of UV-absorbing compounds in the epidermis, and correlations have been found between the concentrations of absorbing compounds in the epidermis and the depth of penetration of UV radiation in a wide variety of species (Bornman and Vogelmann, 1988, 1991; Day, 1993; Day *et al.*, 1993). Rather less attention has been paid to the role played by the cuticle and waxes (Barnes, 1994). Yet, recent reports indicate that the cuticles isolated from a range of leaves and fruits exhibit strong absorbance in the UV; transmission ranging from 1 to 55% at 300 nm (Krauss *et al.*, 1996; Scherbatskoy, 1994). Reports of UV absorbance by leaf wax extracts are more variable, with some authors reporting strong absorbance peaks in the UV (Bornman and Vogelmann, 1988; Günthardt-Goerg, 1996), while others find that the waxes absorb either very weakly (Krauss *et al.*, 1996; Wuhrmann-Meyer and Wuhrmann-Meyer, 1941) or not at all (Day *et al.*, 1992; Robinson *et al.*, 1993) at UV wavelengths. The contradictory nature of these findings may reflect genuine differences between the UV absorbing properties of the waxes extracted from different species or may reflect differences in experimental protocol. It is of interest to note in this regard, that phenylpropanoid compounds (e.g. ferulic acid and flavonoid aglycones) are common constituents of the cuticle, and are known to be extruded on to the surfaces of the leaves of a number of species (Wollenweber, 1982). There is, however, a clear need for further research to establish the significance of the cuticle, and associated EW, in protecting against potentially damaging levels of UV radiation.

Visible radiation (400–700 nm). It has been recognized since the turn of the century that many plants native to high radiation environments develop a thick layer of wax on the leaf surface (e.g. species of *Aloe*, *Kalanchoë*, *Cotyledon*, *Yucca*, *Sedum* and *Kleinia*), and early investigators such as Haberlandt (1914) suggested that the presence of such a wax layer might play a significant role in photoprotection. The literature abounds with reports that the surface waxes can increase the reflectance of visible radiation, substantially decreasing the portion of incident radiation absorbed by leaves (Blum, 1975; Cameron, 1970; Eller, 1979; Eller and Willi, 1977; Johnson *et al.*, 1983; Pearman, 1966; Radgg *et al.*, 1988; Reicosky and Hanover, 1975; Thomas and Barber, 1974, Uddin and Marshall, 1988). However, it was not until the recent work of Robinson and her associates (Robinson and Osmond, 1994; Robinson *et al.*, 1993) on *Cotyledon orbiculata*, a succulent which exhibits crassulacean acid metabolism (CAM), that the wax layer was clearly demonstrated to be important in protecting against photoinhibition. The leaves of this species, like many other CAM plants, develop copious quantities of highly reflective EW when plants are grown under high photon flux densities (PFD). Removal of this superficial wax layer increases the absorption of incident radiation by as much as 50% and subsequent exposure of these leaves to full sunlight causes a dramatic reduction in the variable to maximum chlorophyll fluorescence ratio (F_V/F_M), a marked decline in leaf chlorophyll concentrations and an increase in the antheraxanthin and zeaxanthin content (Figure 7.2), effects that are much more pronounced in leaves where the wax layer is left intact. This work provides strong evidence that in some species the EW layer plays an important role in the avoidance of photoinhibition.

It is generally purported that the leaves of only a few plant species preferentially reflect light of certain wavelengths from their surface. However, this conclusion seems to be at odds with the everyday observation that the wax bloom on the surface of leaves and fruits of many species has a blueish cast. The physical mechanisms underlying this optical phenomenon are discussed elsewhere (see Section 7.2.2) and there are several reports of blue-shifts in leaf reflectance spectra in species exhibiting a prominent wax bloom, for example, *Dudleya britonii* (Mulroy, 1979), *Picea pungens* (Bornman and Vogelmann, 1988; Clark and Lister, 1975) and wild-type *Sorghum bicolor* (Grant *et al.*, 1995). Some of these studies, as with those of Johnson *et al.* (1983) on near-isogenic lines of wheat (*Triticum durum* and *Triticum aestivum*) differing in the degree of glaucousness, have sought to link this blue-shift in spectral reflectance with a decline in the rate of net CO_2 assimilation. Invariably, a slight decline in assimilation rate under light-limiting conditions has been found, but the magnitude of these effects is arguably far outweighed by the advantages to be gained from the presence of a superficial wax layer in high UV environments.

Infrared radiation (700–3000 nm). The photosynthetic pigments are transparent to both near (700–1350 nm) and far infrared (1350–3000 nm) radiation so absorption at these wavelengths is mainly determined by leaf structure and water content and this exerts a strong influence on leaf energy budgets (Gausman, 1985). In the previous section we have described how the removal of the reflective wax layer can increase the amount of visible radiation absorbed by leaves, and in many species these effects extend into the infrared region (Figure 7.1). Differences in the absorption of visible and infrared radiation due to the presence of a reflective wax layer would be expected to result in lower leaf temperatures and thus reduced rates of transpiration, since the saturating water vapour pressure at leaf temperature is a major determinant of the forces driving transpiration (Ehleringer, 1981). In support of these conclusions, data are available to demonstrate that plants with an intact or particularly well-developed reflective wax layer exhibit lower leaf and canopy temperatures, reduced rates of transpiration and improved leaf water status

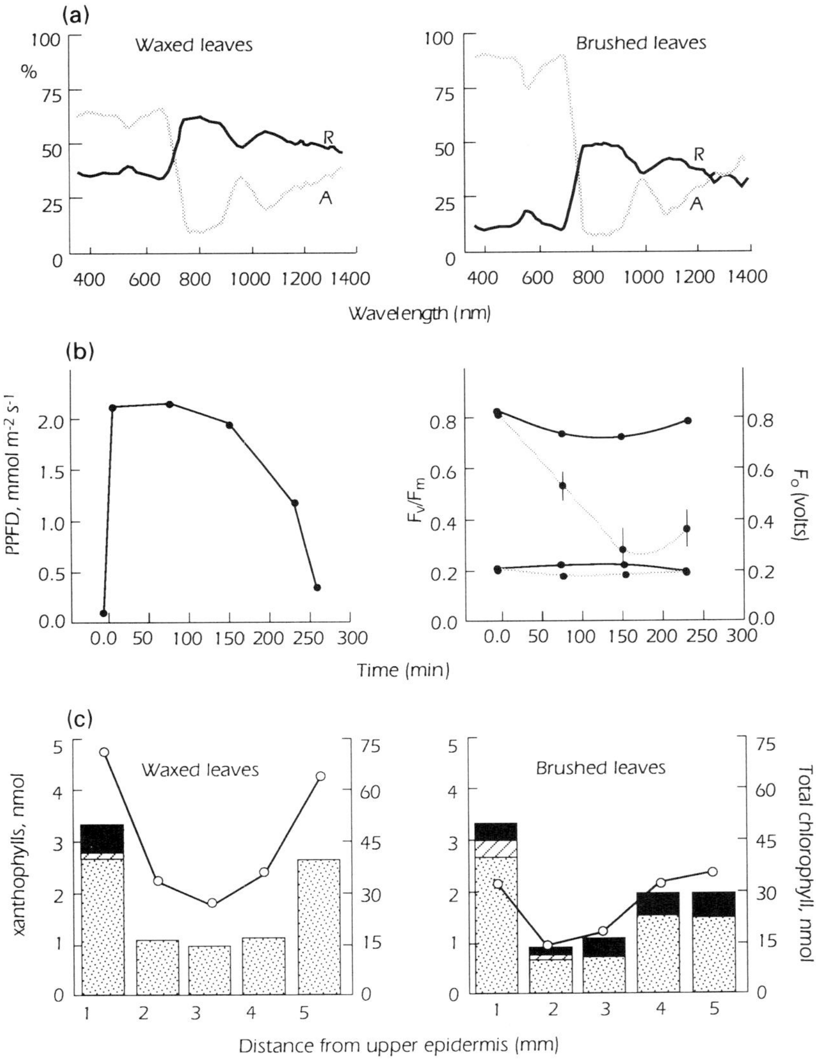

Figure 7.2. *Significance of epicuticular wax (EW) layer in protecting against photoinhibition in* Cotyledon orbiculata. *(a) Reflectance (R)/absorbance (A) spectra of leaves before and after brushing to remove surface wax layer. (b) Photon flux density (PFD) of incident sunlight on experimental plots during the period that chlorophyll fluorescence measurements were made on waxed (solid lines) and brushed (broken lines) leaves in the field. (F_V/F_M, the ratio of variable to maximum chlorophyll fluorescence, F_O, initial level of fluorescence.) (c) Changes in the concentration of violaxanthin (▦), antheraxanthin (▨), zeaxanthin (■) and total chlorophyll (o) (nmols) in successive layers of waxed and brushed leaves after 12 days' exposure to natural conditions. ((a) Redrawn from Sinclair and Thomas (1970). (b,c) Reproduced from Robinson* et al. *(1993) Wax as a mechanism for protection against photoinhibition: a study of* Cotyledon orbiculata. Bot. Acta ***106:*** *307–312, with permission from Georg Thieme Verlag.)*

(Jefferson *et al.*, 1989; Johnson *et al.*, 1983; Premachandra *et al.*, 1994; Robinson *et al.*, 1993). This means that the presence of a reflective EW layer may result in increases in leaf water-use efficiency (WUE; molar ratio of CO_2 fixed:water transpired) without necessarily invoking a role for the EW on the permeability of the cuticle to water (see Chapter 4). In combination with the added protection afforded against high PFDs, heat stress and damaging levels of UV radiation this may help to explain why glaucousness is a trait linked with improved growth and productivity of crops such as wheat (Johnson *et al.*, 1983; Richards *et al.*, 1986) and sorghum (Blum 1975; Jordan *et al.* 1984; Premachandra *et al.*, 1994) in environments where soil water is limiting.

7.4 Influence of solar radiation on the cuticular membrane and epicuticular wax

7.4.1 *Ultraviolet radiation (280–400 nm)*

It has long been known that the physicochemical nature of the leaf surfaces of glasshouse-grown, and controlled environment-grown, plants are different from those of plants in the field, but the reasons for these differences are not well understood (Cape and Percy, 1993; cf. Section 12.4.2). The materials used to roof glasshouses, and controlled-environment cabinets, do not generally transmit UV-B radiation and some authors have suggested that this may be one factor contributing to the differences in leaf-surface properties observed (Martin and Juniper, 1970). However, despite the recent upsurge in interest in the impacts of UV-B radiation on vegetation, promoted by concerns over depletion of the stratospheric ozone layer, relatively little attention has been paid to the impacts of UV-B radiation on the cuticle and EW or the way in which subtle UV-B-induced modifications in the cuticle and EW may influence subsequent responses to other environmental stresses (Barnes, 1994; Barnes *et al.*, 1994). Recent studies indicate that UV-B radiation may have marked effects on the production and chemical composition of the EW in a range of crop plants (Barnes *et al.*, 1994, 1996; Gonzalez *et al.*, 1996; Steinmüller and Tevini, 1985; Tevini and Steinmüller, 1987). Wax production is generally stimulated by elevated levels of UV-B radiation (Figure 7.3), but the extent of the effect seems to depend upon the biologically effective dose of UV-B radiation, plant species and leaf orientation (Barnes *et al.*, 1994, 1996; Gonzalez *et al.*, 1996). UV-B radiation is also known to cause subtle changes in the chemical composition of the waxes extruded on to the leaf surface, with greater effects generally observed on the adaxial vs. the abaxial surface of leaves (Figure 7.4), and consistent shifts in alkane and ester homologue distribution patterns (Barnes *et al.*, 1994, 1996; Gonzalez *et al.*, 1996; Tevini and Steinmüller, 1987); UV-B exposure generally leading to an increase in the ratio of short to long chain-length wax homologues (Table 7.2). The pattern of these changes has been interpreted as evidence for a direct and highly specific effect of UV-B radiation on microsomal-based elongases responsible for the addition of C_2 units within the elongase-decarboxylase pathway (*see* Chapter 3). Moreover, it is of interest to note that elevated levels of UV-B radiation have also been shown to influence cuticular properties, for example leaf surface wettability (Barnes *et al.*, 1996) and cuticular water permeance (Kerstiens *et al.*, 1996) which may, in turn, have subtle but far-reaching physiological consequences for plants *in vivo*.

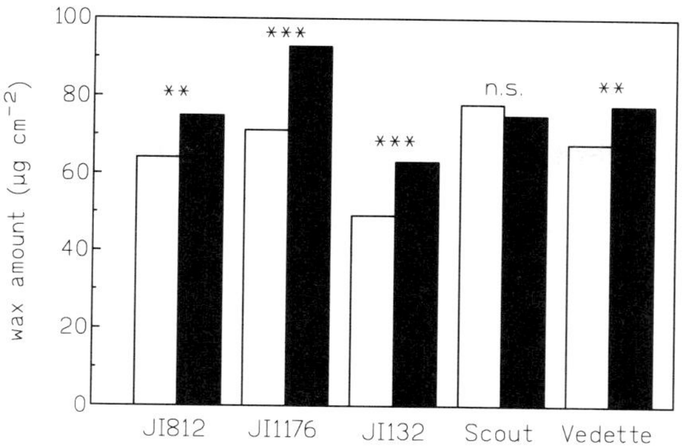

Figure 7.3. *Wax amount extracted from the leaf surfaces of five genotypes of pea (*Pisum sativum *L.) grown under controlled conditions (maximum irradiance of 950 (μmol m^{-2} day^{-1}) at two daily biologically effective doses of ultraviolet-B (UV-B) radiation: 0 (□) and 6.5 (■) kJ m^{-2} day^{-1}. The elevated treatment corresponds to an ozone depletion of approximately 20% using the current modelled clear-sky dose for July in Southern England of 4.3 kJ m^{-2} day^{-1} as a guideline. (n.s., not significant; **p<0.01; ***p<0.001.) (Recalculated from data presented in Gonzalez* et al.*, 1996.)*

7.4.2 *Visible radiation (400–700 nm)*

Cuticle thickness, and the amount of EW deposited on the external surfaces of leaves, varies between plants grown in different light environments and even between leaves on the same plant dependent on their orientation and position in the canopy. There is a great deal of evidence to show that plants grown under high PFDs produce more EW than those grown at low PFD or under shade screens (Baker, 1974; Berg, 1985; Cameron, 1970; Eglinton and Hamilton, 1967; Giese, 1975; Macey, 1970; Reed and Tukey, 1982; Whitecross and Armstrong, 1972) and it is generally considered that PFD is one of the most important factors determining wax production. One of the most elegant demonstrations of the impact of PFD on wax production was conducted by Juniper (1960). He grew peas (*Pisum sativum* L. cv. Alaska) in total darkness at 18°C for 8 days before transferring them to a PFD approximately equivalent to full English summer daylight. Plants grown in the dark were almost totally devoid of the crystalline waxes characteristic of field-grown plants, but within 24 h of transfer to the light wax crystals became visible and within a matter of days the leaf surface closely resembled that of control plants grown continuously in high light. In similar experiments, but this time using shade screens placed over a species of *Eucalyptus*, Hallam (1970) demonstrated that plants needed to be exposed to approximately 20% of full sunlight to produce normal amounts of wax on the leaf surface. The effects of radiant energy have generally been found to affect the geometric configuration of the wax on the surfaces of leaves rather than its chemical composition (Baker, 1974, 1982; Giese, 1975), indicating that the intensity of incident radiation influences wax production rather than its biosynthesis (Tevini and Steinmüller, 1987).

It has been known for many years that the cuticular membrane is directly influenced by the intensity of incident radiation and there is a substantial body of evidence to show that plants grown under high PFDs produce thicker cuticles, with the adaxial cuticle usually more substantial than the abaxial (Baker, 1974; Juniper and Jeffree, 1983; Martin and Juniper, 1970). Indeed, differences in cuticle thickness

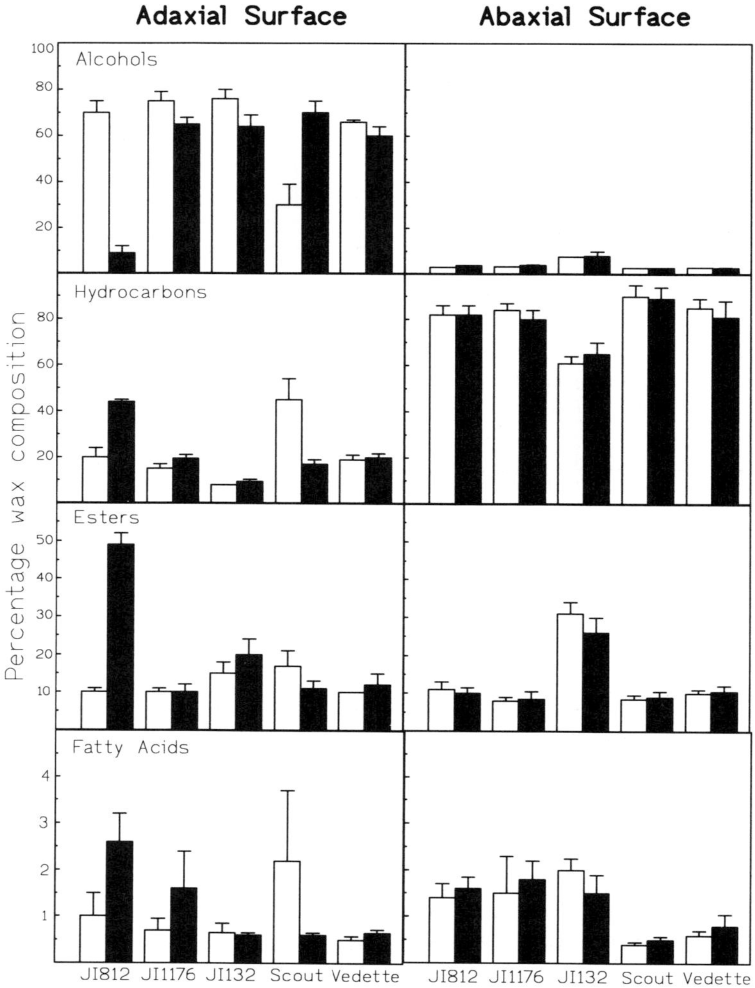

Figure 7.4. *Leaf surface epicuticular wax (EW) chemical composition of five genotypes of pea (*Pisum sativum *L.) grown under controlled conditions (maximum irradiance of 950 µmol m^{-2} day^{-1}) at two daily biologically effective doses of ultraviolet-B (UV-B) radiation: 0 (□) and 6.5 (■) kJ m^{-2} day^{-1}. The elevated treatment corresponds to an ozone depletion of approximately 20% using the current modelled clear-sky dose for July in Southern England of 4.3 kJ m^{-2} day^{-1} as a guideline. (Redrawn from Gonzalez* et al.*, 1996.)*

are one of the anatomical characteristics used to distinguish between 'sun' and 'shade' leaves (Givnish, 1988; Osborn and Taylor, 1990). It may also be worthy of note that, in some studies, a strong correlation has been found between cuticular thickness and the light-saturated rate of CO_2 assimilation of individual leaves (Ashton and Berlyn, 1992) and that cuticle thickness is generally influenced more by leaf temperature than PFD (Juniper and Jeffree, 1983).

Table 7.2. *The influence of controlled levels of UV-B radiation on the straight-chain (n-alkane) and branched-chain (br-alkane) homologue ratios[1] of wax recovered from the adaxial and abaxial surfaces of two tobacco (*Nicotiana tabacum* L. cv. Samsun) genotypes: wild-type plants (WT) and GR32-3, a mutant transgenic for chimeric glutathione reductase based on a pea GR cDNA sequence*

Daily UV-B_{BE} dose[2] ($kJ\ m^{-2}$)	Fatty acids		*n*-Alkanes		*br*-Alkanes	
	WT	GR32-3	WT	GR32-3	WT	GR32-3
Adaxial surface						
0.00	1.178[a]	2.908[b]	0.174[f]	0.268[k]	0.414[w]	0.449[wz]
4.54	1.348[a]	3.371[b]	0.321[gk]	0.313[k]	0.57[x]	0.57[x]
5.66	2.760[b]	4.515[c]	0.397[g]	0.620[l]	0.588[x]	0.790[y]
Abaxial surface						
0.00	1.915[a]	3.459[b]	0.321[gk]	0.345[gk]	0.424[w]	0.454[wz]
4.54	2.323[ab]	1.940[a]	0.293[k]	0.377[g]	0.443[wz]	0.486[z]
5.66	1.417[a]	2.899[b]	0.576[h]	0.441[g]	0.545[x]	0.459[wz]
Main effects						
UV-B	ns		***		***	
Surface	ns		ns		**	
Genotype	*		ns		ns	
UV-B x surface	ns		ns		*	
UV-B x genotype	ns		ns		ns	
UV-B x surface x genotype	ns		*		ns	

Reproduced from Barnes *et al.* (1996) The influence of UV-B radiation on the physicochemical nature of tobacco (*Nicotiana tabacum* L.) leaf surfaces. J. *Exp. Bot.* **47:** 99–109, with permission from Oxford University Press.
Treatment means bearing the same letter are not significantly different at the 5% level. Significance of main effects: ns, not significant; $^{*}p<0.05$; $^{**}p<0.01$; $^{***}p<0.001$.

[1] Ratios were calculated by dividing total proportions of short-chain homologues ($C_{<30}$) by long-chain components ($C_{>30}$).

[2] Daily biologically effective UV-B dose calculated using the generalized plant action spectrum normalized at 300 nm (Caldwell, 1971). The daily UV-B_{BE} dose of 4.54 $kJ\ m^{-2}$ is approximately equivalent to that received at Lancaster, England (54°N, 2°W) under present day clear-sky conditions in midsummer, and the UV-B_{BE} dose of 5.66 $kJ\ m^{-2}$ represents a *c.*25% increase in the midsummer UV-B flux at this location.

7.5 Conclusions

Leaves are neither perfectly diffuse (i.e. Lambertian) nor perfectly specular (i.e. non-Lambertian) reflectors of incident radiation. The diffuse component of leaf reflectance emanates primarily from the interior of the leaf through multiple scattering, although there may be a small contribution from scattering due to rough elements at the leaf surface, while the specular component originates from reflectance at the air–cuticle interface, the first refractive discontinuity encountered by incident radiation. The magnitude and angular dependence of this surface reflectance is strongly influenced by the size, orientation and distribution of wax crystals (and other particles) on the leaf surface and much of the surface reflectance is linearly polarized. This provides a means to separate the reflectance at the air–cuticle interface from that originating from the leaf interior. Consequently, measurements of the degree of polarized reflected radiation contain inherent information about the leaf surface. They also provide potentially useful additional discriminatory information to aid vegetation classification using conventional remote sensing methods.

There is clear evidence that the cuticle and EW play an important role in determining the quantity and quality of radiation reaching the leaf mesophyll, but many important gaps still remain in our knowledge, particularly in relation to the impact of the cuticle on the penetration of UV-B radiation. The development of a thicker cuticle and the production of copious quantities of highly reflective wax on the surface of leaves, and other plant components, render increased protection against high PFDs and UV radiation, while improving leaf water relations through effects on leaf temperature. This type of external photoprotection also applies to other features which render leaves reflective, for example, epidermal salt crusts (Ehleringer, 1981) and leaf hairs (Ehleringer and Mooney, 1978) and it is generally considered that waxy leaves confer a selective advantage to plants growing in high radiation environments. It is, however, important to recognize that the cuticle and its associated wax layer not only influence the quantity and quality of the radiation absorbed by leaves, but also that, vice versa, light quantity and quality influence the physicochemical nature and properties of the cuticle and waxes *per se*.

Acknowledgements

The authors thank Professor A.W. Davison and Dr. A.M. Borland for their assistance and comments on an earlier draft of the manuscript and Mrs. J. Preece for reproducing illustrations. This manuscript is dedicated to J.D.B.'s father who died suddenly during the preparation of this transcript. J.D.B. is indebted to the Royal Society for continued financial support.

References

Antyufeev, V.S. and Marshak, A.L. (1990) Inversion of Monte Carlo model for estimating vegetation canopy parameters. *Remote Sens. Environ.* **33**, 201–209.

Ashton, P.M.S. and Berlyn, G.P. (1992) Leaf adaptations of some *Shorea* species to sun and shade. *New Phytol.* **121**, 587–596.

Baker, E.A. (1974) The influence of the environment on leaf wax development in *Brassica oleracea* var. *Gemmifera*. *New Phytol.* **73**, 955–966.

Baker, E.A. (1982) Chemistry and morphology of plant EW. In: *The Plant Cuticle* (eds D.F. Cutler, K.L. Alvin and C.E. Price). Academic Press, London, pp. 139–166.

Baret, F. (1995) Use of spectral reflectance variation to retrieve canopy biophysical characteristics. In: *Advances in Remote Sensing* (eds F.M. Danson and S.E. Plummer). John Wiley & Sons, New York, pp. 33–52.

Barnes, J.D. (1994) Session V: Rapporteur's Report. In: *Air Pollutants and the Leaf Cuticle,* NATO ASI Series Vol. 36, (eds K.E. Percy, J.N. Cape, R. Jagels and C.J. Simpson). Springer-Verlag, Berlin. pp. 385–386.

Barnes, J.D., Paul, N.D., Percy, K.E. *et al.* (1994) Effects of UV-B radiation on wax biosynthesis. In: *Air Pollutants and the Leaf Cuticle,* NATO ASI Series Vol. 36, (eds K.E. Percy, J.N. Cape, R. Jagels and C.J. Simpson). Springer-Verlag, Berlin, pp. 195–205.

Barnes, J.D., Percy, K.E., Paul, N.D. *et al.* (1996) The influence of UV-B radiation on the physicochemical nature of tobacco (*Nicotiana tabacum* L.) leaf surfaces. *J. Exp. Bot.* **47**, 99–109.

Berg, V.S. (1985) Physical, chemical and spectral properties of leaf surfaces of *Berberis aquifolium*. *Plant Cell Environ.* **8**, 631–638.

Blum, A. (1975) Effect of the *Bm* gene on epicuticular wax deposition and the spectral characteristics of sorghum leaves. *SABRAO J.* **7,** 45–52.

Bornman, J.F. and Vogelmann, T.C. (1988) Penetration of blue and UV radiation measured by fiber optics in spruce and fir needles. *Physiol. Plant.* **72,** 699–705.

Bornman, J.F. and Vogelmann, T.C. (1991) Effect of UV-B radiation on leaf optical properties measured with fibre optics. *J. Exp. Bot.* **42,** 547–554.

Brakke, T.W., Wergin, W.P., Erbe, E.F. *et al.* (1993) Seasonal variation in the structure and red reflectance of leaves from yellow poplar, red oak and red maple. *Remote Sens. Environ.* **43,** 115–130.

Breece HT, Holmes RH. (1971) Bidirectional scattering characteristics of healthy green soybeans and corn *in vivo*. *Appl. Opt.* **10,** 119–127.

Caldwell, M.M. (1971) Solar UV irradiation and the growth and development of higher plants. In: *Photophysiology* (ed. A.C. Giese). Academic Press, London, pp. 131–177.

Caldwell, M.M. (1981) Plant response to solar ultraviolet radiation. In: *Encyclopedia of Plant Physiology: Physiological Plant Ecology I,* New Series Vol. 12A, (eds O.L. Lange, P.S. Nobel, C.B. Osmond and H. Ziegler). Springer-Verlag, Berlin, pp. 169–197.

Cameron, R.J. (1970) Light intensity and the growth of *Eucalyptus* seedlings. II. The effect of cuticular waxes on light absorption in leaves of *Eucalyptus* species. *Aust. J. Bot.*, **18,** 275–284.

Cape, J.N. and Percy, K.E. (1993) Environmental influences on the development of spruce needle cuticles. *New Phytol.* **125,** 787–799.

Cen, Y.-P. and Bornman, J.F. (1993) The effect of exposure to enhanced UV-B radiation on the penetration of monochromatic and polychromatic UV-B radiation in leaves of *Brassica napus*. *Physiol. Plant.* **87,** 249–255.

Clark, J.B. and Lister, G.R. (1975) Photosynthetic action spectra of trees. *Plant Physiol.* **55,** 407–413.

Danson, F.M. and Plummer, S.E. (1995) *Advances in Remote Sensing*. John Wiley & Sons, New York.

Day, T.A. (1993) Relating UV-B radiation screening effectiveness of foliage to absorbing compound concentration and anatomical characteristics in a diverse group of plants. *Oecologia* **95,** 542–550.

Day, T.A., Vogelmann, T.C. and DeLucia, E.E. (1992) Are some plant life forms more effective than others in screening out ultraviolet-B radiation? *Oecologia* **92,** 513–519.

Day, T.A., Martin, G. and Vogelmann, T.C. (1993) Penetration of UV-B radiation in foliage: evidence that the epidermis behaves as a non-uniform filter. *Plant Cell Environ.* **16,** 735–741.

Day, T.A., Howells, B.W. and Rice, W.J. (1994) Ultraviolet absorption and epidermal-transmittance spectra in foliage. *Physiol. Plant.* **92,** 207–218.

Egan, W.G. and Hallock, H.B. (1969) Coherence-polarization phenomenon in remote sensing. *Proc. IEEE* **57,** 621–628.

Eglinton, G. and Hamilton, R.J. (1967) Leaf epicuticular waxes. *Science* **156,** 1322–1335.

Eguchi, H., Hamakoga, M. and Matsui, T. (1982) Digital image processing in polarized light for evaluation of foliar injury. *Environ. Exp. Bot.* **22,** 277–283.

Ehleringer, J. (1981) Ecology and ecophysiology of leaf pubescence in North American desert plants. In: *Biology and Chemistry of Plant Trichomes* (eds E. Rodriguez, P.L. Healey and I. Mehta). Plenum Press, New York, pp. 113–132.

Ehleringer, J. and Mooney, H.A. (1978) Leaf hairs: effects on physiological activity and adaptive value to a desert shrub. *Oecologia* **37,** 183–200.

Eller, B.M. (1979) Die Strahlungsökologische Bedeutung von Epidermisauflagen. *Flora* **168,** 146–192.

Eller, B.M. and Willi, P. (1977) Die Bedeutung der Wachsausblühungen auf Blättern von *Kalanchoë pumila* Baker für die Absorption der Globalstrahlung. *Flora* **166,** 461–474.

Fukshansky, L. (1981) Optical properties of plants. In: *Plants and the Daylight Spectrum* (ed. H. Smith). Academic Press, New York, pp. 21–40.

Fung, A.K. (1984) Surface scattering effects at different spectral regions. In: IIe *Coll. Int. Signatures Spectrales d'Objets en télédétection* (eds G. Guyot and M. Verbrugghe). INRA Publication, Nancy, France, pp. 693–707.

Gates, D.M. (1980) *Biophysical Ecology*. Springer-Verlag, New York.

Gates, D.M., Keegan, H.J., Schleter, J.C. *et al.* (1965) Spectral properties of plants. *Appl. Opt.* **4,** 11–20.

Gausman, H.W. (1974) Leaf reflectance of near-infrared. *Photogramm. Eng.* **15,** 183–191.

Gausman, H.W. (1985) *Plant Leaf Optical Properties in Visible and Near-Infrared Light*. Texas Tech Press, Lubbock, TX.

Giese, B.N. (1975) Effects of light and temperature on the composition of epicuticular wax of barley leaves. *Phytochemistry* **14**, 921–929.

Givnish, T.J. (1988) Adaptation to sun and shade: a whole plant perspective. *Aust. J. Plant Physiol.* **15**, 63–92.

Gonzalez, R., Paul, N.D., Percy, K.E. *et al.* (1996) Responses to ultraviolet-B radiation (280–315 nm) of pea (*Pisum sativum* L.) lines differing in leaf surface wax. *Physiol. Plant.* (in press).

Grant, L. (1987) Diffuse and specular characteristics of leaf reflectance. *Remote Sens. Environ.* **22**, 309–312.

Grant, L., Daughtry, C.S.T. and Vanderbilt, V.C. (1987) Variations in the polarized leaf reflectance of *Sorghum bicolor*. *Remote Sens. Environ.* **21**, 333–339.

Grant, L., Daughtry, C.S.T. and Vanderbilt, V.C. (1993) Polarized and specular reflectance: variation with leaf surface features. *Physiol. Plant.* **88**, 1–9.

Grant, R.H., Jenks, M.A., Rich, P.J. *et al.* (1995) Scattering of ultraviolet and photosynthetically active radiation by *Sorghum bicolor*: influence of epicuticular wax. *Agric. For. Meteorol.* **75**, 263–281.

Günthardt-Goerg, M.S. (1996) Leaf surface features (*Populus x euramericana* and *Populus tremula*) developed under lowered or ambient UV-B radiation. *J. Exp. Bot.* **47** (suppl.), 58.

Haberlandt, G. (1914) *Physiological Plant Anatomy*, 4th edition. Macmillan, London.

Hadley, N.F. (1989) Lipid water barriers in biological systems. *Prog. Lipid Res.* **28**, 1–33.

Hallam, N.D. (1970) Growth and regeneration of waxes on the leaves of *Eucalyptus*. *Planta* **93**, 257–268.

Hecht, E. and Zajac, A. (1974) *Optics*. Addison-Wesley Publishing Company, London.

Jacquemoud, S. and Baret, F. (1990) PROSPECT: A model of leaf optical properties spectra. *Remote Sens. Environ.* **34**, 75–91.

Jefferson, P.G., Johnson, D.A. and Asay, K.H. (1989) Epicuticular wax production, water status and leaf temperature in *Triticaceae* range grasses of contrasting visible glaucousness. *Can. J. Plant Sci.* **69**, 513–519.

Johnson, D.A., Richards, R.A. and Turner, R.C. (1983) Yield, water relations, gas exchange, and surface reflectances of near-isogenic wheat lines differing in glaucousness. *Crop Sci.* **23**, 318–325.

Jordan, W.R., Shouse, P.J., Blum, A. *et al.* (1984) Environmental physiology of sorghum. II. Epicuticular wax load and cuticular transpiration. *Crop Sci.* **24**, 1168–1173.

Juniper, B.E. (1960) Growth, development, and effect of the environment on the ultrastructure of plant surfaces. *J. Linn. Soc. (Bot.)* **56**, 413–418.

Juniper, B.E. and Jeffree, C.E. (1983) *Plant Surfaces*. Edward Arnold, London.

Kerstiens, G., Moody, S.A. and Paul, N.D. (1996) Effects of elevated UV-B radiation on cuticular water permeability of ivy grown under natural or artificial light. *J. Exp. Bot.* **47** (suppl.), 58.

Knipling, E.B. (1970) Physical and physiological basis for the reflectance of visible and near-infrared radiation from vegetation. *Remote Sens. Environ.* **1**, 155–159.

Krauss, P.S., Markstädter, C. and Riederer, M. (1996) Attenuation of ultraviolet radiation by plant cuticles. *J. Exp. Bot.* **47** (suppl.), 58.

Kumar, R. and Silva, L. (1973) Light ray tracing through a leaf cross section. *Appl. Opt.* **12**, 2950–2954.

Macey, M. (1970) The effect of light on wax synthesis in leaves of *Brassica oleracea*. *Phytochem*istry **9**, 757–761.

Martin, J.T. and Juniper, B.E. (1970) *The Cuticles of Plants*. Edward Arnold, London.

McClendon, J.H. (1984) The micro-optics of leaves.1. Patterns of reflection from the epidermis. *Am. J. Bot.* **71**, 1391–1397.

Mestre, H. (1935) The absorption of radiation by leaves and algae. *Cold Spring Harbor Symp. Quant. Biol.* **3**, 191–209.

Mulroy, T. (1979) Spectral properties of heavily glaucous and non-glaucous leaves of a succulent rosette plant. *Oecologia* **38**, 349–357.

Myers, V.I. (1983) Remote sensing applications in agriculture. In: *Manual of Remote Sensing. Vol. II. Interpretation and Application*, 2nd Edn (ed. R.N. Colwell). American Society of Photogammetry, Falls Church, VA, pp. 2111–2218.

Osborn, J.M. and Taylor, T.N. (1990) Morphological and ultrastructural studies of plant cuticular membranes. I. Sun and shade leaves of *Quercus velutina* (Fagaceae). *Bot. Gaz.* **151:** 465–476.

Osborne, B.A. and Raven, J.A. (1986) Light absorption by plants and its implications for photosynthesis. *Biol. Rev.* **61,** 1-61.
Pearman, G.I. (1966) The reflection of visible radiation from leaves of some western Australian species. *Aust. J. Biol. Sci.* **19,** 97-103.
Premachandra, G.S., Hahn, D.T., Axtell, J.D. *et al.* (1994) Epicuticular wax load and water-use efficiency in bloomless and sparse-bloom mutants of *Sorghum bicolor* L. *Environ. Exp. Bot.* **34,** 293-301.
Radgg, C.S., Rao, G.G., Devakumar, A.S. *et al.* (1988) Clonal variation in leaf epicuticular waxes and reflectance: possible role in drought tolerance in *Hevea. Ind. J. Natural Rubber Res.* **1,** 84-87.
Reed, D.W. and Tukey, H.B. (1982) Light intensity and temperature effects on epicuticular wax morphology and internal cuticle ultrastructure of carnation and brussel sprouts leaf cuticles. *J. Am. Soc. Hort. Sci.* **107,** 417-420.
Reicosky, D.A. and Hanover, J.W. (1978) Physiological effects of surface waxes. I. Light reflectance for glaucous and nonglaucous *Picea pungens. Plant Physiol.* **62,** 101-104.
Richards, R.A., Rawson, H.M. and Johnson, D.A. (1986) Glaucousness in wheat: Its development and effect on water-use efficiency, gas exchange and photosynthetic tissue temperatures. *Aust. J. Plant Physiol.* **13,** 465-473.
Robberecht, R., Caldwell, M.M. and Billings, W.D. (1980) Leaf ultraviolet optical properties along a latitudinal gradient in the arctic-alpine life zone. *Ecology* **61,** 612-619.
Robinson, S.A. and Osmond, C.B. (1994) Internal gradients of chlorophyll and carotenoid pigments in relation to photoprotection in thick leaves of plants with Crassulacean acid metabolism. *Aust. J. Plant Physiol.* **21,** 497-506.
Robinson, S.A., Lovelock, C.E. and Osmond, C.B. (1993) Wax as a mechanism for protection against photoinhibition: a study of *Cotyledon orbiculata. Bot. Acta* **106,** 307-312.
Ross, J. and Marshak, A. (1989) The influence of leaf orientation and the specular component of leaf reflectance on the canopy bidirectional reflectance. *Remote Sens. Environ.* **27,** 251-260.
Scherbatskoy, T. (1994) Leaf cuticles as mediators of environmental influences: new developments in the use of isolated cuticles. In: *Air Pollutants and the Leaf Cuticle,* NATO ASI Series, Vol. 36,(eds K.E. Percy, J.N. Cape, R. Jagels and C.J. Simpson). Springer-Verlag, Berlin, pp. 149-164.
Sinclair, R. and Thomas, D.A. (1970) Optical properties of leaves from some species in arid South Australia. *Aust J. Bot.* **18,** 261-273.
Sinclair, T.R., Schreiber, M.M. and Hoffer, R.M. (1973) Diffuse reflectance hypothesis for the pathway of solar radiation through leaves. *Agron. J.* **65,** 276-283.
Steinmüller, D. and Tevini, M. (1985) UV-B induced effects upon cuticular waxes of cucumber, bean and barley leaves. In: *Stratospheric Ozone Reduction, Solar UV Radiation and Plant Life* (eds R.C. Worrest and M.M. Caldwell). Springer-Verlag, Berlin, pp. 271-285.
Teramura, A.H. (1983) Effects of ultraviolet-B radiation on the growth and yield of crop plants. *Physiol. Plant.* **58,** 415-427.
Tevini, M. and Steinmüller, D. (1987). Influence of light, UV-B radiation, and herbicides on wax biosynthesis of cucumber seedlings. *J. Plant Physiol.* **131,** 111-121.
Thomas, D.A. and Barber, H.N. (1974) Studies on leaf characteristics of a cline of *Eucalyptus urnigera* from Mount Wellington, Tasmania. II. Reflection, transmission and absorption of radiation. *Aust. J. Bot.* **22,** 701-707.
Torrance, K.E. and Sparrow, E.M. (1967) Theory for off-specular reflection from roughened surfaces. *J. Opt. Soc. Am.* **57,** 1105-1119.
Uddin, M.N. and Marshall, D.R. (1988) Variation in epicuticular wax content in wheat. *Euphytica* **38,** 3-9.
Vanderbilt, V.C. and Grant, L. (1985) Plant canopy specular reflectance model. *IEEE Trans. Geosci. Remote Sens.* **23,** 722-730.
Vanderbilt, V.C., Grant, L. and Daughtry, C.S.T. (1985) Polarization of light scattered by vegetation. *Proc. IEEE* **73,** 1012-1024.
Vogelmann, T.C. (1993) Plant tissue optics. *Annu. Rev. Plant Physiol. Plant Mol. Biol.* **44,** 231-251.
Vogelmann, T.C., Knapp, A.K., McClean, T.M. *et al.* (1988) Measurement of light within thin plant tissues with fiber optic microprobes. *Physiol. Plant.* **72,** 623-630.
Walter-Shea, E.A., Norman, J.M. and Blad, B.L. (1989) Leaf bidirectional reflectance and transmittance in corn and soybean. *Remote Sens. Environ.* **29,** 161-174.
Walter-Shea, E.A., Norman, J.M., Blad, B.L. *et al.* (1991) Leaf reflectance and transmittance in soybean and corn. *Agron. J.* **83,** 631-636.

Whitecross, M.I. and Armstrong, D.J. (1972) Environmental effects on epicuticular waxes of *Brassica napus* L. *Aust. J. Bot.* **20,** 87–95.
Wollenweber, E. (1982) Flavones and flavonols. In: *The Flavonoids: Advances in Research* (eds J.B. Harborne and T.J. Mabry). Chapman and Hall, London, pp. 189–260.
Woolley, J.T. (1971) Reflectance and transmittance of light by leaves. *Plant Physiol.* **47,** 656–662.
Wuhrmann-Meyer, K. and Wuhrmann-Meyer, M. (1941) Untersuchungen über die Absorption Ultravioletter Strahlen durch Kuticular- und Wachsschichten von Blättern I. *Planta* **32,** 43–50.
Ziska, L.H., Teramura, A.H. and Sullivan, J.H. (1992) Physiological sensitivity of plants along an elevational gradient to UV-B radiation. *Am. J. Bot.* **79,** 863–871.
Ziska, L.H., Teramura, A.H., Sullivan, J.H. *et al.* (1993) Influence of ultraviolet-B (UV-B) radiation on photosynthetic and growth characteristics in field-grown cassava (*Manihot esculentum* Crantz.). *Plant Cell Environ.* **16,** 73–79.

8

Fungal attachment and penetration

K. Mendgen

8.1 Introduction

Many plant pathogenic fungi infect their hosts without using wounds or natural openings, such as stomata. They can penetrate the plant epidermis directly, although its structure is composed in such a way as to deter microorganisms (Juniper, 1991). The outermost layer of the epidermis, the cuticle, is a mixture of the polyester cutin and wax (Holloway, 1993; see Chapter 2) anchored to the wall layers below which consist mainly of carbohydrates and proteins (Carpita and Gibeaut, 1993). The way fungi manage to infect the epidermis has been a matter of interest since the earliest studies on plant diseases. DeBary (1884) and Chester (1946) were the first to review the literature on fungal penetration. More recently, Martin (1964) and, some years later, Van den Ende (1974) concentrated on the degradation of the cuticle and its role in disease resistance. Kolattukudy (1985) was the first to summarize the exciting new facts on the molecular qualities of the enzyme cutinase and the mechanism of cutin degradation. In most recent years, reviews have concentrated on the different tasks of cutinase isozymes (Köller, 1991; Köller *et al.*, 1995) and their importance as a pathogenicity factor during the early stages of infection (Schäfer, 1994). Since cutin is only one hurdle in the obstacle race of the fungus to gain access to the epidermal protoplast, more general reviews may also be of interest to the reader. Elad and Evensen (1995) discuss the mode of infection by *Botrytis* and other reviews deal with the different types of fungal infection structures produced by biotrophic or necrotrophic pathogens (Hardham, 1992; Mendgen and Deising, 1993; Mendgen *et al.*, 1996).

This chapter concentrates on recent results explaining the factors that contribute to the initiation of the penetration process. In addition, observations demonstrating modifications of the plant cell wall during fungal penetration are included.

8.2 Initial contact of the fungus with the cuticle

The leaf surface influences adhesion and growth direction of various fungi by providing factors that can be recognized by spores and germ tubes (Allen *et al.*, 1991; Hoch and Staples, 1991). Cuticles may be smooth (Figure 8.1) or covered with a dense array of wax crystals (Figure 8.2). Water may run off more or less easily from

Plant Cuticles, edited by G. Kerstiens.

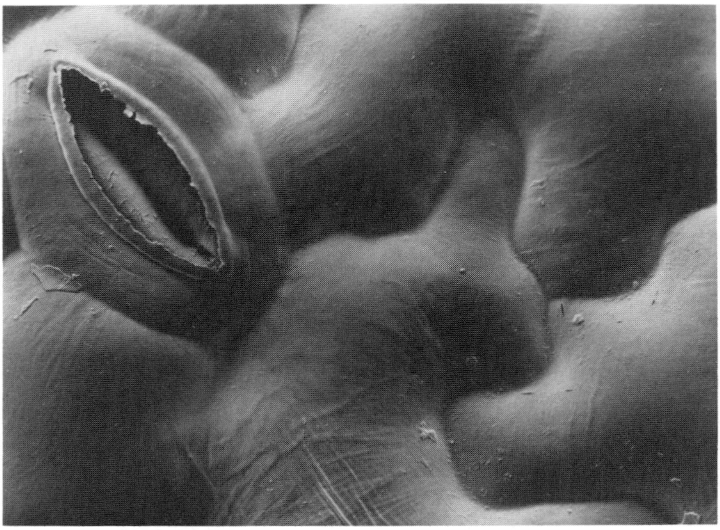

Figure 8.1. *The upper leaf surface of* Vicia faba, *as seen with a low temperature scanning electron microscope (LTSEM; × 200).*

such cuticles. Depending on the plant, the water droplets on the leaf may keep a neutral or reach a more basic pH after a few minutes (Harr and Guggenheim, 1995; see Chapter 16). Fungi have adapted to the characteristic features of their host plant, irrespective of whether the fungi are endophytic (Viret *et al.*, 1993) or pathogenic. They are able to attach under dry and under wet conditions. Therefore, quite a variety of mechanisms exist that help fungi to keep in contact with their host and to penetrate the cuticle. The different mechanisms of adhesion have recently been reviewed by Nicholson (1996) and cannot be dealt with here in detail. In addition, different mechanisms may prevail in adhesion of aquatic fungi (Hyde *et al.*, 1993; Jones, 1994), nematophagous fungi (Tunlid *et al.*, 1992), insect pathogens (Hajek and St. Leger, 1994) or during human pathogenesis (Dykstra and Bradley-Kerr, 1994).

For urediospores of the rust fungi, after landing on a dry leaf, the most important factor in attachment seems to be the hydrophobic interaction between the cuticles of the plant and of the fungal spore (Clement *et al.*, 1994). During this initial stage of infection, the surface contact between urediospores and the cuticle is mediated by the spines emerging from the spore surface (Figure 8.3). Such a urediospore would have less contact on the rice leaf (Figure 8.4) because of the humps produced by the leaf cuticle. Pycnidiospores of *Phyllosticta ampelicida* germinate only on hydrophobic substrata on which they are firmly attached (Kuo and Hoch, 1996).

8.3 Slime production in advance of and during penetration

Conidia of *Uncinuliella australiana*, a powdery mildew fungus, are coated with a networklike layer of extracellular mucilage. This material spreads from conidial surfaces on to a dialysis membrane and forms an adhesion pad between the underside of a conidium and the membrane surface (Mims *et al.*, 1995).

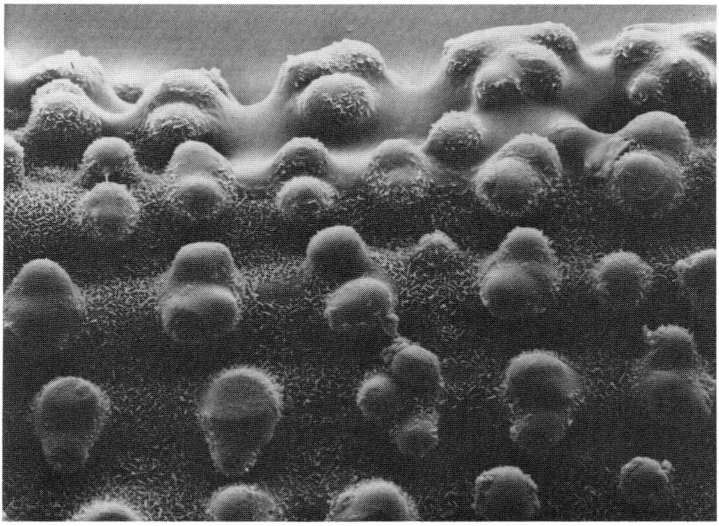

Figure 8.2. *The adaxial leaf surface of* Oryza sativa *as seen with the LTSEM (× 400).*

Conidia of *Erysiphe graminis* release a liquid within 5 min of contact of the spore with the leaf surface (Nicholson *et al.*, 1993). This liquid not only erodes a portion of the leaf cuticle, but also makes the surface of the conidium hydrophilic and the surface of the leaf less hydrophobic. Thus, the fungus prepares the infection court (see also Chapter 13).

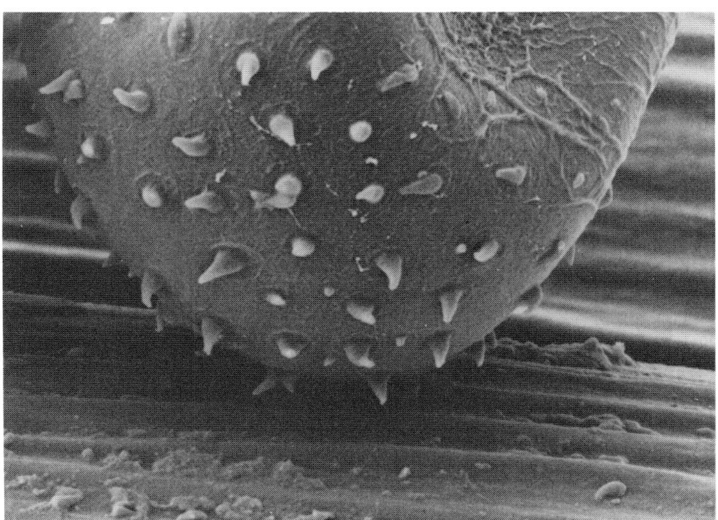

Figure 8.3. *Contact zone between the urediospore of* Uromyces viciae-fabae *and the upper leaf cuticle of* Vicia faba. *Note the spines on the spore surface. (LTSEM micrograph, × 500.)*

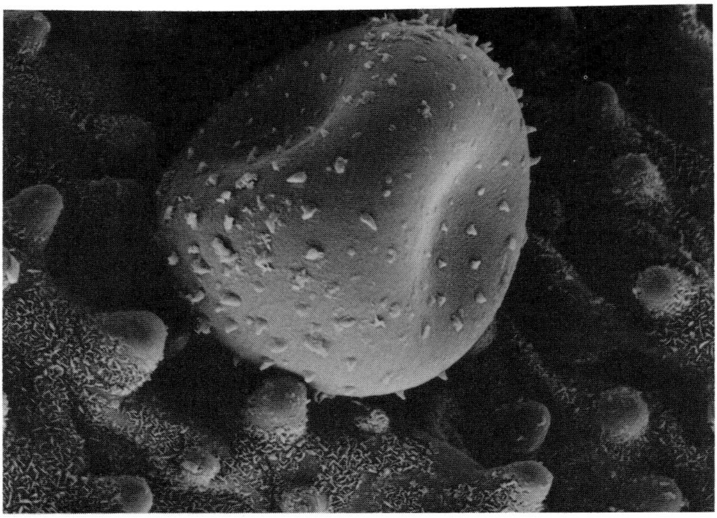

Figure 8.4. *Urediospore of* Uromyces viciae-fabae *on the adaxial cuticle surface of* Oryza sativa, *a non-host. The bumps on the cuticle obviate good contact. (LTSEM micrograph, × 500.)*

In the presence of water, spores of *Cochliobolus heterostrophus* and *Colletotrichum graminicola,* produce a matrix within 20–30 min that accumulates between spores and the plant cuticle as a result of metabolic activities of the fungus (Braun and Howard, 1994; Mercure *et al.*, 1995). A similar situation is observed with fungal endophytes; for example, the beech endophyte *Discula umbrinella* (Viret *et al.*, 1993). This matrix or slime seems to swell with the availability of water on the cuticle. The matrix components are quite heterogenous (Vreeland and Epstein, 1996). In *Nectria haematococca* macroconidia, the mucilage and a 90 kDa glycoprotein are associated with adhesion (Kwon and Epstein, 1993). In *Botrytis* species and in *Colletotrichum lindemuthianum*, glycoproteins typical for spores or infection structures have been characterized with monoclonal antibodies (Cole *et al.*, 1996; O'Connell *et al.*, 1996). In urediospores of *Uromyces viciae-fabae* (Figure 8.5), cutinases and esterases in the matrix of the adhesion pad contributed, among other factors, to the attachment of the fungus to the cuticle (Deising *et al.*, 1992).

A matrix of slime was also observed between European hazelnut leaves and ascospores of *Anisogramma anomala* (Pinkteron *et al.*, 1995). These spores are released from perithecia during periods of rain. Under such conditions, they must adhere rapidly after deposition to prevent being washed off the host. A preformed adhesive attaches these spores and, a few hours later, a more durable adhesive matrix is secreted at the point where the germ tube emerges from the spore. This germ tube immediately penetrates the host while still below the spore. Only the second step in adhesion depends on spore metabolism. Also germlings of *Phyllosticta amphicida* produce a matrix with a specific pattern during their development (Kuo and Hoch, 1995). Similarly, the matrix compostion of *C. lindemuthianum* varies with the development of infection structures and fimbriae project out (O'Connell *et al.,* 1996).

The recent discovery of hydrophobins (Wessels, 1994), which are hydrophobic proteins that mediate the contact of the fungus with hydrophobic parts of the plant

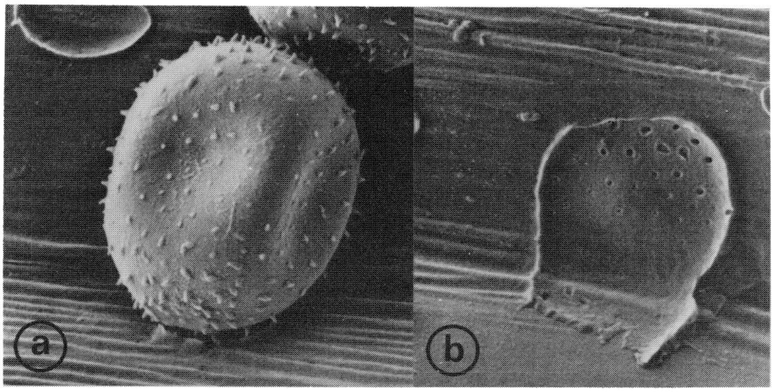

Figure 8.5. *(a) Urediospore of* Uromyces viciae-fabae *with an adhesion pad that mediates contact between plant cuticle and spore about 20 min after wetting the leaf. (b) Removal of the spore leaves the adhesion pad behind. (LTSEM micrographs, × 250.)*

by forming an amphipathic protein layer between fungal hyphae and the hydrophobic plant cuticle, may explain the adhesion process to a certain extent. Coating of glass slides with hydrophobic silicon enhanced adhesion of germ tubes of *Uromyces appendiculatus* to such surfaces (Terhune and Hoch, 1993).

The matrix of slime around spores appears not only typical for plant pathogens. Figure 8.6 shows one of many non-pathogenic microorganisms on an apple leaf. Since all observed microorganisms were surrounded by a matrix, the production of slime is a phenomenon common to many phylloplane fungi, and is not restricted to pathogenic fungi.

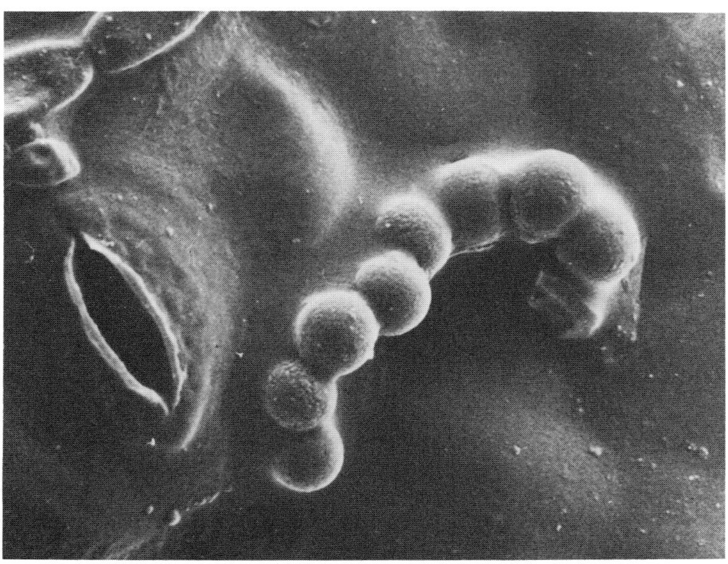

Figure 8.6. *Unknown microorganism on the upper surface of a* Malus *spp. leaf. (LTSEM, × 800.)*

8.4 The influence of water on the infection process

Most infection studies are performed under conditions that do not control the amount of water on the leaf surface. Raspberry cultivars resistant to *Botrytis cinerea* have hairy, waxy canes and their resistance is attributed to greater run off of water (Jarvis, 1994). Thus, the amount of water on the plant surface may influence the interaction of host and pathogens (see also Chapter 14). To study this effect, Williamson *et al.* (1995) dusted rose flowers with dry conidia of *B. cinerea* and incubated them with air of different relative humidities. Conidia germinated in the absence of surface water at 94% relative humidity (r.h.) or above. Germ tubes were smooth, without slime extending over the cuticle and penetrated the epidermis without the formation of appressoria or any structural changes in the cuticle around the germ tube. The authors suggested that no adhesion of the spore may be needed for subsequent infection. Since some germ tubes also had limited contact with the plant surface, penetration of this fungus may require good cuticle contact of the hyphal tip only.

8.5 Morphogenesis of appressorium and penetration hypha

The penetration process of most plant pathogenic fungi is correlated with a series of morphological changes (Figure 8.7), e.g. the differentiation of appressorium and penetration hypha (Mendgen and Deising, 1993). In its simplest form, the germling widens in diameter before penetration occurs and subsequently differentiates a new type of hypha, the penetration hypha with a smaller diameter compared to the germ tube (Figure 8.7). Alternatively, fully developed appressoria may be differentiated (e.g. *Colletotrichum* spp.). This morphogenetic event is preceded by a host signal.

8.5.1 *Signal perception*

Spore germination can be induced by a number of chemical signals (French, 1992). For appressorial development, signals include hardness and hydrophobicity of the substrate (Freytag *et al.*, 1988; Gilbert *et al.*, 1996; Warwar and Dickmann, 1996), heat shock, salts, cyclic AMP (Hoch *et al.*, 1987), starvation (Clay *et al.*, 1994; Jelitto *et al.*, 1994), ethylene (Flaishman and Kollattudkudy, 1994) and components of the cuticle (Kollattukudy *et al.*, 1995). In *Magnaporthe grisea*, the most effective inducers of appressorium development were found to be 1,16-hexadecanedial and 1,16-hexadecanediol (Gilbert *et al.*, 1996). *Uredinales* may recognize topographical signals such as the outer ledge of the stoma or furrows of the cuticle (Hoch and Staples, 1991). Corrêa *et al.*, (1996) suggest that integrin-like proteins are involved in the transmission of such signals from the leaf surface.

8.5.2 *Development of an appressorium*

In a fully differentiated appressorium, the cytoplasm of the germ tube accumulates and a septum separates the appressorial cytoplasm from the highly vacuolated germ tube cytoplasm. The wall of the appressorium may be thickened, or as in the case of *Colletotrichum* spp. and *M. grisea,* enforced with melanin (see below). The appressorium glues to the host surface (Mims *et al.*, 1995). In the case of *M. grisea*, a ring on the base of the appressorium appears to seal the appressorial base to the leaf surface (Howard *et al.*, 1991a).

Deletion of genes expressed during appressorium development of *Colletotrichum gloeosporioides* may interrupt the infection process (Hwang and Kolattukudy, 1995). In *M. grisea,* such a gene coded for a hydrophobin (Talbot *et al.*,

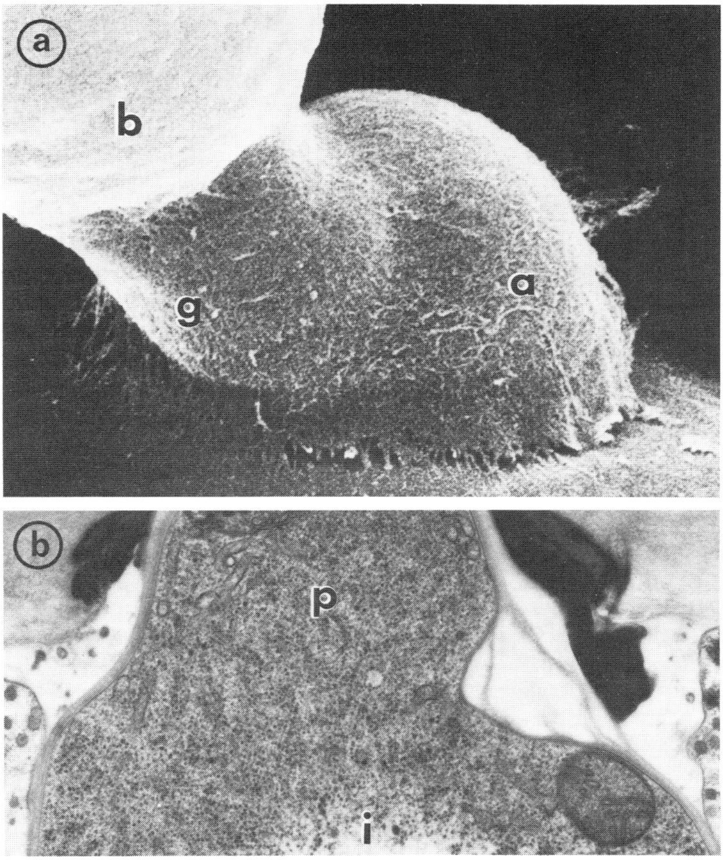

Figure 8.7. *(a) Surface view of a* Uromyces appendiculatus *basidiospore, infecting* Phaseolus vulgaris *(× 5500). (Modified from Gold and Mendgen, 1984.) (b) Cross-section of the* Uromyces vignae *basidiospore, infecting* Vicia faba *(× 5500). (Reproduced from Xu and Mendgen (1994) Endocytosis of 1,3-β-glucans by broad bean cells at the penetration site of the cowpea rust fungus (haploid stage).* Planta ***195****, 282–290, with permission from Springer-Verlag.) (b, basidiospore; g, germ tube; a, appressorium; p, penetration hypha; i, vesicle or infection hypha.)*

1993). Therefore, hydrophobins not only may support adhesion of spores and germ tubes, but also of appressoria.

8.5.3 *Development of the penetration hypha*

In the middle of the appressorial base, the initials of the penetration hypha are characterized by an extremely thin wall, very often surrounded by a cone-like structure. Thus, pressure built up within the appressorium exerts the major force in this place. Vesicles accumulate where penetration hypha development starts. The growing penetration hypha of *Colletotrichum trifolii* does not always grow straight into the host epidermis, as if governed by pressure only, but may turn off in order to avoid the anticlinal walls of the epidermis (Mould *et al.*, 1991). Also, penetration hyphae of *Uredinales* formed in the absence of the host plant, that is on artificial surfaces, still keep their typical shape (Deising *et al.*, 1991). These observations indicate that the penetration hypha itself is able to define its own morphology and direction of growth. The wall of the penetration hypha also is unique. It seems to be modified in such a way as to reduce the elicitation of host defences by masking or modifying the fungal chitin (Deising and Siegrist, 1995).

8.6 Cuticle and cell wall degradation during penetration

The complexity in structure of the cuticle and the plant wall layers below is described in Chapters 2 and 3 by Jeffree and Kolattukudy, respectively. Cuticle and epidermal wall can vary in thickness. In cultivars of rose resistant to *Botrytis cinerea*, the cuticle was thicker as compared to the susceptible cultivar. Under most conditions, this factor seemed to contribute to host plant resistance (Hammer and Evensen, 1994). Studies with transgenic plants having modified wall compositions are not available yet. They would help to elucidate the contribution of the host wall to resist penetration.

The role of fungal enzymes is also unresolved. Since plant pathogenic fungi range from facultative (mostly saprophytic) parasites to obligate parasites entirely dependent on a living host, cell wall-degrading enzymes may enable growth on dead matter and/or contribute to localized penetration into the living cell. This problem was most obvious during the elucidation of the importance of cutinase in pathogenicity. The enzyme is produced by many soil-inhabiting fungi that degrade dead plant material (Heinen, 1963). In plant pathogenic fungi different isoforms of cutinase obviously have different tasks. Conidia of *E. graminis* release cutinase as a component of the conidial exudate to modify the plant cuticle (Pascholati *et al.*, 1993) and in *Uromyces viciae-fabae*, cutinases and esterases contribute to the attachment of the adhesion pad of the spore (Deising *et al.*, 1992). In *B. cinerea* dissolution of the cuticle at the penetration site is suggested by the erosion of the cuticle (Rijkenberg *et al.*, 1980). The enzyme also accumulates around appressoria of *Colletotrichum* spp. (Podila *et al.*, 1995). There, the enzyme may soften the host cuticle in the area of the appressorium. Subsequently the penetration hypha may penetrate by force. Köller *et al.* (1995) have found distinct cutinase isoforms in *Alternaria brassicola,* which may contribute either to saprophytic or pathogenic proliferation. It remains a challenging task to demonstrate a special cutinase isoform at the tip of the penetration hypha. The discussion of this problem is exciting (Kolattukudy *et al.*, 1995; Köller *et al.*, 1995; Schäfer, 1994; Stahl *et al.*, 1994) since the importance of cutinase for infection varied with the conditions used for the infection experiments. We learn from this debate that infection conditions need to be controlled rigorously and have to be as natural as possible.

Work on enzymes that dissolve the wax in the cutin is still completely missing. This work is needed since 'waxases' and cutinases might cooperate during cuticle degradation.

Many components of the plant cell wall are in close contact with or reach into the cutin layer (see Chapter 2). Degradation of polysaccharides may occur by targeted secretion of enzymes. Cellulase production by *Ustilago maydis* occurs during pathogenic development of the fungus only (Schauwecker *et al.*, 1995). *Venturia inaequalis* produces cellulases after contact with cellulose sheets or solvent-extracted apple leaves. Any mechanical disintegration of the leaf substrate resulted in a complete loss of inductive capacity (Kollar, 1994). Presumably, topographic features of the plant surface can induce enzyme production.

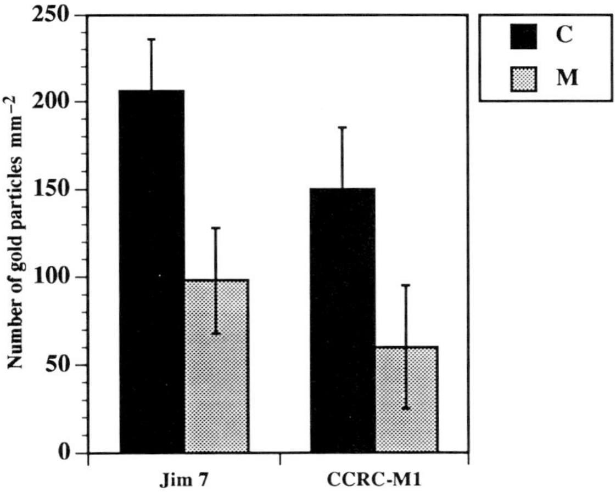

Figure 8.8. *Localized degradation of pectin (detected with the monoclonal antibody Jim 7) and xyloglucan (detected with the monoclonal antibody CCRC-M1) in a host wall area, 0.2 µm wide along the penetration hypha of* Uromyces vignae. *Degradation is indicated by the number of gold particles bound to the antibodies. (M, enzymatically modified or degraded cell wall; C, normal, non-modified cell wall.) (H. Xu and K. Mendgen, unpublished data.)*

Basidiospores of *Uromyces vignae* obviously penetrate the cuticle with the help of enzymes (Figures 8.7 and 8.8). With antibodies against typical wall constituents, it has been possible to discover localized degradation of pectin and xyloglucans just below the cuticle in the contact zone with the penetration hypha of *Uromyces vignae*. The amount of epitopes for these wall constituents was reduced by 50% (Figure 8.8, H. Xu and K. Mendgen, unpublished data). However, enzymes may be secreted without a role in host wall penetration. Uredospores of the rust fungi germinate without saprophytic phase and produce a number of cell wall-degrading enzymes during stomatal penetration. Only later, during differentiation of haustorial mother cells, do these biotrophic fungi dissolve host walls at very small areas at the tip of the penetration hypha that will give rise to a haustorium. The parasite must be able to handle these enzymes in a way that they are restricted to a tiny area of action. This might be performed by regulated secretion and by binding of the enzyme to the host wall. The high isoelectric point of some cell wall-degrading enzymes (Deising *et al.*, 1995; Deising *et al.*, 1996) suggests that they are bound to the host wall during penetration.

Thus, it is very difficult to prove the role of single wall-degrading enzymes during penetration and subsequent development of plant pathogenic fungi. Although Walton (1994) produced a series of mutants deficient in different enzymes, none of these differed from the wild type in their pathogenicity; for example, deletion of exo β-1,3-glucanase in *Cochliobolus carbonum* led to a restriction in saprophytic growth but did not affect pathogenic development (Schaeffer *et al.*, 1994).

8.7 Turgor pressure

Many fungi such as *Colletotrichum, Magnaporthe, Cochliobolus* or the dikaryon of rusts develop highly differentiated appressoria. *M. grisea* and *C. lindemuthianum* are known to penetrate inert materials such as teflon or epoxy resin (Howard *et al.*, 1991a). A turgor pressure of 8 MPa in appressoria of *M. grisea* (Howard *et al.*, 1991b) enables the fungus to penetrate inert surfaces. Melanization of the appressorial wall seems to be a prerequisite for the ability of the fungus to build up this pressure. Chemically induced melanin-deficient mutants were both unable to form melanized appressoria and apathogenic (Chumley and Valent, 1990; Kubo and

Furusawa, 1991). These observations support the idea of a decisive role of turgor during penetration of host walls by some fungi.

8.8 The cytoskeleton

Since growth of the hyphal tips of *Achlya bisexualis* and *Saprolegnia ferax* proceeds under conditions where turgor could no longer be measured, other factors must contribute to this type of movement (Harold *et al.*, 1995; Money and Harold, 1993). In this case, actin filaments may play the crucial role. Microfilaments grow by the addition of actin monomers to the barbed end that abuts upon the plasma membrane. Elongation is thought to involve detachment of the filament from the membrane and the insertion of a new monomer.

For the acrosomal process of the Thyone sperm, with a diameter of 0.1 μm that contains 140 microfilaments, Cooper (1991) has calculated a force of 60×10^4 dyn cm^{-2} corresponding to a pressure of 0.06 MPa. If the density of filaments would be the same in a fungal hyphal tip, Harold *et al.* (1995) have calculated an actin pressure of 0.004 MPa for a fungal hypha with a larger radius. Since penetration hyphae generally have a smaller diameter, the pressure might differ to some extent at the tip, but the contribution of the actin cytoskeleton to the tip pressure is still only in the range of 1% of considered standard osmotic pressure of 0.4 MPa in fungal hyphae.

8.9 Initial plant defence: cell wall alteration during penetration

Nothing is known about the events during and after cuticle penetration when the fungus contacts the wall layers immediately below. Studies with abraded epidermal surfaces of cucumber (Fauth *et al.*, 1996) and observations with the primary germ tube of *E. graminis* f.sp. *hordei,* which penetrates the epidermal wall of barley only to a certain degree, but never completely, suggest that signals reach the host cell immediately after breach of the cuticle (Kunoh *et al.*, 1991). The time schedule of the first plant cell reactions, measured in cell cultures, is very fast. A few seconds after elicitor binding to parsley cells, Ca^{2+} influx and anion efflux is stimulated that would cause depolarization (Nürnberger *et al.*, 1994). Further reactions include the release of oxygen radicals and peroxide, a phenomenon termed the oxidative burst (Mehdy, 1994). It leads to cross-linking of pre-existing hydroxyproline-rich structural proteins (Brisson *et al.*, 1994). Other plant cell responses in a number of different host–parasite systems include callose synthesis (Kauss, 1995), accumulation of ions (Kunoh, 1990) and the production of hydroxycinnamoyl esters (Matern *et al.*, 1995). The esterification of wall polysaccharides with hydroxycinnamic acids modifies the cell to resist the action of lytic enzymes produced by fungal pathogens (Matern *et al.*, 1995). In addition, the plant may produce proteins that inhibit the action of cell wall-degrading enzymes, such as polygalacturonase-inhibiting proteins (Yao *et al.*, 1995).

Not only the existing cell wall is reinforced. As soon as *Fusarium oxysporum* f.sp. *vasinfectum* starts to penetrate the cells in the meristematic zone of cotton (*Gossypium barbadense*), roots react with the formation of wall appositions. These wall appositions consist of normal cell wall material such as cellulose, xyloglucans, pectin and polygalacturonic acid (Rodriguez-Gálvez and Mendgen, 1995). In addition, large amounts of callose are incorporated. The mechanisms by which pathogenic fungi overcome such wall appositions are mostly unknown.

8.10 Conclusions

We now know most tools a fungus uses to penetrate a plant epidermis. It remains to be elucidated for each host-parasite system which factor contributes to what extent to the penetration process. During these studies, it will be very important to use tests for pathogenicity that are as close as possible to the natural conditions in order to elucidate their importance (Schäfer, 1994).

Acknowledgements

I am indebted to R.J. Howard (Du Pont, Wilmington) for the use of the LT-scanning electron microscope, Christine Struck, Holger Deising, Matthias Hahn and Tim Carver for reading the manuscript. The Deutsche Forschungsgemeinschaft and the Fonds der chemischen Industrie gave financial support.

References

Allen, E.A., Hoch, H.C., Stavely, J.R. *et al.* (1991) Uniformity among races of *Uromyces appendiculatus* in response to topographic signaling for appressorium formation. *Phytopathology* **81,** 883-887.

Braun, E.J. and Howard, R.J. (1994) Adhesion of *Cochliobolus heterostrophus* conidia and germlings to leaves and artificial surfaces. *Exp. Mycol.* **18,** 211-220.

Brisson, L.F., Tenhaken, R. and Lamb, C. (1994) Function of oxidative cross-linking of cell wall structural proteins in plant disease resistance. *Plant Cell* **6,** 1703-1712.

Carpita, N.C. and Gibeaut, D.M. (1993) Structural models of primary cell walls in flowering plants: consistency of molecular structure with the physical properties of the walls during growth. *Plant J.* **3,** 1-30.

Chester, K.S. (1946) *The Cereal Rusts.* Chronica Botanica Co., Waltham, MA.

Chumley, F.G. and Valent, B. (1990) Genetic analysis of melanin-deficient, nonpathogenic mutants of *Magnaporthe grisea. Mol. Plant-Microbe Interact.* **3,** 135-143.

Clay, R.P., Enkerli, J. and Fuller, M.S. (1994) Induction and formation of *Cochliobolus sativus* appressoria. *Protoplasma* **178,** 34-47.

Clement, J.A., Porter, R., Butt, T.M. *et al.* (1994) The role of hydrophobicity in attachment of urediniospores and sporelings of *Uromyces viciae-fabae*. *Mycol. Res.* **98,** 1217-1228.

Cole, L., Dewey, F.M. and Hawes, C.R. (1996) Infection mechanisms of *Botrytis* species: pre-penetration and pre-infection processes of dry and wet conidia. *Mycol. Res.* **100,** 277-286.

Cooper, J.A. (1991) The role of actin polymerization in cell motility. *Annu. Rev. Physiol.* **53,** 585-605.

Corrêa, J.A., Staples, R.C. and Hoch, H.C. (1996) Inhibition of thigmostimulated cell differentiation with RGD-peptides in *Uromyces* germlings. *Protoplasma* (in press).

DeBary, A. (1884) *Vergleichende Morphologie und Biologie der Pilze, Mycetozoen und Bacterien.* Engelmann, Leipzig.

Deising, H. and Siegrist, J. (1995) Chitin deacetylase activity of the rust *Uromyces viciae-fabae* is controlled by fungal morphogenesis. *FEMS Microbiol. Letts* **127,** 207-212.

Deising, H., Jungblut, P.R. and Mendgen, K. (1991) Differentiation-related proteins of the broad bean rust fungus *Uromyces viciae-fabae*, as revealed by high resolution two-dimensional polyacrylamide gel electrophoresis. *Arch. Microbiol.* **155,** 191-198.

Deising, H., Nicholson, R.L., Haug, M. *et al.* (1992) Adhesion pad formation and the involvement of cutinase and esterases in the attachment of uredospores to the host cuticle. *Plant Cell* **4,** 1101-1111.

Deising, H., Frittrang, A.K., Kunz, S. *et al.* (1995) Regulation of pectin methylesterase and polygalacturonate lyase activity during differentiation of infection structures in *Uromyces viciae-fabae*. *Microbiology* **141,** 561-571.

Deising, H., Heiler, S., Rauscher, M. *et al.* (1996) Cellular aspects of rust infection structure differentiation: Spore adhesion and fungal morphogenesis. In: *Histology, Ultrastructure and Molecular Cytology of Plant-Microorganism Interactions* (eds M. Nicole and V. Gianinazzi-Pearson). Kluwer Academic Publishers, Dordrecht, pp.135-156.

Dykstra, M.J. and Bradley-Kerr, B. (1994) The adhesive droplet of capilliconidia of *Basidiobolus ranarum* exhibits unique ultrastructural features. *Mycologia* **86,** 336–342.

Elad, Y. and Evensen, K. (1995) Physiological aspects of resistance to *Botrytis cinerea*. *Phytopathology* **85,** 637–643.

Fauth, M., Merten, A., Hahn, M.G. *et al.* (1996) Competence for elicitation of H_20_2 in hypocotyls of cucumber is induced by breaching the cuticle and is enhanced by salicylic acid. *Plant Physiol.* **110,** 347–354.

Flaishman, M.A. and Kolattukudy, P.E. (1994) Timing of fungal invasion using host's ripening hormone as a signal. *Proc. Natl Acad. Sci. USA* **91,** 6579–6583.

French, R.C. (1992) Volatile chemical germination stimulators of rust and other fungal spores. *Mycologia* **84,** 277–288.

Freytag, S., Bruscaglioni, L., Gold, R.E. *et al.* (1988) Basidiospores of rust fungi (*Uromyces* species) differentiate infection structures *in vitro*. *Exp. Mycol.* **12,** 275–283.

Gilbert, R.D., Johnson, A.M. and Dean, R.A. (1996) Chemical signals responsible for appressorium formation in the rice blast fungus *Magnaporthe grisea*. *Physiol. Mol. Plant Pathol.* **48,** 335–346.

Gold, R.E. and Mendgen, K. (1984) Cytology of basidiospore germination, penetration, and early colonization of *Phaseolus vulgaris* by *Uromyces appendiculatus* var. *appendiculatus*. *Can. J. Bot.* **62,** 1989–2002.

Hajek, A.E. and St. Leger, R.J. (1994). Interactions beween fungal pathogens and insect hosts. *Annu. Rev. Entomol.* **39,** 293–322.

Hammer, P.E. and Evensen, K.B. (1994) Differences between rose cultivars in susceptibility to infection by *Botrytis cinerea*. *Phytopathology* **84,** 1305–1312.

Hardham, A.R. (1992) Cell biology of pathogenesis. *Annu. Rev. Plant Physiol. Plant Mol. Biol.* **43,** 491–526.

Harold, F.M., Harold, R.L. and Money, N.P. (1995) What forces drive cell wall expansion? *Can. J. Bot.* **75,** 379–383.

Harr, J. and Guggenheim, R. (1995) *The Leaf Surface of Major Crops.* Friedrich Reinhardt Verlag, Basle.

Heinen, W. (1963) Über den enzymatischen Cutin-Abbau. III. Mitteilung: Die enzymatische Ausrüstung von *Penicillium spinulosum* zum Abbau der Cuticularbestandteile. *Arch. Mikrobiol.* **41,** 268–281.

Hoch, H.C. and Staples, R.C. (1991) Signaling for infection structure formation in fungi. In: *The Fungal Spore and Disease Initiation in Plants and Animals* (eds G. T. Cole and H. C. Hoch). Plenum Press, New York, pp.25–46.

Hoch, H.C., Staples, R.C., Whitehead, B. *et al.* (1987) Signaling for growth orientation and cell differentiation by surface topography in *Uromyces. Science* **235,** 1659–1662.

Holloway, P.J. (1993). Structure and chemistry of plant cuticles. *Pestic. Sci.* **37,** 203–232.

Howard, R.J., Bourett, T.M. and Ferrari, M.A. (1991a) Infection by *Magnaporthe:* An *in vitro* analysis. In: *Electron Microscopy of Plant Pathogens* (eds K. Mendgen and D.-E. Lesemann). Springer-Verlag, Berlin, pp. 251–264.

Howard, R.J., Ferrari, M.A., Roach, D.H. *et al.* (1991b) Penetration of hard substrates by a fungus employing enormous turgor pressures. *Proc. Natl Acad. Sci. USA* **88,** 11281–11284.

Hwang, C.-S. and Kolattukudy, P.E. (1995) Isolation and characterization of genes expressed uniquely during appressorium formation by *Colletotrichum gloeosporioides* conidia induced by the host surface wax. *Mol. Gen. Genet.* **247,** 282–294.

Hyde, K.D., Greenwood, R. and Jones, E.B.G. (1993) Spore attachment in marine fungi. *Mycol. Res.* **97,** 9–14.

Jarvis, W.R. (1994) Latent infections in pre- and postharvest environment. *Hortscience* **29,** 749–751.

Jelitto, T.C., Page, H.A. and Read, N.D. (1994) Role of external signals in regulating the pre-penetration phase of infection by the rice blast fungus, *Magnaporthe grisea*. *Planta* **194,** 471–477.

Jones, E.B.G. (1994) Fungal Adhesion. *Mycol. Res.* **98,** 961–981.

Juniper, B. (1991) The leaf from the inside and the outside: A microbe's perspective. In: *Microbial Ecology of Leaves* (eds J. H. Andrews and S. S. Hirano). Springer-Verlag, Heidelberg, pp.21–42.

Kauss, H. (1996) Callose synthesis. In: *Membranes: Specialized Functions in Plants* (eds M. Smallwood, J.P. Knox and D. J. Bowles). BIOS Scientific Publishers, Oxford, pp. 77–92.

Kolattukudy, P.E. (1985) Enzymatic penetration of the plant cuticle by fungal pathogens. *Annu. Rev. Phytopathol.* **23,** 223–250.

Kolattukudy, P.E., Li, D., Hwang, C.-S. *et al.* (1995) Host signals in fungal gene expression involved in penetration into the host. *Can. J. Bot.* **73**, 1160–1168.

Kollar, A. (1994) Characterization of specific induction, activity, and isozyme polymorphism of extracellular cellulases from *Venturia inaequalis* detected *in vitro* and on the host plant. *Mol. Plant-Microbe Interact.* **7**, 603–611.

Köller, W. (1991) The plant cuticle: a barrier to be overcome by fungal plant pathogens. In: *The Fungal Spore and Disease Initiation in Plants and Animals* (eds G. T. Cole and H. C. Hoch). Plenum Press, New York, pp.219–246.

Köller, W., Yao, C., Trial, F. *et al.* (1995) Role of cutinase in the invasion of plants. *Can. J. Bot.* **73**, S1109–S1118.

Kubo, Y. and Furusawa, I. (1991) Melanin biosynthesis. Prerequisite for successful invasion of the host by appressoria of *Colletotrichum* and *Pyricularia*. In: *The Fungal Spore and Disease Initiation in Plants and Animals* (eds G.T. Cole and H.C. Hoch). Plenum Press, New York, pp.205-218.

Kunoh, H. (1990). Ultrastructure and mobilisation of ions near infection sites. *Annu. Rev. Phytopathol.* **28**, 93–111.

Kunoh, H., Nicholson, R.L. and Kobayashi, I. (1991) Extracellular materials of fungal structures: Their significance at prepenetration stages of infection. In: *Electron Microscopy of Plant Pathogens* (eds K. Mendgen and D.-E. Lesemann). Springer-Verlag, Berlin, pp.223–234.

Kuo, K. and Hoch, H.C. (1995) Visualization of the extracellular matrix surrounding pycnidiospores, germlings, and appressoria of *Phyllosticta ampelicida*. *Mycologia* **87**, 759–771.

Kuo, K. and Hoch, H.C. (1996) Germination of *Phyllosticta ampelicida* pycnidiospores: prerequisite of adhesion to the substratum and the relationship of substratum wettability. *Fungal Genet. Biol.* **20**, 18–29.

Kwon, Y.K. and Epstein, L. (1993). A 90 kDa Glycoprotein associated with adhesion of *Nectria haematococca* macroconidia to substrata. *Mol. Plant-Microbe Interact.* **6**, 481-487.

Martin, T.J. (1964) Role of the cuticle in the defense against plant diseases. *Annu. Rev. Phytopathol.* **2**, 81–100.

Matern, U., Grimmig, B. and Kneusel, R.E. (1995) Plant cell wall reinforcement in the disease-resistance response: molecular composition and regulation. *Can. J. Bot.* **73**, 511–517.

Mehdy, M.C. (1994). Active oxygen species in plant defense against pathogens. *Plant Physiol.* **105**, 467–472.

Mendgen, K. and Deising, H. (1993) Infection structures of fungal plant pathogens — a cytological and physiological evaluation. *New Phytol.* **124**, 193–213.

Mendgen, K., Hahn, M. and Deising, H. (1996) Mechanisms and morphogenesis of penetration by plant-pathogenic fungi. *Annu. Rev. Phytopathol.* **34**, 367–386.

Mercure, E.W., Kunoh, H. and Nicholson, R.L. (1995) Visualization of materials released from adhered, ungerminated conidia of *Colletotrichum graminicola*. *Physiol. Mol. Plant Pathol.* **46**, 121–135.

Mims, C.W., Liljebjelke, K.A. and Richardson, E.A. (1995) Surface morphology, wall structure, and initial adhesion of conidia of the powdery mildew fungus *Uncinuliella australiana*. *Phytopathology* **85**, 352-358.

Money, N.P. and Harold, F.M. (1993) Two water molds can grow without measurable turgor pressure. *Planta* **190**, 426–430.

Mould, R.J.M., Boland, G.J. and Robb, J. (1991) Ultrastructure of the *Colletotrichum trifolii-Medicago sativa* pathosystem. 2. Post penetration events. *Physiol. Mol. Plant Pathol.* **38**, 195–210.

Nicholson, R.L. (1996) Adhesion of fungal propagules. In: *Histology, Ultrastructure and Molecular Cytology of Plant-Microorganism Interactions* (eds M. Nicole and V. Gianinazzi-Pearson). Kluwer Academic Publishers, Dordrecht, pp.117–134.

Nicholson, R.L., Kunoh, H., Shiraishi, T. *et al.* (1993) Initiation of the infection process by *Erysiphe graminis*: Conversion of the conidial surface from hydrophobicity to hydrophilicity and influence of the conidial exudate on the hydrophobicity of the barley leaf surface. *Physiol. Mol. Plant Pathol.* **43**, 307–318.

Nürnberger, T., Nennstiel, D., Jabs, T. *et al.* (1994) High affinity binding of a fungal oligopeptide elicitor to parsley plasma membranes triggers multiple defense responses. *Cell* **78**, 449–460.

O'Connell, R J., Pain, N.A., Hutchinson, K.A. *et al.* (1996) Ultrastructure and composition of the cell surfaces of infection structures formed by the fungal plant pathogen *Colletotrichum lindemuthianum*. *J. Microsc.* **181**, 204–212.

Pascholati, S.F., Deising, H., Leite, B. *et al.* (1993) Cutinase and non-specific esterase activities in the conidial mucilage of *Colletotrichum graminicola*. *Physiol. Mol. Plant Pathol.* **42**, 37–51.

Pinkteron, J.N., Stone, J.K., Neslon, S.J. *et al.* (1995) Infection of European hazelnut by *Anisogramma anomala*: Ascospore adhesion, mode of penetration of immature shoots, and host response. *Phytopathology* **85**, 1260–1268.

Podila, G.K., Rosen, E., San Francisco, M.J.D. *et al.* (1995) Targeted secretion of cutinase in *Fusarium solani* f. sp. *pisi* and *Colletotrichum gloeosporioides*. *Phytopathology* **85**, 238–242.

Rijkenberg, F.H.J., De Leeuw, G.T.N. and Verhoeff, K. (1980) Light and electron microscopy studies on the infection of tomato fruits by *Botrytis cinerea*. *Can. J. Bot.* **58**, 1394–1404.

Rodriguez-Gálvez, E. and Mendgen, K. (1995) Cell wall synthesis in cotton roots after infection with *Fusarium oxysporum*. The deposition of callose, arabinogalactans, xyloglucans, and pectic components into walls, wall appositions, cell plates and plasmodesmata. *Planta* **197**, 535–545.

Schaeffer, H.J., Leykam, J. and Walton, J.D. (1994) Cloning and targeted gene disruption of *EXG1*, encoding exo-β-1,3-glucanase, in the phytopathogenic fungus *Cochliobolus carbonum*. *Appl. Environ. Microbiol.* **60**, 594–598.

Schäfer, W. (1994) Molecular mechanisms of fungal pathogenicity to plants. *Annu. Rev. Phytopathol.* **32**, 461–477.

Schauwecker, F., Wanner, G. and Kahmann, R. (1995) Filament-specific expression of a cellulase gene in the dimorphic fungus *Ustilago maydis*. *Biol. Chem. Hoppe-Seyler* **376**, 617–625.

Stahl, D.J., Theuerkauf, A., Heitefuss, R. *et al.* (1994) Cutinase of *Nectria haematococca* (*Fusarium solani* f. sp. *pisi*) is not required for fungal virulence or organ specificity on pea. *Mol. Plant-Microbe Interact.* **7**, 713–725.

Talbot, N.J., Ebbole, D.J. and Hamer, J.E. (1993) Identification and characterisation of mpg1, a gene involved in pathogenicity from the rice blast fungus *Magnaporthe grisea*. *Plant Cell* **5**, 1575–1590.

Terhune, B.T. and Hoch, H.C. (1993) Substrate hydrophobicity and adhesion of *Uromyces* urediospores and germlings. *Exp. Mycol.* **17**, 241–252.

Tunlid, A., Jansson, H.-B. and Nordbring-Hertz, B. (1992) Fungal attachment to nematodes. *Mycol. Res.* **96**, 401–412.

Van den Ende, G. and Linskens, H.F. (1974) Cutinolytic enzymes in relation to pathogenesis. *Annu. Rev. Phytopathol.* **12**, 247–258.

Viret, O., Scheidegger, C. and Petrini, O. (1993) Infection of beech leaves (*Fagus sylvatica*) by the endophyte *Discula umbrinella* (teleomorph: *Apiognomonia errabunda*): low-temperature scanning electron microscopy studies. *Can. J. Bot.* **71**, 1520–1527.

Vreeland, V. and Epstein, L. (1996) Analysis of plant-substratum adhesives. In: *Modern Methods of Plant Analysis* (eds J. F. Jackson and H. F. Linskens). Springer-Verlag, Berlin, pp.95–116.

Walton, J.D. (1994) Deconstructing the cell wall. *Plant Physiol.* **104**, 1113–1118.

Warwar, V. and Dickman, M.B. (1996) Effects of calcium and calmodulin on spore germination and appressorium development in *Colletotrichum trifolii*. *Appl. Environ. Microbiol.* **62**, 74–79.

Wessels, J.G.H. (1994) Developmental regulation of fungal cell wall formation. *Annu. Rev. Phytopathol.* **32**, 439–459.

Williamson, B., Duncan, G.H., Harrison, J.G. *et al.* (1995) Effect of humidity on infection of rose petals by dry-inoculated conidia of *Botrytis cinerea*. *Mycol. Res.* **99**, 1303–1310.

Xu, H. and Mendgen, K. (1994) Endocytosis of 1,3-β-glucans by broad bean cells at the penetration site of the cowpea rust fungus (haploid stage). *Planta* **195**, 282–290.

Yao, C., Conway, W.S. and Sams, C.E. (1995) Purification and characterization of a polygalacturonase-inhibiting protein from apple fruit. *Phytopathology* **85**, 1373–1377.

9

Cuticular waxes: a critical assessment of current knowledge

M. Riederer and C. Markstädter

9.1 Introduction

Cuticular waxes are the main functional components of the primary aerial surfaces of higher plants. They are embedded within the cutin matrix of plant cuticular membranes (CM) and make up epicuticular films and aggregates. Their main function is to limit the diffusional flow of water and solutes across the cuticle (Riederer and Schreiber, 1995; Schönherr, 1982). Removing cuticular waxes from isolated plant cuticles leads to an increase of permeability by up to several orders of magnitude (Figure 9.1). The composition and fine structure of epicuticular waxes (EW) determine the wettability of primary plant surfaces (see Chapter 15) and thus may control leaching (Holloway, 1969a,b, 1970), epiphyllic microflora (Jacques and Morris, 1995; Schreiber and Schönherr, 1993), and foliar uptake of pesticides (Schönherr *et al.*, 1991). Components of cuticular waxes have been shown to function as semiochemicals in the interactions of herbivores (Eigenbrode and Espelie, 1995; Espelie *et al.*, 1991; Juniper, 1995; Chapters 10–12) and fungi (Carver and Thomas, 1990; Flaishman *et al.*, 1995; Podila *et al.*, 1993; Chapters 12–13) with their host plants. The physical structure and chemical composition of cuticular waxes may also contribute to the attenuation of photosynthetically active radiation (PAR) in order to avoid light inhibition (Eller and Flach, 1990; Flach and Eller, 1990, 1991; Robinson *et al.*, 1993) and of ultraviolet radiation (Bornman and Vogelmann, 1988; Cappelletti, 1961; Krauss *et al.*, 1996; Wuhrmann-Meyer and Wuhrmann-Meyer, 1941; Chapter 7).

Understanding the role cuticular waxes play in any of these functions necessitates the knowledge of their chemical composition. Composition of cuticular waxes, physical structure and the functions of plant cuticles are closely related. This fact was recognized decades ago and much work and effort has since been invested in developing methods for analysing cuticular waxes (Hamilton, 1995; Holloway, 1984; Holloway and Challen, 1966). However, as the insight into the functions and physical structure of plant cuticles advances, the limitations of the knowledge currently available about the chemical composition of cuticular waxes become apparent. It is, therefore, the purpose of this chapter critically to evaluate the state of the art of cuticular wax analysis and the current information on the

Plant Cuticles, edited by G. Kerstiens.

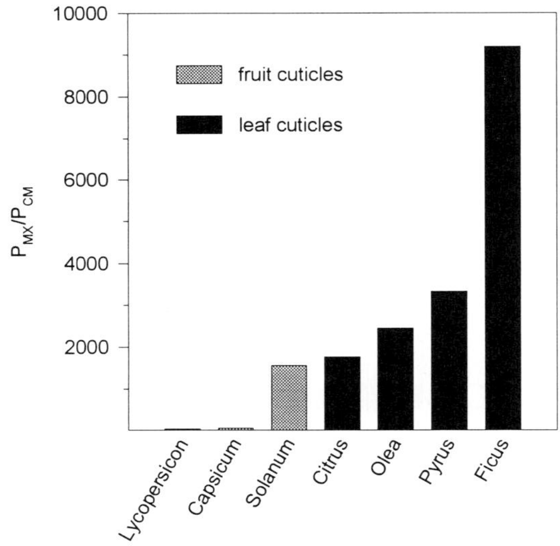

Figure 9.1. *The effect of removal of cuticular waxes on the permeability of cuticular membranes (CM) to 2,4-dichlorophenoxy acetic acid. (P_{MX}, permeance of dewaxed cuticle; P_{CM}, permeance of intact cuticle. Data from Riederer and Schönherr, 1985.)*

composition of cuticular waxes. Comprehensive reviews of the chemical composition and biosynthesis of plant cuticular waxes have been published recently (Avato *et al.*, 1990; Baker, 1982; Bianchi, 1995; Bianchi and Bianchi, 1990; Kolattukudy, 1980; Kolattukudy and Espelie, 1994; Kolattukudy *et al.*, 1976; Tulloch, 1976; Walton, 1990).

9.2 Methods for analysing cuticular waxes

9.2.1 *Extraction of cuticular waxes*

Cuticular waxes are generally obtained either by dipping intact organs (leaves, fruits, stems) into an organic solvent or by extracting isolated cuticular membranes. Whenever the cuticular surface is covered by an extensive bloom of epicuticular aggregates a problem arises: current experimental evidence suggests that epicuticular aggregates are neat crystals consisting only of one compound (Jetter, 1994; Jetter and Riederer, 1994) or a mixture of closely related positional isomers (Jetter and Riederer, 1995). As there are no methods available which are suitable for separating reliably and quantitatively epicuticular aggregates from the rest of cuticular waxes both fractions will combine upon extraction with an organic solvent. Analyses performed on such samples will give results biased in favour of the component(s) making up the epicuticular aggregates. This fact will limit the usefulness of such data for correlations with the physical structure and functions associated with the wax fraction not involved in aggregate formation. The literature available on the composition of cuticular waxes does not pay sufficient attention to this potential problem.

Extracting cuticular waxes from intact organs by dipping is prone to several sources of experimental error. The kinetics of extraction yields may be extended and vary from species to species; for instance, the leaf blades of *Citrus aurantium* had to be immersed in chloroform at 20°C for 30 min in order to reach a plateau of wax yield (Figure 9.2), while with herbaceous species dips lasting only for several tens of seconds were fully sufficient (M. Riederer, A. Liebe, M. Burghardt, unpublished results). As a consequence the kinetics of extraction should be checked for the material to be studied and documented in the resulting publication. Otherwise, the reliability of coverage data cannot be assessed.

A further problem arising from simply dipping plant material into a solvent is that waxes originating from the adaxial and abaxial surfaces of a leaf become amalgamated. Wax composition may vary considerably between leaf surfaces (Riederer and Schneider, 1989) and thus samples obtained in this way may, at best, give a rough impression of the actual qualitative and quantitative composition of the wax covering any of the two surfaces.

Finally, extracting whole leaves may lead to discrimination of certain compound classes and/or chain-lengths of wax constituents. Using again *Citrus aurantium* leaves it has been shown (Riederer and Schneider, 1989) that dipping under optimum conditions (30 min in chloroform at 30°C) results in underestimating the amounts of longer-chain alkanes and alkan-1-ols and practically a complete discrimination of alkyl esters (Figure 9.3). The conclusion from this evidence is that, whenever possible, isolated CM should be used as sources for wax samples. Isolated cuticles can be exhaustively extracted in order to achieve maximum yield and to avoid discriminations, and cuticles from the adaxial and abaxial sides of a leaf can be studied separately. Enzymatic isolation of cuticles may lead to contamination with saturated and unsaturated hexa- and octadecanoic acids which, however, can be removed by a subsequent purification step (Schönherr and Riederer, 1986).

9.2.2 *Chemical analysis of cuticular waxes*

Much progress has been made in the chemical analysis of plant cuticular waxes since the advent of powerful analytical instrumentation like capillary gas

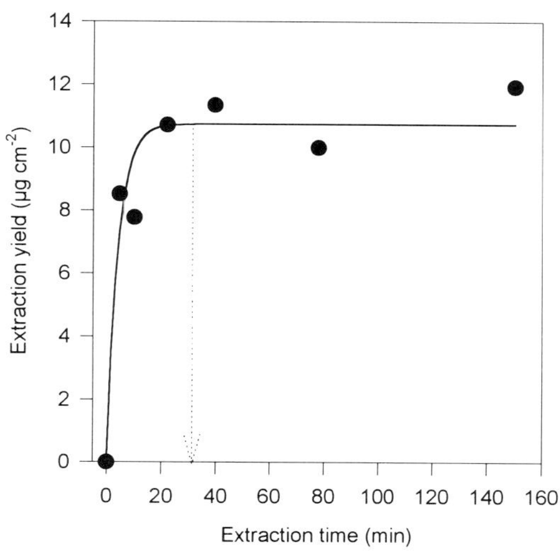

Figure 9.2. *Kinetics of extraction of cuticular waxes from leaf blades of* Citrus aurantium. *Leaves were extracted with chloroform at 20°C. (Data from Riederer and Schneider, 1989.)*

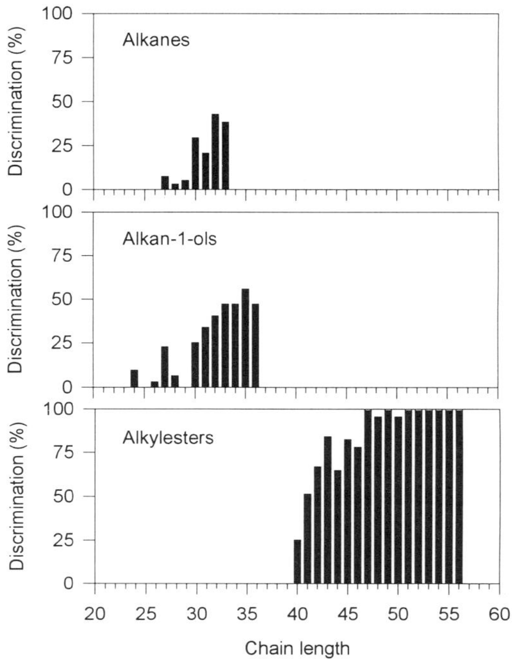

Figure 9.3. *Discrimination of cuticular wax constituents during extraction of* Citrus aurantium *leaf blades (30 min in chloroform at 30°C). After this treatment, the cuticles were isolated enzymatically and extracted again in hot chloroform. For each chain-length, the fraction recovered only by the second extraction step is given. (Data from Riederer and Schneider, 1989.)*

chromatography and mass spectrometry (Holloway, 1984). Nevertheless, the large number of components present in a single wax sample (which may exceed a total of 100) and the diversity of their functional groups still represent a challenge to the workers in this field. The objective is a qualitatively and quantitatively correct and precise determination of the composition of a wax sample. As a rule, samples should be submitted to quantitative analysis after a minimum number of pre-separation steps in order to reduce cumulative errors.

A main problem associated with wax analysis is the determination of compounds with hydroxy and/or carbonyl groups. The reactivity of these compounds may lead to discriminations during sample preparation and the chromatographic process and to inadequate separation. Each of these factors will cause the quality of the analytical results to deteriorate to an unknown degree if their impact is not systematically studied and quantified. The prerequisite for doing this is the use of adequate internal standards added to the sample at the earliest possible step, preferentially during wax extraction. Correction factors relating peak areas to the actual amounts of a compound present can be established on the basis of an internal standard. These factors can also be used for assessing the sensitivity of the whole analytical procedure and its reproducibility.

An evaluation of this type leads to the conclusion that alkanals are especially difficult to analyse when compounds having hydroxy and carboxy groups are present in the sample. In this case, these reactive groups have to be transformed into their corresponding inactive esters or ethers. All derivatization agents studied so far in the authors' laboratory had adverse effects on the quantification of the alkanals present (Figure 9.4). Peak areas of alkanals were up to threefold smaller than expected from a linear proportionality with the internal standard and, among consecutive analyses of identical samples, correction factors varied by more than a factor of two (M. Riederer, C. Markstädter, R. Naß, unpublished results). The introduction of various protective groups for alkanals did not improve overall analytical performance.

These analytical problems can be overcome by separating samples of cuticular waxes containing alkanals into a non-polar and a polar fraction by a solid-phase extraction step in the microscale. When applied to the cuticular wax of *Fagus sylvatica* leaves the non-polar fraction contains alkanes, alkyl and benzyl esters and alkanals. This fraction can be analysed without further derivatization, producing reliable and reproducible results for all compound classes present. The polar fraction containing all compounds with hydroxy and carboxy functions can be treated with silylating reagents and subsequently analysed. The procedure of solid-phase pre-separation should be optimized for each type of wax sample. Then, adequate separation and yields not significantly different from 100% can be achieved, which has been shown by adding internal standards for all major compound classes (M. Riederer, C. Markstädter, R. Naß, unpublished results).

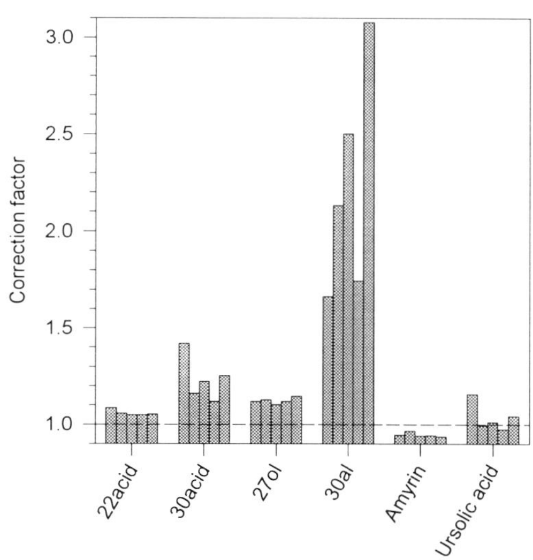

Figure 9.4. *Variability of correction factors for the components of a standard containing representative wax constituents. Five aliquots of a standard solution were silylated in parallel and analysed by gas chromatography. The correction factor is the ratio of the peak areas of the analyte and an internal standard (tetracosane) when equal amounts are injected. (22acid, docosanoic acid; 30acid, triacontanoic acid; 27ol, heptacosan-1-ol; 30al, triacontanal; amyrin, α-amyrin.)*

9.3 Ontogenetic factors influencing amounts and composition of cuticular waxes

Amounts and composition of plant cuticular waxes are subject to the influence of endogenous and exogenous factors. Numerous studies have shown that light, temperature and humidity may influence the coverage and composition of cuticular waxes and, in some cases, the fine structure of epicuticular aggregates (Günthardt-Goerg *et al.*, 1994; Riederer and Schneider, 1990; Shepherd *et al.*, 1995; Sümmchen *et al.*, 1995; von Wettstein-Knowles *et al.*,1980). Another factor exerting an even more pronounced impact on wax amounts and compositions is ontogenetic development. During the development of a *Fagus sylvatica* leaf, wax coverage increases from 20 μg $leaf^{-1}$ just after bud break to more than 700 μg $leaf^{-1}$ at the time when the leaf reaches its final size (Markstädter, 1994). Simultaneously, coverage per unit area doubles from approximately 6 to 12 μg cm^{-2} (Figure 9.5). Thus, wax coverages determined at an arbitrary developmental stage may be misleading as they most likely are not representative.

A refined analysis of wax composition at different stages of the ontogenetic development of a *Fagus sylvatica* leaf underlines this statement. Chain-lengths of the aliphatic wax constituents shift systematically from a bimodal distribution with maxima in the range of C_{28} and C_{52} in the bud state to a distribution dominated by a large single maximum at C_{28} when the leaf has reached final size (Figure 9.6). An even more complex picture is obtained when the mass fraction contributed by different compound classes is estimated at characteristic points during leaf development. The wax of the very young *Fagus* leaf just emerging from the bud is dominated by long-chain alkenyl esters (> 40 mass%) which disappear due to dilution during leaf development. The surface wax of unfolding leaves is characterized by a predominance of alkan-1-ols, while during the phase of maximum leaf expansion *n*-alkyl-*p*-coumarates and alkanes are the most abundant compound classes each contributing approximately 25% of total mass. In the mature

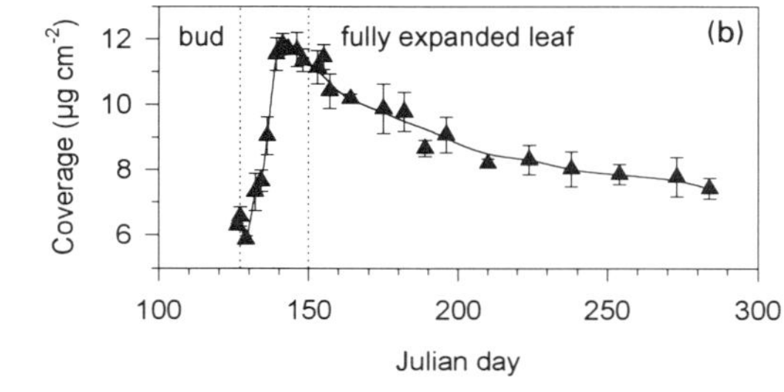

Figure 9.5. *Variation of total wax coverage of* Fagus sylvatica *leaves during the vegetation period. Coverages were determined by gas chromatography and are given related to (a) unit leaf area and (b) per leaf. (Data from Markstädter, 1994.)*

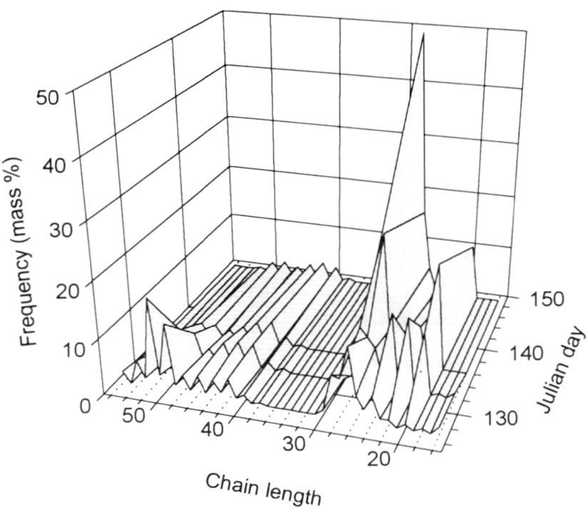

Figure 9.6. *Variation of the chain-length distribution of cuticular wax of* Fagus sylvatica *leaves during ontogenesis. Distributions are shown for four characteristic states of leaf development: end of bud dormancy (day 127), unfolding of leaf (day 132), maximum rate of leaf expansion (day 136) and maximum wax coverage (day 149). (Data from Markstädter, 1994.)*

state alkanals are the most important compound class making up more than 35% of the total mass (Figure 9.7).

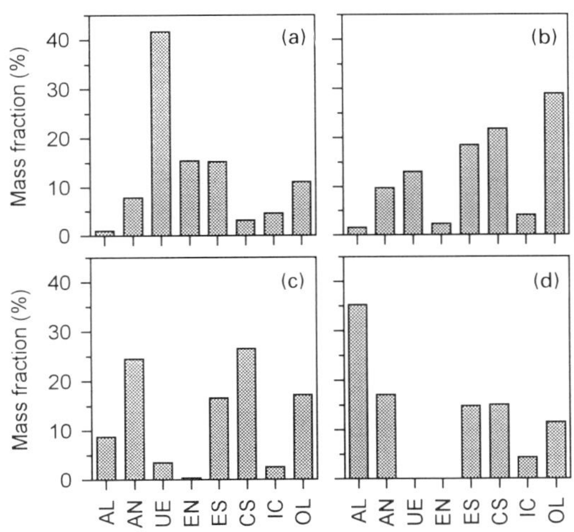

Figure 9.7. *Variation of the composition of cuticular wax of* Fagus sylvatica *leaves during ontogenesis. Contributions of substance classes are shown for four characteristic states of leaf development: (a) end of bud dormancy (day 127), (b) unfolding of leaf (day 132), (c) maximum rate of leaf expansion (day 136), (d) maximum wax coverage (day 149). (AL, alkanals; AN, alkanes; UE, long-chain alkenyl esters; EN, alkenes; ES, long-chain alkyl esters; CS,* n-*alkyl*-p-*coumarates; IC, alkanoic acids; OL, alkan-1-ols. (Data from Markstädter, 1994.)*

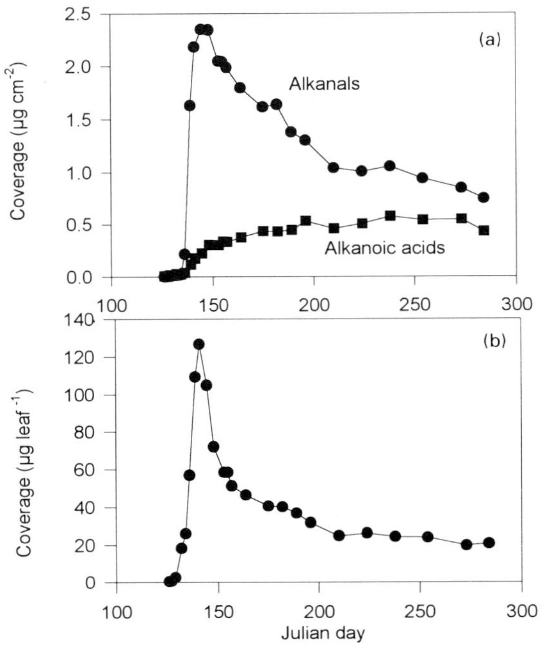

Figure 9.8. *Seasonal variation of the coverages of reactive constituents of the cuticular wax of* Fagus sylvatica *leaves. (a) Alkanal and alkanoic acid coverages; (b) coverages of* n-*alkyl*-p-*coumarates. (Data from Markstädter, 1994.)*

This example shows that wax compositions at different developmental states of the leaf of a single species (in the present case even of a single branch) may vary to the same extent as observed between mature leaves of different species. Thus, a wax sample arbitrarily taken without regard to the developmental state of the plant or organ may well be a snapshot not at all characteristic for the species. In general, the composition of cuticular wax should be determined only for clearly defined developmental states. Otherwise, such data are of poor value when chemical composition and any of the cuticular functions are to be compared.

9.4 Abiotic factors influencing wax composition

Cuticular waxes are deposited on the outermost surface of the aerial parts of higher plants. Most of these organs are of considerable longevity and, thus, cuticular waxes may be exposed to the impacts of the physical environment for extended periods of time. This may lead to a mechanical abrasion of epicuticular waxes by the action of wind, rain, snow and the friction between leaves (Baker and Hunt, 1986; Simoneit *et al.*, 1991) or to volatilization losses.

Recently, evidence has been obtained that certain major constituents of plant cuticular waxes may also undergo spontaneous chemical reactions, while others (e.g. alkan-1-ols, alkyl esters) are not affected (Markstädter, 1994). The alkanal coverages of *Fagus sylvatica* leaves increase as long as leaf expansion and wax biosynthesis proceed, while they decline exponentially during the mature state of the leaf. This decline is paralleled by an increase of alkanoic acids of the same chain-lengths as the alkanals suggesting that they are the products of a (photo)oxidative conversion of alkanals to alkanoic acids (Figure 9.8a). However, this increase of alkanoic acid coverages cannot fully account for the disappearance of the alkanals, suggesting that further reaction products are formed which cannot be extracted or analysed by the methods used. The coverages of the *n*-alkyl-*p*-coumarates present in *Fagus sylvatica* leaf wax follow a time course similar to that exhibited by the alkanal fraction (Figure 9.8b). No reaction products have been detected so far.

These examples show that amounts and composition of cuticular waxes covering longevous parts of plants may not only be affected by environmental factors acting on biosynthesis or by removal but may also be subject to spontaneous abiotic transformations. Such changes can only be detected when the composition of cuticular waxes is documented as a function of the age of the organ studied. As it is unknown whether these transformations influence any functional properties of plant cuticular waxes it is not advisable to ignore them.

9.5 What do we know about cuticular waxes?

The preceding discussion clearly demonstrates that numerous sources of systematic errors may affect the analytical results. There is only a minute number of reports in the literature identifying and eliminating at least some of the uncertainties related to wax extraction, sample work-up, chromatography and ontogenetic variation. The effects of abiotic reactions in cuticular waxes have been overlooked completely. Therefore, the reliability of the analytical results available is fairly limited and attempts to correlate these data on chemical composition with functions should be dealt with very cautiously.

Another important question is whether all relevant wax constituents are detected when standard analytical protocols are used. As long as there is no good reason for excluding a given compound class or chain-length because it is proven not to contribute to any function, maximum completeness must be the ultimate goal of wax analysis. Reviewing the literature on wax composition published during the last three decades suggests that the number of substances and substance classes described for a given wax depends on technical progress and scientific effort invested. Therefore, we are presently far from a position where we could tell what cuticular waxes are. From a functional point of view, is it legitimate to exclude the aliphatics with chain-lengths $< C_{16}$ and $> C_{64}$ simply because they do not fit into the conventional notion of a 'wax' or because standard techniques are not suitable? And can we be sure that alicyclic constituents like triterpenoids, diterpenoids, sterols and flavonoids do not contribute to any function of the cuticle even though they may make up between 50 and 80% of the total mass of the cuticular waxes of some species? The same is true for monoterpenes and other volatile components.

These problems related more or less directly to analytical difficulties or deficiencies limit our current view of cuticular waxes to a few snapshots taken at mostly arbitrary developmental states through a narrow window. However, this window can be gradually opened when the sources of error will be identified and eliminated and analytical methods available in other fields applied to the analysis of plant cuticular waxes.

In contrast to this, there is a much more fundamental problem concerning the empirical basis of our knowledge on cuticular waxes. A survey of the pertinent literature indicates that the cuticular waxes of not much more than 100 plant species have been analysed so far. This count even considers all analyses covering more than just one substance class. In comparison to the total number of known species of higher plants (Wilson, 1988) this is only a fraction of 0.04%. In addition, most of the species investigated so far are crops and/or growing in the temperate zones of the Northern hemisphere. Analyses dealing with species from the diverse tropical and subtropical biomes or from extreme habitats are almost completely lacking. It is fairly unlikely that the incredibly small and biased sample from which we derive our notion of plant cuticular waxes is, by chance, representative even in the widest sense.

Acknowledgements

The work reported in this chapter has been supported by grants from the Deutsche Forschungsgemeinschaft, the Sonderforschungsbereich 251 and the Bayerisches Klimaforschungsprogramm. The authors gratefully acknowledge the permission by Reinhold Naß for using unpublished data, and skilful technical assistance by Daniela Seifert.

References

Avato, P., Bianchi, G. and Pogna, N. (1990) Chemosystematics of surface lipids from maize and some related species. *Phytochemistry* **29,** 1571–1576.

Baker, E.A. (1982) Chemistry and morphology of plant epicuticular waxes. In: *The Plant Cuticle* (eds D.F. Cutler, K.L. Alvin and C.E. Price). Academic Press, London, pp. 139–165.

Baker, E.A. and Hunt, G.M. (1986) Erosion of waxes from leaf surfaces by simulated rain. *New Phytol.* **102,** 161–173.

Bianchi, A. and Bianchi, G. (1990) Surface lipid composition of C-3 and C-4 plants. *Biochem. Syst. Ecol.* **18,** 533–537.

Bianchi, G. (1995) Plant waxes. In: *Waxes: Chemistry, Molecular Biology and Functions* (ed. R.J. Hamilton). The Oily Press, Dundee, pp. 175–222.

Bornman, J.F. and Vogelmann, T.C. (1988) Penetration of blue and UV radiation measured by fiber optics in spruce and fir needles. *Physiol. Plant.* **72,** 699–705.

Cappelletti, C. (1961) Ricerche sulla permeabilità delle cuticole alle radiazioni ultraviolette. - I. Piante di duna. *Rend. Sc. fis. mat. nat.* **30,** 331–342.

Carver, T.L.W. and Thomas, B.J. (1990) Normal germling development by *Erysiphe graminis* on cereal leaves freed of epicuticular wax. *Plant Pathol.* **39,** 367–375.

Eigenbrode, S.D. and Espelie, K.E. (1995) Effects of plant epicuticular lipids on insect herbivores. *Annu. Rev. Entomol.* **40,** 171–194.

Eller, B.M. and Flach, B.M.-T. (1990) Solar energy input to plant surfaces: II. Leaf dimorphism of *Aloe dichotoma* Masson and diurnal absorption of global radiation. *Bot. Helvet.* **100,** 239–248.

Espelie, K.E., Bernays, E.A. and Brown, J.J. (1991) Plant and insect cuticular lipids serve as behavioral cues for insects. *Arch. Insect Biochem. Physiol.* **17,** 223–233.

Flach, B.M.T. and Eller, B.M. (1990) Solar energy input to plant surfaces: I. Modeling of spectral power distribution with the computing program ECOSOL. *Bot. Helvet.* **100,** 225–238.

Flach, B.M.T. and Eller, B.M. (1991) Solar energy input to plant surfaces: III. Influence of leaf curvature and self-shading. *Bot. Helvet.* **101,** 121–134.

Flaishman, M.A., Hwang, C.S. and Kolattukudy, P.E. (1995) Involvement of protein phosphorylation in the induction of appressorium formation in *Colletotrichum gloeosporioides* by its host surface wax and ethylene. *Physiol. Mol. Plant Pathol.* **47,** 103–117.

Günthardt-Goerg, M.S., Keller, T., Matyssek, R. *et al.* (1994) Environmental effects on Norway spruce needle wax. *Eur. J. For. Pathol.* **24,** 92–111.

Hamilton, R.J. (1995) Analysis of waxes. In: *Waxes: Chemistry, Molecular Biology and Functions* (ed. R.J. Hamilton). The Oily Press, Dundee, pp. 311–342.

Holloway, P.J. (1969a) Chemistry of leaf waxes in relation to wetting. *J. Sci. Food Agric.* **20,** 124–128.

Holloway, P.J. (1969b) The effects of superficial wax on leaf wettability. *Ann. Appl. Biol.* **63,** 145–153.

Holloway, P.J. (1970) Surface factors affecting the wetting of leaves. *Pestic. Sci.* **1,** 156–163.

Holloway, P.J. (1984) Surface lipids of plants and animals. In: *CRC Handbook of Chromatography. Lipids*, Vol. 1 (eds H.K. Mangold, G. Zweig and J. Sherma). CRC Press, Boca Raton, FL, pp. 347–380.

Holloway, P.J. and Challen, S.B. (1966) Thin layer chromatography in the study of natural waxes and their constituents. *J. Chromatogr.* **25**, 336–346.

Jacques, M.A. and Morris, C.E. (1995) A review of issues related to the quantification of bacteria from the phyllosphere. *FEMS Microbiol. Ecol.* **18**, 1–14.

Jetter, R. (1994) Chemische Zusammensetzung, Struktur und Bildung röhrenförmiger Wachskristalle auf Pflanzenoberflächen. Doct. Diss., Universität Kaiserslautern.

Jetter, R. and Riederer, M. (1994) Epicuticular wax crystals of nonacosan-10-ol: In vitro reconstitution and factors influencing crystal habits. *Planta* **195**, 257–270.

Jetter, R. and Riederer, M. (1995) In vitro reconstitution of epicuticular wax crystals: Formation of tubular aggregates by long-chain secondary alkanediols. *Bot. Acta* **108**, 111–120.

Juniper, B.E. (1995) Waxes on plant surfaces and their interactions with insects. In: *Waxes: Chemistry, Molecular Biology and Functions* (ed. R.J. Hamilton). The Oily Press, Dundee, pp. 157–174.

Kolattukudy, P.E. (1980) Cutin, suberin, and waxes. In: *The Biochemistry of Plants,* Vol. 4 (ed. P.K. Stumpf). Academic Press, New York, pp. 571–654.

Kolattukudy, P.E. and Espelie, K.E. (1994) Chemistry, biochemistry, and function of suberin and associated waxes. In: *Natural Products of Woody Plants* (ed. J.W. Rowe). Springer-Verlag, Berlin, pp. 304–367.

Kolattukudy, P.E., Croteau, R. and Buckner, J.S. (1976) Biochemistry of plant waxes. In: *Chemistry and Biochemistry of Natural Waxes* (ed. P.E. Kolattukudy). Elsevier, Amsterdam, pp. 289–347.

Krauss, P., Markstädter, C. and Riederer, M. (1996) Attenuation of ultraviolet radiation by plant cuticles. *J. Exp. Bot.* **47** (suppl. 4), 58.

Markstädter, C. (1994) Untersuchungen zur jahreszeitlichen Entwicklung der kutikulären Wachse von *Fagus sylvatica* L. Doct. Diss., Universität Kaiserslautern.

Podila, G.K., Rogers, L.M. and Kolattukudy, P.E. (1993) Chemical signals from avocado surface wax trigger germination and appressorium formation in *Colletotrichum gloeosporioides. Plant Physiol.* **103**, 267–272.

Riederer, M. and Schneider, G. (1989) Comparative study of the composition of waxes extracted from isolated leaf cuticles and from whole leaves of *Citrus:* Evidence for selective extraction. *Physiol. Plant.* **77**, 373–384.

Riederer, M. and Schneider, G. (1990) The effect of the environment on the permeability and composition of *Citrus* leaf cuticles. II. Composition of soluble cuticular lipids and correlation with transport properties. *Planta* **180**, 154–165.

Riederer, M. and Schönherr, J. (1985) Accumulation and transport of (2,4-dichlorophenoxy)acetic acid in plant cuticles: II. Permeability of the cuticular membrane. *Ecotoxicol. Environ. Safety* **9**, 196–208.

Riederer, M. and Schreiber, L. (1995) Waxes — the transport barriers of plant cuticles. In: *Waxes: Chemistry, Molecular Biology and Functions* (ed. R.J. Hamilton). The Oily Press, Dundee, pp. 130–156.

Robinson, S.A., Lovelock, C.E. and Osmond, C.B. (1993) Wax as a mechanism for protection against photoinhibition — a study of *Cotyledon orbiculata. Bot. Acta* **106**, 307–312.

Schönherr, J. (1982) Resistance of plant surfaces to water loss: Transport properties of cutin, suberin and associated lipids. In: *Physiological Plant Ecology, part 2, Encyclopedia of Plant Physiology, N.S.*, Vol. 12B (eds O.L. Lange, P.S. Nobel, C.B. Osmond and H. Ziegler). Springer-Verlag, Berlin, pp. 153–179.

Schönherr, J. and Riederer, M. (1986) Plant cuticles sorb lipophilic compounds during enzymatic isolation. *Plant Cell Environ.* **9**, 459–466.

Schönherr, J., Riederer, M., Schreiber, L. *et al.* (1991) Foliar uptake of pesticides and its activation by adjuvants: theories and methods for optimization. In: *Pesticide Chemistry* (ed. H. Frehse). VCH Verlag, Weinheim, pp. 237–253.

Schreiber, L. and Schönherr, J. (1993) Determination of foliar uptake of chemicals —influence of leaf surface microflora. *Plant Cell Environ.* **16**, 743–748.

Shepherd, T., Robertson, G.W., Griffiths, D.W. *et al.* (1995) Effects of environment on the composition of epicuticular wax from kale and swede. *Phytochemistry* **40**, 407–417.

Simoneit, B.R.T., Crisp, P.T., Mazurek, M.A. *et al.* (1991) Composition of extractable organic matter of aerosols from the Blue Mountains and Southeast coast of Australia. *Environ. Int.* **17**, 405–419.

Sümmchen, P., Markstädter, C. and Wienhaus, O. (1995) Composition of the epicuticular wax esters of *Picea abies* (L.) Karst. Z. *Naturforsch.* **50c,** 11–14.

Tulloch, A.P. (1976) Chemistry of waxes of higher plants. In: *Chemistry and Biochemistry of Natural Waxes* (ed. P.E. Kolattukudy). Elsevier, Amsterdam, pp. 235–287.

von Wettstein-Knowles, P., Avato, P. and Mikkelsen, J.D. (1980) Light promotes synthesis of the very long fatty acyl chains in maize wax. In: *Biogenesis and Function of Plant Lipids* (eds P. Mazliak, P. Benveniste, C. Costes and R. Douce). Elsevier/North-Holland Biomedical Press, Amsterdam, pp. 271–274.

Walton, T.J. (1990) Waxes, cutin and suberin. In: *Lipids, Membranes and Aspects of Photobiology* (eds J.L. Harwood and J. Boyer). Academic Press, London, pp. 105–158.

Wilson, E.O. (1988) The current state of biological diversity. In: *Biodiversity* (eds E.O. Wilson and F.M. Peter). National Academy Press, Washington, DC, pp. 3–18.

Wuhrmann-Meyer, K. and Wuhrmann-Meyer, M. (1941) Untersuchungen über die Absorption ultravioletter Strahlen durch Kutikular- und Wachsschichten von Blättern. I. *Planta* **32,** 43–50.

10

Plant surface waxes and insect behaviour

S. D. Eigenbrode

10.1 Introduction

The aerial surfaces of plants are covered with a layer of lipophilic compounds commonly known as waxes (see Table 3.1). Plant surface waxes are obviously important in protecting the plant against water loss and entry (Jeffree, 1986), but their location, chemical and structural complexity are consistent with the variety of ecological roles discussed in this volume, including influencing interactions between insects and plants.

Plant surface waxes determine the structural morphology and physical properties of the plant surface on a scale relevant to phytophagous and entomophagous insects. The wax surface is the 'playing field' for interactions between plants, insect herbivores and their predators and parasitoids. The chemistry and morphology of surface waxes also provide cues important in host plant selection by herbivores (Eigenbrode and Espelie, 1995; Juniper, 1995). In this chapter, particular emphasis will be placed on the influences of wax morphology on both phytophagous and entomophagous insects. The allelochemical activity of plant surface wax components will also be reviewed in light of recent studies.

10.2 Plant surface wax morphology and insect behaviour

10.2.1 *Insect attachment*

The problem of attachment by insects to plants intrigued early naturalists (Power, 1664) and is an obvious force shaping plant-insect interactions. Southwood (1973, 1986) has consistently emphasized the problems of attachment for herbivorous insects, but this aspect of the relationship between insects and their host plants remains somewhat understudied considering its evident and widespread importance. Many supracellular structures, such as trichomes, affect attachment (e.g. Kennedy, 1986), but covering all of these structures are plant waxes; it is these, essentially, to which insects adhere.

Microscopically crystalline waxes (blooms) occur frequently on terrestrial plants (Martin and Juniper, 1970) and can interfere with attachment to plant surfaces by insects. This has been exploited most elegantly by carnivorous pitcher plants in the genus *Nepenthes* (Juniper *et al.*, 1989). Surface waxes on the interiors

Plant Cuticles, edited by G. Kerstiens.

of *Nepenthes* pitchers crystallize as plates, each of which seems to be attached to the cuticle by a thin stalk that should break away readily when disturbed by an insect tarsus. Removal of these crystalline waxes was shown by Knoll (1914) to eliminate the 'slipperiness' of *Nepenthes* to ants, which were otherwise readily trapped by the plant.

It is tempting to speculate that wax blooms on wholly autotrophic plants are adaptations to interfere with insect attachment and thereby provide defence against herbivory. Juniper (Juniper, 1995; Juniper *et al.*, 1989) suggests that the loose granular waxes on the flowering shoots of some plants (e.g. *Salix* spp.) are adaptations to prevent crawling insects from robbing nectar and other concentrated resources intended to attract flying pollinators and dispersers. Waxes on flowering shoots of *Hypenia* and *Eriope* (Ocimeae) exemplify such a 'greasy pole' defensive syndrome (Harley, 1991).

Wax blooms may also defend vegetative tissues from insect herbivory. For example, Mulroy (1979) reported that individuals of the crassulaceous plant *Dudleya britonii* with a waxy bloom were less attacked in the field by the pyralid moth *Rhagea stigmella* than were bloomless plants of this species. Juvenile foliage of both *Eucalyptus nitens* and *E. globulus* has a prominent wax bloom and is not attacked by the cassidine chrysomelid beetle *Paropsis charybdis* (Edwards, 1982). The insects readily feed on mature foliage that lacks a waxy bloom. Experimental observations showed that *P. charybdis* repeatedly slipped from the waxy juvenile leaves, but not from the more mature glabrous leaves. Mechanical removal of the wax crystals eliminated slipping from juvenile leaves.

Subsequent work (Edwards and Wanjura, 1990) showed that the densities of *P. charybdis* and several other chrysomelids were less on the very waxy *E. bridgesiana* and *E. bicostata* than on eucalypts with less prominent blooms.

The prominent waxy blooms on *Brassica* crops also apparently provide protection against feeding by some insects (Anstey and Moore, 1954; Bodnaryk, 1992b; Stoner, 1990; Way and Murdie, 1965). Against chrysomelids, this defence is related to their reduced adhesion to these plants. Stork (1980b) showed that adhesion to a waxy Brussels sprout variety by *Phaedon cochleariae* was much less than to a near-isogenic 'glossy' Brussels sprout lacking a waxy bloom. Beetles placed on leaves of the waxy plants fell off when these were raised to a median inclination of 40°, but beetles remained attached to glossy leaves at angles up to 180°. Wax particles could be seen by scanning electron microscopy (SEM) fouling the tarsal hairs of *Ph. cochleariae* collected from waxy plants. Bodnaryk (1992b) showed that mechanical removal of the wax bloom from a normal variety of *Brassica napus* markedly increased settling and feeding by the chrysomelid *Phyllotreta cruciferae*, as is beautifully illustrated in Figure 10.1.

The mechanism by which wax blooms provide defence against herbivory seems clear. More puzzling is the frequent occurrence of insect resistance in glossy (i.e. non-waxy) crops. Glossy forms of eight crops are reported resistant to 17 insect species, as compared with only five cases of increased susceptibility of glossy forms (Eigenbrode and Espelie, 1995). Although several explanations for this pattern have been offered, as will be discussed later on, it is possible that an important factor is attachment, not by herbivores but by entomophagous insects.

Attachment and multitrophic interactions. Way and Murdie (1965) first suggested that insect resistance in a glossy *Brassica* was related to increased predation on these plants, and this has since occurred to others working with glossy *Brassica* (Eigenbrode, 1990; Eigenbrode *et al.*, 1991a; Stoner, 1990; J.J.A. Van Loon,

Figure 10.1. Phyllotreta cruciferae *flea beetles preferentially attacking* Brassica napus *on part of the leaf surface from which the wax bloom has been removed mechanically, forming the letter 'B' (photograph courtesy of R.P. Bodnaryk).*

unpublished observations). Way and Murdie (1965) reported finding more predators, including coccinellids and anthocorids on glossy plants. In the field in Arizona, ants are found much more often on glossy cabbage than on normal-wax cabbage (S.D. Eigenbrode, unpublished observations).

One explanation for the larger numbers of predators on glossy plants is that these insects can adhere better to these plants, just as do herbivores. Larvae of the lacewing *Chrysoperla carnea*, repeatedly fall from leaves of waxy Brussels sprouts, but remain attached to plants such as *Brassica chinensis* that have less prominent blooms (von Arzet 1973). Similarly, coccinellid larvae (*Adalia bipunctata*) spend 70% of their time in 'unproductive activities' and no time feeding on waxy *Brassica* (Shah, 1982). In contrast they spend only 35% of their time in unproductive activities and 40% of their time feeding on less waxy *B. chinensis.*

Recent experimental work confirms that three predator species are more mobile and consequently more effective on a glossy mutant *Brassica* as compared with a standard variety (Eigenbrode *et al.*, 1996). Adults of the coccinellid beetle *Hippodamia convergens* and the anthocorid bug *Orius insidiosus*, and larvae of the *C. carnea* spent more time walking on a glossy cabbage than on a cabbage with normal bloom. All three predators also spent more time in 'unproductive behaviours', such as grooming, on the waxy plants. Finally, these predators explored more of the leaf surface of glossy plants, including the abaxial surface, while on waxy plants their movements were more confined to upper surfaces and leaf edges. This last tendency is dysfunctional for predation because of the prevalence of prey on abaxial surfaces of leaves.

As a consequence of these differences in mobility, all three predators are more effective at reducing populations of the larvae of the diamondback moth *Plutella xylostella* on glossy cabbage than on waxy cabbages (Eigenbrode *et al.*, 1995). On individually caged plants in the greenhouse, predators always significantly reduced survival of *P. xylostella* larvae on glossy plants, but never on normal-wax plants. On caged plants in the field, predators significantly reduced *P. xylostella* survival on glossy plants, but only *C. carnea* was effective on normal-wax plants.

Some other entomophagous insects, including hymenopteran parasitoids, that depend on agility may be less effective on *Brassica* with waxy blooms. The aphid parasitoid *Diaretiella rapae* was more commonly observed on glossy cabbage plants even though aphid populations were higher on waxy ones (S.D. Eigenbrode, unpublished observations). In a cage experiment, an aphelinid parasitoid *Eretmocerus* sp. produced more progeny per female and parasitized a greater percentage of *Bemisia* sp. whitefly nymphs on glossy cabbage plants than on waxy cabbage plants (O. Minkenberg and S.D. Eigenbrode, unpublished observations; Figure 10.2). Newly emerged parasitoids were caged on individual plants with abundant hosts (18 second-instar nymphs per female on average) for 16 days. Although the specific mechanisms for this effect remain unexplained, the wax bloom on normal-wax plants may have reduced mobility and searching efficiency of the wasps, just as trichomes influence parasitism by the whitefly parasitoid *Encarsia formosa* (van Lenteren, 1990).

Wax blooms on crops other than *Brassica* also potentially interfere with attachment and effectiveness of entomophagous insects; for example, adult coccinellids fall frequently from pea foliage (Kareiva and Sahakian, 1990; S.D. Eigenbrode and Rohde, unpublished observations) apparently because of poor adhesion to the waxy leaves. Kareiva and Sahakian (1990) recorded 45% of *Coccinella septempunctata* and 32% of *Hippodamia variegata* fell from pea plants within a 3-min period. Thus it is possible that 'glossy' crop varieties are 'resistant' to herbivores in part because reduced bloom results in enhanced predation and parasitism. In support of this, on mutant peas lacking a wax bloom, *H. convergens* is more effective at attacking pea aphid *Acyrthosiphon pisum* (S.D. Eigenbrode and

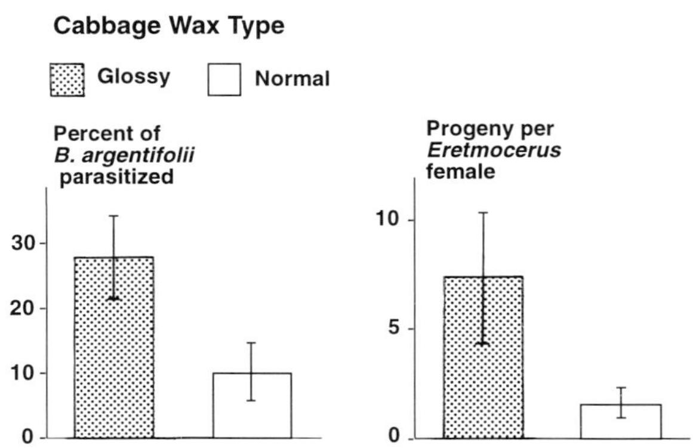

Figure 10.2. *Progeny per female* Eretmocerus *sp. and percent parasitism of* Bemisia *sp. on glossy and normal-wax cabbage (O. Minkenberg and S.D. Eigenbrode, unpublished observations.) (Percent parasitism = parasitized nymphs + exuviae of parasitized nymphs/total nymphs and exuviae; progeny = parasitized nymphs + exuviae of parasitized nymphs/female.)*

Rohde, unpublished observations), just as it is a better predator of *P. xylostella* on glossy *Brassica*. In addition, glossy crop varieties often express insect resistance more strongly on plants tested under field conditions than in greenhouses, clip cages, or other enclosures that would exclude entomophagous insects (Table 10.1). The evidence suggests that, although resistance to insects in glossy varieties of different crops must depend on various mechanisms, enhanced predation is a likely contributing factor.

Predators in natural systems. As they do on crop plants, wax blooms in natural systems also potentially interfere with active entomophagous insects. Predaceous coccinellids, for example, may forage less on very waxy plants. As a result, plants

Table 10.1. *Relative resistance to insect pests of glossy genotypes of three crop species tested in the field and greenhouse*

System	Relative scores*		Reference
	Field	Greenhouse or cage	
P. rapae damage score on PI 234599 glossy cauliflower (*B. oleracea*) vs. average for normal-wax cauliflower	0†	1.32	Dickson and Eckenrode (1975)
Trichoplusia ni damage score on PI 234599 vs. normal-wax cauliflowers	0†	0.90	Dickson and Eckenrode (1975)
P. xylostella survival on PI 234599 vs. normal wax varieties	0.0	0.94	Lin *et al.* (1983)
P. xylostella survival on NY 8329 glossy *B. oleracea* vs. normal-wax variety 'Round-Up'	0.10	0.78	Eigenbrode (1990); Eigenbrode *et al.* (1991b)
Bemisia tabaci counts and survival on 'Greenglaze' vs. 'Vates' collards (*B. oleracea*)	0.45	0.61-1.01	Elsey *et al.* (1994)
Schizaphis graminum counts on bloom and bloomless sorghums Redbine, RCK-60 and RWD 3 × Weskan *Bm* and *bm*	0.01-0.04	0.70-0.94	Starks and Weibel (1981)
Sitobion avenae infestation scores or counts on wheat varieties:			
non-glaucous and glaucous 'Avalon', 'Brigand', 'Virtue'	0.55-1.00	0.86-1.00	Lowe *et al.* (1985)
non-glaucous 320/30 vs. glaucous 'Sandown'	0.26-0.59	0.69-0.79	Lowe (1984)

* Mean value from glossy/mean value from normal wax for an experiment; ranges are for several experiments.

† Data from unreplicated trial.

with waxy blooms may provide some herbivores refuge against these and other active predators and parasitoids (see also Section 12.5.1, p. 242). Smaller coccinellid species that apparently adhere better to various surfaces, including wax blooms (Frazer and McGregor, 1994; Kareiva and Sahakian, 1990) may be more prevalent on waxy plants. As does plant architecture (Frazer and McGregor, 1994), plant surface waxes may in this way drive habitat specialization by coccinellids and other generalist predators.

How do waxes interfere with insect adhesion? Exfoliation of wax crystals appears to be the principal mechanism through which wax blooms interfere with insect attachment. Loose crystalline material on a surface, regardless of its composition, apparently will reduce insect adhesion. Indeed, Knoll (1914) found that talcum powder (microscopic crystals of $Mg_3Si_4O_{10}(OH)_2$) effectively restored the 'slipperiness' to ants of *Nepenthes* pitchers from which the waxes had been mechanically removed. Adhesion by *H. convergens* to glass plates can be greatly reduced by a deposit of crystalline tetracosanoic (C_{24}) acid (a common plant wax constituent), although the crystals of this pure compound are regular parallelograms unlike plant wax crystals (S.D. Eigenbrode, unpublished observations). Plant wax crystals as diverse in composition and structure as the plates and crenate ribbons of *Pisum* and *Eucalyptus* and the rods and filaments of *Brassica* all disrupt adhesion.

Quantitative differences in adhesion to different waxy blooms occur and may be related to crystal density, form, or their tendency to exfoliate readily from the plant. Adhesion by *P. charybdis* differs threefold between two bloomed species of *Eucalyptus* (Edwards and Wanjura, 1990), probably because of crystal density. Crystal density among several *Brassica* mutants is negatively correlated with susceptibility to flea beetles and positively correlated with the apparently predator-mediated resistance to *P. xylostella* (Figure 10.3). The additional variation may be related to the shapes of crystals on these *Brassica* mutants (Figure 10.4) or their tendency to exfoliate. Rigorous studies have not been made of the effects on adhesion of different wax crystal attributes.

In the absence of crystals, insect adhesion will depend on the molecular adhesive forces between the cuticular surfaces of insect and plant. Differences in wettability among plant wax constituents deposited as amorphous films (Holloway, 1969) suggest that adhesive properties of non-crystalline plant waxes differ. No studies address the possibility directly, but the chrysomelid beetle *Chrysolina polita* adheres equally well to smooth glass, gold-coated glass and to 'perspex' plastic (Stork, 1980a), materials with very different molecular adhesive properties (Visser, 1976). Nonetheless, it remains possible that small differences in adhesive properties of amorphous waxes have ecological significance for some insects.

Can insects adapt to improve adhesion to wax blooms? Because wax blooms are common and can hamper insect attachment, we might expect to find evidence of insect adaptations to cope with them. Remarkable structures permit insects to adhere to diverse substrates (Nachtigall, 1974), but loose crystalline surfaces seem to confound most of these. An exception might be suction or adhesion dependent on mucilaginous secretions; for example, *Chrysoperla carnea* possesses an anal adhesive organ that produces such a secretion (Spiegler, 1962). One experiment indicates this organ actually adheres *better* to waxy bloom than to glossy cabbages. Third-instar *C. carnea* were attached to a thread and allowed to walk for a moment on each type of leaf mounted to the pan of a milligram balance. The animal was then pulled vertically off the leaf surface by slowly reeling in the attached thread. The

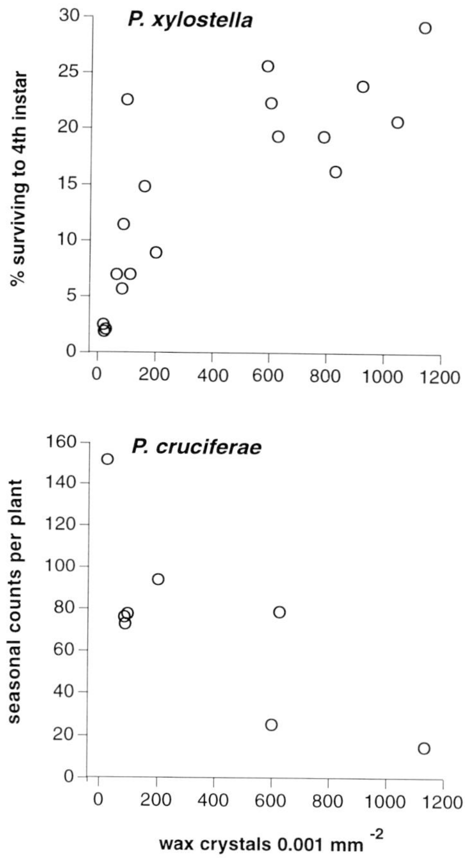

Figure 10.3. *Relationship between crystal densities of* Brassica *lines and infestation levels by* Phyllotreta *flea beetles and survival of* Plutella xylostella. *(Data from Eigenbrode* et al., *1991b and Stoner, 1990.)*

tarsi always broke free first, leaving the animal attached only by its abdominal organ. The 'force' (mg) required to break the abdominal organ free was 11.6 ± 2.4 from glossy plants and 25.6 ± 7.2 from normal-wax plants (n = 6 for each treatment; M.-B. Roux and S.D. Eigenbrode, unpublished observations). This suggests the mucilage is able to wet the wax crystals and incorporate them into an adhesive matrix or colloid (Visser, 1976).

Unfortunately for *C. carnea*, as we have seen, the superior adhesion of the abdominal organ becomes dysfunctional when coupled with reduced tarsal adhesion to the waxy surface of cabbage plants; the animal seems to have difficulty breaking the abdominal attachment, and 'scrambles' (Eigenbrode *et al.*, 1996). But adhesive secretions, such as those produced by some Syrphidae larvae, could be effective for adhering to wax blooms.

Certain structural adaptations of the legs may improve walking on wax blooms; for example, the cabbage aphid *Brevicoryne brassicae*, walks more effectively on waxy leaves of *Brassica* plants than does *Lipaphis erysimi.* Both species tend to

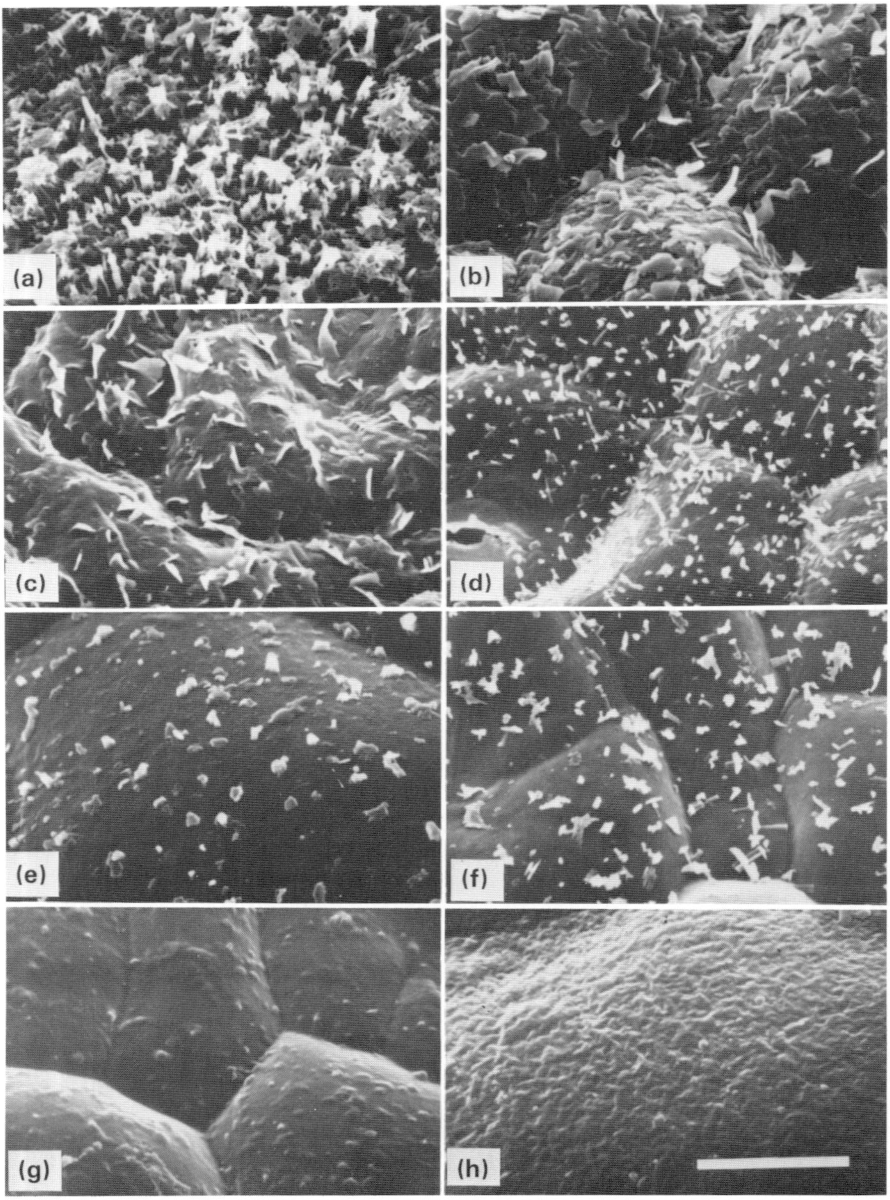

Figure 10.4. *Crystal morphology of several* Brassica *mutations that confer resistance to* Plutella xylostella. *(a) Normal bloom 'Round-Up'. Glossy types: (b) Broc5; (c) PI 261597; (d) Broc3; (e) 'Glossy Andes'; (f) 'Glazed Vates'; (g) PI 234599; (h) 'Green Glaze'. (Bar = 10 μm.) (Reproduced from Eigenbrode* et al. *(1991) Characteristics of leaf waxes of* Brassica oleracea *associated with resistance to diamondback moth.* J. Econ. Entomol. ***83****, 1609–1618, with permission from the Entomological Society of America.)*

move to the underside of a leaf if placed on the upper surface. Åhman (1990) showed that *B. brassicae* completes this manoeuvre equally well on host plants with a range of wax loads, while in contrast *L. erysimi* movement is impeded on very waxy plants; 40% of the insects reach the underside in 2 h on waxy plants (27–48 μg waxes cm^{-2}) and 80% on low-wax plants (3–15 μg waxes cm^{-2}). After mechanical removal of the wax crystals from a waxy cultivar, more than three times as many *L. erysimi* relocated to the leaf underside in 2 h.

Based on measurements from laboratory colonies, the two aphid species differ in the morphology of their legs (S.D. Eigenbrode, unpublished observations). *L. erysimi* body length is 1.92 ± 0.03 mm and lengths of the meta tibia and femur are 0.74 ± 0.02 and 0.44 ± 0.02 mm, respectively. *B. brassicae* body length is 1.55 ± 0.06 mm and the lengths of the meta tibia and femur are 0.93 ± 0.03 and 0.50 ± 0.02 mm, respectively. The other legs of *B. brassicae* are also longer than those of *L. erysimi.* The proportionately longer legs of *B. brassicae* may enable it to apply its tarsi perpendicular to the waxy surface and reduce lateral vectors that could reduce adhesion (e.g. Southwood, 1986).

Insect adaptations to cope with plant surface features are often behavioural rather than morphological (Kennedy, 1986; Southwood, 1986) and the same might be expected for waxes. A chrysomelid beetle, *Paropsis aegrota*, that utilizes very waxy *Eucalyptus* spp., has no obvious tarsal modifications for this, but accommodates by feeding and ovipositing while straddling the leaf edge (Edwards and Wanjura, 1990). This behaviour allows the mechanical advantage of opposing the tarsi and potentially applying more force perpendicular to the surface than supplied by gravity (S.D. Eigenbrode and M. Rohde, unpublished observations). *Phyllotreta* flea beetles and *H. convergens* also exhibit 'edgewalking' on leaves with wax blooms (Bodnaryk, 1992a; Eigenbrode *et al.*, 1996).

The expanded host range of the whitefly *Bemisia argentifolii* Bellows and Perring to include many crops, including brassicas, not exploited by its sister species *B. tabaci* (Gennadius), may also be associated with behavioural adaptations to surface features including waxes. Morphological differences between the adults of these species have not been reported, yet oviposition by *B. tabaci* is reduced on waxy cabbage while *B. argentifolii* accepts this host readily (Blua *et al.*, 1995). Mechanical removal of wax bloom from cabbage renders the plant acceptable for *B. tabaci* as well.

10.2.2 *Leaf colour*

Wax blooms reflect relatively uniformly across the visible spectrum (Mulroy, 1979; Prokopy *et al.*, 1983; Chapter 7), and they can also be brilliantly reflective in the ultraviolet (UV) wavelengths (Clark and Lister, 1975; Mulroy, 1979). Herbivorous insects, however, often are attracted to specific wavelengths in the 'green' near 540 nm. A wax bloom can make the plants less attractive to these insects. Cabbage root maggot flies, *Delia radicum*, alight less frequently on plants with waxy blooms than on the same plant species with reduced blooms and more reflectance in the attractive wavelengths (Prokopy *et al.*, 1983; Figure 10.5). This visual mechanism may account for the observed higher numbers of eggs laid by some insects on or near glossy mutant *Brassica* (Dickson and Eckenrode, 1975; Way and Murdie, 1965).

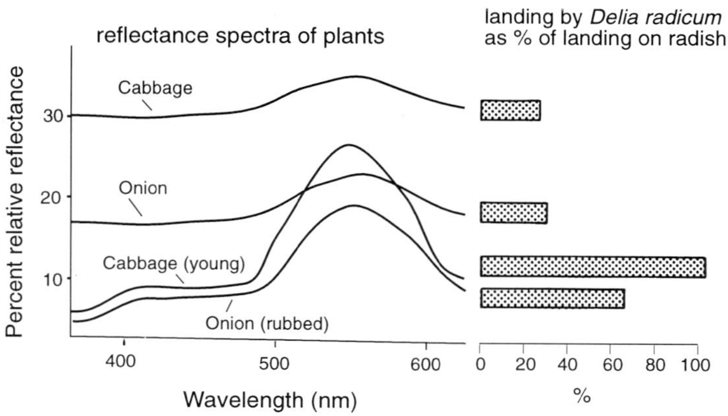

Figure 10.5. *Attractiveness of leaf surfaces to* Delia radicum *in relation to leaf surface waxes and spectral reflectance. (Data from Prokopy* et al., *1983.)*

10.2.3 *Pleiotropic effects of wax blooms on the plant*

Changes in the thickness or amount of crystallization of the surface waxes may have indirect physiological effects on the plant that, in turn, alter suitability for insects. Cole and Riggall (1992) demonstrated that *Brevicoryne brassicae* colony size was reduced on glossy Brussels sprouts by two-thirds relative to normal-bloom isolines in the greenhouse, but only when plants were experimentally water-stressed to approximate field conditions. Phloem ingestion was also reduced on the water-stressed glossy plants but not on water-stressed normal-bloom plants. The glossy Brussels sprouts, which have less surface wax than normal plants, are apparently more susceptible to water stress, and the associated changes in their constituent chemistry are detrimental to the aphids. A similar pleiotropic effect may produce some resistance to other insects in glossy *Brassica;* for example, even in the absence of predators, glossy cabbage is slightly more resistant to *P. xylostella* when grown in the field, but not when grown in the greenhouse (Eigenbrode, 1990; Eigenbrode *et al.*, 1995).

10.3 Surface waxes as semiochemicals

The chemicals on plant surfaces are important cues influencing host plant selection by phytophagous insects (Chapman and Bernays, 1989). Sometimes acceptance or rejection of a potential host occurs solely on the basis of these chemicals; for example, rejection of potential host plants by grasshoppers can occur after palpation of the plant surface and before biting (Chapman and Sword, 1993). Although plant surface chemicals are primarily waxes, minute but potentially semiochemically active amounts of sugars (Derridj *et al.*, 1989), plant volatiles (Spence *et al.*, 1983), amino acids (Soldaat *et al.*, 1996), and polar plant compounds (Städler and Buser, 1984) are also present. Trichome exudates may also 'contaminate' surface waxes and influence insects as well as pathogens on the plants surface (see Chapter 12). The presence of these materials calls into question our traditional view of what constitutes 'leaf surface waxes'. A discussion of all these surface constituents of plants is beyond the scope of a single review. In this chapter, the semiochemical activity of the most common wax constituents (long-chain alkyl components and terpenoids; Walton, 1990) will primarily be considered.

10.3.1 *Semiochemical activity of plant surface lipids*

Removal of plant surface lipids with organic solvents can alter insect host acceptance by phytophagous insects; for example, leaf discs of *Canna generalis* are rejected in choice tests by larvae of the tobacco hornworm *Manduca sexta*, but after their surface lipids are removed these plants become acceptable (de Boer and Hanson, 1988). Similarly, rejection of various less preferred plants at palpation by *Locusta migratoria* is eliminated when surface waxes are removed from the plants (Chapman, 1977; Woodhead and Chapman, 1986). A problem with such experiments is that plant surfaces can be changed considerably by removal of the lipids due to increased water loss and escape of internal constituents. However, the organic extracts themselves are also behaviourally active and may act as biting or oviposition deterrents (Bernays and Chapman, 1975; Woodhead, 1983; Woodhead and Padgham, 1988) or stimulants (Albert and Parisella, 1983; Bernays *et al.*, 1976; Foster and Harris, 1992; Klingauf *et al.*, 1978) for phytophagous insects.

Whether due to deterrents or stimulants, some insects behave differently on wax extracts of host and non-host plants. *P. xylostella* larvae will bite and palpate more on glass slides coated with surface waxes extracted from a susceptible cabbage than on waxes from a less preferred, glossy variety (Eigenbrode *et al.*, 1991b). Similarly, neonates of *Phthorimaea operculella* move more slowly and bite more on dichloromethane extracts of their host plant, potato, than on extracts of non-host plants (Varela and Bernays, 1988).

Subtle but critical herbivore behaviour may be influenced by surface lipid extracts. Successful establishment on sorghum by *Chilo partellus* requires that the neonate larvae migrate up the plant and into the whorl to feed. Successful completion of the orientation behaviour required to gain the whorl depends on surface waxes of the host plant (Bernays *et al.*, 1985).

In most of the above examples, the active components are probably the surface wax components. Usually, the extracts were first treated by evaporation under vacuum, clean-up on silica column, or both. Often, activity is retained in individual chromatographic fractions dominated by particular wax components (e.g. Atkin and Hamilton, 1982b; Foster and Harris, 1992; Klingauf *et al.*, 1978; Woodhead, 1983), but it has recently become apparent that even minute amounts of other components can influence bioassays with surface wax extracts. Roessingh *et al.* (1992) reported that glucosinolate concentrations as low as 10^{-9} M are capable of stimulating tarsal chemoreceptors of *D. radicum*. To establish the activity of the wax components, therefore, particular care is needed to ensure there are no such contaminants in materials used for bioassays.

10.3.2 *Wax components as cues for host acceptance*

The best evidence that waxy constituents rather than volatiles or polar compounds are semiochemicals is provided by studies with purified compounds. As summarized in Table 10.2, representative compounds from most major classes of wax chemicals have been found to be behaviourally active. The most common wax components often stimulate acceptance behaviours in insects, while less common ones, such as shorter-chain free fatty acids and even-chain-length alkanes are deterrents. This pattern is consistent with plant wax constituents signifying 'plant' or 'not-plant' to an exploring insect perhaps as a first step in assessment of the surface of a potential host. However, the ubiquity of most wax components makes

Table 10.2. *Effects of pure plant surface wax components on behaviour of phytophagous insects*

Compound(s)	Insect	Effect	Reference
STIMULATE FEEDING BEHAVIOUR OR OVIPOSITION			
***n*-Alkanes**			
C_{32}	*Acyrthosiphon pisum*	Increases probing into sachets with the alkane in suspension	Klingauf *et al.* (1971)
C_{25}, C_{26}, C_{28}, C_{29}, C_{33}	*Ostrinia nubilalis*	Stimulate oviposition on filter paper	S. Udayagiri and C.E. Mason (unpublished observations)
Fatty alcohols			
C_{26} and C_{28}	*Bombyx mori*	Stimulate feeding on artificial diets, other alcohols and C_{26} C_{28} acids inactive	Mori (1982)
Fatty acids			
C_{14}	*Plutella xylostella*	Stimulates palpation on glass substrate	S.D. Eigenbrode (unpublished observations)
$C_{>14}$	*Myzus persicae*	Stimulate settling	Greenway *et al.* (1978)
DETER FEEDING BEHAVIOUR OR SETTLING			
***n*-Alkanes**			
C_{19}, C_{21}, C_{23}	*Locusta migratoria*	Deter feeding on sucrose-treated glass fibre discs in choice test vs. sucrose controls. Longer chain-length and even chain-length alkanes not deterrent	Woodhead (1983)
Fatty acids			
C_8–C_{13}	*M. persicae*	Deter settling	Greenway *et al.* (1978); Phelan and Miller (1982); Sherwood *et al.* (1981)
C_{12}	*M. persicae*	Limits colonization and virus transmission when applied to sugar beets in the field	Herbach (1987)
$C_{18:1}$	*P. xylostella*	Deters biting on glass substrate	S.D. Eigenbrode (unpublished observations)
Fatty alcohols			
C_{25}	*P. xylostella*	Deters biting when added to cabbage waxes on glass	S.D. Eigenbrode (unpublished observations)

(continued)

Compound(s)	Insect	Effect	Reference
Alkyl esters			
C_{12} acid - C_{24} alcohol	*L. migratoria*	Deter feeding on sucrose-treated glass fibre discs in choice test vs. sucrose controls. Longer chain-length and even chain-length alkanes not deterrent	Woodhead (1983)
C_{12} acid-methyl	*L. migratoria*	Deters feeding on sucrose treated discs	E.A. Bernays (unpublished observations)
Triterpenols			
Amyrins	*L. migratoria*	Deter feeding on wheat flour discs	Bernays *et al.* (1976)
	P. xylostella	Deter biting when added to acceptable waxes	S.D. Eigenbrode (unpublished observations)
Other components			
p-hydroxy-benzaldehyde	*L. migratoria*	Deters feeding	Woodhead (1982)

them poor sign stimuli for host recognition; for example, the C_{26} and C_{28} *n*-alcohols stimulate feeding by the silkworm *Bombyx mori* (Mori, 1982), but these alcohols are so common in surface waxes they cannot unambiguously identify the silkworm's host, mulberry.

Waxes as synergists with other semiochemicals. Rather than acting as sign stimuli for host recognition, dominant wax components may be part of a host–plant-specific chemical 'signature' (e.g. Städler, 1978) of polar and non-polar surface chemicals. Consistent with this idea, waxy compounds have been found to synergize with host-specific secondary compounds to stimulate host acceptance in herbivorous insects. Empirically developed surrogate plant models used to test oviposition stimulants to different species of anthomyiid flies are often most effective if coated with paraffins (Harris *et al.*, 1987; Roessingh *et al.*, 1992). Roessingh and Städler (1990) found that the number of eggs laid by cabbage flies, *D. radicum*, on paper leaves treated with ethanolic extract of cabbage was increased as much as 50-fold when the models were coated with paraffin.

Recent work by Spencer (1996) demonstrates that sinigrin, an established oviposition stimulant for *Plutella xylostella* (Gupta and Thorsteinson, 1960; Reed *et al.*, 1989), is much more stimulating in the presence of alkanes. Individual mated *P. xylostella* females were exposed for 12 h in multiple choice arenas to aluminium foil targets coated with either sinigrin (10^{-4} M), a mixture of alkanes (C_{16}–C_{29}) 80 μg cm^{-2}, both sinigrin and alkanes together, or untreated. The combination of alkanes and sinigrin attracted more than four times as many eggs as sinigrin or alkanes alone (Figure 10.6a). In a no-choice test, only sinigrin or sinigrin + alkanes stimulated oviposition, and the combination was eight times as attractive as sinigrin alone. Spencer also showed that this synergy is apparently a manifestation of enhanced

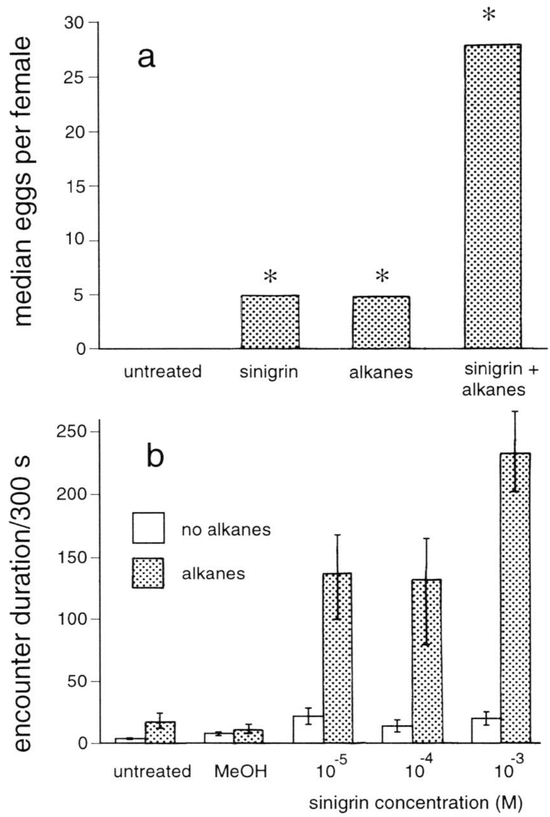

Figure 10.6. *Synergy between alkanes and sinigrin for* Plutella xylostella *oviposition. (a) Median eggs laid per female* P. xylotella *on foil targets treated with sinigrin (10^{-5} M), alkanes (equimolar C_{16}–C_{29} inclusive, 81 μg cm^{-2}) or both. (*, significantly different;* $p \leq 0.001$.*) (b) Mean time spent by females in contact with targets treated with sinigrin at different concentrations with or without alkanes. (Bars are SEMs.) (Spencer, 1996.)*

arrestment of the moths on foil targets treated with alkanes as well as sinigrin. During the first 5 min of exposure to an array of targets, arrestment of moths was dose dependent on sinigrin concentration only in the presence of the alkanes (Figure 10.6b).

In these examples, synergy by alkanes is independent of chain length. The paraffins used in fly experiments are a mixture of different chain-length alkanes, and pilot studies with *P. xylostella* show that individual odd-chain-length alkanes typical of plant waxes are no more effective synergists than are even-chain-length alkanes (J.L. Spencer, unpublished observations). Nevertheless, the ability to detect hydrocarbons in general might permit insects to recognize potential plant surfaces, and increase their responsiveness to host-specific chemical cues on the plant surface. The synergistic potential of other wax constituents such as *n*-alcohols has not been examined.

Theoretically, the relative concentrations of ubiquitous surface wax components alone could be used by insects to identify host plants. The relative concentrations of alkanes, for example, can be an effective chemotaxonomic trait (Harborne and Turner, 1984) and might be expected to serve as well for insects. This potential is suggested by Foster and Harris' (1992) finding that Hessian fly *Mayetiola destructor* is more stimulated by combined chromatographic fractions of the surface waxes of its host plant, wheat, than by any individual fraction. Unfortunately, plant surface wax composition can vary substantially with plant age and environmental conditions (e.g. Atkin and Hamilton, 1982a; Baker, 1974; see Sections 9.3–4 and 11.2) and this can limit their reliability for species identification by chemotaxonomists (e.g. Zygadlo *et al.*, 1994) and potentially by insects. However, insects selecting among host plants within a limited spatial and temporal context may be able to utilize these cues effectively. Surface lipid composition is sufficient to distinguish chemotaxonomically between 19 species of *Chionochloa,* as long as the analysis was performed within localities (Savill *et al.*, 1988).

Wax components as deterrents. Deterrent chemicals in non-hosts are at least as important for host selection by phytophagous insects as are positive cues in host plants (Jermy, 1966), and there are several examples of plant surface wax components acting as deterrents (Table 10.2). Some of these deterrents are so atypical (e.g. very short-chain fatty acids) that they may merely signal 'not plant' to insect herbivores. Others, such as triterpenoids, are more common but not ubiquitous, and may be valuable signals by which insects distinguish non-host and host plants; for example, waxes of some plants are composed of as much as 50% pentacyclic triterpenoids or related saponins (Balsdon *et al.*, 1995; Smith and Severson, 1992) while in other plants these components are scarce or absent. The triterpenoids, α- and β-amyrin are not normally found in the waxes of cultivated *Brassica,* but surface waxes of cabbages with a gene causing glossy wax contain as much as 6% amyrins and are resistant to *Plutella xylostella* (Eigenbrode *et al.*, 1991a; Eigenbrode *et al.*, 1991b; K. Espelie, unpublished data). When extracted and deposited on glass, waxes from these glossy plants elicit less biting and palpating and more walking by neonate *P. xylostella* than do waxes from normal cabbage (Eigenbrode *et al.*, 1991a). This syndrome of reduced acceptance can be reproduced on waxes from normal-wax cabbage by augmenting them with 1% amyrins (S.D. Eigenbrode and S. Pillai, unpublished observations). Thus the amyrins appear to be the principal deterrents influencing larval acceptance of these plants. Larvae of *P. xylostella* foraging in a mixed canopy, through this response to amyrins, may be able to rapidly recognize and avoid inappropriate feeding on non-hosts.

Amyrins are also deterrent to *L. migratoria*, a generalist species that predominantly feeds on Poaceae (Bernays and Chapman, 1976). Although amyrins are reported in grass waxes (Walton, 1990), their concentrations in waxes of hosts of *L. migratoria* are largely unknown. Concentration-dependent deterrence by these compounds may influence host preference behaviour of *L. migratoria*.

Correlations between concentrations of particular wax components and insect resistance suggest possible deterrent activities. These have been reported between amyrins and the azalea lacebug *Stephanitis pyrioides* (Balsdon *et al.*, 1995), the raspberry aphid *Amphorophora idaea* (Robertson *et al.*, 1991), and the fall armyworm *Spodoptera frugiperda* (Yang *et al.*, 1993; see Chapter 11), between esters of the C_{30} *n*-alcohol and the spotted alfalfa aphid *Therioaphis maculata* (Bergman *et al.*, 1991), and between the C_{22} *n*-alcohol and the tobacco budworm *Heliothis virescens* (Johnson and Severson, 1984). Deterrent effects of these components have not, however, been determined.

Other semiochemical phenomena involving waxes. Surface wax chemistry as well as morphology may influence interactions at higher trophic levels. Some predators and parasitoids respond to cuticular hydrocarbons of their prey. It is interesting that hydrocarbon profiles of some herbivores resemble those of their host plants (Espelie and Brown, 1990; Espelie *et al.*, 1991). It may be that these similarities constitute a chemical crypsis defence by a herbivore, making it more difficult to distinguish chemically from the surface of its host plant. The demonstrated ability of some insects to alter their surface hydrocarbons in response to their diet (Espelie and Bernays, 1989) may contribute to this effect by allowing insects to improve the match between their surface lipids and those of a host plant.

Plant surface waxes are excellent solvents for some volatile components from the host plant (Spence *et al.*, 1983). They may also serve as wicks for adsorption and retention of volatiles from the environment, including insect pheromones (Noldus *et al.*, 1991). This phenomenon may be quite widespread in complex canopies and have interesting ecological repercussions.

10.4 Surface wax morphology and chemistry are interdependent

Chemical and morphological aspects of plant surface waxes will act together to influence insect behaviour. Host plant recognition potentially depends on visual, tactile and chemical cues of the surface wax mixture. Certainly all these modalities can be used during host plant selection. Because morphology depends to a large extent on chemical composition (Jeffree, 1986), insects will sometimes encounter particular chemical and morphological attributes together; for example, *n*-alcohols and the plate-like crystals they predominantly form. Wax morphology may provide a tactile component reinforcing chemically based host recognition. In other cases a wax component present at low but semiochemically active concentrations may not influence crystal form. The information content of these combined sensory modalities is large.

Although no studies have attempted to examine interactions between surface wax chemistry and morphology and insect behaviour, some observations suggest them. Neonate *P. xylostella* discriminate between intact surfaces of resistant and susceptible plants as well as between the extracted surface waxes of these plants. But on the plant surface, where crystals are intact, differences in palpation predominate, while on the waxes deposited as an amorphous film, differences in biting predominate (Eigenbrode *et al.*, 1991a). The spectacular synergy between alkanes and polar surface extracts for oviposition by *D. radicum* (Roessingh and Städler, 1990) only occurred if the paraffin was applied to the plant model as a smooth film. If paraffin was sprayed on in a solvent, no synergy occurred, suggesting an overriding importance of tactile qualities of the paraffin. The surfaces were not examined with SEM, but solvent-applied paraffin may have had some structural features.

10.5 Some implications and conclusions

10.5.1 *Plant–insect interactions*

Study of the effects of plant surface waxes on herbivores has been dominated by work with cultivated crops, particularly those with glossy waxes, but what has been discovered indicates that waxes can influence insect–plant ecology in natural systems. Striking variation in wax blooms occurs within natural populations of several plant species (e.g. *Salix lutea* and *S. lasiandra* in the Pacific Northwest, Hitchcock *et al.*, 1964; *Jojoba* spp. and *Ephedrus* spp. in the desert Southwest, S.D. Eigenbrode and E.A. Bernays, unpublished observations; several genera of Crassulaceae, Mulroy, 1979; *Cercis canadensis mexicana*, Tipton and White, 1995). This variation undoubtedly influences phytophagous and entomophagous arthropods associated with these plants, but this has not been very well studied.

Wax components may play a larger role than heretofore recognized as deterrents defining herbivore host ranges. The role of the triterpenoids in waxes as protective deterrents has been surmised (Gershenzon and Croteau, 1991) but not investigated in natural systems. The dominance of these compounds in some fruit waxes (Walton, 1990) is interesting in this regard. As with so many plant secondary compounds, these wax triterpenoids may deter some insects but become arrestants for specialists: for example, the blackmargined aphid, *Monellia caryella* settles preferentially on the surfaces of its host, pecan *Carya illinoiensis*, the waxes of which contain over 30% amyrins (Smith and Severson, 1992).

Finally, the synergistic interactions between surface wax components and other semiochemicals may help explain the chemical basis of insect and plant host associations. The basis of this synergism requires study, and its potential influence must be considered during development of experimental studies of insect host selection behaviour.

10.5.2 *Pest management*

As knowledge of the functions of plant surface waxes in insect–plant interactions grows, so will our ability to manipulate surface waxes genetically to alter these interactions. We know that wax characteristics are influenced by a large number of genes. Loci have been identified in maize, barley, *Brassica* spp., *Pisum sativum*, and *Arabidopsis thaliana* (Baker, 1974; Bianchi *et al.*, 1985; Koorneef *et al.*, 1989; Macey and Barber, 1970; von Wettstein-Knowles, 1979). Mutations at these loci have permitted us to develop models for the mechanisms of wax biosynthesis (von Wettstein-Knowles, 1995).

Recently, molecular techniques have permitted cloning and sequencing of some of the genes involved (Aarts *et al.*, 1995; Tacke *et al.*, 1995). Thus, transgenic crop plants expressing particular wax traits will not be far behind. Indicative of the potential for pest management, recent work by A. Rashotte (unpublished data) documents apparent antibiotic resistance to green peach aphid *Myzus persicae* in *Arabidopsis* with insertion mutation of the CER3 wax gene. Although glossy-leaf resistance has been documented in germplasm of several crops, there are few glossy varieties in wide use because of some of the liabilities of the glossy trait. Transgenic plants conferring resistance but without problems with water relations and pathogen susceptibility are within the realm of possibility. Potentially biotechnology can be used to elevate insect feeding deterrents in waxes, or to produce plant surfaces that provide advantages to entomophagous insects, or both, and provide exciting new possibilities for insect pest management.

References

Aarts, M.G.M., Keijzer, C.J., Stiekema, W.J. *et al.* (1995) Molecular characterization of the CER1 gene of Arabidopsis involved in epicuticular wax biosynthesis and pollen fertility. *Plant Cell* **7**, 2115–2127.

Åhman, I. (1990) Plant-surface characteristics and movements of two *Brassica*-feeding aphids, *Lipaphis erysimi* and *Brevicoryne brassicae*. *Symp. Biol. Hung.* **39**, 119–125.

Albert, P.J. and Parisella, S. (1983) Chemical bases of host-plant selection by eastern spruce budworm (*Choristoneura fumiferana* Clem.) (Lepidoptera: Tortricidae). In: *Proceedings, forest defoliator-host interactions: a comparison between gypsy moth and spruce budworm*. USDA Forest Service, Washington, DC, General Technical Report NE-85, pp. 9–14.

Anstey, T.H. and Moore, J.F. (1954) Inheritance of glossy foliage and cream petals in Green Sprouting broccoli. *J. Heredity* **45**, 39–41.

Atkin, D.S.J. and Hamilton, R.J. (1982a) The changes with age in the epicuticular wax of *Sorghum bicolor*. *J. Nat. Prod.* **45**, 697–703.

Atkin, D.S.J. and Hamilton, R.J. (1982b) The effects of plant waxes on insects. *J. Nat. Prod.* **45**, 694–696.

Baker, E.A. (1974) The influence of environment on leaf wax development in *Brassica oleracea* var. *gemmifera*. *New Phytol.* **73**, 955–966.

Balsdon, J.A., Espelie, K.E. and Braman, S.K. (1995) Epicuticular lipids from azalea (*Rhododendron* spp.) and their potential role in host plant acceptance by azalea lace bug, *Stephanitis pyrioides* (Heteroptera: Tingidae). *Biochem. Syst. Ecol.* **23**, 477–485.

Bergman, D.K., Dillwith, J.W., Zarrabi, A.A. *et al.* (1991) Epicuticular lipids of alfalfa relative to its susceptibility to spotted alfalfa aphids (Homoptera: Aphididae). *Environ. Entomol.* **20**, 781–785.

Bernays, E.A. and Chapman, R.F. (1975) The importance of chemical inhibition of feeding in host-plant selection by *Chorthippus parallelus* (Zetterstedt). *Acrida* **4**, 83–93.

Bernays, E.A. and Chapman, R.F. (1976) Deterrent chemicals as the basis of oligophagy in *Locusta migratoria* (L.). *Ecol. Entomol.* **2**, 1–18.

Bernays, E.A., Blaney, W.M., Chapman, R.F. *et al.* (1976) The ability of *Locusta migratoria* (L.) to perceive plant surface waxes. *Symp. Biol. Hung.* **16**, 35–40.

Bernays, E., Woodhead, S. and Haines, L. (1985) Climbing by newly hatched larvae of the spotted stalk borer *Chilo partellus* to the top of sorghum plants. *Entomol. Exp. Appl.* **39**, 73–79.

Bianchi, A., Bianchi, G., Avato, P. *et al.* (1985) Biosynthetic pathways of epicuticular wax of maize as assessed by mutation, light, plant age and inhibitor studies. *Maydica* **30**, 179–198.

Blua, M.J., Yoshida, H.A. and Toscano, N.C. (1995) Oviposition preference of two *Bemisia* species (Homoptera: Aleyrodidae). *Environ. Entomol.* **24**, 88–93.

Bodnaryk, R.P. (1992a) Distinctive leaf feeding patterns on oilseed rapes and related Brassicacea by flea beetles, *Phyllotreta cruiciferae* (Goeze) (Coleoptera:Chrysomelidae). *Can. J. Plant Sci.* **72**, 575–581.

Bodnaryk, R.P. (1992b) Leaf epicuticular wax, an antixenotic factor in *Brassicaceae* that affects the rate and pattern of feeding in flea beetles, *Phyllotreta cruciferae* (Goeze). *Can. J. Plant Sci.* **72**, 1295–1303.

Chapman, R.F. (1977) The role of the leaf surface in food selection by acridids and other insects. *Colloq. Int. Cent. Nat. Rech. Sci. France* **265**, 133–149.

Chapman, R.F. and Bernays, E.A. (1989) Insect behavior at the leaf surface and learning as aspects of host plant selection. *Experientia* **45**, 215–221.

Chapman, R.F. and Sword, G. (1993) The importance of palpation in food selection by a polyphagous grasshopper (Orthoptera: Acrididae). *J. Insect Behav.* **6**, 79–91.

Clark, J.B. and Lister, G.R. (1975) Photosynthetic action spectra of trees. *Plant Physiol.* **55**, 407–413.

Cole, R.A. and Riggall, W. (1992) Pleiotropic effects of genes in glossy *Brassica oleracea* resistant to *Brevicoryne brassicae*. In: *Proceedings of the 8th International Symposium on Insect-Plant Relationships* (eds S.B.J. Menken, J.H. Visser and P. Harrewjin). Kluwer Academic Publishers, Dordrecht, pp. 313–315.

de Boer, G. and Hanson, F.E. (1988) The role of leaf lipids in food selection by larvae of the tobacco hornworm, *Manduca sexta*. *J. Chem. Ecol.* **14**, 669–682.

Derridj, S., Gregoire, V., Boutin, J.P. *et al.* (1989) Plant growth stages in the interspecific oviposition preference of the European corn borer and relations with chemicals present on the leaf surface. *Entomol. Exp. Appl.* **53**, 267–276.

Dickson, M.H. and Eckenrode, C.J. (1975) Variation in *Brassica oleracea* resistance to cabbage looper and imported cabbage worm in the greenhouse and field. *J. Econ. Entomol.* **68**, 757–760.

Edwards, P.B. (1982) Do waxes on juvenile Eucalyptus leaves provide protection from grazing insects? *Aust. J. Ecol.* **7**, 347–352.

Edwards, P.B. and Wanjura, W.J. (1990) Physical attributes of eucalypt leaves and the host range of chrysomelid beetles. *Symp. Biol. Hung.* **39**, 227–236.

Eigenbrode, S.D. (1990) Diamondback moth resistance mechanisms in *Brassica*. Ph.D. Dissertation, Cornell University, Ithaca, NY.

Eigenbrode, S.D. and Espelie, K.E. (1995) Effects of plant epicuticular lipids on insect herbivores. *Annu. Rev. Entomol.* **40,** 171-194.

Eigenbrode, S.D., Espelie, K.E. and Shelton, A.M. (1991a) Behavior of neonate diamondback moth larvae on leaves and leaf waxes of resistant and susceptible cabbages. *J. Chem. Ecol.* **17,** 1691-1704.

Eigenbrode, S.D., Stoner, K.A., Shelton, A.M. *et al.* (1991b) Characteristics of leaf waxes of *Brassica oleracea* associated with resistance to diamondback moth. *J. Econ. Entomol.* **83,** 1609-618.

Eigenbrode, S.D., Moodie, S. and Castagnola, T. (1995) Generalist predators mediate resistance to a phytophagous pest in cabbage with glossy leaf wax. *Entomol. Exp. Appl.* **77,** 335-342.

Eigenbrode, S.D., Castagnola, T. and Roux, M.B. (1996) Mobility of three generalist predators is greater on cabbage with glossy leaf wax than on cabbage with a wax bloom. *Entomol. Exp. Appl.* (in press).

Elsey, K.D. and Farnham, M.W. (1994) Response of *Brassica oleracea* L. to *Bemisia tabaci* (Gennadius). *HortScience* **29,** 814-817.

Espelie, K.E. and Bernays, E.A. (1989) Diet-related differences in the cuticular lipids of *Manduca sexta* larvae. *J. Chem. Ecol.* **15,** 2003-2017.

Espelie, K.E. and Brown, J.J. (1990) Cuticular hydrocarbons of species which interact on four trophic levels: Apple, *Malus pumila* Mill.; codling moth, *Cydia pomonella* L.; a hymenopteran parasitoid, *Ascogaster quadridentata* Wesmael; and a hyperparasite *Perilampus fulvicornis* Ashmead. *Comp. Biochem. Physiol.* **95B,** 131-136.

Espelie, K.E., Bernays, E.A. and Brown, J.J. (1991) Plant and insect cuticular lipids serve as behavioral cues for insects. *Arch. Insect Biochem. Physiol.* **17,** 223-233.

Foster, S.P. and Harris, M.O. (1992) Foliar chemicals of wheat and related grasses influencing oviposition by Hessian fly, *Mayetiola destructor* (Say) (Diptera, Cecidomyiidae). *J. Chem. Ecol.* **18,** 1965-1980.

Frazer, B.D. and McGregor, R.R. (1994) Searching behaviour of adult female Coccinellidae (*Coleoptera*) on stem and leaf models. *Can. Entomol.* **126,** 389-399.

Gershenzon, J. and Croteau, R. (1991) Terpenoids. In: *Herbivores: Their Interactions with Secondary Plant Metabolites* (eds M.R. Berenbaum and G.A. Rosenthal). Academic Press, San Diego, CA, pp. 165-220.

Greenway, A.C., Griffiths, D.C. and Lloyd, S.L. (1978) Response of *Myzus persicae* to components of aphid extracts and to carboxylic acids. *Entomol. Exp. Appl.* **24,** 369-374.

Gupta, P.D. and Thorsteinson, A.J. (1960) Food plant relationships of the diamondback moth *Plutella maculipennis* (Curt.). III. Sensory regulation of oviposition of the adult female. *Entomol. Exp. Appl.* **3,** 305-314.

Harborne, J.B. and Turner, B.L. (1984) *Plant Chemosystematics.* Academic Press, London.

Harley, R. (1991) The greasy pole syndrome. In: *Ant - Plant Interactions* (eds C.R. Huxley and D.F. Cutler). Oxford University Press, Oxford, pp. 430-433.

Harris, M.O., Keller, J.E. and Miller, J.R. (1987) Responses to *n*-dipropyl disulfide by ovipositing onion flies. *J. Chem. Ecol.* **13,** 1261-1277.

Herbach, E. (1987) Effect of dodecanoic acid on the colonisation of sugar beet by aphids and the secondary spread of virus yellows. *Annu. Appl. Biol.* **111,** 477-482.

Hitchcock, C.L., Conquist, A., Ownbey, M. *et al.* (1964) *Vascular plants of the Pacific Northwest. Part 2: Salicaceae to Saxifragaceae.* University of Washington Press, Seattle, WA.

Holloway, P.J. (1969) Chemistry of leaf waxes in relation to wetting. *J. Sci. Food Agric.* **20,** 124-128.

Jeffree, C.E. (1986) The cuticle, epicuticular waxes and trichomes of plants, with reference to their structure, functions and evolution. In: *Insects and the Plant Surface* (eds B. Juniper and T.R.E. Southwood). Edward Arnold, London, pp. 23-64.

Jermy, T. (1966) Feeding inhibitors and food preference in chewing phytophagous insects. *Entomol. Exp. Appl.* **9,** 1-12.

Johnson, A.W. and Severson, R.F. (1984) Leaf surface chemistry of tobacco budworm resistant tobacco. *J. Agric. Entomol.* **1,** 23-33.

Juniper, B.E. (1995) Waxes on plant surfaces and their interactions with insects. In: *Waxes: Chemistry, Molecular Biology and Functions* (ed. R.J. Hamilton). The Oily Press, Dundee, pp.157-174.

Juniper, B.E., Robins, R.J, and Joel, D,M. (1989) *The Carnivorous Plants.* Academic Press, London.

Kareiva, P. and Sahakian, R. (1990) Tritrophic effects of a simple architectural mutation in pea plants. *Nature* **345,** 433-434.

Kennedy, C.E.J. (1986) Attachment may be a basis for specialization in oak aphids. *Ecol. Entomol.* **11,** 291-300.

Klingauf, F., Nöcker-Wenzel, K. and Klein, W. (1971) Einfluss einiger Wachskomponenten von *Vicia faba* L. auf das Wirtswahlverhalten von *Acyrthosiphon pisum* (Harris) (Homoptera: Aphididae). *Z. Pflanzenkr. Pflanzenpathol. Pflanzenschutz* **78**, 641–648.

Klingauf, F., Nöcker-Wenzel, K. and Röttger, U. (1978) Die Rolle peripherer Pflanzenwachse für den Befall durch phytophage Insekten. *Z. Pflanzenkr. Pflanzenpathol. Pflanzenschutz* **85**, 228–237.

Knoll, F. (1914) Über die Ursache des Ausgleitens der Insectenbeine an wachsbedeckten Pflanzentheilen. *Jahrb. Wiss. Bot.* **54**, 448–497.

Koorneef, M., Hahnhart, C.J. and Thiel, F. (1989) A genetic and phenotypic description of *Eceriferum* (*cer*) mutants in *Arabidopsis thaliana*. *Heredity* **80**, 118–122.

Lin, J., Eckenrode, C.J. and Dickson, M.H. (1983) Variation in *Brassica oleracea* resistance to diamondback moth (Lepidoptera: Plutellidae). *J. Econ. Entomol.* **76**, 1423–1427.

Lowe, H.J.B. (1984) Development and practice of a glasshouse screening technique for resistance of wheat to the aphid *Sitobion avenae*. *Ann. Appl. Biol.* **104**, 297–305.

Lowe, H.J.B., Murphy, B.J.P. and Parker, M.L. (1985) Non-glaucousness, a possible aphid-resistance character of wheat. *Ann. Appl. Biol.* **106**, 555–560.

Macey, M.J.K. and Barber, H.N. (1970) Chemical genetics of wax formation on the leaves of *Pisum sativum*. *Phytochemistry* **9**, 5–12.

Martin, J.T. and Juniper, B.E. (1970) *The Cuticles of Plants*. Edward Arnold, London.

Mori, M. (1982) *n*-Hexacosanol and *n*-octacosonol; feeding stimulants for larvae of the silkworm, *Bombyx mori*. *J. Insect Physiol.* **11**, 969–973.

Mulroy, T.W. (1979) Spectral properties of heavily glaucous and non-glaucous leaves of a succulent rosette plant. *Oecologia* **38**, 349–357.

Nachtigall, W. (1974) *Biological Mechanisms of Attachment*. Springer-Verlag, New York.

Noldus, L.P.J.J., Potting, R.P.J. and Barendregt, H.E. (1991) Moth sex pheromone adsorption to leaf surface: bridge in time for chemical spies. *Physiol. Entomol.* **16**, 329–344.

Phelan, P.L. and Miller, J.R. (1982) Post-landing behavior of alate *Myzus persicae* as altered by (*E*)-β-farnesene and three carboxylic acids. *Entomol. Exp. Appl.* **32**, 46–53.

Power, H. (1664) *Experimental Philosophy, Containing New Microscopical, Mercurial and Magnetic Experiments*. London.

Prokopy, R.J., Collier, R.H. and Finch, S. (1983) Leaf color used by cabbage root flies to distinguish among host plants. *Science* **221**, 190–192.

Reed, D.W., Pivnick, K.A. and Underhill, E.W. (1989) Identification of chemical oviposition stimulants for the diamondback moth, *Plutella xylostella*, present in three species of Brassicaceae. *Entomol. Exp. Appl.* **53**, 277–286.

Robertson, G.W., Griffiths, D.W., Birch, A.N.E. *et al.* (1991) Further evidence that resistance in raspberry to the virus vector aphid, *Amphorophora idaei*, is related to the chemical composition of the leaf surface. *Ann. Appl. Biol.* **119**, 443–449.

Roessingh, P. and Städler, E. (1990) Foliar form, colour and surface characteristics influence oviposition behaviour in the cabbage root fly *Delia radicum*. *Entomol. Exp. Appl.* **57**, 93–100.

Roessingh, P., Städler, E., Fenwick, G.R. *et al.* (1992) Oviposition and tarsal chemoreceptors of the cabbage root fly are stimulated by glucosinolates and host plant extracts. *Entomol. Exp. Appl.* **65**, 267–282.

Savill, M.G., Bickestaffe, R. and Connor, H.E. (1988) Interspecific variation in epicuticular waxes of *Chionochloa*. *Phytochemistry* **27**, 3499–3507.

Shah, M.A.T. (1982) The influence of plant surfaces on the searching behavior of coccinellid larvae. *Entomol. Exp. Appl.* **31**, 377–380.

Sherwood, M.H., Greenway, A.R. and Griffiths, D.C. (1981) Responses of *Myzus persicae* (Sulzer) (Hemiptera: Aphididae) to plants treated with fatty acids. *Bull. Entomol. Res.* **71**, 133–136.

Smith, M.T. and Severson, R.F. (1992) Host recognition by the blackmargined aphid (Homoptera: Aphididae) on pecan. *J. Entomol. Sci.* **27**, 93–112.

Soldaat, L.L., Boutin, J. and Derridj, S. (1996) Species-specific composition of free amino acids on the leaf surface of four *Senecio* species. *J. Chem. Ecol.* **22**, 1–12.

Southwood, T.R.E. (1973) The insect-plant relationship — an evolutionary perspective. *Symp. R. Entomol. Soc. Lond.* **6**, 143–155.

Southwood, T.R.E. (1986) Plant surfaces and insects — an overview. In: *Insects and the Plant Surface* (eds T.R.E. Southwood and B. Juniper). Edward Arnold, London, pp. 1–22.

Spence, R.M., Tucknott, O.G., Baker, E.A. *et al.* (1983) Compounds associated with the surface lipid layer of watercress. *Phytochemistry* **22**, 1753–1756.

Spencer, J.L. (1996) Waxes enhance *Plutella xylostella* (Lepidoptera: Plutellidae) oviposition in response to sinigrin and cabbage homogenates. *Entomol. Exp. Appl.* (in press).

Spiegler, P.E. (1962) The origin and nature of the adhesive substance in larvae of the genus *Chrysopa* (Neuroptera: Chrysopidae). *Ann. Entomol. Soc. Am.*, **55**, 69–72.
Städler E. (1978) Chemoreception of host plant chemicals by ovipositing females of *Delia (Hylemya) brassica*. *Entomol. Exp. Appl.* **24**, 711–720.
Städler, E. and Buser, H.R. (1984) Defense chemicals in leaf surface wax synergistically stimulate oviposition by a phytophagous insect. *Experientia* **40**, 1157–1159.
Starks, K.J. and Weibel, D.E. (1981) Resistance in bloomless and sparse-bloom sorghum to greenbugs. *Environ. Entomol.* **10**, 963–965.
Stoner, K.A. (1990) Glossy leaf wax and host-plant resistance to insects in *Brassica oleracea* L. under natural infestation. *Environ. Entomol.* **19**, 730–739.
Stork, N.E. (1980a) Experimental analysis of adhesion of *Chrysolina polita* (Chrysomelidae: Coleoptera) on a variety of surfaces. *J. Exp. Biol.* **88**, 91–107.
Stork, N.E. (1980b) Role of waxblooms in preventing attachment to brassicas by the mustard beetle, *Phaedon cochleariae*. *Entomol. Exp. Appl.* **28**, 100–107.
Tacke, E., Korfhage, C. and Doring, H.-P. (1995) Transposon tagging of the maize Glossy2 locus with the transposable element En/Spm. *Plant J. Cell Mol. Biol.* **8**, 907–912.
Tipton, J.T. and White, M. (1995) Differences in leaf cuticle structure and efficacy among Eastern redbud and Mexican redbud phenotypes. *J. Am. Soc. Hort. Sci.* **120**, 59–64.
van Lenteren, J.C. (1990) Biological control in a tritrophic system approach. In: *Aphid-Plant Interactions: Populations to Molecules* (eds D.C. Peters, J.A. Webster and C.S. Chlouber). Oklahoma State University, Stillwater, OK, pp. 3–28.
Varela, L.G. and Bernays, E.A. (1988) Behavior of newly hatched potato tuber moth larvae, *Phthorimaea operculella* Zell. (Lepidoptera: Gelechiidae), in relation to their host plants. *J. Insect. Behav.* **1**, 261–275.
Visser, J. (1976) Adhesion of colloid particles. In: *Surface and Colloidal Science* (ed. Matijević). John Wiley and Sons, New York, pp. 3–84.
von Arzet, H.R. (1973) Suchverhalten der Larven von *Chrysopa carnea* Steph. (Neuroptera: Chrysopidae). *Z. Angew. Entomol.* **74**, 64–79.
von Wettstein-Knowles, P. (1979) Genetics and biosynthesis of plant epicuticular waxes. In: *The Biochemistry and Physiology of Plant Lipids* (eds L.-Å. Appelqvist and C. Liljenberf). Elsevier/North Holland Biomedical Press, Amsterdam, pp. 1–26.
von Wettstein-Knowles, P. (1995) Biosynthesis and genetics of plant epicuticular waxes. In: *Waxes: Chemistry, Molecular Biology and Functions* (ed. R.J. Hamilton). The Oily Press, Dundee, pp. 81–129.
Walton, T.J. (1990) Waxes, cutin and suberin. In: *Methods in Plant Biochemistry* (eds P.M. Dey and J.B. Harborne). Academic Press Inc., San Diego, CA, pp. 105–158.
Way, M.J. and Murdie, G. (1965) An example of varietal resistance of Brussels sprouts. *Ann. Appl. Biol.* **56**, 326–328.
Woodhead, S. (1982) *p*-hydroxybenzaldehyde in the surface wax of sorghum: Its importance in seedling resistance to acridids. *Entomol. Exp. Appl.* **31**, 296–302.
Woodhead, S. (1983) Surface chemistry of *Sorghum bicolor* and its importance in feeding by *Locusta migratoria*. *Physiol. Entomol.* **8**, 345–352.
Woodhead, S. and Chapman, R.F. (1986) Insect behaviour and the chemistry of plant surface waxes. In: *Insects and the Plant Surface* (eds B. Juniper and T.R.E. Southwood). Edward Arnold, London, pp. 123–135.
Woodhead, S. and Padgham, D.E. (1988) The effect of plant surface characteristics on resistance of rice to the brown planthopper, *Nilaparvata lugens*. *Entomol. Exp. Appl.* **47**, 15–22.
Yang, G., Wiseman, B.R., Isenhour, D.J. *et al.* (1993) Chemical and ultrastructural analysis of corn cuticular lipids and effect of these lipids on feeding by fall armyworm larvae. *J. Chem. Ecol.* **19**, 2055–2074.
Zygadlo, J.A., Maestri, D.M. and Grosso, N.R. (1994) Alkane distribution in epicuticular wax of some Solanaceae species. *Biochem. Syst. Ecol.* **22**, 203–209.

11

Integrated case study: effects of maize leaf epicuticular lipids on insect pests

K.E. Espelie

11.1 Introduction

R.F. Chapman and his colleagues were among the first to recognize the important role that plant epicuticular lipids (EL) play in insect–plant interactions (Bernays *et al.*, 1976; Chapman, 1977; Chapman and Bernays, 1989; Woodhead and Chapman, 1986). When an insect first arrives on the surface of a plant, it immediately comes into contact with a complex mixture of lipids that are characteristic of that particular plant species (Baker, 1982; Jeffree, 1986). Chapman has suggested that it would be beneficial to the plant for it to have chemicals on its surface that insects could recognize and associate with a specific plant species, and that the presence of these chemicals would discourage most insects from biting or ovipositing (Chapman, 1977). It is now clear that for some insects, on some plants, the chemical and microstructural characteristics of plant EL can affect many aspects of insect behaviour, such as orientation, movement, oviposition and feeding (Bernays *et al.*, 1976; Chapman and Bernays, 1989; Eigenbrode and Espelie, 1995; Espelie *et al.*, 1991; Städler, 1986; Woodhead and Chapman, 1986; see Chapter 10).

Variations in the chemical composition and microstructure of the EL on the plant surface may result in variable plant resistance to insect attack (Klingauf *et al.*, 1978; Woodhead, 1983). If the insects do not like the chemical and structural characteristics of the plant surface, they may exhibit non-acceptance behaviour and move away from that surface (Bernays *et al.*, 1985; Eigenbrode *et al.*, 1991; Stoner, 1990; Widstrom *et al.*, 1979). This chapter will review the information available regarding the effect that the EL of maize, *Zea mays* (L.), have upon herbivorous insects.

11.2 Maize epicuticular lipid composition

Maize EL consist primarily of *n*-alkanes, wax esters, free fatty alcohols and free fatty acids (Bianchi and Avato, 1984; Blaker and Greyson, 1988). The major *n*-alkanes

Plant Cuticles, edited by G. Kerstiens.

range in chain length from C_{23} to C_{35}, while the dominant free fatty alcohols range from C_{24} to C_{32}. Genotypes of maize which vary in their susceptibility to herbivorous insects do not have major differences in their EL composition (Yang *et al.*, 1993d).

There are, however, dramatic differences in the surface lipids of the foliage of maize from one part of the plant to another (Yang *et al.*, 1993d). Younger (upper) leaves of maize plants have much higher proportions of *n*-alkanes (45–52%), while lower (older) leaves have higher amounts of free fatty alcohols (12–18%). There are also changes in the concentrations of the triterpenols, α- and β-amyrin. The concentrations of these compounds are two to three times greater in younger leaves than they are in older leaves (Figure 11.1). The differences in the chemical composition of younger and older leaves result in major variations in microstructure of the leaf surface. Scanning electron microscopy (SEM) has shown that the upper leaves of maize have a smooth, amorphous appearance (characteristic of a high proportion of *n*-alkanes), while the lower leaves have a dense array of plate-like crystals (characteristic of free fatty alcohols; Yang *et al.*, 1993d).

Growth of microorganisms on the leaf surface of maize is very dependent on leaf surface chemistry, especially the amount of low molecular weight carbohydrates (Derridj *et al.*, 1989). The distribution of monosaccharides on the surface of maize foliage can vary dramatically from one part of the plant to another, although the ability of these compounds to penetrate through the cuticle to the leaf surface does not seem to be affected by the EL (Stammitti *et al.*, 1995). Monosaccharides on the leaf surface of maize affect the growth of the yeast, *Sporobolomyces roseus*, which, in turn, results in reduced oviposition by the European corn borer, *Ostrinia nubialis* (Martin *et al.*, 1993).

11.3 Effects of ingestion of epicuticular lipids

The results of a wide variety of plant–insect interaction studies have suggested that ingestion of EL affects the development of some insect species (Eigenbrode and Espelie, 1995). The growth of fall armyworm, *Spodoptera frugiperda*, was

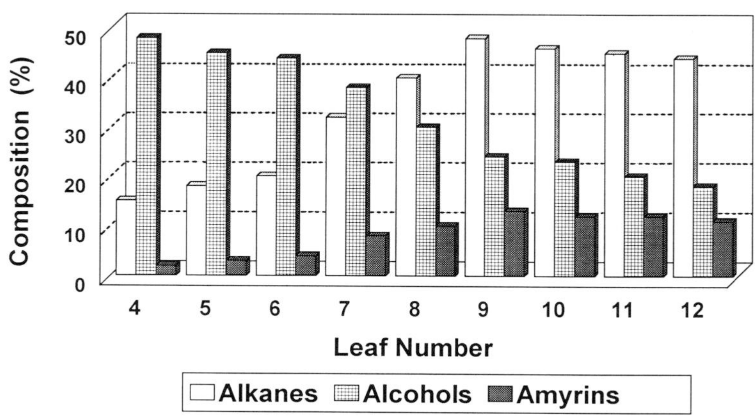

Figure 11.1. *Percent composition of* n-*alkanes, free fatty alcohols, and α- and β-amyrins of the epicuticular lipids (EL) from the lower (4) to the upper (12) leaves of maize.*

enhanced when the surface lipids were extracted from the foliage of bermuda grass (*Cynodon dactylon*) which was added to their diet (Quisenberry *et al.*, 1988). Similarly, fall armyworm larvae grew better when they were fed a diet containing peanut foliage from which the EL had been removed, than when they were reared on a diet containing untreated peanut foliage (Yang *et al.*, 1993a, 1995). Addition of the palmitate ester of α-amyrin to the diet has been shown to interfere with insect development for several species of Lepidoptera (Shankaranarayana *et al.*, 1980).

The effects of ingesting maize EL have been studied extensively with the fall armyworm, *Spodoptera frugiperda*. This insect is a major pest of maize in North and Central America (Sparks, 1986). Ingestion of maize EL inhibited the growth and development of larvae of the fall armyworm (Yang *et al.*, 1991). Larvae were reared in the laboratory on a meridic diet containing either maize foliage or maize foliage from which the EL had been removed by extraction with an organic solvent. Larvae that fed on diet with foliage from which the surface lipids had been extracted weighed much more, at both 7 and 14 days of age, than larvae that fed on diet with untreated foliage (Table 11.1). Yang *et al.* (1993d) compared the growth of fall armyworm larvae on artificial diet containing either young or old foliage that was either untreated (i.e. the EL were present) or was extracted (without surface lipids). In each case, larvae developed better when they were reared on diet containing foliage without surface lipids. However, the differences in growth were more dramatic when larvae were fed younger leaves (Yang *et al.*, 1993d). This result suggests that the higher proportions of *n*-alkanes and amyrins that are present in younger maize leaves (Figure 11.1) may have an adverse effect on fall armyworm development.

When EL, which had been extracted from maize foliage, were added to a meridic diet and fed to fall armyworm, the weight of individual larvae was significantly lower than the weight of insects reared on a control diet (Yang *et al.*, 1991). In an attempt to determine which components of the maize EL inhibit insect development, individual components were added, in a range of concentrations, to artificial diet and fed to fall armyworm larvae. A major alkane of maize surface lipids, *n*-nonacosane (C_{29}), caused increased growth of fall armyworm larvae when it was added to the insect diet (Yang *et al.*, 1993d). Similarly, larvae developed more rapidly when they were reared on diet containing the free fatty alcohol, 1-triacontanol (C_{30}), the most dominant alcohol in maize surface lipids. The free fatty acids, hexadecanoic acid (C_{16}) and dodecanoic acid (C_{22}) also caused increased growth of fall armyworm larvae, when these compounds were added, in separate tests, to the insect diet. When α-amyrin was added to the diet of fall armyworm

Table 11.1. *Effect of ingestion of maize epicuticular lipids (EL) on fall armyworm development**

Insect parameter	Foliage with lipids	Foliage without lipids
7-day larval weight (mg)	15.5	24.0
14-day larval weight (mg)	299.9	373.0
Time to pupation (days)	19.7	18.6
Pupal weight (mg)	198.4	219.0
Adult eclosion (days)	29.0	28.4

*Larvae were reared on a meridic diet containing either untreated maize (foliage with lipids) or maize from which the EL had been extracted (foliage without lipids).

larvae, the growth of the insects was enhanced. The same result was obtained when larvae were fed diet supplemented with β-amyrin (Yang *et al.*, 1993d).

The southwestern corn borer, *Diatraea grandiosella* Dyar, is a pest of maize which feeds preferentially on foliage in the whorl of the plant (Davis *et al.*, 1995). The aromatic allelochemical, 2-hydroxy-4,7-dimethoxy-1,4-benzoxazin-3-one (*N-O*-Me-DIMBOA), is present in the EL of whorl foliage and has been shown to be toxic to the southwestern corn borer (Hedin *et al.*, 1993). Maize genotypes that exhibit resistance to the southwestern corn borer have higher amounts of *N-O*-Me-DIMBOA in their surface lipids than do susceptible genotypes. The resistant genotypes also have 1.5 times as much total EL in the whorl foliage as the susceptible lines. Hedin *et al.* (1993) suggested that the methoxyl moiety attached to the ring nitrogen of *N-O*-Me-DIMBOA may provide this compound with increased stability in the surface wax layer.

Larvae of the corn earworm, *Helicoverpa zea*, often feed on the silks of maize, where they are protected from many predators and parasitoids, as well as from conventional pesticide applications (Wiseman and Davis, 1990). The silks are covered with a layer of EL that are rich in alkenes (3–15%) and aldehydes (14–36%) (Yang *et al.*, 1992). Ingestion of these lipids has an adverse effect on the growth and development of corn earworm larvae. Larvae reared on a meridic diet containing maize silks from which the lipids had been removed were significantly larger (mean average weight of 317 mg) after 8 days than larvae maintained on a diet with untreated silks (284 mg) (Yang *et al.*, 1992). Although corn earworm larvae grew better when they were fed a diet containing maize silks from which the surface lipids had been removed, insect development was not inhibited when larvae were fed a diet to which these lipids had been added (Yang *et al.*, 1994).

Polar lipid components which affect both insect pests and fungal pathogens have been isolated from the outer layer of developing maize kernels. These compounds include the steryl esters of *p*-coumaric acid and ferulic acid (Norton and Dowd, 1996). Ingestion of these esters had no effect on the growth and development of larvae of the corn earworm, *Helicoverpa zea*, but they enhanced the growth of larvae of the driedfruit beetle, *Carpophilus hemipterus*. Maize kernels are covered with a layer of EL and a cutin polymer composed of hydroxyfatty acids (Espelie *et al.*, 1979). These lipids have been reported to play a role in providing resistance in certain genotypes of maize to the fungus *Aspergillus flavus* (Guo *et al.*, 1995). Removal of the EL from maize kernels resulted in increased production of aflatoxins.

11.4 Effect of epicuticular lipids on insect movement

In some cases, when insects exhibit non-acceptance of a plant, they are responding to the composition of EL (Eigenbrode and Espelie, 1995). This response has been documented very nicely by Sanford Eigenbrode and his colleagues working with the diamondback moth, *Plutella xylostella*, on insect-resistant and insect-susceptible varieties of cabbage (see Chapter 10). Variations in the chemical composition of the EL of cabbage foliage dramatically affect the behaviour of larvae of the diamondback moth (Eigenbrode and Shelton, 1990; Eigenbrode *et al.*, 1991). Larvae spent more time crawling, crawled more rapidly, and spent more time searching when they were on resistant cabbage foliage than they did when they were on the foliage of susceptible cabbage plants. The EL extracted from the two types of cabbage were very different (Eigenbrode *et al.*, 1991). The differences in crawling and in searching behaviour of the diamondback moth larvae were also observed when the larvae were placed on the EL extracts of the resistant and susceptible genotypes that had been applied to glass plates. To date, this work is the best demonstration that EL affect the behaviour of herbivorous insects.

Guang Yang and his colleagues have examined the effect that EL of maize have upon the movement of fall armyworm larvae. In a series of laboratory studies, which were recorded by video camera, the behaviour of fall armyworm larvae on untreated, excised maize foliage was compared to the behaviour of larvae on excised foliage from which the EL had been removed (Yang *et al.*, 1993b). Larvae spent more time crawling, crawled more slowly, spent less time searching, and searched less frequently when they were on maize leaves from which the EL had been removed than they did when they were on untreated leaves. When larvae were left for 24 h in a chamber that contained excised, untreated foliage from upper leaves of maize and extracted, upper maize leaves, they chose to be on the extracted foliage more often (68% of the time) than they chose to be on the untreated leaves (32%). Conversely, when fall armyworm larvae were given a choice between untreated and extracted foliage from the lower part of maize plants, the larvae selected both types of leaves with the same frequency (Yang *et al.*, 1993b).

In the same series of laboratory tests, fall armyworm larvae travelled greater distances, crawled faster, spent more time crawling, spent more time searching, and searched more frequently when they were on upper, excised leaves of maize than they did when they were on lower, excised leaves (Figure 11.2; Yang *et al.*, 1993b). When these same behaviours were monitored for larvae placed on foliage from which the EL had been extracted, there was no difference in the behaviour of the larvae on upper leaves versus lower leaves of maize.

In field studies, it was found that there was no significant difference in the movement behaviour of fall armyworm larvae that had been placed on foliage of resistant and susceptible maize genotypes (Yang *et al.*, 1993c). However, neonate fall armyworm larvae travelled more rapidly when they were placed on upper leaves of both resistant and susceptible maize genotypes in the field than they did when placed on lower leaves. Fall armyworm neonates were placed on the adaxial surface

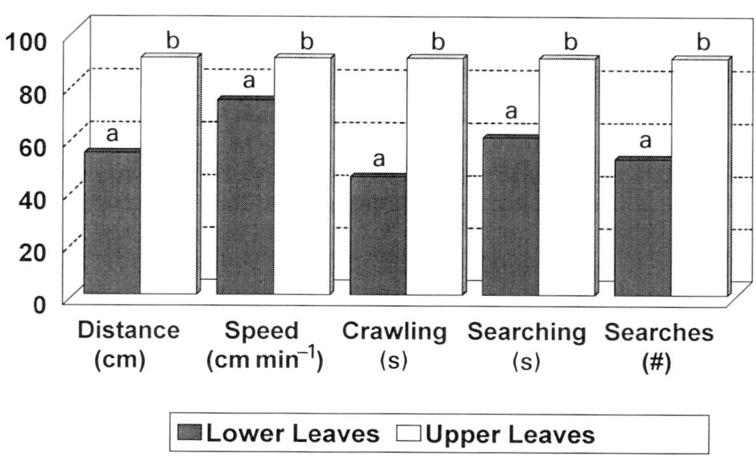

Figure 11.2. *Movement behaviour of fall armyworm larvae, in the laboratory, on lower and upper leaves which had been excised from field-grown maize plants.*

of either upper or lower leaves of maize plants and individual larvae were watched for 30 min. The majority of larvae (65%) that were placed on the upper leaves had arrived at the axil after 30 min, while only 30% of the larvae that were placed on the lower leaves made it to the axil during the 30 min observation time (Yang *et al.*, 1993c).

In a related field study, Yang *et al.* (1993c) placed neonate fall armyworm larvae on maize plants and then determined their location after 24 and 48 h. Larvae were placed on either upper or lower leaves. After 24 h, a significantly higher proportion of larvae that had been placed on upper leaves had arrived at the central, or yellow, whorl than larvae that had been placed on a lower leaf (Figure 11.3). When the location of the fall armyworm larvae was determined 48 h after placement on the leaves, there were twice as many larvae in the yellow whorl on the plants where the larvae had been placed on upper leaves as there were on plants where the larvae had been placed on lower leaves (Yang *et al.*, 1993c).

The fact that fall armyworm larvae move more rapidly on younger maize foliage than they do on older leaves may be due to the large difference in the chemical composition of the EL between younger and older leaves of maize (Figure 11.1). It would be beneficial for the larvae to move quickly on the foliage in the field so that they can more quickly reach their preferred feeding site, the yellow whorl, where they will be less susceptible to desiccation, predators and parasitoids.

11.5 Conclusions

Differences in the chemical composition of EL of maize foliage result in differences in the behaviour of herbivorous insect pests. We are just beginning to gain an understanding of these behavioural differences. When we have a better awareness of the effects that these plant surface lipids have upon insect pests, we should be able to select for plants with host plant resistance much more efficiently (Wiseman *et al.*, 1981).

It is very likely that changes in the chemistry of the plant surface influence interactions through several trophic levels (Eigenbrode and Espelie, 1995). A better understanding of the role(s) that surface lipids play in plant–insect and insect–insect

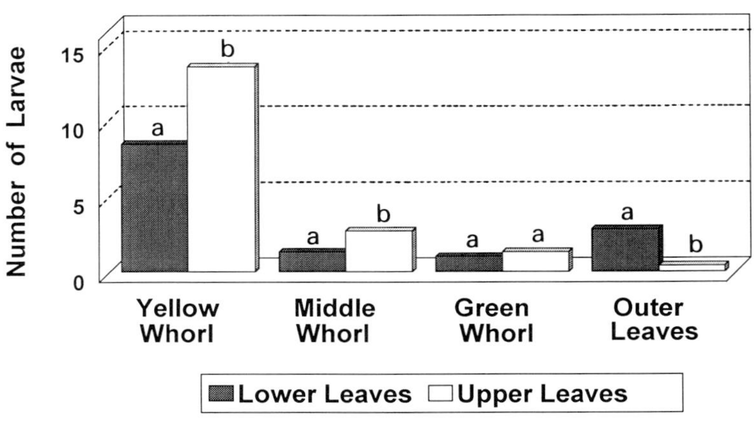

Figure 11.3. *Location of fall armyworm neonates 24 h after 30 larvae had been placed on either a lower or an upper leaf of a maize plant in the field.*

interactions will allow us to incorporate many aspects of biological control more effectively into our pest management systems. It has become clear that our investigations must take a more comprehensive approach in order to better explain structure/function relationships in the plant cuticle (Kerstiens, 1996). As T.R.E. Southwood (1986) stated more than 10 years ago, when he was referring to our understanding of the importance of plant EL in plant–insect interactions, 'we investigators are only beginning to scratch the surface.'

Acknowledgements

Some of the work described in this review was supported by Pioneer Hi-Bred International, Inc. and by HATCH projects allocated to the Georgia Agricultural Experiment Station. I would like to thank Gerhard Kerstiens for helpful comments.

References

Baker, E.A. (1982) Chemistry and morphology of plant epicuticular lipids In: *The Plant Cuticle* (eds D.E. Cutler, K.L. Alvin and C.F. Price). Academic Press, London, pp. 139–166.

Bernays, E.A., Blaney, W.M., Chapman, R.F. *et al.* (1976) The ability of *Locusta migratoria* L. to perceive plant surface waxes. In: *The Host-Plant in Relation to Insect Behaviour and Reproduction* (eds T. Jermy and Á. Szentesi). Akadémiai Kiadó, Budapest, pp. 35–40.

Bernays, E.A., Woodhead, S. and Haines, L. (1985) Climbing by newly hatched larvae of the spotted stalk borer *Chilo partellus* to the top of sorghum plants. *Entomol. Exp. Appl.* **39,** 73–79.

Bianchi, G. and Avato, P. (1984) Surface waxes from grain, leaves, and husks of maize (*Zea mays* L.). *Cereal Chem.* **61,** 45–47.

Blaker, T.W. and Greyson, R.I. (1988) Developmental variation of leaf surface wax of maize, *Zea mays. Can. J. Bot.* **66,** 839–846.

Chapman, R.F. (1977) The role of the leaf surface in food selection by acridids and other insects. *Colloq. Int. C.N.R.S.* **265,** 133–149.

Chapman, R.F. and Bernays, E.A. (1989) Insect behavior at the leaf surface and learning as aspects of host plant selection. *Experientia* **45,** 215–223.

Davis, F.M., Baker, G.T. and Williams, W.P. (1995) Anatomical characteristics of maize resistant to leaf feeding by southwestern corn-borer (Lepidoptera: Pyralidae) and fall armyworm (Lepidoptera: Noctuidae). *J. Agric. Entomol.* **12,** 55–65.

Derridj, S., Gregoire, V., Boutin, J.P. *et al.* (1989) Plant growth stages in the interspecific oviposition preference of the European corn borer and relation with chemicals present on the leaf surface. *Entomol. Exp. Appl.* **53,** 267–276.

Eigenbrode, S.D. and Espelie, K.E. (1995) Effects of plant epicuticular lipids on insect herbivores. *Annu. Rev. Entomol.* **40,** 171–194.

Eigenbrode, S.D. and Shelton, A.M. (1990) Behavior of neonate diamondback moth larvae (Lepidoptera: Plutellidae) on glossy-leaved resistant genotypes of *Brassica oleracea. Environ. Entomol.* **19,** 1566–1571.

Eigenbrode, S.D., Espelie, K.E. and Shelton, A.M. (1991) Behavior of neonate diamondback moth larvae [*Plutella xylostella* (L.)] on leaves and on extracted leaf waxes of resistant and susceptible cabbages. *J. Chem. Ecol.* **17,** 1691–1704.

Espelie, K.E., Dean, B.B. and Kolattukudy, P.E. (1979) Composition of the lipid-derived polymers from different anatomical regions of several plant species. *Plant Physiol.* **64,** 1089–1093.

Espelie, K.E., Bernays, E.A. and Brown, J.J. (1991) Plant and insect cuticular lipids serve as behavioral cues for insects. *Arch. Insect Biochem. Physiol.* **17,** 223–233.

Guo, B.Z., Russin, J.S., Cleveland, T.E. *et al.* (1995) Wax and cutin layers in maize kernels associated with resistance production by *Aspergillus flavus. J. Food Protect.* **58,** 296–300.

Hedin, P.A., Davis, F.M. and Williams, P.M. (1993) 2-Hydroxy-4,7-dimethoxy-1,4-benzoxazin-3-one (*N-O*-ME-DIMBOA), a possible toxic factor in corn to the southwestern corn borer. *J. Chem. Ecol.* **19,** 531–542.

Jeffree, C.E. (1986) The cuticle, epicuticular waxes and trichomes of plants, with reference to their structure, functions and evolution. In: *Insects and the Plant Surface* (eds B.E. Juniper and T.R.E. Southwood). Edward Arnold, London, pp. 23–64.

Kerstiens, G. (1996) Signalling across the divide: a wider perspective of cuticular structure-function relationships. *Trends Plant Sci.* **1,** 125–129.

Klingauf, F., Nöcker-Wenzel, K. and Röttger, U. (1978) Die Rolle peripherer Pflanzenwachse für den Befall durch phytophage Insekten. *Z. Pflanzenkr. Pflanzensch.* **85,** 228–237.

Martin, D., Bedel de Buzareingues, F., Barry, P. *et al.* (1993) An epiphytic yeast (*Sporobolomyces roseus*) influencing in oviposition preference of the European corn borer (*Ostrinia nubilalis*) on maize. *Acta Oecologica* **14,** 563–574.

Norton, R.A. and Dowd, P.F. (1996) Effect of steryl coumaric acid derivatives from corn bran on *Aspergillus flavus*, corn earworm larvae and driedfruit beetle larvae. *J. Agric. Food Chem.* (in press).

Quisenberry, S.S., Caballero, P. and Smith, C.M. (1988) Influence of bermudagrass leaf extracts on development and survivorship of fall armyworm (Lepidoptera: Noctuidae) larvae. *J. Econ. Entomol.* **81,** 910–913.

Shankaranarayana, K.H., Ayyar, K.S. and Khrishna Rao, G.S. (1980) Insect growth inhibitor from the bark of *Santalum album. Phytochemistry* **11,** 579–585.

Southwood, R. (1986) Plant surfaces and insects: an overview. In: *Insects and the Plant Surface* (eds B.E. Juniper and T.R.E. Southwood). Edward Arnold, London, pp. 1–22.

Sparks, A.N. (1986) Fall armyworm (Lepidoptera: Noctuidae): potential for area-wide management. *Florida Entomol.* **69,** 603–614.

Städler, E. (1986) Oviposition and feeding stimuli in leaf surface waxes. In: *Insects and the Plant Surface* (eds B.E. Juniper and T.R.E. Southwood). Edward Arnold, London, pp. 105–121.

Stammitti, L., Garrec, J.P. and Derridj, S. (1995) Permeability of isolated cuticles of *Prunus laurocerasus* to soluble carbohydrate. *Plant Physiol. Biochem.* **33,** 319–326.

Stoner, K.A. (1990) Glossy leaf wax and plant resistance to insects in *Brassica oleracea* under natural infestation. *Environ. Entomol.* **19,** 730–739.

Widstrom, N.W., McMillian, W.W. and Wiseman, B.R. (1979) Ovipositional preference of the corn earworm and the development of trichomes on two exotic corn selections. *Environ. Entomol.* **8,** 833–839.

Wiseman, B.R. and Davis, F.M. (1990) Plant resistance to insects attacking corn and grain sorghum. *Florida Entomol.* **73,** 446–458.

Wiseman, B.R., Williams, W.P. and Davis, F.M. (1981) Fall armyworm: resistance mechanisms in selected corns. *J. Econ. Entomol.* **74,** 622–624.

Woodhead, S. (1983) Surface chemistry of *Sorghum bicolor* and its importance in feeding by *Locusta migratoria. Physiol. Entomol.* **8,** 345–352.

Woodhead, S. and Chapman, R.F. (1986) Insect behavior and the chemistry of plant surface waxes. In: *Insects and the Plant Surface* (eds B.E. Juniper and T.R.E. Southwood). Edward Arnold, London, pp. 123–135.

Yang, G., Isenhour, D.J. and Espelie, K.E. (1991) Activity of maize leaf cuticular lipids in resistance to leaf-feeding by the fall armyworm. *Florida Entomol.* **74,** 229–236.

Yang, G., Wiseman, B.R. and Espelie, K.E. (1992) Cuticular lipids from silks of seven corn genotypes and their effect on development of corn earworm larvae [*Helicoverpa zea* (Boddie)]. *J. Agric. Food Chem.* **40,** 1058–1061.

Yang, G., Espelie, K.E., Todd, J.W. *et al.* (1993a) Cuticular lipids from wild and cultivated peanuts and the relative resistance of these peanut species to fall armyworm and thrips. *J. Agric. Food Chem.* **41,** 814–818.

Yang, G., Espelie, K.E., Wiseman, B.R. *et al.* (1993b) Effect of corn foliar cuticular lipids on the movement of fall armyworm (Lepidoptera: Noctuidae) neonate larvae. *Florida Entomol.* **76,** 302–316.

Yang, G., Wiseman, B.R. and Espelie, K.E. (1993c) Movement of neonate fall armyworm (Lepidoptera: Noctuidae) larvae on resistant and susceptible genotypes of corn. *Environ. Entomol.* **22,** 547–553.

Yang, G., Wiseman, B.R., Isenhour, D.J. *et al.* (1993d) Chemical and ultrastructural analysis of corn cuticular lipids and their effect on feeding by fall armyworm larvae. *J. Chem. Ecol.* **19,** 2055–2074.

Yang, G., Wiseman, B.R. and Espelie, K.E. (1994) Effect of cuticular lipids from silks of selected corn genotypes on the development of corn earworm larvae. *J. Entomol. Sci.* **29,** 239–246.

Yang, G., Espelie, K.E., Todd, J.W. *et al.* (1995) Characterization of cuticular lipids from cultivated and wild peanut species and their effect on feeding by fall armyworm (Lepidoptera: Noctuidae). *Peanut Sci.* **22,** 49–54.

12

Integrated case study: *Nicotiana* leaf-surface components and their effects on insect pests and diseases

D.M. Jackson and D.A. Danehower

12.1 Introduction

In the genus *Nicotiana* (Solanaceae), chemical components on the leaf surface fall into two groups: waxes that are an integral part of the epidermis, and exudates that arise primarily from glanded trichomes (Goodspeed, 1954; Wolf, 1946). Leaf-surface (LS) components from *Nicotiana* species include both primary and secondary metabolites.

LS compounds from commercial tobacco, *Nicotiana tabacum* L., are important primarily because they contribute to leaf quality. Components such as acyl saccharide esters (sugar esters) and diterpenes contribute to tobacco flavour and aroma (Enzell and Wahlberg, 1980; Roberts, 1988). The LS components from *Nicotiana* species also have an array of allelochemical activities (Cutler *et al.*, 1986b; Severson *et al.*, 1985b). Secondary metabolites may be allomones, which benefit the plant as pre-formed chemical inhibitors, toxicants, or repellents to pests. Or they may serve as kairomones, which benefit other organisms as attractants, feeding or ovipositional stimulants for insects (both pests and beneficials), or as germination and/or growth stimulants for microorganisms (including disease organisms) or the seed of competing plants (weeds).

This chapter will review the physical structure of the leaf surface; the composition, production, biochemistry, and methods of analysis of the LS components; the allelochemical effects of these compounds on microorganisms, insects, and other plants; and the potential for utilizing LS components from *Nicotiana* species for control of agricultural pests.

Plant Cuticles, edited by G. Kerstiens.

12.2 Leaf-surface structure

Leaf trichomes from *Nicotiana* species have been classified as simple, branched, glandular-headed, hydathodes, and specialized stalk cells (Goodspeed, 1954; Johnson *et al.*, 1985). Only glandular-headed trichomes with well defined chloroplasts produce significant amounts of exudates. Non-secreting glanded trichomes lack chloroplasts (Nielsen *et al.*, 1991). Tobacco accessions possessing only simple (glandless) trichomes produce very low levels of secondary metabolites, but normal levels of primary metabolites (not produced by trichomes; Komari *et al.*, 1986; Severson *et al.*, 1983, 1984). The size of the glandular trichomes as well as their density are important in the net production of exudates (Nielsen and Severson, 1990; Roberts *et al.*, 1981).

The phylloplane (epicuticle, trichomes and glands) has been associated with a plant's ability to defend itself from leaf-borne diseases. LS waxes provide a physiochemical barrier to penetration by microorganisms. Berry and McKeen (1985) showed that resistance to blue mould, *Peronospora tabacina* (Adam), in tobacco was partially due to infection droplets not adhering to the leaf surface. In addition, topological features on plant surfaces can stimulate the germination of fungal spores and signal differentiation of germ tubes into appressoria (Kolattukudy *et al.*, 1995; Roberts *et al.*, 1984).

12.3 Leaf-surface chemistry and biochemistry

The LS chemistry of most *Nicotiana* species consists of a complex mixture of primary and secondary metabolites. Alkanes are the principal primary metabolites found in the epicuticular waxes (EW) of *Nicotiana* species (Mold *et al.*, 1963; Severson *et al.*, 1985b). Other primary metabolic products include fatty alcohols, sterols, and wax esters. Diterpenes, sugar esters and acylnornicotines are the predominant secondary metabolites, but small amounts of flavonoids, acetogenins and pyridine alkaloids (especially nicotine) also occur. Severson (1990) comprehensively reviewed the LS chemistry of tobacco, and the reader is referred to this work for additional information on the epicuticular composition of *Nicotiana*.

12.3.1 *Chemistry and biochemistry of primary metabolites*

To simplify chromatographic analyses, crude LS extracts are typically partitioned against aqueous-alcohol and hexane phases to separate polar and non-polar constituents, respectively. Approximately 15% of the total LS lipids from *N. tabacum* are found in the non-polar fraction, and about 50% of this fraction consists of long-chain alkanes (Reid, 1974). LS alkanes from tobacco are typically composed of a mixture of normal, iso (2-methyl), and anteiso (3-methyl) branched hydrocarbons, whose chain lengths vary from C_{25} to C_{36} (Gamou and Kawashima, 1981; Mold *et al.*, 1963). Other primary metabolites in the LS wax of *N. tabacum* include C_{16}–C_{30} fatty alcohols, wax esters and sterols (Reid, 1974; Severson *et al.*, 1984, 1985a,b). *N*-docosanol is the predominant fatty alcohol. The wax esters consist of C_{12}–C_{30} saturated fatty acids esterified to fatty alcohols (Arrendale *et al.*, 1988). Docosanyl myristate is the most abundant wax ester (Severson *et al.*, 1985b).

12.3.2 *Chemistry and biochemistry of secondary metabolites*

Diterpenoids are the principal LS secondary metabolites of most *Nicotiana* species (Gamou and Kawashima, 1979). The major cembranoid diterpenes are

(1*S*,4*S*,6*R*,2*E*,7*E*,11*E*)-4,6-dihydroxycembra-2,7,11-triene (α-CBT-diol) and its (4*R*) epimer (β-CBT-diol), trivially known as α- and β-4,8,13-duvatriene-1,3-diols (Chang and Grunwald, 1976a; Reid, 1974; Roberts and Rowland, 1962; Severson *et al.*, 1984). Other cembranes from *Nicotiana* species include (1*S*,4*S*,6*R*,2*E*,7*E*,11*E*)-4-hydroxycembra-2,7,11-triene (α-CBT-monol) and its corresponding (4*R*) epimer (β-CBT-monol), trivially known as α- and β-4,8,13-duvatrien-1-ols (Wahlberg *et al.*, 1981).

Whereas virtually all accessions of *N. tabacum* produce cembranoid diterpenes, only a few, including oriental-type tobaccos, produce labdanoid diterpenes (Sato *et al.*, 1981, 1982; Severson *et al.*, 1984, 1985b). The principal labdanes are (12*Z*)-labda-12,14-diene,8α-ol (*Z*- or *cis*-abienol) and (13*E*)-labda-13-ene-8α,15-diol (labdenediol) (Colledge *et al.*, 1974; Reid, 1974). *N. glutinosa* L. produces five additional labdanes, principally sclareol and its 13*S* epimer, episclareol. Minor metabolites are manool, 2-ketoepimanool and its corresponding enol (Colledge *et al.*, 1974). Another labdane diterpene, 2-hydroxy manool, is found in certain accessions of *N. glutinosa* (Jackson *et al.*, 1991a). Noma *et al.* (1982) identified raimonol and isoraimonol, two unique labdane diterpenes in the LS extracts of *N. raimondii* Macbride.

Glucose esters (GEs) and sucrose esters (SEs) represent a unique class of natural products referred to as the sugar esters. These compounds have been reported from the LS extracts of solanaceous plants, including species in the genera *Nicotiana, Solanum, Lycopersicon, Datura* and *Petunia* (Burke *et al.*, 1987; Kays *et al.*, 1994; King and Calhoun, 1988; King *et al.*, 1986; Severson *et al.*, 1984, 1985a, 1991). For the SEs of *N. tabacum*, hydroxyl groups at the 2, 3, 4 and 6 positions of the glucosyl moiety of sucrose are esterified with C_2–C_8 aliphatic (fatty) acids (Severson *et al.*, 1985b). To date, 19 different structural types of SE and five types of GE have been reported from the *Nicotiana* species (Arrendale *et al.*, 1990; Matsuzaki *et al*, 1988a, 1989a,b, 1991, 1992; Severson *et al.*, 1985b, 1991, 1994a,b; Shinozaki *et al.*, 1992). Only the Repandae section of *Nicotiana* does not have species that produce sugar esters (Severson *et al.*, 1991). Many of the *Nicotiana* species lack the acetyl group on the C_6 carbon that is present in the SE of *N. tabacum*. These SEs possess only C_3–C_8 aliphatic acids on the 2, 3 and 4 carbons of the glucosyl moiety and the 1′, 3′ and 4′ carbons of the fructosyl moiety (Chortyk *et al.*, 1993; Severson *et al.*, 1994a).

A third class of secondary metabolites, *N*-acylnornicotines and *N*-hydroxyacylnornicotines, occurs in the three species of the Repandae section of *Nicotiana* (*N. nesophila* Johnston, *N. repanda* Willdenow ex Lehmann, and *N. stocktonii* Brandegee; Matsuzaki *et al.* 1988b; Severson *et al.*, 1988a,b). Replacement of the methyl group of the pyrrolidine ring of nicotine by acyl or hydroxyacyl moieties gives rise to this series of natural products. The *N*-acyl chain lengths range from C_{12} to C_{16}, with C_{13} the most prominent. These components appear to be localized mainly, if not entirely, on the leaf surface of the plant (Huesing and Jones, 1987). Zador and Jones (1986) indicated that the *N*-hydroxyacylnornicotines are found exclusively in the trichome exudate of *N. stocktonii*. They suggested a biosynthetic route by which nicotine is converted to nornicotine in the leaves followed by *N*-acylation to the corresponding *N*-hydroxyacylnornicotine in the trichomes.

The biosynthesis of CBT-diols is localized in the glandular head of trichomes (Kandra and Wagner, 1988; Keene and Wagner, 1985; Michie and Reid, 1968). The CBT-diols are exuded from the glanded trichomes and eventually become embedded in the LS wax (Chang and Grunwald, 1980). *Z*-abienol and CBT-monols can be synthesized by cell-free extracts prepared from trichome-bearing tissues, but not from sub-epidermal or epidermal tissue (Guo and Wagner, 1995; Guo *et al.*, 1994). *N. tabacum* SEs are also produced by glanded trichomes (Kandra and Wagner, 1988). The biosynthesis of the branched-chain acids on SEs occurs via utilization and extension of the biochemical pathways for production of branched-chain amino acids (Kandra *et al.*, 1990).

12.3.3 *Analysis of LS components*

Generally, organic solvents of low to medium polarity have been used for the extraction of *Nicotiana* LS compounds. Severson *et al.* (1984) and Danehower (1987) showed that short, sequential dips in methylene chloride removed over 95% of the LS waxes from fresh tobacco leaves. Keene and Wagner (1985) and Zador and Jones (1986) used acetonitrile to extract LS materials. Isopropanol has been used to remove sugar esters for production of biorational insecticides (Jackson *et al.*, 1996).

Liquid chromatography (LC) has been used for preparative isolations of LS components. Typically, solvent partitioning is combined with column chromatography to obtain enriched or purified fractions (Arrendale *et al.*, 1988; Jackson *et al.*, 1986). Classes of compounds isolated using LC include hydrocarbons and waxes, cembranes, labdanes, SEs and acylnornicotines (Arrendale *et al.*, 1990; Bailey *et al.*, 1974; Chang and Grunwald, 1976a; Danehower *et al.*, 1987; Gamou and Kawashima, 1979; Keene and Wagner, 1985; Matsuzaki *et al.*, 1989a,b, 1991; Roberts and Rowland, 1962; Severson *et al.*, 1985a).

Gas chromatography (GC) with derivatization is commonly used to analyse LS extracts (Chang and Grunwald, 1976a,b). Severson *et al.* (1984) developed a silylation/capillary GC procedure that provides nearly complete resolution of most *Nicotiana* LS mixtures. Danehower (1987) also developed a procedure for GC analysis of SEs.

Thin layer chromatography (TLC) systems for *Nicotiana* LS constituents have been used primarily to gather qualitative information (Bailey *et al.*, 1974; Reid, 1974). TLC has been used to measure LS diterpenes (Komari *et al.*, 1986; Kubo *et al.*, 1982; Sato *et al.*, 1981). Similarly, silica gel was used for separation and tentative identification of purified fractions of SE from 57 *Nicotiana* species in a study of the chemotaxonomic relationships between species (Matsuzaki *et al.*, 1989a). Zador and Jones (1986) separated *N. stocktonii* hydroxyacylnornicotines on Kieselgel 60 TLC plates. Lawson and Danehower (1989) described a quantitative TLC technique for tobacco LS components. Their method involved the use of silica gel followed by densitometry at 200 nm for detection of the diterpene constituents. Detection of other LS components (hydrocarbons, wax esters, SEs) required charring of the plates using fuming sulphuric acid followed by scanning in white light. Lin and Wagner (1994b) reported on a novel quantitative assay for SEs involving staining with Rhodamine B followed by spectrophotometric detection; minimum detection levels were 0.1 μg using fluorescence.

Mass spectroscopy and proton or carbon-13 nuclear magnetic resonance (NMR) techniques have been used for the structural elucidation of LS constituents (Arrendale *et al.*, 1990; Severson *et al.* 1985a; Wahlberg *et al.*, 1986). Einolf and Chan (1984) reported on the quantification of SEs using direct chemical ionization mass spectrometry. Himmelsbach *et al.* (1990) reported on the ^{1}H and ^{13}C-NMR spectra of the 2, 3, 4-tri-β-methylpentanoyl 6-acetyl SE (Group 5; Severson *et al.*, 1985b) of *N. tabacum*. This paper should prove to be an invaluable aid to the interpretation of SEs from *Nicotiana* and other solanaceous species.

12.4 Factors affecting production of leaf-surface compounds

Numerous factors influence the production, secretion and loss of chemicals from the leaf surface of *Nicotiana* species. At a basic level, genes encode for the enzymes in metabolic pathways, and the level of expression of these genes can be affected by both biotic and abiotic stresses. Environmental factors such as light, temperature and rainfall can influence secretion, degradation, or physical removal of compounds from the leaf surface. Nutritional and physiological factors also influence the quantity of components found on the leaf. The sum of these factors ultimately determines the amount of these components present at a given time during the plant's life cycle. Much of our understanding of the importance of these factors arises from studies with *N. tabacum*.

12.4.1 *Intrinsic factors affecting LS metabolites*

Upper leaves of *N. tabacum* yield more total LS compounds per unit area than leaves at lower stalk positions. This is most likely due to the greater trichome density of the upper leaves before they have fully expanded (Chang and Grunwald, 1976b; Gamou and Kawashima, 1979). Once leaf senescence begins, the levels of LS compounds start to decline (Danehower *et al.*, 1988).

The production of LS components depends on the inheritance of specific biosynthetic capabilities as well as trichome morphology. Nielsen and Severson (1992) found that single gene control explained the inheritance of CBT-diols in tobaccos with glanded, secreting trichomes. Likewise, Sisson *et al.* (1993) reported that production of α- and β-CBT-monols was inherited as a single recessive trait, and that production of CBT-monols was negatively correlated with production of CBT-diols.

Inheritance of the glandless trait appears to be controlled by three recessive genes (Burk *et al.*, 1982). However, the presence of glanded trichomes *per se* is not sufficient for diterpene production; for example, the tobacco cultivar Enshu has glanded trichomes but produces abnormally low levels of CBT-diols (Sato *et al.*, 1981). The production of low levels of CBT-diols in Enshu is controlled by a single recessive gene (Komari *et al.*, 1986; Nielsen *et al.*, 1982).

Most *Nicotiana* species in the sections Paniculatae, Undulatae and Repandae do not produce sugar esters, which correlates well with their botanical classifications (Matsuzaki *et al.*, 1989a; Severson *et al.*, 1991). A single dominant gene located on chromosome A controls SE inheritance in *N. tabacum* (Danehower *et al.*, 1989; Gwynn *et al.*, 1985). Production of *Z*-abienol in *N. tabacum* is also controlled by a single dominant gene located on chromosome A (Kubo *et al.*, 1982; Tomita *et al.*, 1980), and it is closely linked with the gene for SE production (Danehower *et al.*, 1989).

12.4.2 *Extrinsic factors affecting LS metabolites*

Fertilization is an important factor influencing production of LS components by *N. tabacum*. Court *et al.* (1984) and Simpson *et al.* (1985) found that CBT-diol concentrations of tobacco decreased with increased nitrogen fertilization. Soil type also affects LS chemical composition (Gamou and Kawashima, 1979). Severson *et al.* (1985b) reported that drought-stressed tobacco produced higher levels of CBT-diols than non-stressed tobacco. Wilkinson and Kasperbauer (1972, 1980) found that photoperiod and temperature influence concentrations of alkanes, fatty acids and fatty alcohols. Due to reduced sunlight in the greenhouse, tobacco grown there produces only about 50% of the total LS components that are found on field-grown

Table 12.1. *Allelochemical activities of LS components from* Nicotiana *species*

Class of LS component	Materials bioassayed	Tobacco budworm ovipositional response[a]	Topical toxicity to tobacco aphids[b]	Inhibition of wheat coleoptiles[c]	Inhibition of bacteria[d]	Inhibition of fungi[e]
Hydrocarbons	Hexane solubles[f] from NC 2326	-	-	nt	nt	nt
Alkaloids	Nicotine	nt	+	nt	nt	nt
	Acylnornicotines	nt	nt[g]	nt	+	nt
Fatty alcohols	Docosanol	-	nt	-	nt	nt
Wax esters	Docosanyl myristate	-	nt	-	nt	nt
Sugar esters	Sucrose esters	+	+	+[h]	±	+
	Glucose esters	+	+	+/-	±	nt
Labdanoid diterpenes	*Z*-abienol	-	+	+	+	+
	2-Hydroxy manool	-	nt	+	nt	nt
	Sclareols	-	nt	+	+	nt
	Manool	+	nt	+	nt	nt
	Labdenediol	+	nt	+	+	nt
Cembranoid diterpenes	α-CBT-diols	+	+	+[i]	+/-[j]	+/-[k]
	β-CBT-diols	+	+	+	-	+/-
	CBT-monols	+	+	+	nt	nt
	CBT-oxy-diols	-	nt	nt	nt	nt

[a] +, significant response (paired *t* test, $p < 0.05$) to material applied to a non-preferred surface in choice tests versus an untreated control; -, no significant response; nt, not tested (data from Cutler *et al.*, 1986b; Jackson *et al.*, 1984, 1986, 1989, 1991a; Severson *et al.*, 1991).

[b] +, significant response (probit analysis) to material applied to the dorsum of apterous aphids; -, no significant response; nt, not tested (data from Eckel *et al.*, 1991, unpublished; Jackson *et al.*, 1991b; Severson *et al.*, 1991, 1994a,b).

[c] +, significant inhibition of wheat coleoptiles at 10^{-4} M concentration; -, no significant response; +/-, significant in some tests, not significant in others; nt, not tested (data from Cutler and Cole, 1974; Cutler *et al.*, 1977; 1986a,b, 1991, 1992, 1994; Matsuzaki *et al.*, 1988a, 1989a; Severson *et al.*, 1985b; Shinozaki *et al.*, 1991, 1992; Springer *et al.*, 1975).

[d] +, significant inhibition of bacteria; -, no significant response; +/-, significant in some tests, not significant in others; nt, not tested (data from Chortyk *et al.*, 1993; Cutler *et al.*, 1986a,b; Severson *et al.*, 1985b, 1988a,b).

tobacco (Gamou and Kawashima, 1979). Supplemental greenhouse lighting markedly increases the quantity of LS compounds on tobacco (Severson *et al.*, 1985b). This is particularly noteworthy because many insect and disease screening trials are conducted in greenhouse environments. Failure to use proper supplemental lighting could obscure the role of LS chemicals in resistance phenomena (cf. Section 7.4).

Rainfall can physically reduce the amount of LS compounds. Severson *et al.* (1985b) reported that losses of diterpenoid constituents from *N. tabacum* varied from 10 to 96%, while losses of SEs varied from 23 to 57%. Danehower *et al.* (1988) observed substantially reduced *Z*-abienol levels after rainfall. Rainfall that had occurred at least two days before sampling did not substantially reduce *Z*-abienol levels, indicating a 'rebound effect'.

High levels of atmospheric ozone can greatly reduce LS chemical levels (Heggestad and Middleton, 1959). Court and Hendel (1983) found a negative correlation between levels of LS components and ozone. Tobacco plants grown under elevated ozone conditions had lower levels of cembranoid diterpenes, but elevated ultraviolet-B (UV-B) radiation in these same experiments had no effect on tobacco growth or LS chemistry (Jackson *et al.*, 1992).

12.5 Allelochemistry of *Nicotiana* leaf-surface compounds

The LS components from *Nicotiana* species have many biological activities, including plant growth regulation, antimicrobial properties and effects on behaviour, development and survival of insect pests and beneficials (Table 12.1). Studies of the allelochemistry of *N. tabacum*, in particular, have provided scientists with some of the best information available on the role of plant LS chemistry on plant–insect, plant–pathogen and plant–plant interactions.

12.5.1 Nicotiana*–insect allelochemical interactions*

Tobacco is attacked by a variety of insect pests (Akehust, 1968; Chaplin *et al.*, 1976), ranging from the highly polyphagous *Heliothis* species to the nearly monophagous *Manduca* species that feed primarily on Solanaceae (Yamamoto and Fraenkel, 1960). This review will highlight the major insect pests of tobacco in the USA: the tobacco hornworm, *Manduca sexta* (L.); the tobacco budworm, *Heliothis virescens* (F.); and the tobacco aphid, *Myzus nicotianae* Blackman.

Tobacco seedlings generally are pest-free when they are transplanted. It is crucial that adult females place their eggs (or nymphs) on or near a suitable host plant,

---

[e] +, significant inhibition of fungi; -, no significant response; +/-, significant in some tests, not significant in others; nt, not tested (data from Cruickshank *et al.*, 1977; Cutler *et al.*, 1986a,b; Lawson *et al.*, 1988; Menetrez *et al.*, 1990).

[f] Contains low levels of fatty alcohols and wax esters in addition to C_{25}–C_{36} hydrocarbons (Severson *et al.*, 1984, 1985b).

[g] Toxic to tobacco hornworm larvae, *Manduca sexta*, but not to tobacco budworm larvae, *Heliothis virescens* (Huesing and Jones, 1987; Severson *et al.*, 1988a,b).

[h] Also inhibits seed germination of barnyard grass and tobacco seed (Matsuzaki *et al.*, 1988a; 1989b; Shinozaki *et al.*, 1991; 1992).

[i] Also inhibits seed germination of barnyard grass and hemp sesbania (Lawson *et al.*, 1988).

[j] Inhibits Gram-positive bacteria, *Bacillus subtilis* and *B. cereus*, but not Gram-negative bacteria, *Escherichia coli* (Cutler *et al.*, 1986a,b, Severson *et al.*, 1988a,b).

[k] Antibiotic against blue mould, *Peronospora tabacina*, but not against *Curvularia lunata* or *Aspergillus flavus* (Cruickshank *et al.*, 1977; Cutler *et al.*, 1986a,b, Severson *et al.*, 1988a,b).

because immature insects generally cannot move off a tobacco plant until they are fully developed. Little is known about how insects find tobacco plants over a distance, but chemical cues are probably very important. Insects are affected throughout their lives by chemicals in their environment. Chemical cues mediate many aspects of insect behaviour, including host finding and acceptance, avoidance of predators and mate location. A particular chemical may act as an attractant, repellent, arrestant, excitant, stimulant, deterrent, or have another function. Once an insect locates a plant using visual and olfactory cues, contact with the plant surface is usually essential before feeding or oviposition is initiated. Contact chemoreceptors and mechanoreceptors are used to evaluate the plant surface. Typically, insect pests of tobacco utilize secondary plant metabolites (stimulants or deterrents) to identify potential hosts without perception of the leaf interior. After acceptance of the host plant, secondary LS components and internal leaf constituents, including nutrients, become important for determining host suitability for the insect's survival, growth, development and reproduction.

Manduca sexta. Tobacco hornworm moths lay their eggs exclusively on solanaceous plants (Yamamoto and Fraenkel, 1960). Willis and Arbas (1991) showed that hornworm moths fly upwind in response to odours emanating from fresh tobacco leaves. Contact chemoreception is then required to elicit oviposition. Water-soluble fractions from leaves of several solanaceous plants, including tobacco, elicited ovipositional behaviour in *M. sexta* (Bordner *et al.*, 1983; Tichenor *et al.*, 1981). The ovipositional preferences of *M. sexta* and *H. virescens* for 70 accessions from 64 of the 66 *Nicotiana* species were evaluated in field-cage experiments (Jackson *et al.*, 1989; Severson *et al.*, 1991). Forty-one species were significantly less attractive to *M. sexta* than standard tobacco cultivars, and no species was more attractive than tobacco. Few eggs of either insect species were oviposited on *Nicotiana* species that had narrow or glabrous leaves, or leaves devoid of LS diterpenes or SEs (Jackson *et al.*, 1989; Severson *et al.*, 1991).

Stimulation of either olfactory or gustatory receptors may result in host acceptance by hornworm larvae, but gustatory reception alone is responsible for rejection of unacceptable plants (Hanson and de Boer, 1987). Larvae may use their subtle sense of taste to select parts of the plant without high concentrations of deterrents. Hornworm larvae feed exclusively on solanaceous plants in the wild, but if their taste organs are removed, they will feed on other plants (Waldbauer and Fraenkel, 1961). Both polar and non-polar LS compounds stimulate feeding by larvae of *M. sexta* (Bordner *et al.*, 1983; de Boer and Hanson, 1987).

Acyl- and hydroxyacylnornicotines are toxic to *M. sexta* larvae upon contact or through feeding (Huesing and Jones, 1987, 1988; Severson *et al.*, 1988a,b). Tobacco hornworm larvae cannot detoxify these unique nicotine derivatives as they can nicotine itself. Interestingly, acylnornicotines are not toxic to larvae of *Heliothis* species (Severson *et al.*, 1988a).

Heliothis virescens. Visual, tactile, olfactory and contact chemosensory cues are all involved in host finding and acceptance by *H. virescens* (Mitchell *et al.*, 1990; Navasero and Ramaswamy, 1991; Ramaswamy *et al.*, 1987). Tobacco budworm moths orient toward nectar sources and ovipositional sites by positive anemotaxis in response to airborne volatiles emitted by plants, including tobacco (Lingren and Wolf, 1982; Mitchell *et al.*, 1991; Tingle and Mitchell, 1992). Mitchell *et al.* (1991) concluded that flight responses in *H. virescens* were affected by plant-specific volatiles and not just by ubiquitous green plant odours.

Tobacco budworm moths prefer to oviposit on or near the flowering portions of a wide range of host plants, including many *Nicotiana* species (Jackson *et al.*, 1989; Severson *et al.*, 1991). In the absence of flowers, vegetative structures are chosen for oviposition sites, and eggs are concentrated on the first few leaves below the apical leaf bud of tobacco (Jackson *et al.*, 1983). Before oviposition, the surface texture, moisture and chemistry of a leaf are analysed by mechanoreceptors and chemosensilla on the tarsi and ovipositor (Mitchell *et al.*, 1990; Navasero and Ramaswamy, 1993; Ramaswamy *et al.*, 1987).

A primary mode of resistance to the tobacco budworm is ovipositional antixenosis (non-preference) associated with the lack of trichome exudates on some tobacco introductions (TIs; Johnson and Chaplin, 1982); for instance, TI 1112, which is resistant to tobacco budworm infestations, has simple (glandless) trichomes and is nearly devoid of most LS components found on commercial tobaccos (Severson *et al.*, 1984). In field plots and in field-cage experiments, 75–90% of the tobacco budworm eggs were oviposited on flue-cured tobacco cultivars in choice situations with TI 1112 (Elsey and Chaplin, 1978; Jackson *et al.*, 1983). However, budworm larvae placed directly on TI 1112 foliage survived as well as they did on foliage of flue-cured cultivars (Elsey and Chaplin, 1978).

Whole-leaf extracts from flue-cured tobacco (NC 2326) stimulate oviposition by tobacco budworm moths when sprayed onto TI 1112 plants (Jackson *et al.*, 1984) or cotton cloths (Mitchell *et al.*, 1990), but extracts from TI 1112 are inactive. The methanol-water-soluble fraction (MWS) of the whole-leaf extracts from NC 2326 is active, but the hexane-soluble fraction (HS) does not enhance oviposition by *H. virescens* (Jackson *et al.*, 1986). The α- and β-CBT-diols are present primarily (99.5+%) in the NC 2326 MWS fraction, while the HS fraction contains fatty alcohols, wax esters, hydrocarbons and other non-polar compounds. NC 2326 produces very low levels of α- and β-CBT-monols and SEs (Johnson and Severson, 1982). Other tobaccos with higher levels of α- and β-CBT-monols (TI 1223 and TI 1341) and/or SEs (TI 165) are heavily oviposited upon in cage experiments (Jackson *et al.*, 1988; Severson *et al.*, 1985b). In studies with pure components isolated from green-leaf extracts from several *Nicotiana* species, tobacco budworm oviposition was stimulated by α-CBT-diol, β-CBT-diol, α+β-CBT-monols, SEs, manool and labdenediol, but not by docosanyl myristate, *Z*-abienol, 2-hydroxymanool, sclareols, fructose, glucose, sucrose or CBT-oxy-diols (a cembrane degradation product; Jackson *et al.*, 1984, 1986, 1988, 1989, 1991a). The relatively non-volatile, polar cembranoid diterpenes and sugar esters act as contact stimuli and are perceived by tarsal sensilla (Ramaswamy and Hanson, 1990).

Some tobacco entries having high levels of CBT-diols and SEs, such as TI 165, exhibit antibiotic resistance to tobacco budworm larvae (Johnson and Severson, 1984; Johnson *et al.*, 1992). The sticky trichome exudates from these tobaccos may act as a physical deterrent that prevents small larvae from becoming established by entangling them on leaf surface or by gluing their mouth parts shut. However, this mechanism does not explain all of the observed resistance in TI 165 (Jackson *et al.*, 1989; Johnson and Severson, 1984; Severson *et al.*, 1985b). Recently, Snook *et al.* (1995) identified hydroxygeranyllinolool rhamnosides and glucosides as resistance factors in TI 165, however these are internal leaf components.

Myzus nicotianae. For the most part, alate aphids are passively carried to tobacco fields, and orientation is primarily to colour and light reflections (Van Emden *et al.*, 1969). After landing, aphids probe plants to determine acceptability, and they may minimize the intake of deleterious plant compounds by withdrawing their stylets from cells containing allelochemicals (Campbell *et al.*, 1986). If tobacco aphids encounter an unfavourable plant, they may abandon it.

Aphids confined on tobacco entries with low levels of LS components, such as TI 1112, survive and reproduce normally in laboratory experiments (Jackson *et al.*, 1991b; Johnson, 1978). Resistance in TI 1112 to aphids is due to antixenosis that may be associated with low levels of CBT-diols (Johnson, 1978; Johnson and Severson, 1982). Eckel *et al.* (1991) found a positive correlation between levels of CBT-diols and aphid infestations for several tobacco entries in the field. Electronic feeding-monitor studies also suggest that CBT-diols are important for host acceptance by tobacco aphids (R.V.W. Eckel and D.M. Jackson, unpublished).

Although many aphid-resistant tobacco entries have low levels of LS components, some tobaccos with heavy trichome exudates are also resistant (Jackson *et al.*, 1991b; Johnson 1980; Johnson and Severson, 1982; Severson *et al.*, 1994b). During their searching and leaf probing behaviours, aphids often come into contact with trichome exudates, and toxic components may pass through their soft cuticle. Thurston and Webster (1962) showed that a toxic factor from *N. gossei* Domin could be transferred from the leaf surface, and that aphids that contacted this material were killed. Aphids also may become entrapped in the gummy trichomes exudates.

Eckel *et al.* (1991) reported significant negative correlations between field infestations of tobacco aphids and levels of CBT-monols, sugar esters and *Z*-abienol for several tobacco entries. Furthermore, the topical toxicity of various *Nicotiana* LS components (nicotine, α- and β-CBT-monols, α- and β-CBT-diols, labdenediol, SEs and *Z*-abienol) to tobacco aphids was investigated (Severson *et al.*, 1994b). All of the components except labdenediol were toxic to tobacco aphids. An SE mixture from *N. tabacum* (PD 1097) had the highest toxicity, with an LC_{50} (dose required to kill 50% of the test population) of 0.25 μg $aphid^{-1}$ in 1 μl acetone solution. The relative toxicity of each chemical was greater when applied in acetone than in methanol. This was particularly true of *Z*-abienol (LC_{50} 7.5 μg $aphid^{-1}$ with acetone, >500 μg $aphid^{-1}$ with methanol). These LS components also had deleterious effects on the reproductive ability of tobacco aphids (Severson *et al.*, 1994b). The toxicities of the LS components of *N. tabacum* to tobacco aphids suggest their possible enhancement in breeding programmes or their potential for use as biopesticides.

Beneficial species. Beneficial natural enemies of insect pests represent the third trophic level. Populations of these entomophagous species have been shown to vary according to the physical and chemical characteristics of the leaves of *Nicotiana* species (Jackson *et al.*, 1989). Plant-produced chemicals may affect beneficial parasitoids, predators or pathogens directly or they may act through their insect host.

Lepidopteran eggs are seldom parasitized on commercial tobacco, probably because the minute parasitic wasps become trapped in the sticky trichome exudates (Rabb and Bradley, 1968). However, tobacco budworm and hornworm eggs laid on tobacco introductions without glanded trichomes, such as TI 1112 and TI 1298, have significant egg parasitism (Elsey and Chaplin, 1978; Katanyukul and Thurston, 1979). Parasitic wasps that attack tobacco aphids and hornworm larvae may also be hindered by trichome exudates on tobacco (Marcovitch and Stanley, 1937; Takada and Takenaka, 1982).

Predatory insects, such as *Chrysopa carnea* Stephens and ladybird beetles, have difficulty searching for prey on tobacco with heavy exudates (Belcher and Thurston, 1982; Elsey, 1974). Elsey and Chaplin (1978) and Jackson *et al.* (1989) found more ladybird beetles on TI 1112 than on flue-cured tobacco. In general, ladybird beetles are found in higher numbers on tobacco entries and *Nicotiana* species with fewer trichomes or exudates regardless of the size of aphid populations, whereas stilt bugs, *Jalysus wickhami* Van Duzee, are seldom found on tobaccos that are not hairy and sticky (Jackson *et al.*, 1989).

Parasitoids and predators use both plant-produced and host-produced chemicals for orientation. The initial phase of orientation, host habitat finding, is often mediated by chemical cues emanating from plants, regardless of the presence of a herbivore host; for example, *Cardiochiles nigriceps* Viereck, a nearly obligatory larval parasitoid of *H. virescens*, is attracted to plants in the absence of its larval hosts. Over a 10-year period, 60 times as many *C. nigriceps* adults were counted in plots of *N. noctiflora* Hooker than were found in commercial tobacco, even though *N. noctiflora* is an extremely poor host for tobacco budworm larvae (Jackson *et al.*, 1995). *C. nigriceps* adults are attracted to volatile components from *N. noctiflora* leaves and flowers. After field-grown *N. noctiflora* flowers and leaves were extracted with methylene chloride, GC analysis showed only four volatile peaks of interest. These LS components were identified as nitrogen-sulphur heterocycles (Jackson *et al.*, 1995; Schlotzhauer *et al.*, 1993).

12.5.2 Nicotiana–*plant allelochemical interactions*

The LS constituents of *Nicotiana* also play a role in controlling the growth of axillary buds and competing plant species. Using the etiolated wheat (*Triticum aestivum* L., cv. Wakeland) coleoptile assay, Cutler and Cole (1974) determined that crude LS extracts of tobacco inhibited coleoptile elongation by 54%. Growth inhibitory activity was attributed to two compounds in this mixture, later identified as α- and β-CBT-diols (Springer *et al.*, 1975).

Additional compounds from *Nicotiana* species have been evaluated using the wheat coleoptile bioassay (Cutler *et al.*, 1977, 1986a,b). Enriched SE mixtures significantly inhibited coleoptile growth at 10^{-3} and 10^{-4} M. Pure 6-*O*-acetyl-2,3,4-tri-*O*-3-methylpentanoyl sucrose (Group 5 SE; Severson *et al.*, 1985b) demonstrated the highest activity, inhibiting growth 100% and 81% at 10^{-3} and 10^{-4} M, respectively (Cutler *et al.*, 1986b). Activity was strongly correlated with the proportion of 3-methylvalerate residues in all SE mixtures tested. Cutler *et al.* (1986b, 1991, 1992, 1994) conducted structure/activity experiments with the group 5 SEs to determine which characteristics of the molecule were important for activity. They reported that neither 3-methylvaleric acid alone nor its potassium salt caused inhibition. Furthermore, the related GE, 6-*O*-acetyl-2,3,4-tri-*O*-3-methylvaleryl glucose, was inactive in this bioassay. They concluded that the entire SE molecule is required for biological activity, and that minor differences in acyl acid composition have a profound effect on their expression of biological activity.

Matsuzaki *et al.* (1988a) surveyed the germination and growth-inhibiting activities of the LS components from 54 *Nicotiana* species against tobacco seed. They identified SEs as the principal components in these extracts that inhibited growth. The effects of tetra- and triacyl sugar esters from *N. umbratica* Burbidge, *N. miersii* Remy and *N. glutinosa* on the growth of barnyard grass, *Echinochloa crusgalli* L. Beauvoir, have also been reported (Matsuzaki *et al.*, 1988a, 1989b; Shinozaki *et al.*, 1991,1992). Lawson *et al.* (1988) reported that CBT-diols inhibited seed germination in barnyard grass and hemp sesbania, *Sesbania exaltata* (Raf) Cory.

Cutler *et al.* (1986a) first reported that hydroxyacylnornicotines had antimicrobial and plant-inhibitory properties. Matsuzaki *et al.* (1988b) analysed the effects of 10 *N*-hydroxyacylnornicotines and hydroxyanatabine from the Repandae section of *Nicotiana,* as well as synthetic derivatives, against germination of tobacco seed. They found that chain length of the acyl moiety was important for biological activity. The greatest inhibitory activity was seen with *N*-acylnornicotines having C_5–C_9 aliphatic acyl groups and with *N*-hydroxyacylnornicotines with C_9–C_{12} acyl groups and hydroxylation at either the 4 or 5 positions.

12.5.3 *Nicotiana–microbe allelochemical interactions*

Plants have developed an array of biochemical defences against fungal and bacterial invaders. Cumulatively, these chemical defence systems make the successful infection of plants the exception rather than the rule. A growing body of evidence suggests that interactions between LS allelochemicals and pathogens may play an important role in the resistance of *Nicotiana* and other plant species to manifestation of disease symptoms (Table 12.2). Interestingly, little attention has been given to the hypothesis that plant prohibitins (Mahadevan, 1970) or preformed chemical inhibitors on the leaf surface might play a significant role in defending plants against foliar pathogens. However, much of the limited research in this area has been conducted with solanaceous plants.

It should be emphasized that the entire ontogeny of a pathogen should be considered before one draws conclusions about the importance of LS components in foliar disease resistance. Most of the studies described below looked only at specific effects of LS compounds on germination or growth *in vitro*. It is equally important to know how LS components affect germ-tube growth, germ-tube differentiation, the penetration of appressorial pegs, and subsequent growth, development and reproduction of foliar fungal pathogens. Furthermore, chemical components may inhibit conidial growth or colony formation *in vitro*, but they may not be active in field situations. Environmental factors in the field may reduce activity or rain may wash LS compounds off the plant. Finally, concentrations of specific trichome-produced components, such as SEs, vary within a plant and even over a leaf surface (Lin and Wagner, 1994a), resulting in localized areas where the concentration of LS chemicals may be insufficient to suppress pathogen growth.

Cruickshank *et al.* (1977) identified α- and β-CBT-diols as two inhibitors of blue mould sporangia and conidia germination. These chemicals accounted for approximately 80% of the inhibitory activity against blue mould, with ED_{50}s of 24 μg ml^{-1} and 17 μg ml^{-1}, respectively. Germination of blue mould sporangia is completely inhibited by 10 μg cm^{-2} of CBT-diols (Menetrez *et al.*, 1990). Reuveni *et al.* (1987) found that removal of the CBT-diols from the leaf surface of tobacco plants by a one-second dip in acetone resulted in significantly increased susceptibility to *P. tabacina* infection. Reapplication of the extracts to the leaves restored some resistance to the plant. In contrast to these findings, Rao *et al.* (1989) reported that removal of the CBT-diols from older leaves resulted in little increase in blue mould infection. Rao *et al.* (1989) also found little relationship between CBT-diol content of burley tobacco and susceptibility to blue mould in the field. This suggests that other age-related or environmental factors account for additional aspects of resistance to the blue mould disease.

Z-abienol also inhibits germination of *P. tabacina* spores, while SEs from *N. tabacum* were inactive against this disease (Menetrez *et al.*, 1990). The activity of *Z*-abienol (ED_{50} of 10 μg cm^{-2}) in inhibiting spore germination was somewhat higher than for CBT-diols. Interestingly, low concentrations of either CBT-diols or *Z*-abienol (< 0.1 μg ml^{-1}) resulted in slight, but significant, stimulation of germination of spores (Menetrez *et al.*, 1990). Kennedy *et al.* (1992) developed an *in vivo* leaf disc assay to measure the effects of LS chemicals on the germination and growth of blue mould. Results of assays using LS chemicals isolated from 25 *Nicotiana* species indicated that α-CBT-diol had an IC_{50} (concentration at which conidia germination was reduced 50%) of 3.0 μg cm^{-2}, while β-CBT-diol had a value of 2.9 μg cm^{-2}. Two labdanes, labdenediol and sclareols, had IC_{50}s of 0.4 μg cm^{-2} and 4.7 μg cm^{-2}, respectively (Kennedy *et al.*, 1992).

Cohen *et al.* (1983) investigated the effects of LS diterpenes from *N. glutinosa* on conidial germ-tube emergence of *Erysiphe cichoracearum* DC, tobacco powdery mildew. The fungus failed to grow beyond the single germ-tube stage on *N. glutinosa* and immune entries of *N. tabacum*. In susceptible *N. tabacum* cultivars, 50% of the conidia proceeded to the normal, three-germ-tube stage and produced colonies. Susceptible *N. tabacum* cultivars sprayed *in vivo* with 10 mg ml^{-1} LS exudates from *N. glutinosa* were fully protected from powdery mildew infection. At 5 mg ml^{-1}, inhibition ranged from 55 to 99%. A single component, 2-ketoepimanool, inhibited *in vitro* germination beyond the first-germ-tube stage 15, 83 and 92% of the time at 10, 50 and 100 μg cm^{-2}, respectively. The ED_{90} for this compound for colony development was 1 μg cm^{-2}. This work emphasizes the importance of fully examining not only germination but also developmental ontogeny.

An epimeric mixture of sclareol and 13-episclareol from *N. glutinosa* at 100 μg ml^{-1} provided over 90% protection against French bean rust, *Uromyces appendiculatus*, broad bean rust, U. *viciae-fabae*, and wheat leaf rust, *Puccina recondita* f. sp. *tritici* (Bailey *et al.*, 1974). The sclareol/episclareol mixture incorporated into agar at 20 μg ml^{-1} increased the degree of hyphal branching of 16 fungal species, resulting in reduction of colony size (Bailey *et al.*, 1974). Concentrations of the epimeric mixture in *N. glutinosa* ranged from 200 μg g^{-1} fresh weight in the oldest leaves to 600 μg g^{-1} fresh weight in the youngest leaves, suggesting that endogenous levels may be adequate to influence significantly the growth of many fungi.

We (D.L. Miner, D.A. Danehower and A.R. Butler, unpublished data) recently found that crude extracts from the leaves of *N. glutinosa*, *N. repanda, N. exigua* Wheeler, *N. amplexicaulis* Burbidge and *N. gossei* strongly inhibited germination and growth of two strains of *Rhizoctonia solani*, the causal agent of target leaf spot. In other recent work, α-CBT-diol, β-CBT-diol, *Z*-abienol, and a mixture of Group 1–6 *N. tabacum* SEs (Severson *et al.*, 1985b) showed no inhibitory activity against the brown spot disease of tobacco (*Alternaria alternata* f. sp. *tabaci* Simmons; M.L. Menetrez, H.W. Spurr, Jr, D.A. Danehower and D.R. Lawson, unpublished data). However, LS extracts from *N. glutinosa* strongly inhibited growth of *A. alternata*, and extracts from *N. gossei* moderately inhibited this pathogen. Bioassays against *Phytophthora parasitica* pv. *nicotianae,* the black shank disease of tobacco, indicated excellent inhibitory activity in the crude LS washes from *N. debneyi* Domin, *N. gossei*, *N. glutinosa* and *N. tabacum* (D.A. Danehower, unpublished data).

Kennedy *et al.* (1995) bioassayed LS compounds from 23 *Nicotiana* species against conidia of *Colletotrichum lagenarium* (Pass.) Ell. and Halst, the anthracnose pathogen of cucumber. IC_{50}s were 6.3 μg cm^{-2} for CBT-diols, 19.3 μg cm^{-2} for sclareols, 1.0 μg cm^{-2} for labdenediol and 1.2 μg cm^{-2} for SEs from *N. gossei*. At concentrations of 100 μg cm^{-2}, CBT-diols and a sclareol-labdenediol mixture protected cucumbers against lesion development by 93% and 98%, respectively. Sugar esters from 19 of 23 *Nicotiana* species inhibited *C. lagenarium* conidial germination to some degree, and sugar esters from nine *Nicotiana* species completely inhibited conidial germination (Kennedy *et al.*, 1995).

Table 12.2. *Biological activities of crude LS extracts and purified chemicals from*

	Pseudomonas solanacearum (Bacterial wilt; tobacco)	*Colletotrichum gloeosporioides* (Anthracnose; tobacco)	*Peronospora tabacina* (Blue mould; tobacco)	*Erysiphe cichoracearum* (Powdery mildew; tobacco)	*Rhizoctonia solani* (Target spot; tobacco)
Nicotiana species					
N. occidentalis	-[a]	-	nt	nt	+/-
N. glauca	-	nt	nt	nt	-
N. debneyii	-	-	nt	nt	+/-
N. suaveolens	-	-	nt	nt	+/-
N. repanda	-	-	+	nt	+/-
N. rosulata	-	nt	nt	nt	-
N. exigua	-	-	nt	nt	+
N. ingulba	-	nt	nt	nt	-
N. glutinosa	-	+	nt	++	++
N. hesperis	-	-	nt	nt	-
N. velutina	-	-	nt	nt	-
N. amplexicaulis	-	+	nt	nt	+/-
N. gossei	-	-	nt	nt	+
N. fragrans	nt	-	nt	nt	-
N. kawakamii	nt	-	nt	nt	+
N. otophora	nt	-	nt	nt	+
N. tabacum (Galpao)	-	nt	++	nt	+
N. tabacum (NFT)	nt	-	nt	nt	+
N. tabacum (TI 454)	nt	-	nt	nt	+
N. tabacum (TI 536)	nt	-	nt	nt	+
N. tabacum (TI 1068)	nt	-	++	nt	+
N. tabacum (TI 1341)	nt	-	nt	nt	+
LS chemical					
SEs[b] Group 1-6 (*N. tabacum*)	-	nt	-	nt	+
Z-abienol (*N. tabacum*)	-	nt	++	nt	-
α-CBT-diol (*N. tabacum*)	-	nt	++	nt	-
β-CBT-diol (*N. tabacum*)	-	nt	++	nt	-
2-Hydroxy manool (*N. glutinosa*)	++	++	nt	nt	++
CBT-monols (*N. tabacum*)	++	-	-	nt	-
sclareols (*N. glutinosa*)	-	++	nt	nt	-
manool (*N. glutinosa*)	-	-	nt	nt	+/-
2-ketoepimanool (*N. glutinosa*)	nt	nt	nt	++	nt
SEs Group 1	++	++	-	nt	++
SEs Group 2	++	++	-	nt	-

[a] -, no activity; +/-, variable activity, dependent on assay; +, good activity; ++, excellent activity; nt, not
1988; Menetrez *et al.*, 1990; Severson *et al.*, 1985b, 1988a; D.A. Danehower, D.L. Miner, A.R. Butler,
[b] SEs, sucrose esters. For group designations see Severson *et al.* (1985b).

Nicotiana *species against pathogens and non-pathogens*

Phytophthora parasitica (Black shank; tobacco)	*Alternaria alternata* (Brown spot; tobacco)	*Phytophthora infestans* (Late blight; potato)	*Alternaria solani* (Early blight; tomato)	*Clavibacter michiganensis* (Bacterial canker; tomato)	*Bacillus cereus* (Saphrophyte)	*Bacillus subtilis* (Saphrophyte)
+/-	-	+	+/-	-	-	-
-	-	-	-	+	nt	nt
-	+/-	+/-	+/-	+	++	+/-
-	+/-	+/-	-	++	++	+
+	+	+	+	++	++	++
-	-	++	-	-		-
-	-	++	-	++	++	-
-	-	-	-	-	nt	-
++	+	+	+	++	++	++
-	-	+	-	-	-	-
-	-	-	-	-	-	-
-	-	+	-	+	-	-
-	-	nt	-	++	-	+
-	+	-	-	+	-	-
+	++	-	+	++	++	+
+/-	++	+	+	++	-	-
+	+	+	+	++	nt	+
++	+	-	+	++	++	++
++	++	++	+	++	+	+
+	++	++	+	++	++	++
+	++	+	+	++	++	++
+	++	+/-	+	++	++	++
nt	nt	nt	nt	-	nt	-
nt	nt	nt	nt	-	nt	-
nt	nt	nt	nt	-	nt	-
nt	nt	nt	nt	-	nt	-
nt	nt	++	++	++	++	++
-	nt	+	+	++	-	+
+	nt	++	+	++	++	++
+	nt	-	-	+	+	++
nt	nt	nt	nt	nt	nt	nt
-	nt	++	+	++	++	++
-	nt	++	+	++	++	++

tested (data from Chortyk *et al.*, 1993; Cruickshank *et al.*, 1977; Cutler *et al.*, 1986a,b; Lawson *et al.*, M.L. Menetrez, and G.D. Goins, unpublished).

Nicotiana LS components also exhibit selective inhibition of Gram-positive and Gram-negative bacteria. Using zone inhibition studies, Cutler *et al.* (1986a) showed that α-CBT-diol inhibits *Bacillus subtilis* and *B. cereus* (Gram-positive bacteria), but not *Escherichia coli* (Gram-negative). The β-CBT-diol isomer was inactive against Gram-positive bacteria, indicating the importance of stereochemistry in expression of biological activity. SEs from several *Nicotiana* species were effective against three Gram-positive bacteria (*B. subtilis, B. cereus* and *Mycobacterium thermosphactum*) and two Gram-negative bacteria (*Citrobacter freundii* and *Escherichia cloacae*), but not against *E. coli* (Chortyk *et al.*, 1993; Cutler *et al.* 1986b, 1991, 1992, 1994). As in the wheat coleoptile bioassays, activity was directly correlated with chain length of the acyl groups. SEs with C_7 or C_8 groups tended to have higher activities than SEs with C_5 or C_6 groups (Chortyk *et al.*, 1993).

We (D.L. Miner, D.A. Danehower and A.R. Butler, unpublished data) have recently observed that the epicuticular components from *N. exigua* and *N. gossei* have good activity against *Claverbacter michiganensis*, a bacterial rot of tomato. Sclareols, 2-hydroxymanool, and SEs (Group 1 and 2, Severson *et al.*, 1985b) also had significant activity against this pathogen. Manool had activity against *B. subtilis, B. cereus* and *C. michiganensis*.

Activity against saprophytes, such as the *Bacillus* species, poses an interesting ecological problem. If LS allelochemicals suppress growth of biological control agents or naturally occurring saprophytes on the leaf surface, while having no activity against disease organisms, then these components may actually encourage the growth of disease organisms. This possibility should not be overlooked by those attempting to manipulate LS allelochemistry to impart disease resistance.

12.6 Utilization and manipulation of leaf-surface components

Breeding chemically based, insect-resistance factors into tobacco holds promise for reducing problems associated with the use of pesticides. The international tobacco research organization (CORESTA) recently conducted a symposium on the genetic resources available for breeding for resistance to insects and diseases (Johnson *et al.*, 1992; Nielsen, 1992; Sisson, 1992a,b; Smeeton and Ternouth, 1992). Many of the tobacco germplasm sources reported in this symposium have resistance factors based on LS components. Whereas, there has been a long history of breeding for resistance to tobacco diseases (Stavely, 1979), efforts to breed for resistance to insects are fairly recent (Severson *et al.*, 1985b). Interspecific hybridization has been used repeatedly for incorporating disease-resistance genes into tobacco breeding lines (Stavely, 1979), but only limited attempts have been made to use wild *Nicotiana* species when breeding for insect resistance. However, Harlow *et al.* (1995) found that several interspecific hybrids with *N. tabacum* as a parent were resistant to tobacco aphids in the field. Palakarcheva *et al.* (1995) crossed *N. gossei* with an oriental tobacco cultivar (*N. tabacum*) to produce F_1 hybrid tobacco lines with resistance to tobacco aphids, powdery mildew, tobacco mosaic virus and blue mould. Partially sterile regenerates and their progenies also were resistant to tobacco aphids, and these materials could be used in tobacco breeding programmes.

The *Nicotiana* species are a rich source of potentially valuable agrochemicals; for example, the diterpenoids produced by *N. tabacum* have potential uses in the flavour and fragrance industries (Tornatzky *et al.*, 1996). LS components also affect the behaviour of pestiferous and beneficial insects (Jackson *et al.*, 1989), and/or they have insecticidal properties (Buta *et al.*, 1993; Neal *et al.*, 1994; Severson *et al.*, 1994a,b). Recent field trials have documented the efficacy of sugar esters for control of soft-bodied insects such as whiteflies and aphids (Liu and Stansly, 1995; McPherson *et al.*, 1995; Severson *et al.*, 1994a,b), and there is interest in the commercial development of natural and synthetic sugar esters as biorational insecticides. The sugar esters from *N. gossei* have been patented as a biopesticide (Pittarelli *et al.*, 1993), and they have been shown not to harm beneficial biological control agents, such as *Encarsia formosa* Gahan (Bentz and Neal, 1995). Chortyk (1996) produced inexpensive synthetic analogues of SEs that are structurally similar and nearly as active against whiteflies as natural SEs from *Nicotiana*.

Tobacco and *Nicotiana* species have been used widely in genetic and biotechnology research (Gadani *et al.*, 1995; Nielsen and Chappell, 1995; Wagner, 1992). In fact, tobacco is among the easiest plants to manipulate through biotechnology, and it is sometimes colloquially referred to as the 'white rat' of plant biotechnology (Tornatzky *et al.*, 1996). Besides being highly amenable to genetic transformation, tobacco also has been the subject of a vast amount of basic and applied research. Thus, it is likely that tobacco will continue to be an important plant for future applications of biotechnology. Some of this work will undoubtedly involve manipulation of the epicuticular chemistry of *Nicotiana* species (Nielsen and Chappell, 1995, Tornatzky *et al.*, 1996).

Acknowledgements

This paper is dedicated in loving memory to Dr. Ray F. Severson, Research Chemist, U.S. Department of Agriculture, Agricultural Research Service, who worked steadfastly in the pursuit of knowledge on the chemistry and biology of the tobacco leaf surface. His innovative contributions to science will be deeply missed. We thank Jennifer Grob and Wilant van Giessen for their efforts in helping us prepare this chapter.

References

Akehurst, B.C. (1968) *Tobacco.* Tropical Agriculture Series, Humanities Press, New York.

Arrendale, R.F., Severson, R.F., Chortyk, O.T. *et al.* (1988) Isolation and identification of the wax esters from the cuticular waxes of green tobacco leaf. *Beitrag. Tabakforsch. Int.* **14,** 67–76.

Arrendale, R.F., Severson, R.F., Sisson, V.A. *et al.* (1990) Characterization of the sucrose ester fraction from *Nicotiana glutinosa. J. Agric. Food Chem.* **38,** 75–85.

Bailey, J.A., Vincent, G.G. and Burden, R.S. (1974) Diterpenes from *Nicotiana glutinosa* and their effects on fungal growth. *J. Gen. Microbiol.* **85,** 57–64.

Belcher, D.W. and Thurston, R. (1982) Inhibition of movement of larvae of the convergent lady beetle by leaf trichomes of tobacco. *Environ. Entomol.* **11,** 91–94.

Bentz, J.-A. and Neal, J.W, Jr (1995) Effect of a natural insecticide from *Nicotiana gossei* on the whitefly parasitoid *Encarsia formosa* (Hymenoptera: Aphelinidae). *J. Econ. Entomol.* **88,** 1611–1615

Berry, J.W. and McKeen, W.E. (1985) Development of *Peronospora hyoscyami* f. sp. *tabacina* in the resistant host *Nicotiana exigua. Can. J. Plant Pathol.* **7,** 262–269.

Bordner, J., Danehower. D.A., Thacker, J.D. *et al.* (1983) Chemical basis for host plant selection. In: *Plant Resistance to Insects* (ed. P.A. Hedin), Am. Chem. Soc. Symp. Ser. No. 208. American Chemical Society, Washington, DC, pp. 245–264.

Burk, L.G., Chaplin, J.F. and Jackson, D.M. (1982) Inheritance of the glandless trichome trait in *Nicotiana tabacum* L. *Tobacco Sci.* **26,** 51–53.

Burke, B.A., Goldsby, G. and Mudd, J.B. (1987) Polar epicuticular lipids of *Lycopersicon pennellii. Phytochemistry* **26,** 2567–2571.

Buta, J.G., Lusby, W.R., Neal, J.W., Jr *et al.* (1993) Sucrose esters from *Nicotiana gossei* active against the greenhouse whitefly *Trialeuroides vaporariorum. Phytochemistry* **32,** 859–864.

Campbell, B.C., Jones, K.C. and Dreyer, D.L. (1986) Discriminative behavioral responses by aphids to various plant matrix polysaccharides. *Entomol. Exp. Appl.* **41,** 17–24.

Chang, S.Y. and Grunwald, C. (1976a) Duvatrienediols in cuticular wax of Burley tobacco leaves. *J. Lipid Res.* **17,** 7–11.

Chang, S.Y. and Grunwald, C. (1976b) Duvatrienols, alkanes, and fatty acids in cuticular wax of tobacco leaves of various physiological maturity. *Phytochemistry* **15,** 961–963.

Chang, S.Y. and Grunwald, C. (1980) Structural organization of tobacco leaf polar cuticular lipids. *Bot. Gaz.* **141,** 360–365.

Chaplin, J.F., Baumhover, A.H., Bortner, C.E. *et al.* (1976) Tobacco Production. *U.S. Dept. Agric., Agric. Res. Serv., Agric. Info. Bull.* No. 245.

Chortyk, O.T. (1996) New pesticidal sucrose esters for whitefly control. In: Fourth Annual Progress Review of the 5-Year National Research and Action Plan for the Development of Management and Control Methodology for the Silverleaf Whitefly. *U.S. Dept. Agric., Agric. Res. Serv.* (in press).

Chortyk, O.T., Severson, R.F., Cutler, H.C. *et al.* (1993) Antibiotic activities of sugar esters isolated from selected *Nicotiana* species. *Biosci. Biotech. Biochem.* **57,** 1355–1356.

Cohen, Y., Eyal, H., Goldschmidt, Z. *et al.* (1983) A preformed chemical inhibitor of tobacco powdery mildew on leaves of *Nicotiana glutinosa. Physiol. Plant Pathol.* **22,** 143–150.

Colledge, A., Reid, W.W. and Russell, R.A. (1974) A survey of surface diterpenoids of green leaves. *Ann. du Tabac., Sec. 2,* **11,** 159–164.

Court, W.A. and Hendel, J.G. (1983) Chemical composition of flue-cured tobacco cultivars with different sensitivities to weather fleck. *Can. J. Plant Sci.* **63,** 293–297.

Court, W.A., Elliot, J.M. and Hendel, J.G. (1984) Influence of applied nitrogen fertilization on certain lipids, terpenes, and other characteristics of flue-cured tobacco. *Tobacco Sci.* **28,** 69–72.

Cruickshank, I.A., Perrin, D.R. and Mandryk, M. (1977) Fungitoxicity of duvatrienediols associated with the cuticular wax of tobacco leaves. *Phytopathology* **90,** 243–249.

Cutler, H.G. and Cole, R.J. (1974) Properties of a plant growth inhibitor extracted from immature tobacco leaves. *Plant Cell Physiol.* **15,** 19–28.

Cutler, H.G., Reid, W.W. and Delatang, J. (1977) Plant growth inhibiting properties of diterpenes from tobacco. *Plant Cell Physiol.* **18,** 711–814.

Cutler, H.G., Severson, R.F., Cole, P.D. *et al.* (1986a) Plant growth inhibitory and antimicrobial properties of hydroxyacylnornicotines from select *Nicotiana* species. *Proc. - Plant Growth Regulator Soc. Am.* **13,** 188–200.

Cutler, H.G., Severson, R.F., Cole, P.D. *et al.* (1986b) Secondary metabolites from higher plants: Their possible role as biological control agents. In: *Natural Resistance of Plants to Pests: Roles of Allelochemicals* (eds M.B. Green and P.A. Hedin), Am. Chem. Soc. Symp. Ser. No. 296. American Chemical Society, Washington, DC, pp. 178–196.

Cutler, H.G., Severson, R.F., Montemurro, N. *et al.* (1991) Plant growth inhibitory and antimicrobial properties of sucrose esters from *Nicotiana tabacum* cultivars and species. *Plant Growth Regul. Soc. Am. Q.* **19,** 69–81.

Cutler, H.G., Severson, R.F., Cole, P.D. *et al.* (1992) The biological activity of some mixed sucrose and glucose esters from exotic tobacco cultivars. *Plant Growth Regul. Soc. Am. Q.* **20,** 18–28.

Cutler, H.G., Severson, R.F., Sisson, V.A. *et al.* (1994) *Nicotiana gossei:* A source of biologically active diversely mixed glucose and sucrose esters. *Plant Growth Regul. Soc. Am. Q.* **22,** 116–129.

Danehower, D.A. (1987) A rapid method for the isolation and quantification of the sucrose esters of tobacco. *Tobacco Sci.* **31,** 32–35.

Danehower, D.A., Izac, R.R., Grubbs, H.J. *et al.* (1987) A rapid method for the analysis of *Z*-abienol. *Tobacco Sci.* **31,** 48–51.

Danehower, D.A., Long, R.C., Core, M.T. *et al.* (1988) Changes in *Z*-abienol levels during growth and curing of Oriental and South American flue-cured tobacco. *Tobacco Sci.* **32,** 49–52.

Danehower, D.A., Reed, S.M. and Wernsman, E.A. (1989) Identification of the chromosome carrying the gene for production of β-methylvaleryl sucrose esters in *Nicotiana tabacum. Agric. Biol. Chem.* **53,** 2813–2815.

De Boer, G. and Hanson, F.E. (1987) Feeding responses to solanaceous allelochemicals by larvae of the tobacco hornworm, *Manduca sexta. Entomol. Exp. Appl.* **45,** 123–131.

Eckel, R.V.W., Jackson, D.M., Severson, R.F. *et al.* (1991) Correlations between cuticular *Nicotiana tabacum* components and resistance to the tobacco aphid, *Myzus nicotianae. 45th Tobacco Chem. Res. Conf.* Asheville, NC, 20–23 October 1991, Program Book, **45,** 16.

Einolf, W.N. and Chan, W.G. (1984) Estimation of sucrose esters in tobacco by direct chemical ionization mass spectrometry. *J. Agric. Food Chem.* **32,** 785.

Elsey, K.D. (1974) Influence of plant host on searching speed of two predators. *Entomophaga* **19**, 3–6.

Elsey, K.D. and Chaplin, J.F. (1978) Resistance of tobacco introduction 1112 to the tobacco budworm and green peach aphid. *J. Econ. Entomol.* **71**, 723–725.

Enzell, C.R. and Wahlberg, I. (1980) Leaf composition in relation to smoking quality and aroma. *Rec. Adv. Tobacco Sci.* **6**, 64–122.

Gadani, G., Ayers, D. and Hempfling, W. (1995) Tobacco: A tool for plant genetic engineering research and molecular farming. *Agro-Food Indus. Hitech* **6**, 1–19; **7**, 3–6.

Gamou, K. and Kawashima, N. (1979) Studies on leaf surface lipid of tobacco I. Changes in leaf surface lipid and duvatrienediol during growth, senescence, and curing of tobacco leaves. *J. Agric. Biol. Chem.* **43**, 2163–2168.

Gamou, K. and Kawashima, N. (1981) Studies on leaf surface lipid of tobacco IV. Comparisons of epicuticular alkane composition in *Nicotiana* species. *Agric. Biol. Chem.* **45**, 645–652.

Goodspeed, T.H. (1954) *The Genus Nicotiana: Origins, Relationships, and Evolution of its Species in the Light of Their Distribution, Morphology, and Cytogenetics.* Chronica Botanica Co. Waltham, MA.

Guo, Z. and Wagner, G.J. (1995) Biosynthesis of cembratrienols in cell-free extracts from trichomes of *Nicotiana tabacum. Plant Sci.* **110**, 1–10.

Guo, Z., Severson, R.F. and Wagner, G.J. (1994) Biosynthesis of the diterpene *cis*-abienol in cell-free extracts of tobacco trichomes. *Arch. Biochem. Biophys.* **308**, 103–108.

Gwynn, G.R., Severson, R.F., Jackson, D.M. *et al.* (1985) Inheritance of sucrose ester containing beta-methylvaleric acid in tobacco. *Tobacco Sci.* **29**, 79–81.

Hanson, F.E. and de Boer, G. (1987) What sensory organs control selection of leaf-chewing caterpillars? *Ser. Entomol.* **41**, 391.

Harlow, C.D., Jackson, D.M., Chortyk, O.T. *et al.* (1995) Development of insect pest management strategies using leaf surface components of *Nicotiana* species and hybrids. *Proc. 36th Tobacco Workers Conf.,* Tampa, FL, 9–12 January 1995 (in press).

Heggestad, H.E. and Middleton, J.T. (1959) Ozone in high concentrations as cause of tobacco leaf injury. *Science* **129**, 208–210.

Himmelsbach, D.S., van Halbeck, H., Arrendale, R.F. *et al.* (1990) Assignment of proton and carbon-13 spectra of a sucrose ester from tobacco by inverse detected two-dimensional heteronuclear correlated NMR spectroscopy. *Mag. Res. Chem.* **28**, 682–687.

Huesing, J.E. and Jones, D. (1987) A new form of antibiosis in *Nicotiana. Phytochemistry* **26**, 1381–1384.

Huesing, J. and Jones, D. (1988) Antibiosis to the tobacco hornworm (*Manduca sexta*) in wild species of *Nicotiana. Entomol. Exp. Appl.* **47**, 267–274.

Jackson, D.M., Cheatham, J.S., Pitts, J.M. *et al.* (1983) Ovipositional response of tobacco budworm moths (Lepidoptera: Noctuidae) to tobacco introduction 1112 and NC 2326 in cage tests. *J. Econ. Entomol.* **76**, 1303–1308.

Jackson, D.M., Severson, R.F., Johnson, A.W. *et al.* (1984) Ovipositional response of tobacco budworm moths (Lepidoptera: Noctuidae) to cuticular chemical isolates from green tobacco leaves. *Environ. Entomol.* **13**, 1023–1030.

Jackson, D.M., Severson, R.F., Johnson, A.W. *et al.* (1986) Effects of cuticular duvane diterpenes from green tobacco leaves on tobacco budworm (Lepidoptera: Noctuidae) oviposition. *J. Chem. Ecol.* **12**, 1349–1359.

Jackson, D.M., Severson, R.F., Johnson, A.W. *et al.* (1988) Host plant resistance in tobacco to *Heliothis* species. In: *Theory and Tactics of Heliothis Population Management: III. Emerging Control Tactics and Techniques* (eds G.A. Herzog, S. Ramaswamy, G. Lentz, J.L. Goodenough, and J.J. Hamm). *Southern Coop. Ser. Bull.* No. 337, pp. 31–49.

Jackson, D.M., Severson, R.F. and Johnson, A.W. (1989) Effects of natural tobacco constituents on insect survival, development, and behavior. *Rec. Adv. Tobacco Sci.* **15**, 26–116.

Jackson, D.M., Severson, R.F., Sisson, V.A. *et al.* (1991a) Ovipositional response of tobacco budworm moths (Lepidoptera: Noctuidae) to cuticular labdanes and sucrose esters from the green leaves of *Nicotiana glutinosa* L. (Solanaceae). *J. Chem. Ecol.* **17**, 2489–2506.

Jackson, D.M., Sisson, V.A. and Severson, R.F. (1991b) Mechanisms of resistance in *Nicotiana tabacum* to the tobacco aphid. In: *Aphid–Plant Interactions: Populations to Molecules* (eds D.C. Peters, J.A. Webster, and C.S. Chlouber), *Proc. Int. Symp. MP-132*. Agric. Exp. Stn, Div. Agric., Oklahoma State Univ., Stillwater, OK, p. 281.

Jackson, D.M., Rufty, T.W., Eckel, R.V.W. *et al.* (1992) Survival and development of insect pests on tobacco plants grown under elevated levels of ozone and ultraviolet radiation. *46th Tobacco Chem. Res. Conf.,* Montreal, Canada, 27–30 September 1992, Program Book, **46,** 43.

Jackson, D.M., Nottingham, S.F., Schlotzhauer, W.S. *et al.* (1995) Attractiveness of *Nicotiana* species to *Cardiochiles nigriceps. Proc. 36th Tobacco Workers Conf.,* Tampa, FL, 9–12 January 1995.

Jackson, D.M., Simmons, A.M., Chortyk, O.T. *et al.* (1996) Production of biorational insecticides by *Nicotiana* species. In: Fourth Annual Progress Review of the 5-Year National Research and Action Plan for the Development of Management and Control Methodology for the Silverleaf Whitefly. *U.S. Dept. Agric., Agric. Res. Serv.* (in press).

Johnson, A.W. (1978) Evaluation of several *Nicotiana tabacum* entries for resistance of two tobacco insect pests. *Tobacco Sci.* **22,** 41–43.

Johnson, A.W. (1980) Resistance in tobacco to the green peach aphid. *J. Econ. Entomol.* **73,** 707–709.

Johnson, A.W. and Chaplin, J.F. (1982) Sources of tobacco budworm resistance in tobacco. *Tobacco Sci.* **26,** 157–158.

Johnson, A.W. and Severson, R.F. (1982) Physical and chemical leaf surface characteristics of aphid resistant and susceptible tobacco. *Tobacco Sci.* **26,** 98–102.

Johnson, A.W. and Severson, R.F. (1984) Leaf surface chemistry of tobacco budworm resistant tobacco. *J. Agric. Entomol.* **1,** 23–32.

Johnson, A.W., Severson, R.F., Hudson, J. *et al.* (1985) Tobacco leaf trichomes and their exudates. *Tobacco Sci.* **29,** 67–72.

Johnson, A.W., Severson, R.F. and Jackson, D.M. (1992) Sources of resistance to insect pests of tobacco. *CORESTA Info. Bull.* **1992(3/4),** 151–175.

Kandra, L. and Wagner, G.J. (1988) Studies of the site and mode of biosynthesis of tobacco trichome exudate components. *Arch. Biochem. Biophys.* **265,** 425–432.

Kandra, L., Severson, R.F. and Wagner, G.J. (1990) Modified branched-chain amino acid pathways give rise to acyl acids of sucrose esters exuded from tobacco leaf trichomes. *Eur. J. Biochem.* **188,** 385–391.

Katanyukul, W. and Thurston, R. (1979) Mortality of eggs and larvae of the tobacco hornworm on Jimson weed and various tobacco cultivars in Kentucky. *Environ. Entomol.* **8,** 802–807.

Kays, S.J., Severson, R.F., Nottingham, S.F. *et al.* (1994) Possible biopesticide from *Petunia* for the control of the sweetpotato whitefly (*Bemisia tabaci*) on vegetable crops. *Proc. Fla. State Hort. Soc.* **107,** 163–167.

Keene, C.K. and Wagner, G.J. (1985) Direct demonstration of duvatrienediol biosynthesis in glandular heads of tobacco trichomes. *Plant Physiol.* **79,** 1026–1032.

Kennedy, B.S., Nielsen, M.T., Severson, R.F. *et al.* (1992) Leaf surface chemicals from *Nicotiana* affecting germination of *Peronospora tabacina* (Adam) Sporangia. *J. Chem. Ecol.* **18,** 1467–1479.

Kennedy, B.S., Nielsen, M.T. and Severson, R.F. (1995) Biorationals from *Nicotiana* protect cucumbers against *Colletotrichum lagenarium* (Pass.) Ell. & Halst disease development. *J. Chem. Ecol.* **21,** 221–231.

King, R.R. and Calhoun, L.A. (1988) 2,3-Di-O- and 1,2,3,-Tri-O-Acylated glucose esters from the glandular trichomes of *Datura metel. Phytochemistry* **27,** 3761–3763.

King, R.R., Pelletier, Y., Singh, R.P. *et al.* (1986) 3,4-Di-O-Isobutryl-6-0-carprylsucrose: The major component of a novel sucrose ester complex from the type B glandular trichomes of *Solanum berthaultii* Hawkes (PL 473340). *J. Chem. Soc. Chem Commun.* **1986,** 1078–1079.

Kolattukudy, P.E., Rogers, L.M., Li, D. *et al.* (1995) Surface signaling in pathogenesis. *Proc. Natl Acad. Sci. USA* **92,** 4080–4087.

Komari, T., Kubo, T. and Sato, M. (1986) Inheritance of the low-duvatrienediol trait in *Nicotiana tabacum* L. *Tobacco Sci.* **30,** 159–162.

Kubo, T., Sato, M., Tomita, H. *et al.* (1982) Identification of the chromosome carrying the gene for *cis*-abienol production by the use of monosomics in *Nicotiana tabacum.* L. *Tobacco Sci.* **26,** 126–128.

Lawson, D.R. and Danehower, D.A. (1989) Quantitative thin-layer chromotagraphy of *Nicotiana tabacum* leaf surface components. *J. Chromatogr.* **463,** 429–440.

Lawson, D.R., Danehower, D.A., Shilling, D.G. *et al.* (1988) Allelochemical properties of *Nicotiana tabacum* leaf surface compounds. In: *Biologically Active Natural Products: Potential Use in Agriculture* (ed. H.G. Cutler), Am. Chem. Soc. Symp. Ser. No. 380. American Chemical Society, Washington, DC, pp. 363–377.

Lin, Y. and Wagner, G.J. (1994a) Surface disposition and stability of pest-interactive, trichome-exuded diterpenes and sucrose esters of tobacco. *J. Chem. Ecol.* **20,** 1907–1921.

Lin, Y. and Wagner, G.J. (1994b) Rapid and simple method for estimation of sugar esters. *J. Agric. Food Chem.* **42,** 1709–1712.

Lingren, P.D. and Wolf, W.W. (1982) Nocturnal activity of the tobacco budworm and other insects. In: *Biometerology in Integrated Pest Management* (eds J.L. Hatfield and I.J. Thomason). Academic Press, New York pp. 211–228.

Liu, T.-X. and Stansly, P.A. (1995) Toxicity and repellency of some biorational insecticides to *Bemisia argentifolia* on tomato plants. *Entomol. Exp. Appl.* **74,** 137–143.

Mahadevan, A. (1970) Prohibitins and disease resistance. *Phytopathol. Z.* **68,** 73–80.

Marcovitch, S. and Stanley, W.W. (1937) Control of tobacco insects in Tennessee. *Univ. Tenn. Agric. Exp. Stn Bull.* No. 160.

Matsuzaki, T., Koseki, K. and Kowai, A. (1988a) Germination and growth inhibition of surface lipids from *Nicotiana* species and identification of sucrose esters. *Agric. Biol. Chem.* **52,** 1889–1897.

Matsuzaki, T., Miyano, M., Yasumatsu, N. *et al.* (1988b) Germination and growth inhibition of acylnornicotines from section Repandae of the genus *Nicotiana* and synthetic acylnornicotines. *Agric. Biol. Chem.* **52,** 1899–1903.

Matsuzaki, T., Shinozaki, Y., Suhara, S. *et al.* (1989a) Isolation of glycolipids from the surface lipids of *Nicotiana bigelovii* and their distribution in *Nicotiana* species. *Agric. Biol. Chem.* **53,** 3079–3082.

Matsuzaki, T., Shinozaki, Y., Suhara, S. *et al.* (1989b) Isolation and characterization of tetra- and tri-acylglucose from the surface glycolipids of *Nicotiana miersii. Agric. Biol. Chem.* **53,** 3343–3345.

Matsuzaki, T., Shinozaki, Y., Suhara, S. *et al.* (1991) Leaf surface glycolipids from *Nicotiana acuminata* and *N. pauciflora. Agric. Biol. Chem.* **55,** 1417–1419.

Matsuzaki, T., Shinozaki, Y., Haagimori, M. *et al.* (1992) Novel glycerolipids and glycolipids from the surface of *Nicotiana benthamiana. Biosci. Biotech. Biochem.* **56,** 1565–1569.

McPherson, R.M., Hall, E.K., Lambert, A.L. *et al.* (1995) Effects of a natural pesticide from *Nicotiana gossei* on controlling whiteflies (Homoptera: Aleyrodidae) on a susceptible and resistant soybean line. *J. Entomol. Sci.* **30,** 312–317.

Menetrez, M.L., Spurr, H.W., Danehower, D.A. *et al.* (1990) Influence of tobacco leaf surface chemicals on germination of *Peronospora tabacina* Adam sporangia. *J. Chem. Ecol.* **16,** 1565–1576.

Michie, M.J. and Reid, W.W. (1968) Biosynthesis of complex terpenes in the leaf cuticle and trichomes of *Nicotiana tabacum. Nature* **218,** 578.

Mitchell, E.R., Tingle, F.C. and Heath, R.R. (1990) Oviposition response of three *Heliothis* species (Lepidoptera: Noctuidae) to allelochemicals from cultivated and wild host plants. *J. Chem. Ecol.* **16,** 1817–1827.

Mitchell, E.R., Tingle, F.C. and Heath, R.R. (1991) Flight activity of *Heliothis virescens* (F.) females (Lepidoptera: Noctuidae) with reference to host-plant volatiles. *J. Chem. Ecol.* **17,** 259–266.

Mold, J.D., Stevens, R.K., Means, R.E. *et al.* (1963) The paraffin hydrocarbons of tobacco: Normal, iso-, and anteiso-homologs. *Biochemistry* **2,** 605–610.

Navasero, R.C. and Ramaswamy, S.B. (1991) Morphology of leaf surface trichomes and its influence on oviposition by *Heliothis virescens* (F.) (Lepidoptera: Noctuidae). *Crop Sci.* **31,** 342–353.

Navasero, R.C. and Ramaswamy, S.B. (1993) Influence of plant age, water stress, larval damage, and presence of conspecific eggs on oviposition by *Heliothis virescens* (F.) on cotton. *J. Appl. Entomol.* **115,** 97–106.

Neal, J.W., Buta, J.G., Pittarelli, G.W. *et al.* (1994) Novel sucrose esters from *Nicotiana gossei:* Effective biorationals against selected horticultural insect pests. *J. Econ. Entomol.* **87,** 1600–1607.

Nielsen, M.T. (1992) Sources of resistance to black shank and black root rot diseases. *CORESTA Info. Bull.* **1992(3/4),** 144–150.

Nielsen, M.T. and Chappell, J.C. (1995) Biotechnology in tobacco improvement: Applications and current status in the industry. *Rec. Adv. Tobacco Sci.* **21,** 134–164.

Nielsen, M.T. and Severson, R.F. (1990) Variation of flavor components on leaf surfaces of tobacco genotypes differing in trichome density. *J. Agric. Food. Chem.* **38,** 467–471.

Nielsen, M.T. and Severson, R.F. (1992) Inheritance of a diterpene constituent in tobacco trichome exudate. *Crop Sci.* **32,** 1148–1150.

Nielsen, M.T., Jones, G.A. and Collins, G.B. (1982) Inheritance pattern for secreting and nonsecreting glandular trichomes in tobacco. *Crop. Sci.* **22,** 1051–1053.

Nielsen, M.T., Akers, C.P., Järlfors, U.E. *et al.* (1991) Comparative ultrastructural features of secreting and nonsecreting glandular trichomes of two genotypes of *Nicotiana tabacum* L. *Bot. Gaz.* **152,** 13–22.

Noma, M., Suzuki, F., Gamou, K. *et al.* (1982) Two labdane diterpenoids from *Nicotiana raimondii. Phytochemistry* **21,** 395–397.

Palakarcheva, M., Staneva, M. and Tsanova, E. (1995) Hybridization between *Nicotiana gossei* Domin and *N. tabacum* L. for development of oriental tobacco lines resistant to tobacco aphids and diseases. *Tobacco Sci.* **39,** 38–42.

Pittarelli, G.W., Buta, J.G., Neal, J.W. *et al* (1993) Biological pesticide derived from *Nicotiana* plants. United States Patent No. 5,260,281.

Rabb, R.L. and Bradley, J.R. (1968) The influence of host plants on parasitism of eggs of the tobacco hornworm. *J. Econ. Entomol.* **61,** 1249–1252.

Ramaswamy, S.B. and Hanson, F.E. (1990) Electrophysiological responses of tarsal chemoreceptors of *Heliothis virescens* (F.) (Lepidoptera: Noctuidae). *Symp. Biol. Hungarica* **39,** 111–117.

Ramaswamy, S.B., Ma, W.K. and Baker, G.T. (1987) Sensory cues and receptors for oviposition by *Heliothis virescens. Entomol. Exp. Appl.* **43,** 159–168.

Rao, M.N., Siegel, M.R., Ferriss, R.S. *et al.* (1989) Relationships between susceptibility of field-grown burley tobacco to blue mold and contents of duvatrienediols. *Phytopathology* **79,** 267–270.

Reid, W.W. (1974) Part I. The cuticular and cytoplasmic lipids of *Nicotiana tabacum. Ann. du Tabac., Sec.* 2, **11,** 151–159.

Reuveni, M., Tuzun, S., Cole, J.S. *et al.* (1987) Removal of duvatrienediols from the surface of tobacco leaves increases their susceptibility to blue mold. *Physiol. Mol. Plant Pathol.* **30,** 441–451.

Roberts, D.L. (1988) Natural tobacco flavor. *Rec. Adv. Tobacco Sci.* **14,** 49–81.

Roberts, D.L. and Rowland, R.L. (1962) Macrocyclic diterpenes. α- and β-4,8,13-duvatriene-1,3-diols from tobacco. *J. Org. Chem.* **27,** 3989–3995.

Roberts, D.W., Aist, J.R., Kolattukudy, R.E. *et al.* (1984) *Infection Processes of Fungi*. Rockefeller Foundation, New York.

Roberts, K.L., Thurston, R. and Jones, G.A. (1981) Density of types of trichomes on leaf lamina of various tobacco cultivars. *Tobacco Sci.* **25,** 68–69.

Sato, M.T., Komari, H., Tomita, H. *et al.* (1981) Survey of leaf surface diterpenoids in flue-cured, cigar and Japanese domestic varieties of tobacco. *Sabrao J.* **13,** 47–54.

Sato, M.T., Komari, H. and Asaine, K. (1982) Varietal differences in the composition of leaf surface diterpenoids in tobacco. *Iwata Tobacco Exp. St. Bull.* **14,** 59–70.

Schlotzhauer, W.S., Horvat, R.J., Chortyk, O.T. *et al.* (1993) Volatile constituents from flowers of selected *Nicotiana* species. *47th Tobacco Chem. Res. Conf.,* Gatlinburg, TN, 18–21 October 1993, Program Book, **47,** 42.

Schoonhoven, L.M. (1987) What makes a caterpillar eat? The sensory code underlying feeding behavior. In: *Perspectives in Chemoreception and Behavior* (eds R.F. Chapman, E.A. Bernays, and J.G. Stoffolano, Jr). Springer-Verlag, New York, pp. 69–97.

Severson, R.F. (1990) The cuticular chemistry of *N. tabacum*. *CORESTA Info. Bull.* **1990,** 34–54.

Severson, R.F., Gwynn, G.R., Chaplin, J.F. *et al.* (1983) Leaf trichome exudate associated with insect resistance in *Nicotiana tabacum* L. *Tobacco Sci.* **27,** 82–83.

Severson, R.F., Arrendale, R.F., Chortyk, O.T. *et al.* (1984) Quantitation of the major cuticular components from green leaf of different tobacco types. *J. Agric. Food Chem.* **32,** 566–570.

Severson, R.F., Arrendale, R.F., Chortyk, O.T. *et al.* (1985a) Isolation and characterization of the sucrose esters of the cuticular waxes of green tobacco. *J. Agric. Food Chem.* **33,** 870–875.

Severson, R.F., Johnson, A.W. and Jackson, D.M. (1985b) Cuticular constituents of tobacco: Factors affecting their production and their role in insect and disease resistance and smoke quality. *Rec. Adv. Tobacco Sci.* **11,** 105-174.

Severson, R.F., Arrendale, R.F., Cutler, H.G. *et al.* (1988a) Chemistry and biological activity of acylnornicotines from *Nicotiana repandae.* In: *Biologically Active Natural Products: Potential Use in Agriculture* (ed. H.G. Cutler), Am. Chem. Soc. Symp. Ser. No. 380. American Chemical Society, Washington, DC, pp. 335–362.

Severson, R.F., Huesing, J.E., Jones, D. *et al.* (1988b) Identification of tobacco hornworm antibiosis factor from cuticulae of Repandae section of *Nicotiana* species. *J. Chem. Ecol.* **14,** 1485–1494.

Severson, R.F., Jackson, D.M., Johnson, A.W. *et al.* (1991) Ovipositional behavior of tobacco budworm and tobacco hornworm: Effects of cuticular components from *Nicotiana* species. In: *Naturally Occurring Pest Bioregulators* (ed. P. A. Hedin), Am. Chem. Soc. Symp. Ser. No. 449. American Chemical Society, Washington, DC, pp. 264–277.

Severson, R.F, Chortyk, O.T., Stephenson, M.G. *et al.* (1994a) Characterization of natural pesticide from *Nicotiana gossei*. In: *Bioregulators for Crop Protection and Pest Control* (ed. P.A. Hedin), Am. Chem. Soc. Symp. Ser. No. 557. American Chemical Society, Washington, DC, pp. 107–121.

Severson, R.F., Eckel, R.V.W., Jackson, D.M. *et al.* (1994b) Aphicidal activity of cuticular components from *Nicotiana tabacum*. In: *Natural and Engineered Pest Management Agents* (eds P.A. Hedin, J.J. Menn and R.M. Hollingworth), Am. Chem. Soc. Symp. Ser. No. 551. American Chemical Society, Washington, DC, pp. 172–179.

Shinozaki, Y., Matsuzaki, T., Suhara, S. *et al.* (1991) New types of glycolipids from the surface lipids of *Nicotiana umbratica*. *Agric. Biol. Chem.* **55**, 751–756.

Shinozaki, Y., Matsuzaki, T., Suzuki, H. *et al.* (1992) Surface glycolipids in the shoot-forming culture of *Nicotiana umbratica*. *Biosci. Biotech. Biochem.* **56**, 1482–1483.

Simpson, J.R., Johnson, A.W. and Severson, R.F. (1985) Leaf surface chemistry of tobacco budworm resistant and susceptible tobacco grown in the field and greenhouse at different fertilization rates. *Tobacco Sci.* **29**, 44–46.

Sisson, V.A. (1992a) Sources of resistance to bacterial wilt. *CORESTA Info. Bull.* **1992(3/4)**, 136–143.

Sisson, V.A. (1992b) The United States Tobacco germplasm collection. *CORESTA Info. Bull.* **1992(3/4)**, 107–112.

Sisson, V.A., Severson, R.F. and Jackson, D.M. (1993) Inheritance of high duvatrienol production in tobacco. *Tobacco Sci.* **37**, 25–29.

Smeeton, B.W. and Ternouth, R.A.F. (1992) Sources of resistance to powdery mildew, wildfire, angular leaf spot, and *Alternaria*. *CORESTA Info. Bull.* **1992(3/4)**, 127–135.

Snook, M.E., White, R.A., Jr, Chortyk, O.T. *et al.* (1995) Terpene glycosides from tobacco with antibiosis activity towards the tobacco budworm (*Heliothis virescens*). *49th Tobacco Chem. Res. Conf.*, Lexington, KY, 24–27 September 1995, Program Book, **47**, 27–28.

Springer, J.P., Clardy, J., Cox, R.H. *et al.* (1975) The structure of a new type of plant growth inhibitor extracted from immature tobacco leaves. *Tetrahedron Lett.* **32**, 2737–2740.

Stavely, J.R. (1979) Disease resistance. In: *Nicotiana: Procedures for Experimental Use* (ed. R.D. Durbin). *U.S. Dept. Agric. Tech. Bull.* No. 1586, pp. 87–110.

Takada, H. and Takenaka, Y. (1982) Parasite complex of *Myzus persicae* (Sulzer) (Homoptera, Aphididae) on tobacco. *Kontyu* **50**, 556–568

Thurston, R. and Webster, J.A. (1962) Toxicity of *Nicotiana gossei* Dominot to *Myzus persicae* (Sulzer). *Entomol. Exp. Appl.* **5**, 233–238.

Tichenor, L.H., Seigler, D.S. and Wei, L. (1981) Oviposition responses of *Manduca sexta* to solanaceous volatile components. *Trans. Ill. State Acad. Sci.* **74**, 35–42.

Tingle, F.C. and Mitchell, E.R. (1992) Attraction of *Heliothis virescens* (F.) (Lepidoptera: Noctuidae) to volatiles from extracts of cotton flowers. *J. Chem. Ecol.* **18**, 907–914.

Tomita, H., Sato, M. and Kawashima, N. (1980) Inheritance of labdanoid producing ability in *N. tabacum*. *Agric. Biol. Chem.* **44**, 2517–2518.

Tornatzky, L., Batts, Y., Casson, L. *et al.* (1996) *Prospects for Plant-Based Biotechnological Products: Capitalizing on the Southern Advantage*. Southern Technology Council, Research Triangle Park, N.C.

Van Emden, H.F., Eastop, V.F., Hughes, R.D. *et al.* (1969) The ecology of *Myzus persicae*. *Annu. Rev. Entomol.* **14**, 197–270.

Wagner, G.J. (1992) Improving tobacco through metabolic engineering. *Rec. Adv. Tobacco Sci.* **18**, 3–43.

Wahlberg, I., Wallin, I., Narborne, C. *et al.* (1981) Note on the stereostructures of thunbergol (isocembrol) and 4-Epiisocembrol. *Acta Chem. Scand. Ser. B Org. Chem. Biochem.* **35**, 65–68.

Wahlberg, I., Walsh, E.B., Forsblom, I. *et al.* (1986) Tobacco chemistry 64*. A new sucrose ester from Greek tobacco. *Acta Chem. Scand. Ser. B Org. Chem. Biochem.* **40**, 724–730.

Waldbauer, G.P. and Fraenkel, G. (1961) Feeding on normally rejected plants by maxillectomized larvae of the tobacco hornworm, *Protoparce sexta* (Lepidoptera, Sphingidae). *Ann. Entomol. Soc. Am.* **54**, 477–485.

Wilkinson, R.E. and Kasperbauer, M.J. (1972) Epicuticular alkane content of tobacco as influenced by photoperiod, temperature and leaf age. *Phytochemistry* **11**, 2439–2442.

Wilkinson, R.E. and Kasperbauer, M.J. (1980) Effect of light and temperature on epicuticular fatty acid and fatty alcohol of tobacco. *Phytochemistry* **19**, 1379–1383.

Willis, M.A. and Arbas, E.A. (1991) Odor-modulated upwind flight of the sphinx moth, *Manduca sexta L. J. Comp. Physiol. A* **169**, 427–440.

Wolf, F.A. (1946) Further consideration of glandular leaf hairs of tobacco and of their significance. *Bull. Torrey Bot. Club* **73**, 224–234.

Yamamoto, R.T and Fraenkel, G. (1960) The specificity of the tobacco hornworm, *Protoparce sexta* to solanaceous plants. *Ann. Entomol. Soc. Am.* **53**, 503–507.

Zador, E. and Jones D. (1986) The biosynthesis of a novel nicotine alkaloid in the trichomes of *Nicotiana stocktonii. Plant Physiol.* **82**, 479–484.

13

Influences of host surface features on development of *Erysiphe graminis* and *Erysiphe pisi*

T.L.W. Carver, S.M. Ingerson and B.J. Thomas

13.1 Introduction

The cuticle-cell wall complex forms the first line of plants' defence against attack by leaf pathogens. Some pathogens avoid these defences; for example by invading through wounds or stomata. Others, like the powdery mildew fungi, form a specialized infection structure, the appressorium, from which a penetration peg grows to penetrate the host cuticle and cell wall directly. This chapter considers evidence that appressorium formation by *Erysiphe* spp. represents the culmination of fungal growth and differentiation processes which occur in an ordered and predictable sequence driven by the perception of, and response to, characteristics of the host leaf surface (substratum). We consider influences of substrata on appressorial development by two different powdery mildew fungi, that is *Erysiphe graminis* DC (*Blumeria graminis* DC), which causes powdery mildew of cereals (including: oat, *Avena sativa* L.; barley, *Hordeum vulgare* L.; wheat, *Triticum aestivum* L.) and other Gramineae, and *Erysiphe pisi* DC, which causes powdery mildew of peas (*Pisum sativum* L.) and other legumes. For *E. graminis* we draw on information from studies of various *formae speciales* (f.sp.; forms of the fungus), which are indistinguishable in their development although they are specialized to attack different genera of Gramineae.

13.2 Spore germination, germling growth and appressorium differentiation by *E. pisi* on pea leaves and *E. graminis* on cereal leaves

Figure 13.1 shows that germlings of both *E. graminis* and *E. pisi* develop through morphologically distinct stages, but that the two fungi differ in certain important aspects. *E. graminis* has been studied more extensively than *E. pisi*, but the chronological details of germination and germling development have been

Plant Cuticles, edited by G. Kerstiens.

described for both *E. graminis* on cereals (e.g. Carver *et al.*, 1995a,b) and *E. pisi* on pea (Kunoh *et al.*, 1979; Singh and Singh, 1983).

Appressorium formation by *E. pisi* follows a relatively simple developmental sequence whereby a single germ tube emerges from the asexual spores, or conidia, between 2 and 4 h after inoculation. This elongates (to 25–30 μm), forms a septum and swells at the tip forming an appressorium 6–8 h after inoculation. The appressorium of *E. pisi* is shaped irregularly, usually with one to five lobes, but lobe number and germ tube length are influenced to some degree by light (Ayres, 1983). One of the appressorial lobes produces an infection peg which attempts to infect the host epidermal cell.

E. graminis differs from *E. pisi* and all other powdery mildew fungi in that conidia produce not one but two germ tubes (Kunoh *et al.*, 1979b). Prior to emergence of either tube, conidia release material containing enzymes including cutinase (Pascholati *et al.*, 1992) onto the underlying surface, and this may help prepare the surface to favour subsequent fungal development (Kunoh *et al.*, 1988; Nicholson *et al.*, 1993). The first germ tube, termed the primary germ tube (PGT; Kunoh *et al.*, 1979a), emerges 0.5–2.0 h after inoculation; it remains short (5–10 μm) and aseptate, and secretes extracellular material, possibly adhesive, soon after making contact with the host surface (Carver *et al.*, 1995a). PGTs never form haustoria, but produce a short penetration peg which enters the host surface to an unknown depth (Carver *et al.*, 1995a; Kunoh *et al.*, 1978) inducing localized host cell responses. PGTs may thus gain access to host water (Carver and Bushnell, 1983), which can be vital to continued germling development under low relative humidity conditions (see below). The second germ tube emerges 3.0–3.5 h after the PGT (Kunoh *et al.*, 1979a), elongates (ultimately 30–40 μm), secretes extracellular material, becomes septate, and swells at the tip to form a single, hook-shaped appressorial lobe (by about 9–10 h). A penetration peg beneath the appressorial lobe attempts to penetrate the underlying host cell cuticle and cell wall, probably aided by the secretion of enzymes (Aist, 1976; Bushnell, 1972; Kunoh, 1981). Penetration is attempted about 12–15 h after inoculation. If, and only if, attempted penetration from the first appressorial lobe fails, a second lobe may be formed distal to the first, and usually on the opposite side of the germ tube. If the second lobe fails to penetrate, further lobes may be formed in distal progression down the appressorial germ tube.

When penetration succeeds, both *E. graminis* and *E. pisi* form feeding structures, haustoria, in their host's epidermal cells. Haustoria export nutrients to the surface where hyphal growth accelerates. Within days, hyphal appressoria produce a succession of further haustoria, and within a week a new generation of conidia is produced from conidiophores, completing the asexual life cycle of the fungi.

13.3 Does spore germination depend upon contact with a firm surface?

Several lines of evidence suggest that extensive surface contact is not a prerequisite for germination. *E. graminis* conidia germinate rapidly when held in a simulated airborne state on 0.5 μm diameter spider's suspension threads (Figure 13.2; Carver and Ingerson, 1987; Nicholson *et al.*, 1993), or when floating on liquid surfaces (Gay *et al.*, 1985; Nicholson *et al.*, 1993). We have recently found similar behaviour by *E. pisi* conidia (Figure 13.2 and T.L.W. Carver and S.M. Ingerson, unpublished observations).

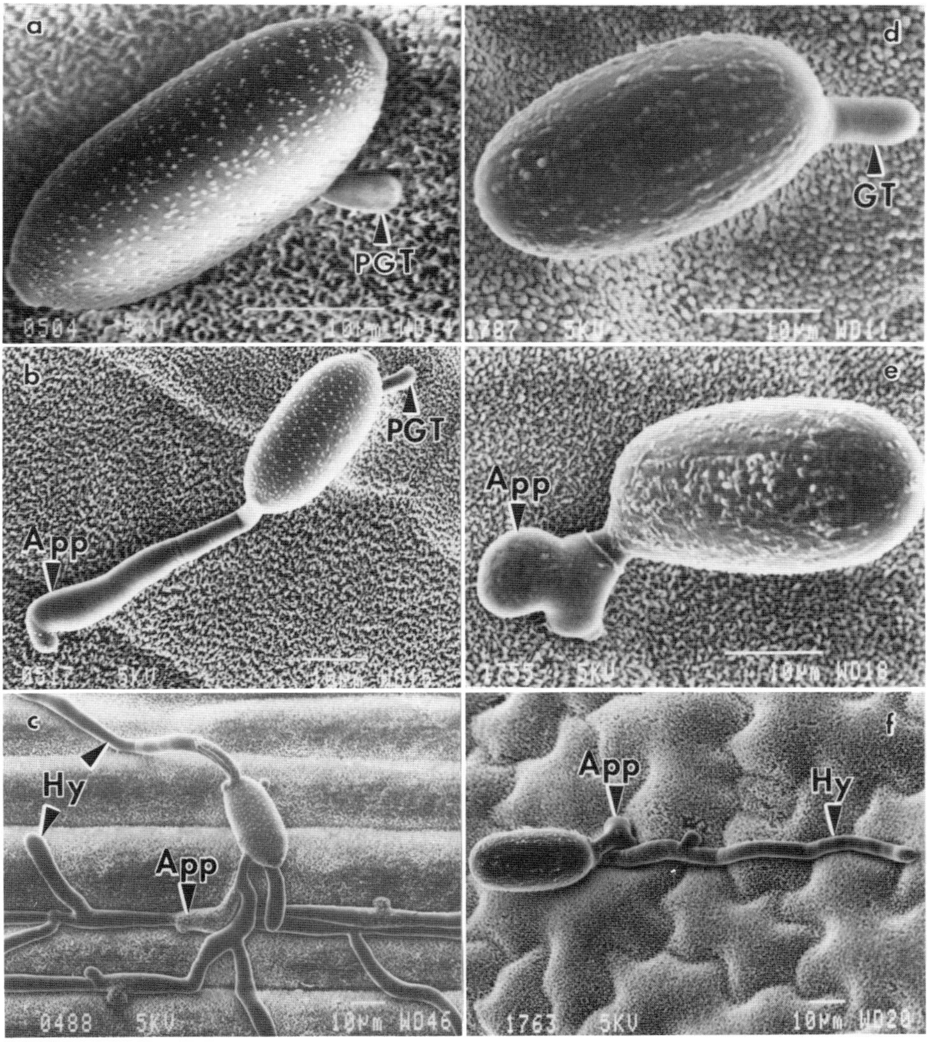

Figure 13.1. *Stages in germination and germling development by* Erysiphe graminis *(a,b,c) and* Erysiphe pisi *(d, e, f) on the surface of oat and pea leaves respectively. (a)* E. graminis *produces a primary germ tube (PGT) 0.5–2.0 h after inoculation. (b) A second germ tube emerges, elongates and differentiates a simple appressorium (App) with an apical lobe by 9–10 h. (c) If infection succeeds, secondary hyphae (Hy) develop. (d)* E. pisi *produces only one germ tube (GT) 2–4 h after inoculation. (e) This same tube elongates and differentiates an appressorium with multiple lobes. (f) Again, secondary hyphae develop if infection succeeds. ((c) Reproduced from Carver* et al. *(1995) The surface of* Erysiphe graminis *and the production of extracellular material at the fungus-host interface during germling and colony development.* Can. J. Bot. **73**, *272–287, with permission from the NRC Research Press.)*

13.4 *E. graminis* development following germination

13.4.1 *Appressorial germ tube elongation and differentiation by* E. graminis *is influenced by substratum characteristics*

In contrast to their behaviour on cereal leaves where most (~ 95%) germinated, *E. graminis* conidia form appressoria (Carver and Thomas, 1990), conidia held in a simulated airborne state, or on clean glass or on agar generally produce only short germ tubes and rarely form a germ tube which elongates or forms an appressorium-like structure (Carver and Ingerson, 1987). Apparently, germinating conidia respond to leaf surface characteristics. Perception of, and response to, these characteristics appears vital to growth and differentiation of the appressorial germ tube.

For *E. graminis* there are at least three stages in the processes leading from germination to appressorium formation that are influenced by the substratum surface (Carver and Ingerson, 1987). The first stage involves perception of the surface by the PGT. This somehow triggers the second stage which involves elongation of the second germ tube to emerge from the conidium. The third stage appears to depend upon the elongating tube itself responding to substratum characteristic(s) and differentiating the appressorial structure.

13.4.2 *Is epicuticular leaf wax involved in stimulating appressorial germ tube development by* E. graminis?

From work with barleys carrying eceriferum mutations (affecting epicuticular wax (EW) synthesis and or excretion), Yang and Ellingboe (1972) suggested that *E. graminis* responds to the topography of epicuticular leaf waxes. However, this now seems unlikely since removal of cereal leaf EW has almost no effect on appressorium formation (Carver and Thomas, 1990). Nevertheless, Staub *et al.* (1974) suggested that leaf waxes may be dissolved beneath *E. graminis* germ tubes, and Kunoh *et al.* (1990) have shown that factors in conidial washings degrade amorphous material (possibly non-crystalline wax or polymeric cutin) surrounding and partially covering the crystalline wax plates on barley leaves. Thus leaf waxes (epicuticular and intracuticular), or their breakdown products, may provide information important to the fungus, either through their chemistry or because of their hydrophobicity.

13.4.3 *The role of substratum hydrophobicity in appressorial germ tube development by* E. graminis

Nicholson *et al.* (1993) found that *E. graminis* conidia hardly ever formed appressorial germ tubes on polystyrene, and from this suggested that hydrophobic substrata may be inhibitory to appressorium formation. However, Nicholson *et al.* incubated their material at 70% relative humidity (r.h.), which is below the optimal

---

Figure 13.2. *Typical characteristics of* Erysiphe graminis *(a,b,c,d) and* E. pisi *(e,f,g,h) germlings developing: (a,e) suspended on spider's thread in a simulated airborne state; (b,f) on clean glass; (c,g) on Perspex; (d,h) on cellulose dialysis membrane. On spider's thread and glass* E. graminis *generally produced only multiple short germ tubes while apparently normal appressorial germ tubes were frequent on Perspex or cellulose membrane.* E. pisi *on spider's thread produced a single germ tube that elongated greatly but never differentiated an appressorium. On glass, Perspex and cellulose membrane,* E. pisi *often formed appressoria. ((a) and (e) light micrographs. (b–d) and (f–g) scanning electron micrographs (SEM) of specimens either viewed from above (d,g,h) or side (b,c,f).)*

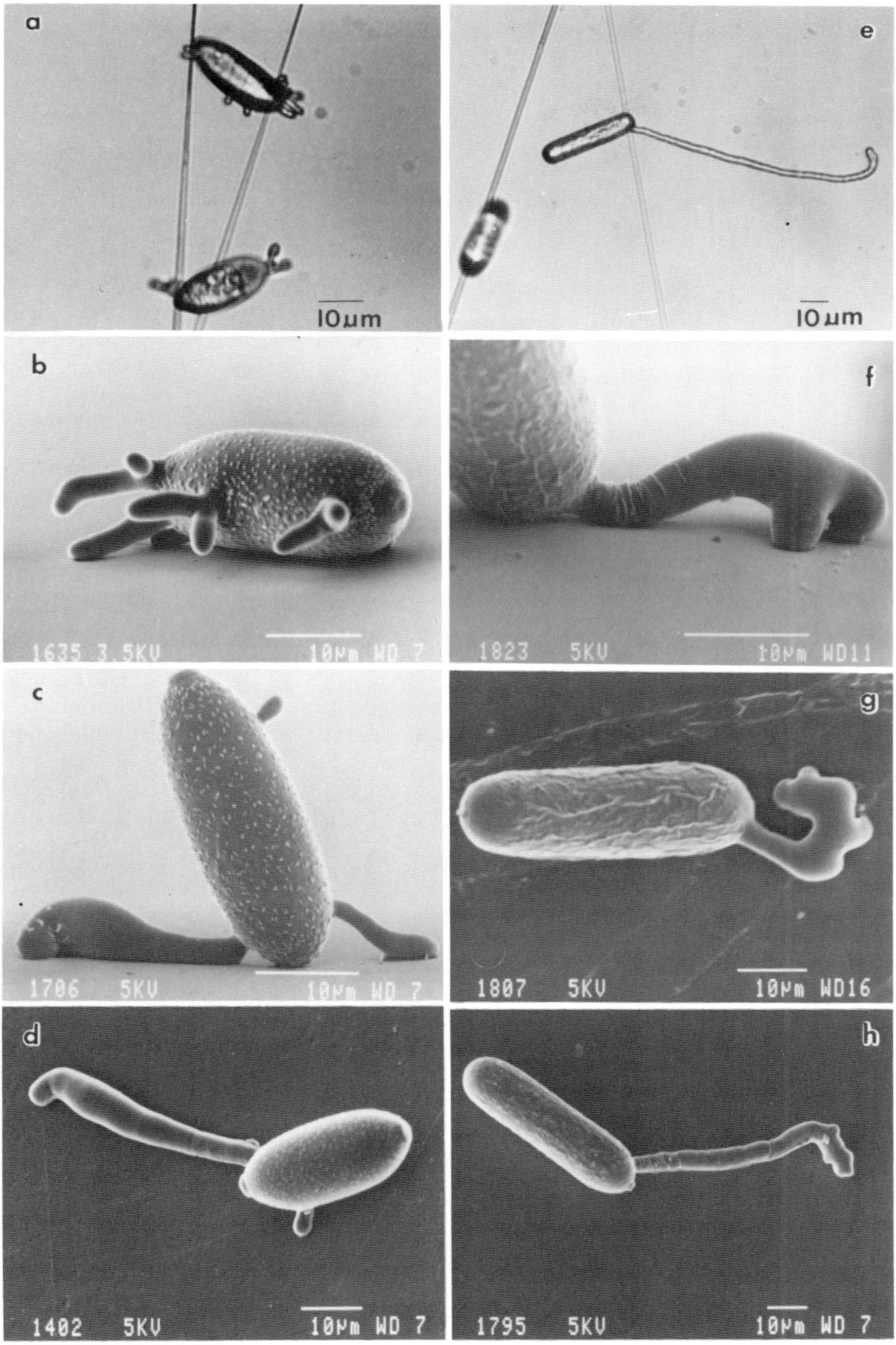
a
10µm
e
10µm
b
1635 3.5KV 10µm WD 7
f
1823 5KV 10µm WD11
c
1706 5KV 10µm WD 7
g
1807 5KV 10µm WD16
d
1402 5KV 10µm WD 7
h
1795 5KV 10µm WD 7

for *E. graminis* development (e.g. Manners and Hossain, 1963). We (unpublished data) confirmed that appressorial germ tubes form very rarely if conidia are incubated at 95% r.h. (or less) on artificial surfaces. By contrast, if incubated at 100% r.h., conidia on hydrophobic artificial surfaces frequently formed appressoria with apparently normal morphology (see below). Therefore, when incubation r.h. is below 100%, failure to form appressoria on hydrophobic substrata may be due to water shortage rather than to inhibitory effects of the substratum. On leaves incubated at low r.h., host water may be taken up via the PGT penetration peg to support appressorial development (Carver and Bushnell, 1983); this would not be possible on artificial surfaces (such as polystyrene) where fungal growth potential may only be realized at 100% r.h.

In a series of experiments, we compared frequencies of appressorium formation on substrata with different surface hydrophobicity incubated at 100% r.h. (Table 13.1). The relative hydrophobicity of surfaces was assessed by goniometer measurements of contact angles formed between water droplets (0.5–0.6 μl) and surface (the greater the angle, the greater the hydrophobicity). After incubation for 24–30 h, 100 germinated conidia were examined on each of at least three replicate samples of each substratum to determine whether they had formed an apparently normal appressorium. The experiment was repeated at least three times for each surface, and results between experiments were very consistent. Table 13.1 gives the results for a single experiment involving different surfaces.

The highest proportion of germinated conidia formed appressoria on leaf surfaces, which also had the highest surface hydrophobicity. On all artificial substrata some conidia produced structures which we classified as appressoria because of their morphology (swollen, lobed apex to elongated germ tube), although whether these structures are fully functional remains unknown. The lowest proportion formed appressoria on agar. Agar was the most hydrophilic of our test substrata, but under the conditions used the agar surface itself may have supported a film of free water (plus solutes) and in this way it may not be directly comparable to other substrata (H. Kunoh, personal communication). However, clean glass was also relatively hydrophilic and here too only a small proportion of conidia formed appressoria while the majority formed multiple short tubes only

Table 13.1. *The relative hydrophobicity of various substrata and percentages of germinated* E. graminis *f.sp.* avenae *conidia forming morphologically normal appressoria on those substrata 24–30 h after inoculation and incubation at 100% r.h.*

Substratum	Contact angle	% Germinated conidia forming appressorium
Host leaf (oat)	137°	94.9 a
Perspex	67°	77.3 b
Polystyrene	94°	50.0 c
Agar	10°	1.0 e
Glass	38°	15.3 d
Glass-silanized	71°	69.0 b

Relative hydrophobicity indicated by the contact angle made between a 0.5–0.6 μl water droplet and substratum surface (the greater the angle, the greater the hydrophobicity). Means (based on observation of at least 300 germinated conidia) followed by different letters differ significantly ($p < 0.01$) as determined by χ^2 analysis of raw count data.

(Figure 13.2). By contrast, when glass was silanized (Terhune and Hoch, 1993) by treatment with 'Repelcote' (2% dimethyldichlorosilane in 1,1,1-trichloroethane; BDH, Poole, UK), increased surface hydrophobicity was associated with a marked increase in the proportion of conidia forming appressoria. A similar proportion formed appressoria on Perspex (Figure 13.2), which had similar hydrophobicity to the silanized glass. Also on polystyrene (relatively hydrophobic) about 50% of germinated conidia formed appressoria. We assume that Perspex, silanized glass and polystyrene surfaces were chemically inert in relation to possible activity of fungal enzymes. Thus, these results indicate that, when water is not limiting, association with a relatively hydrophobic substratum may in itself be sufficient to stimulate germinating conidia to form apparently normal appressoria. Clearly, leaves provide a highly hydrophobic, inductive surface both when the EW is present (Table 13.1) or when the EW is removed (contact angle = 103°) by use of cellulose acetate glue (Carver and Thomas, 1990).

13.4.4 *The role of other substratum characteristics in appressorial germ tube development by* E. graminis

Although the relatively hydrophilic surfaces provided by glass and agar are not conducive to appressorium formation, apparently normal and functional appressoria of *E. graminis* formed with high frequency on hydrated cellulose-based membranes such as cellophane and cellulose dialysis tubing (Kobayashi *et al.*, 1991; Nicholson *et al.*, 1993). Our recent work supports these observations (see Figure 13.2) and goniometer measurement showed that these membranes have highly hydrophilic surfaces (contact angles = 12° and 27° for cellophane and dialysis tubing respectively; compare with Table 13.1 data). We suggested previously (Carver *et al.*, 1995b) that cellulose, or cellulose breakdown products released by the action of fungal enzymes, may provide the stimulus necessary for germling growth and differentiation of the appressorium. This remains a possibility worth further investigation.

13.4.5 *A tentative explanation for* E. graminis *germling development and differentiation on leaves*

E. graminis conidia germinating on leaves are in contact with a hydrophobic surface. Contact with this surface by the fungal PGT may in itself be sufficient to trigger processes involved in recognition of the surface and subsequent elongation of the appressorial germ tube which, in turn, responds to the hydrophobic surface by differentiating an appressorium. This would explain why germling development proceeds normally on inert hydrophobic surfaces such as Perspex. However, on leaves, penetration of the EW and cuticle by the PGT penetration peg, possibly facilitated by cutinases released by the fungus (Pascholati *et al.*, 1992), may allow direct fungal contact with underlying cuticular components and then with the cellulose matrix of the host cell wall. This contact may act as a second or back-up stimulus to trigger fungal development. Alternatively, breakdown products of leaf waxes, cutin or cellulose may act as signal molecules which augment the stimulus due to contact with the hydrophobic leaf surface.

13.5 *E. pisi* development following germination

The results reported in this section summarize some recent investigations in our laboratory which have not yet been reported elsewhere.

13.5.1 E. pisi *does not depend upon perception of substratum characteristics to stimulate germ tube elongation*

When conidia of *E. pisi* were held in a simulated airborne state suspended on spiders' threads (as described previously for *E. graminis*; Carver and Ingerson, 1987), a very high proportion (~ 83%) of germinating conidia produce a germ tube that elongated to around 100 μm (Figure 13.2). This is dramatically higher than the proportion of *E. graminis* conidia that form a long tube on spiders' thread (~ 7%; Carver and Ingerson, 1987) and closely approaches the proportion of conidia which produced an elongated germ tube on host pea leaves (~ 94%). Thus the processes driving germ tube elongation by *E. pisi* appear to be engaged as a simple consequence of germination and do not depend for their stimulation on contact with any form of solid surface. This is in clear contrast to the situation for *E. graminis*.

13.5.2 E. pisi *depends upon perception of substratum characteristics to stimulate appressorial germ tube differentiation*

While germ tube elongation was common for *E. pisi* germinating on spiders' threads, the elongated germ tubes never differentiated appressoria but remained as hypha-like developments without terminal swelling or lobe formation, and they were far longer (~ 100 μm) than those that formed appressoria on pea leaves (~ 25 μm; Figure 13.1). In contrast, when conidia were incubated on various artificial substrata (Table 13.2) many formed structures which we classified as appressoria because of their morphology (swollen, lobed apex to elongated germ tube), although whether these structures are fully functional remains unknown.

Table 13.2 shows that the highest proportion of germinated conidia formed appressoria on pea leaves. As for *E. graminis* on cereal leaves, this proportion was not significantly affected by removal of pea leaf EW by cellulose acetate glue treatment (as described by Carver and Thomas, 1990). Thus for *E. pisi*, as for *E. graminis*, the physical architecture of epicuticular leaf waxes was unimportant in stimulating appressorium formation. Pea leaves with or without their EW in place, were highly hydrophobic (contact angles with 0.5–0.6 μl water droplets = 136° and 102° respectively). High proportions of conidia also formed appressoria on Perspex (Figure 13.2) and silanized glass suggesting that hydrophobicity of the substratum (Table 13.1) *per se* stimulates appressorium differentiation. On clean glass, significantly fewer conidia formed appressoria than on hydrophobic artificial surfaces. However, even on this relatively hydrophilic substratum a substantial proportion of germinated conidia produced apparently normal appressoria (Figure 13.2), broadly confirming earlier findings (Kunoh *et al.*, 1992). This contrasts with the behaviour of *E. graminis* on clean glass where very few appressoria were formed (Table 13.1) and suggests that for *E. pisi* mere contact with a relatively firm surface has a stimulatory influence on appressorium formation. On agar, *E. pisi* conidia frequently formed a germ tube that elongated considerably (up to 100 μm) but, like the germ tubes formed on spiders' thread, they rarely (about 6%) formed an appressorium. However, like *E. graminis*, incubation of *E. pisi* on artificial cellulose membranes (dialysis membrane or cellophane) significantly increased the proportion of conidia forming an appressorium-like structure (Figure 13.2) to levels above those seen on glass or agar, even though all of these surfaces were relatively hydrophilic. This again indicates the possibility of a chemical involvement of cellulose-containing substrata on appressorium formation, possibly through the perception of cellulose breakdown products released by the activity of fungal

Table 13.2. *Percentages of germinated* E. pisi *conidia forming morphologically normal appressoria on various substrata* 24–30 h after inoculation and incubation at 100% r.h.*

	% Germinated conidia forming appressoria		
Substratum	Experiment 1	Experiment 2	Experiment 3
Pea leaf (EW intact)	91.3 a	–	–
Pea leaf (EW removed)	90.7 a	–	–
Perspex	76.3 b	69.0 a	60.3 a
Glass-silanized	77.0 b	–	–
Glass	53.0 c	36.3 b	47.3 b
Cellophane	–	70.3 a	–
Dialysis membrane	–	74.0 a	58.2 a
Agar (2% Technical No. 3)	–	–	6.2 c

Means (based on observation of at least 300 germinated conidia) followed by different letters differ significantly ($p < 0.01$) as determined by χ^2 analysis of raw count data.
–, substratum not included in a particular experiment.
* For information on relative hydrophobicity of substrata, see Table 13.1 and text.

enzymes. Curiously, Kunoh *et al.* (1992) reported that, although the apex of *E. pisi* germ tubes often became swollen on cellulose membranes, appressoria formed very rarely. We cannot explain this discrepancy between their result and ours.

13.5.3 *A tentative explanation for* E. pisi *germling development and differentiation on leaves*

Contact with a leaf surface is not required either for germination of *E. pisi* or to stimulate elongation of the single germ tube produced by conidia. However, differentiation of an appressorial structure by the elongated germ tube is apparently stimulated by contact with a firm surface and further favoured if the surface is hydrophobic. Leaves, where appressoria form with the highest frequency, clearly fulfil these requirements. Appressorium formation is also favoured by contact with cellulose-containing substratum. It may be that fungal enzymes released by germinating conidia erode leaf waxes, allowing the fungus access to underlying cuticular and cell wall components. This may provide an additional stimulus to appressorium differentiation, as suggested earlier for *E. graminis*.

13.6 Abnormal germling behaviour on abaxial leaf surfaces of *Lolium*

We have found that while *E. graminis* germlings develop normally on the adaxial surface of ryegrass leaves (*Lolium* spp.), conidia on the abaxial leaf surface generally produced only multiple short germ tubes, none of which elongated, and very few formed normal appressoria (Carver *et al.*, 1990). The EW on the abaxial leaf surface form amorphous sheets, quite unlike the plate form of wax found on the adaxial leaf surface. If the abaxial sheet waxes were removed prior to inoculation, second-formed germ tubes elongated perfectly normally and formed appressoria. Analysis showed that abaxial wax had a higher content of long-chain aldehydes than adaxial wax which was rich in alkyl esters and 1-alkanols (P.J. Holloway and T.L.W. Carver, unpublished observations). The nature of the amorphous sheet wax present on the abaxial surface of *Lolium* leaves appears to interfere with processes necessary to normal germling development. We had previously suggested that the chemistry or thickness of the abaxial wax may have prevented fungal PGTs making contact with some unknown 'inductive factor' lying beneath the wax (Carver *et al.*, 1990; Carver *et al.*, 1995b). In the light of our recent studies this explanation no longer seems adequate. Appressoria form normally on inert, hydrophobic, artificial surfaces (see Table 13.1), and the abaxial surface of *Lolium* leaves with EW present is hydrophobic (contact angle = 97°). Thus, we have a situation on the abaxial surface of *Lolium* leaves where a high proportion of *E. graminis* conidia should form appressoria if the leaf surface is acting like an inert, hydrophobic artificial surface. The failure of many *E. graminis* conidia to form appressoria on this surface suggests that the EW may in fact contain components which are inhibitory to germling development by *E. graminis*.

Recently, we observed the behaviour of *E. pisi* on the abaxial surface of *Lolium* leaves. More than 94% of germinated *E. pisi* conidia formed apparently normal appressoria in this situation; that is, early fungal development was as successful as on pea leaves. Whatever the factor that inhibits *E. graminis* development, it has no apparent effect on *E. pisi*.

13.7 Concluding remarks

Host leaf surface characteristics promote developmental processes leading from germination to infection structure differentiation by both *E. graminis* and *E. pisi*. Hydrophobicity of the surface may be important to development by both organisms. While surface hydrophobicity has no apparent influence on the development of some fungi, for others it is known to stimulate both the adhesion of spores to surfaces (see e.g. Braun and Howard, 1994; Nicholson, 1996; Terhune and Hoch, 1993), and appressorium differentiation by several plant pathogens including *Botrytis cinerea* (Hill *et al.*, 1980), *Uromyces appendiculatus* (Terhune and Hoch, 1993), and *Magnaporthe grisea* (Lee and Dean, 1993), and the entomopathogenic fungus *Metarhizium anisopliae* (St Leger *et al.*, 1989). For *E. graminis* and *E. pisi* the situation is complicated by the fact that appressoria also differentiate on certain hydrophilic surfaces. However, it is well established for a number of fungi that a variety of apparently different 'signals' (e.g. chemical, topographical and temperature) can stimulate infection structure differentiation (see Hoch and Staples, 1991). Thus, the ability of *E. pisi* to differentiate modestly well on glass may reflect stimulation due simply to contact with a firm surface. Nevertheless, enhancement of appressorium formation seen when either *E. graminis* or *E. pisi* germinate on cellulose membranes, suggests that chemical signals may be derived from the contact, possibly through the action of degradative fungal enzymes.

It was apparent that appressoria were formed more frequently by germlings of both *E. graminis* and *E. pisi* when the fungi were on host leaves than when on any artificial surfaces. This suggests that leaves may provide a combination of signals, acting in an additive or interactive fashion to stimulate fungal development. The layered nature of the leaf surface (EW overlying cuticular membrane (CM) overlying cell wall), may require penetration by the fungus through the various layers in order to access different potential signalling sites. This may be facilitated by:

(i) the early release of conidial exudates containing enzymes including cutinase, which help 'prepare the infection court' (Kunoh *et al.*, 1988; Nicholson *et al.*., 1988; Pascholati *et al.*, 1992);
(ii) in the case of *E. graminis* by production of the PGT penetration peg;
(iii) possibly by the concentration of degradative fungal enzymes in extracellular material secreted by fungal germ tubes (Carver *et al.*, 1995a, b).

It is interesting to speculate which of the two fungi is more primitive. Does the complexity of *E. graminis* development indicate greater sophistication? What could be the evolutionary advantage of this complexity? On some artificial surfaces *E. pisi* forms an appressorial germ tube when *E. graminis* generally fails to do so. Perhaps this gives *E. graminis* conidia that are originally deposited on surfaces other than leaves an additional opportunity to be re-transported to a host leaf.

It has been argued (e.g. Hoch and Staples, 1991) that by understanding the processes that drive pathogen recognition of a potential host and lead to the differentiation of infection structures, we may be able to devise lasting disease control strategies. In this context it is somewhat discouraging to find that *Erysiphe* recognizes such ubiquitous characteristics as hydrophobicity and the presence of cellulose. However, it is encouraging, for *E. graminis* at least, that characteristics such as those present in the abaxial surface of *Lolium* leaves, can, and do, impede fungal development. Perhaps such characteristics could be exploited in plant breeding.

Acknowledgements

We thank Professors Hitoshi Kunoh and Ralph Nicholson for their helpful advice and discussions during the execution of this work and writing of the paper. Work partly funded by the UK Ministry of Agriculture Fisheries and Food.

References

Aist, J.R. (1976) Cytology of penetration and infection-fungi. In: *Physiological Plant Pathology* (eds R. Heitefuss and P.H. Williams). Springer-Verlag, Berlin, pp. 197–221.

Ayres, P.G. (1983) Conidial germination and germ-tube growth of *Erysiphe pisi* in relation to visible light and its transmission through pea leaves. *Trans. Br. Mycol. Soc.* **81**, 269–274.

Braun, E.J. and Howard, R.J. (1994) Adhesion of *Cochliobolus heterosporus* conidia and germlings to leaves and artificial surfaces. *Exp. Mycol.* **18**, 211–220.

Bushnell, W.R. (1972) Physiology of fungal haustoria. *Annu. Rev. Phytopathol.* **10**, 151–176.

Carver, T.L.W. and Bushnell, W.R. (1983) The probable role of primary germ tubes in water uptake before infection by *Erysiphe graminis*. *Physiol. Plant Pathol.* **23**, 229–240.

Carver, T.L.W. and Ingerson, S.M. (1987) Responses of *Erysiphe graminis* germlings to contact with artificial and host surfaces. *Physiol. Mol. Plant Pathol.* **30**, 359–372.

Carver, T.L.W. and Thomas, B.J. (1990) Normal germling development by *Erysiphe graminis* on cereal leaves freed of epicuticular wax. *Plant Pathol.* **39**, 367–375.

Carver, T.L.W., Thomas, B.J., Ingerson-Morris, S.M. *et al.* (1990) The role of the abaxial leaf surface waxes of *Lolium* spp. in resistance to *Erysiphe graminis*. *Plant Pathol.* **39**, 573–583.

Carver, T.L.W., Thomas, B.J. and Ingerson-Morris, S.M. (1995a) The surface of *Erysiphe graminis* and the production of extracellular material at the fungus–host interface during germling and colony development. *Can. J. Bot.* **73**, 272–287.

Carver, T.L.W., Ingerson-Morris, S.M., Thomas, B.J. *et al.* (1995b) Early interactions during powdery mildew infection. *Can. J. Bot.* **73** (Suppl. 1): S632–S639.

Gay, J.K., Martin, M. and Ball, E. (1985) The impermeability of powdery mildew conidia and their germination in arid environments. *Plant Pathol.* **34**, 353–362.

Hill, G., Stellwaag, F. and Schlosser, E. (1980) Substrate surface and appressoria formation by *Botrytis cinerea*. *Phytopathol. Z.* **99**, 186–191.

Hoch, H.C. and Staples, R.C. (1991) Signalling for infection structure formation in fungi. In: *The Fungal Spore and Disease Initiation in Plants and Animals* (eds G.T. Cole and H.C. Hoch). Plenum Press, New York, pp. 25–46.

Kobayashi, I., Tanaka, C., Yamaoka, N.*et al.* (1991) Morphogenesis of *Erysiphe graminis* conidia on artificial membranes. *Trans. Mycol. Soc. Jpn.* **32,** 187–198.

Kunoh, H. (1981) Early stages of infection process of *Erysiphe graminis* on barley and wheat. In: *Microbial Ecology of the Phylloplane* (ed. J.P. Blakeman). Academic Press, London, pp. 85–101.

Kunoh, H., Tsuzuki, T. and Ishizaki, H. (1978) Cytological studies of the early stages of powdery mildew in barley and wheat. IV. Direct ingress from superficial primary germ tubes and appressoria of *Erysiphe graminis hordei* on barley leaves. *Physiol. Plant Pathol.* **13,** 327–333.

Kunoh, H., Aist, J.R. and Israel, H.W. (1979a) Primary germ tubes and host cell penetrations from appressoria of *Erysiphe graminis hordei. Ann. Phytopathol. Soc. Jpn* **45,** 326–332.

Kunoh, H., Itoh, O., Kohno, M. *et al.* (1979b) Are primary germ tubes of conidia unique to *Erysiphe graminis? Ann. Phytopathol. Soc. Jpn* **45,** 675–682.

Kunoh, H., Yamaoka, N., Yoshioka, H. *et al.* (1988) Preparation of the infection court by *Erysiphe graminis*. II. Contact-mediated changes in morphology of the conidium surface. *Exp. Mycol.* **12,** 325–335.

Kunoh, H., Nicholson, R.L., Yoshioka, H. *et al.* (1990) Preparation of the infection court by *Erysiphe graminis*: Degradation of the host cuticle. *Physiol. Plant Pathol.* **36,** 397–407.

Kunoh, H., Toyoda, K., Yamaoka, N. *et al.* (1992) Morphogenesis of *Erysiphe pisi* conidia on artificial substrata. *Trans. Mycol. Soc. Jpn* **33,** 87–93.

Lee, Y.-H. and Dean, RA. (1993) cAMP regulates infection structure formation in the plant pathogenic fungus *Magnaporthe grisea. Plant Cell* **5,** 693–700.

Manners, J.G. and Hossain, S.M. (1963) Effects of temperature and humidity on conidial germination of *Erysiphe graminis. Phytopathology* **53,** 1378–1380.

Nicholson, R.L. (1996) Adhesion of Fungal Propagules: Significance to the success of the fungal infection process. In: *Histology, Ultrastructure and Molecular Cytology of Plant-Micro-organism Interactions* (eds M. Nicole and V. Gianinazzi-Pearson). Kluwer Academic Publishers, Dordrecht (in press).

Nicholson, R.L., Yoshioka, H., Yamaoka, N. *et al.* (1988) Preparation of the infection court by *Erysiphe graminis*. II. Release of esterase enzyme from conidia in response to a contact stimulus. *Exp. Mycol.* **12,** 336–349.

Nicholson, R.L., Kunoh, H., Shiraishi, T. *et al.* (1993) Initiation of the infection process by *Erysiphe graminis*: Conversion of the conidial surface from hydrophobicity to hydrophilicity and influence of the conidial exudate on the hydrophobicity of the barley leaf surface. *Physiol. Mol. Plant Pathol.* **43,** 307–318.

Pascholati, S.F., Yoshioka, H., Kunoh, H. *et al.* (1992) Preparation of the infection court by *E. graminis* f.sp. *hordei:* cutinase is a component of the conidial exudate. *Physiol. Mol. Plant Pathol.* **41,** 53–59.

Singh, U.P. and Singh, H.B. (1983) Development of *Erysiphe pisi* on susceptible and resistant cultivars of pea. *Trans. Br. Mycol. Soc.* **81,** 275–278.

Staub, I., Dahmen, H. and Schwinn, F.J. (1974) Light and scanning electron microscopy of cucumber and barley powdery mildew on host and non host plants. *Phytopathology* **64,** 364–372.

St.Leger, R.J., Butt, T.M., Goettel, M.S. *et al.* (1989) Production *in vitro* of appressoria by the entomopathogenic fungus *Metarhizium anisopliae. Exp. Mycol.* **13,** 274–288.

Terhune, B.T. and Hoch, H.C. (1993) Substrate hydrophobicity and adhesion of *Uromyces* urediospores and germlings. *Exp. Mycol.* **17,** 241-252

Yang, S.L. and Ellingboe, A.H. (1972) Cuticle layer as a determining factor for the formation of mature appressoria of *Erysiphe graminis* on wheat and barley. *Phytopathology* **62,** 708–714.

14

The presence of water on leaf surfaces and its importance for microbes and insects

D.R. Butler

14.1 Introduction

Plant pathologists have long recognized the importance of surface water on leaves to disease development. The need of water for spore germination of *Septoria graminum* was described by Voglino (1904) and several disease warning schemes which used weather variables related to leaf wetness were proposed more than 50 years ago (e.g. Everdingen, 1926; Mills, 1944). While schemes such as these (and many more subsequently) provided a basis for spray advice to growers over many years, their reliability has been questioned. Clear limitations have been the accuracy with which leaf wetness periods can be estimated, and a need to quantify the effect of wetness on the response of pathogens.

This situation led to an interest in measuring leaf wetness periods and a variety of methods have been reported over the last 60 years, using mechanical devices (e.g. Hirst, 1957; Stephan and Mildebrath, 1938), electronic sensors (e.g. Davis and Hughes, 1970; Huband and Butler, 1984) and electrodes on leaves (e.g. Häckel, 1974; Winters and Small, 1934). Reviews of reports on leaf wetness sensors and their design are given by Huber (1992) and Sutton *et al.* (1988).

Models have also been used to estimate leaf wetness periods. These range from simple empirical relationships with weather variables such as relative humidity, to physically based simulation models to calculate evaporation of surface water. These include models which relate to single drops (Butler, 1985; Zhang and Gillespie, 1990), populations of drops (Brain and Butler, 1985), dew on leaves (Gillespie and Barr, 1984; Pedro and Gillespie, 1982), wetness on fruit (Butler, 1980; Huber *et al.*, 1991; Monteith and Butler, 1979) and complete crop canopies (Butler, 1986; Huber, 1988; Huber and Itier, 1990; Norman, 1982; Thompson, 1981). A thorough review of both empirical and physical models is available (Huber and Gillespie, 1992), so this paper will only consider certain aspects of physically based simulation models and highlight some of the difficulties of assessing leaf wetness.

Plant Cuticles, edited by G. Kerstiens.

The response of pathogens to surface water is highly varied and, to utilize information on leaf wetness duration, the response of organisms needs to be quantified. In recent years a large number of reports have been published which include the response of pathogens to leaf wetness; most with respect to infection and an increasing number on sporulation. In this chapter a selection of these reports are discussed with a view to highlighting the diverse nature of plant pathogens.

Reports on the effect of surface water on insect pests are much less common than those for pathogens. There is evidence however, of leaf wetness playing a crucial role in the life cycle of certain insects, especially in the tropics. Such knowledge is important for integrated pest management approaches to pest control.

14.2 Wetness variability

The distribution of water on leaf surfaces varies greatly, depending on leaf morphology, the wettability of the cuticle, and leaf orientation. The degree to which a given volume of water spreads over the surface depends on the contact angle. This is defined as the angle between the tangent plane to the surface of the water and the tangent plane to the surface of the leaf at any point along their line of contact (see also Section 15.3.1). Contact angles vary greatly between plant species and, within a species, there is often a strong effect of leaf age. Young leaves commonly have larger contact angles than older ones, the change being caused by damage of superficial wax (Holloway, 1970). Published values of contact angles for pure water on leaves range from less than 30° (Fogg, 1947) to 170° (Holloway, 1969). Crops such as faba beans and potato with characteristically small contact angles, have wettable leaves and water spreads over their surface as a continuous film. Others such as onions, wheat and barley have characteristically large contact angles, so surface water forms discrete and often tightly formed drops.

Ward (1988) compared surface wetness after rain on six adjacent crops to assess the variability in drying times. She used a visual scale to assess leaf wetness, from 1 for complete dryness to 5 for leaves saturated with water. She recorded wetness on the youngest fully expanded leaves at the top of the canopy and the oldest green leaves at the base of the canopy. The observations were started within 1 h of a rainfall event and, in all cases, less water was held on the young leaves than the old ones from the outset. The young leaves were usually more hydrophobic, but they were also well exposed to radiation and air movement, leading to relatively fast evaporation rates and rapid drying.

Observations were made on two days with contrasting weather conditions. On the first day (22 July 1987) it was cloudy (the mean solar radiation was 177 W m^{-2}) and the vapour pressure deficit was 0.27 kPa, so evaporation rates would have been slow. There were large differences between crops in the drying times both for young and old leaves (Table 14.1). Beans and potatoes dried most rapidly and peas and beet dried more than 5 h later. The leaves of both beans and potatoes are wettable, so intercepted water spread to provide a large surface area for evaporation. Pea leaves have a large contact angle (158°) so water accumulated as tightly formed drops, minimizing the exposed surface area for evaporation. Young leaves of beet are vertically orientated and more hydrophobic than old ones. Water was held as drops mainly in hollows and around the leaf margins. The old leaves are large and prostrate, with an extremely undulating surface in which pools of water collected and persisted for a long time.

On the second day (3 August 1987) it was sunny (the mean solar radiation was 570 W m^{-2}) and the vapour pressure deficit was 0.47 kPa, so the conditions favoured

Table 14.1. *Crop wetness periods following rain (Ward, 1988)*

	Youngest leaf (h)		Oldest leaf (h)	
	22/7/87	3/8/87	22/7/87	3/8/87
Onions (*Allium cepa*)	4.0	3.7	2.5	4.5
Beet (*Beta vulgaris*)	9.0	4.3	>10.0	5.3
Linseed (*Linum usitatissimum*)	4.5	3.5	9.0	4.7
Beans (*Vicia faba*)	<1.0	3.5	4.3	4.3
Peas (*Pisum arvense*)	4.5	4.3	>10.0	5.5
Potatoes (*Solanum tuberosum*)	<1.3	2.5	4.3	3.7

fast evaporation rates. On this occasion all the crops dried within 2 h of each other, and the difference in the drying times of the top and bottom leaves was about 1 h.

These two examples demonstrate the magnitude of variation of leaf wetness periods between crops. Differences depend not only on the crop characteristics but also on the weather conditions. It is also clear that the amount of intercepted water plays an important role in determining the leaf wetness duration; for example, the persistence of water on mature beet leaves was due mainly to the large quantity of water held.

To calculate the duration of leaf wetness it is convenient to divide the wetness period into wetting and drying phases. For dew the wetting phase occurs when the leaf temperature is less than the dew point temperature of the surrounding air, so water condenses on the leaf surface. For rain (or sprinkler irrigation) the wetting phase starts at the onset of precipitation and the amount of water held on the crop increases until canopy saturation is reached. When dew formation or precipitation ends, the quantity of water held on the leaves and the evaporation rate determine the length of the drying phase.

14.3 Dew

In humid climates dew is usually confined to the uppermost leaves in crop canopies which are exposed to radiative cooling at night. Wetness from dew is normally restricted to a thin layer (either a film of water or small uniformally sized drops), but in the humid tropics the amount of dew on wettable leaves commonly exceeds their water-holding capacity. Large water drops then fall from the top of the canopy and wet the lower foliage. In more arid climates, the main source of water for dew formation is from the soil. At night water evaporates from the soil and condenses onto foliage in the canopy, sometimes on lower leaves which dry more slowly than those at the top (Weiss *et al.*, 1989).

A model to estimate accumulation of dew and its evaporation from the leaf energy balance can predict wetness periods to within 1 h of observed values (Pedro and Gillespie, 1982). This model has been adapted to different crops (Gillespie and Barr, 1984) and gave satisfactory predictions of dew occurrence and duration in locations of contrasting climates (Scherm and van Bruggen, 1993). The model was found to be most sensitive to cloud cover and humidity in coastal areas and to humidity and wind speed in an interior valley of California. A simplified multi-layer model for dew suggested by Weihong and Goudriaan (1991) restricted energy balance calculations to the night-time, but handled distillation from the soil. The quantity of dew formed at night was related empirically to the duration of the drying phase after sunrise. Both these energy balance models are sensitive to cloud cover (which may change during the night), so their adoption will be limited by the ease with which cloud cover data can be obtained.

14.4 Rainfall interception

The way that water is intercepted during rain depends on the rainfall intensity. Mist or fog can be thought of as extremely low-intensity rain, when numerous small drops collect on leaves throughout the crop canopy. Initially there would be little redistribution of the water, but with continued interception the small drops would coalesce sufficiently for water to drip from the leaves. With extremely light rain (as opposed to mist) the wetting pattern for individual leaves would be similar, but exposed leaves at the top of the canopy would intercept more water than lower (sheltered) leaves since the rain would be falling.

During heavy rain, drops that collide with leaves break up and are dispersed as numerous droplets, some of which adhere to the leaf, some are carried by the wind and may leave the crop canopy, and others fall to lower leaves or the ground. The impact of large drops falling at terminal velocity causes leaves to vibrate, so water may be shaken from the leaf surface, limiting the water-holding capacity of leaves in these conditions.

To permit calculations of leaf wetness periods due to rain, there is a need to quantify interception. Many studies of rainfall interception by vegetation have been made in forests (Rutter, 1975) and there are some reports for grasses and cereals (e.g. Beard, 1962; Burgy and Pomeroy, 1958; Butler and Huband, 1984; Butler and King, 1982; Couturier and Ripley, 1973; Merriam, 1961). Couturier and Ripley (1973) proposed the equation of an exponential asymptote to describe the wetting of a canopy. This assumed total capture of rain by the canopy initially, with an asymptote equal to the canopy storage capacity. Results from field measurements in wheat however, indicated the need for an additional term to account for an initial interception efficiency of less than one (Butler, 1983). The resulting equation is:

$$I = I_c(1 - \exp(-E_I R/I_c)), \tag{14.1}$$

where I is the amount of intercepted water, I_c is the canopy storage capacity, E_I is the interception efficiency and R is the amount of rainfall. Values of $E_I = 0.7$ and $I_c = 2.3$ mm were suitable for wheat canopies with a leaf area index of about 3 (Royle and Butler, 1986; see Figure 8.1). This implies that the depth of water per unit leaf area was 0.8 mm, held largely on the adaxial surface. This value is consistent with estimates for single leaves, obtained from measurements of drop size and number, using photographs of saturated leaves, which ranged from 0.5 to 1.0 mm (Butler, 1983).

14.4.1 *The effect of rainfall intensity*

As rainfall intensity increases, the size of raindrops and their kinetic energy increases. When small, low-energy drops fall onto a leaf, they are likely to remain on the surface provided it is not already saturated with water. Large, high-energy drops however will break up and be redistributed, and could shake previously intercepted water from the leaf. One would expect therefore, the efficiency of interception to change with rainfall intensity. Butler (1983) monitored throughfall beneath a barley

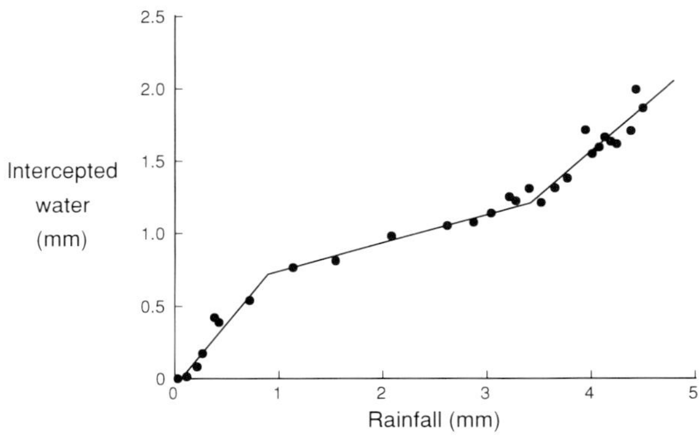

Figure 14.1. *The effect of rainfall intensity on interception for a barley canopy. Intensities were 0.12-0.71 mm h^{-1} (<1 mm rainfall); 1.80 mm h^{-1} (1-3 mm rainfall) and 0.47 mm h^{-1} (> 3 mm rainfall).*

canopy to determine the dynamics of interception from the difference between incident rainfall and throughfall. Measurements on 12–13 July 1980 provided an example with contrasting periods of rainfall intensity (Figure 14.1). The data fall into three groups, when the cumulated rainfall was (a) less than 1.0 mm, (b) between 1.0 and 3.0 mm and (c) greater than 3.0 mm. The efficiency of interception for each group, given by the slopes of fitted lines varies between 0.2 and 0.75 and the values are inversely related to rainfall intensity (Table 14.2).

Table 14.2. *Rainfall interception efficiencies in barley at different rainfall intensities*

Group	Interception efficiency	Rainfall intensity (mm h^{-1})
a	0.75	0.12–0.71
b	0.20	1.80
c	0.56	0.47

The groups refer to data in Figure 14.1 when the rainfall totals were: (a) <1.0 mm; (b) 1.0–3.0 mm; (c) > 3.0 mm.

14.4.2 *The effect of wind speed*

When water collects on leaves as discrete drops, much larger amounts can be held in calm conditions than when it is windy. Strong wind buffets the crop and intercepted water would be shaken from the leaves. This effect is illustrated with measurements (Butler, 1983) which compare the pattern of interception in similar wheat canopies on two occasions with slow and fast wind speeds (Figure 14.2). In this example the effective storage capacity of the canopy was reduced from about 2.3 mm with slow wind (open circles) to 0.5 mm when the mean wind speed was 3.9 m s^{-1} (closed circles). Butler (1983) also observed that the mean interception efficiency in cereals on five occasions when the wind speed was greater than 2 m s^{-1}, was 60–70% lower than the mean efficiency on nine occasions when the wind speed was less than 2 m s^{-1}. The effect of increasing wind speed therefore, was to

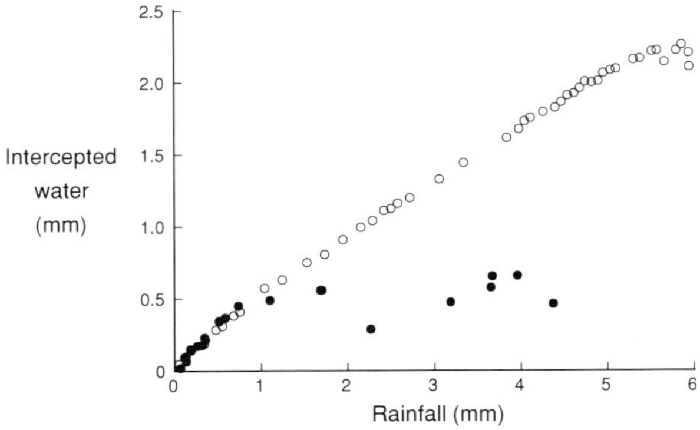

Figure 14.2. *The effect of wind speed on interception in wheat canopies with a leaf area index of 3. On both occasions rain fell at night when evaporation from the canopy would have been small. (Open circles, wind speed < 2 m s^{-1}; closed circles, wind speed = 3.9 m s^{-1} (from 0.5 to 4.5 mm rainfall).)*

increase the drainage rate from the cereal canopies. Hall (1985) found no evidence of any change in the drainage rate from heather at wind speeds of 2.8 and 4.8 m s^{-1}. This was because the leaves of heather are completely wettable and water is held firmly as a film which is not easily disturbed by wind.

It is clear from the discussion above that both the interception efficiency and the effective canopy storage capacity may vary considerably with the prevailing weather conditions. This should be taken into account when estimating leaf wetness duration, but there are few data available to quantify the effects illustrated above. To compare the storage capacity of different canopies it is convenient to express the values as equivalent depth of water per unit leaf area. Values for vegetation (other than forests) vary by a factor of 10 (Table 14.3).

Table 14.3. *Estimates of storage capacities for different canopies expressed as equivalent depth of water per unit leaf area*

Canopy	Depth of water (mm)	Reference
Wheat (*Triticum aestivum*)	0.80	Butler (1983)
Maize (*Zea mays*)	0.15	Norman and Campbell (1983)
Apple (*Malus sylvestris*)	0.06	Miranda and Butler (1986)
Annual ryegrass (*Lolium rigidum*)	0.25	Merriam (1961)
Big bluestem (*Andropogon furcatus*)	0.39	Clark (1940)
Beans (*Vicia faba*)	0.07	Huber (1988)
Heather (*Calluna vulgaris*)	0.60	Hall (1985)

14.5 Evaporation of intercepted water from canopies

A number of simulation models have been written to calculate evaporation of surface water from different heights within plant canopies (Butler, 1986; Huber, 1988; Norman, 1982; Thompson, 1981). Results from these models not only provide a good insight into the extent that leaf wetness varies with height for a variety of canopies in different environments, but also highlight some of the difficulties of estimating leaf wetness periods. Norman and Campbell (1983) emphasized the influence of soil evaporation on the canopy microclimate (and therefore on leaf wetness), and observed that it is highly variable, depending on atmospheric conditions, as well as soil and crop characteristics. The most comprehensive model is called CUPID (Norman, 1982), where the process of rainfall interception is considered, including stem flow and wetting of lower leaves in the canopy by drainage from upper leaves. Tests with maize showed that evaporation of intercepted water was underestimated by CUPID. The model works well for the simulation of dew however, and predicts condensation of water on leaves low in the canopy (by distillation of soil water) which agree with measured surface wetness (Weiss *et al.*, 1989).

The conduction of heat to discrete drops from the surrounding leaf significantly affects their evaporation rate (Butler,1985). This was therefore included in a model to simulate evaporation of drops from non-wettable leaves in a cereal canopy (Butler, 1986). The model was intended for use after rain to assess the persistence of raindrops on leaves. Comparisons between the model predictions and observed drying times of leaves, gave good agreement for leaves near the top of the canopy, but evaporation in the lower half of the canopy was often underestimated (Butler, 1983). Exceptions to this were thought to be associated with an uncertainty of the distribution of intercepted water at the end of a rainfall event. The difficulty of estimating the amount of rainwater held on leaves is a major constraint to the use of this type of model. When water is held as discrete drops, the effect of wind may dominate rainfall interception (e.g. see Figure 14.2) so that realistic estimates are not possible if wind speed data are not available.

Results from a canopy evaporation model for films of water on leaves of faba beans (Huber, 1988), illustrated different patterns of leaf wetness at the top and bottom of the canopy. When this model was modified to include conduction of heat from dry to wet parts of the leaf (Huber and Itier, 1990), there was a statistical improvement in the agreement between observed and predicted wetness periods. It is likely that the underestimation of evaporation of intercepted water by the CUPID model (Norman and Campbell, 1983) would be reduced by incorporating heat conduction into the equations.

To relate leaf wetness to disease in practice, the maximum duration of water drops deep in the canopy is of value (Barr and Gillespie, 1987). For this a multi-layer canopy model is not needed, and a much simpler model using inputs from a nearby weather station was suggested by Zhang and Gillespie (1990). They obtained good agreement between observed and calculated drying times (within 14%) for the majority of drops studied. However, the success of routine applications with this type of model would still depend on accurate estimates of rainfall interception and knowledge of its distribution within the crop canopy.

14.6 Surface water and plant pathogens

14.6.1 *Infection*

Bacteria infect vegetation either through stomata, lenticels, nectaries or hydrathoids. They require the presence of surface water to infect because they are unicellular and, if motile, are only motile in water (Manners, 1982).

Foliar fungal pathogens are dispersed as asexual or sexual spores which must germinate and penetrate the host to cause infection. A source of water is needed for spore germination and germ tube growth and, for the majority of foliar pathogens, this requirement is met by leaf wetness. Powdery mildews are an exception to this general rule; in this case spore germination gradually increases from 0% relative humidity (r.h.) to saturation, but in liquid water germination is less than 1%. Germ tube growth however, is only substantial at 98–100% r.h. (Manners and Hossain, 1963). The validity of other reports of spore germination without liquid water has been questioned because of the difficulty in avoiding transient condensation in controlled conditions when the air is near saturation (Schein, 1964). However, the careful experimentation of Clayton (1942) distinguished spores which germinated on a dry surface (*in vitro*) at relative humidities of 99% and above (e.g. *Venturia inaequalis*) from those which require liquid water (e.g. *Sclerotinia fructicola*). For *in vivo* studies, the humidity of air within the leaf boundary is even more difficult to control, but a recent report provides convincing evidence that *Botrytis cinerea* on rose petals germinates with a dry surface at a high relative humidity (Williamson *et al.*, 1995). The following discussion will be restricted to those pathogens which require liquid water for infection.

When airborne spores are deposited on dry leaves, they will remain inactive until the leaf surface is wetted by dew or precipitation. Leaf wetness can be thought of as a switch to start the infection process (germination, germ tube growth and host penetration). For many pathogens, the relationship between infection and surface wetness duration at a given temperature can be described by an asymptotic exponential (Figure 14.3). There is a minimum period (m) within which no infection occurs followed initially by a rapid increase in disease and then by a slower increase as the asymptote is approached. The equation for this response curve given by Butler and Jadhav (1991) is:

$$D = D_{max}(1 - \exp(-b(W - m))), \quad (14.2)$$

where D_{max} is the maximum amount of disease and b is the rate of disease increase with increasing wetness period (this determines the time when the disease approaches the asymptote). Values of the parameters m and b vary widely for different pathogens, and for a particular pathogen may vary with temperature. Values of m and the time taken to reach 90% of D_{max} at the optimum temperature for a selection of pathogens are given in Table 14.4. For groundnut rust (*Puccinia arachidis*), m was shown to increase and D_{max} was shown to decrease at sub-optimal or super-optimal temperatures (Butler and Jadhav, 1991).

After spore germination and prior to host penetration, leaf surface conditions may be critical to survival of the pathogen. With groundnut rust, for example, an

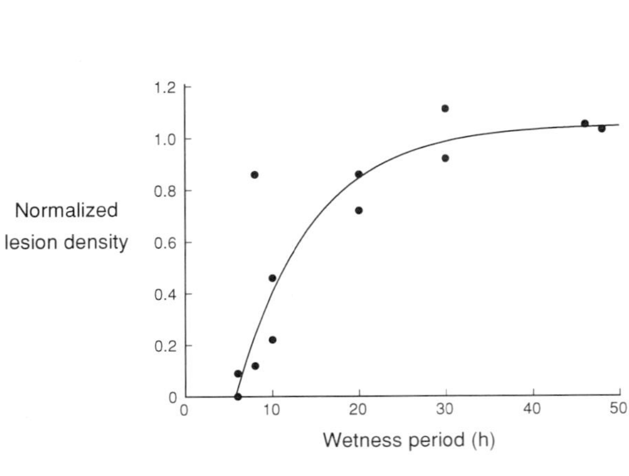

Figure 14.3. *The relationship between groundnut rust lesion density and leaf wetness periods at 20°C. (Redrawn from Butler and Jadhav (1991). Requirements of leaf wetness and temperature for infection of groundnut by rust.* Plant Pathol. ***40:*** *395–400, with permission from Blackwell Science Ltd.)*

Table 14.4. *The minimum wetness period* (m) *for infection to occur and the period for 90% of maximum infection* D_{max} *for selected pathogens*

Pathogen	m	90% D_{max}	Source
Phytophthora cactorum	< 0.5	1	Grove *et al.* (1985)
Puccinia striiformis	3	10	Dennis (1987)
Phytophthora infestans	4	11	Bashi and Rotem (1974)
Colletotrichum coccodes	7	47	Dillard (1989)
Pyrenopeziza brassicae	13	51	Figueroa *et al.* (1995)
Cercospora carotae	20	96	Carisse and Kushalappa (1990)
Phaeoisariopsis personata	20	130	Butler *et al.* (1994)

interruption in leaf wetness of just 1 h after inoculation substantially reduced infection (Butler and Jadhav, 1991). Infection by *Botrytis squamosa* on onion was also reduced with a dry period of 2 h or less, but the extent of reduction was very sensitive to the timing of the dry period (Alderman *et al.*, 1985). Pathogens of this type which are sensitive to desiccation must be able to germinate and penetrate the host rapidly, since periods of continuous leaf wetness much longer than 12 h are not common in nature.

Other pathogens require longer wetness periods for substantial infection and, to survive, must be able to withstand interruptions in leaf wetness; for example, *Septoria tritici* on wheat requires wetness periods in excess of 48 h for maximum infection, and Shaw (1991) found that substantial infection could still occur when wet periods were interrupted by 48 h of dryness. Breaks in leaf wetness usually reduced infection, but the effects varied with the host genotype, humidity during the dry period and there was a light interaction. For *Stemphyllium botryosum* on tomato, similar amounts of infection occurred with continuous and interrupted wetness (Bashi and Rotem, 1974).

The *Cercospora* species grow extensively on the leaf surface in wet conditions, but substantial host penetration may not occur for several days (Beckman and Payne, 1982). Such slow growth is exemplified by late leaf spot of groundnut (*Phaeoisariopsis personata*), where the minimum wetness period for infection (m in Equation 14.2) is about 20 h and the number of lesions reaches 90% of D_{max} after 130 h of wetness (Butler *et al.*, 1994). When leaf wetness is interrupted by 4 h day^{-1} or more, three times as many lesions develop than with continuous wetness. The increase in lesion number results from increased branching of germ tubes, tropism towards stomata and more host penetration (Wadia and Butler, 1994). Evidence of similar germ tube branching and tropism has been shown for *Cercospora beticola* on sugar beet (Rathaiah, 1977), *Cercospora zeae-maydis* on maize (Beckman and Payne, 1982) and *Cercospora arachidicola* on groundnut (D.R. Butler, personal observation). Extensive germ tube growth with continuous wetness and stomatal tropism associated with interrupted wetness has also been reported for *Cercospora musae* on banana (Goos and Tschirch, 1963). In experiments with *Cercospora carotae* on carrot, there was more infection with interrupted wetness than continuous wetness when the total wetness period was 36 h and the dry period was between 3 and 24 h. After an initial wet period of 6 h however, infection occurred without further surface water, and increased with r.h. from 84 to 100% (Carisse and Kushalappa, 1992).

The discussion above alludes to the wide variation in strategies of pathogens to adapt to wetness in the natural environment. Fungal pathogens could be classified into three groups. In one, spores germinate and penetrate the host very quickly to avoid desiccation. The second group are able to withstand prolonged dry periods, so take advantage of leaf wetness when it occurs. The third group depends on alternate wet and dry periods to provide a stimulus for invasion of the host.

14.6.2 *Sporulation and dispersal*

Many plant pathogens produce spores in mucilage which swells in contact with water and is extruded on to the surface of the host. Mucilaginous spores, which are dispersed predominantly by rain (Fitt *et al.*, 1989), include pycnidia (e.g. *Septoria nodorum* on wheat, Griffiths and Ao, 1976; *Ascochyta rabiei* on chickpea, Nene and Reddy, 1987), conidia on stromata (e.g. *Pseudocercosporella herpotrichoides* (eyespot) on cereals, Glynne, 1953; *Rhynchosporium secalis* on barley, Steadman, 1980) or conidia in acervuli (e.g. *Colletotrichum graminicola* on cereals, Nicholson and Moraes, 1980). Some fungi, such as *Phytophthora megakarya* on cocoa release motile zoospores into surface water and these may also be splash-spread.

Splash-dispersal occurs most commonly during rain (or overhead irrigation), but large drops falling from leaves higher in the canopy may also be important, especially in conditions of light rain, mist or fog. Spores are carried in secondary droplets formed at the time of impact of primary drops. Most spores are found in the largest droplets (> 1 mm diameter) and very few in droplets less than 100 μm diameter (Fitt *et al.*, 1989). A small proportion of spores, carried in extremely small droplets become airborne and may be carried long distances by wind (Carnegie, 1980).

Spores that are characteristically dispersed by wind may also be affected by rain. Large increases in airborne spores are commonly observed at the onset of rain or during short-lived showers (e.g. Savary and Janeau, 1986; Smith and Crosby, 1973). This occurs when leaves are shaken by raindrops, or when spores are dislodged by the puff of air preceding a falling raindrop (Hirst and Steadman, 1963).

Surface wetness may be required for spore production and liberation of airborne spores. Dry ascospores of *Venturia inaequalis* are actively liberated when the asci absorb water and swell (Hirst *et al.*, 1955). McCartney and Lacey (1990) found that rain was essential to the production and release of ascospores in *Pyrenopeziza brassicae*, but suggested that dew may sustain production for several days after rain. Airborne spores of *Crinipellis perniciosa* (the cause of witches' broom disease of cocoa) are released from turgid basidiocarps on dead brooms. These fruiting bodies are only produced after days with rain and with intermittent broom wetness (Rudgard, 1986).

Other pathogens may sporulate when the leaf surface remains dry, but only when the r.h. exceeds a threshold value. Harrison and Lowe (1989) found large numbers of sporangia of *Phytophthora infestans* on potato leaves at 100% r.h., but numbers at lower humidities depended on the speed of airflow in their chamber. With *Phaeoisariopsis personata* on groundnut, there was a linear increase in the number of conidia from 94.5% (the threshold) to 100% r.h. (Butler *et al.*, 1995). With constant conditions for 3 days, three times as many conidia were produced at 98–99% r.h. than with wet leaves, but there were similar numbers with intermittent high r.h. and intermittent leaf wetness. Sosa-Alvarez *et al.* (1995) found that 2.5 days of continuous wetness was needed to initiate sporulation of *Botrytis cinerea* on strawberry, and conidial production increased each day from 3 to 7 days. Fewer conidia were produced with interrupted wetness.

Some pathogens, such as rusts and powdery mildews, sporulate over a wide range of humidities (Rotem *et al.*, 1978) and free water may greatly reduce sporulation (Ward and Manners, 1974). Sporulation of *Sphaerotheca fuliginea* on squash increases at low r.h. (Reuveni *et al.*, 1971) and that of *Erysiphe cichoracearum* on tobacco is not affected by humidity (Cole, 1971).

The different mechanisms by which pathogens produce spores strongly influence the requirement of surface water for spore production and release. Liquid water is essential for the release of zoospores, and is normally required for the extrusion of mucilaginous spores. In the case of spores for air dispersal, there are a whole range of responses to surface water. Spores produced when leaves are wet are not normally liberated into the air until the surface dries.

14.7 Insects

Studies on effects of leaf surface wetness on insects are scarce, perhaps because surface water is less critical for insect pests than it is for pathogens. Insects commonly respond to humidity however, and may lose significant mass when exposed to dry air. Weissling and Giblin-Davis (1993) observed that partially desiccated weevils (*Rhynchophorus cruentatus*) regained their body weight when exposed to liquid water, but not in near-saturated air. They suggested a connection between rainwater trapped at the leaf base of *Sabal palmetto* palms and adult weevils harboured between stem internodes and the leaf base.

14.7.1 *Diapause termination*

Water is absorbed or imbibed by many insect species at the end of diapause, and this is thought to be necessary for pupation (Beck, 1980). Examples of this phenomenon are the rice borer, *Chilo simplex* (Koizumi, 1952), and the southwestern corn borer, *Diatraea grandiosella* (Reddy and Chippendale, 1973). However, in the case of the stem borer of sorghum and maize, *Busselo fusca*, it has been shown that contact with surface water is more important to the termination of diapause than water uptake (Okuda, 1990). Larvae given water orally failed to pupate. This reaction to contact with water appears to be a good adaptation to tropical semi-arid conditions, because continuous contact is required for more than 1 week to terminate diapause completely (Okuda, 1988). Sporadic rains would therefore not lead to untimely pupation.

14.7.2 *Larval movement and survival*

Sorghum shoot fly (*Atherigona soccata*) lays its eggs on the abaxial surface of an expanded leaf at the five-leaf growth stage, and the freshly emerged larva migrates to the growing point at the base of the (unexpanded) central whorl leaf. Movement of the larva has been associated with dew (Rivnay, 1960) and Blum (1963) reported that larvae fell from the plant unless the leaves were moistened. This finding was confirmed by Raina (1981), who pointed out that first instar larvae are prone to desiccation but, in the presence of dew, they glide through the film of water.

Nwanze *et al.* (1990) reported differential surface wetness on the central whorl leaf in sorghum genotypes, with consistently more water on those susceptible to shoot fly damage. They found that larval movement was more rapid in susceptible (with more surface wetness) than resistant genotypes, and the larvae often did not reach the growing point in resistant plants. Subsequent studies showed that surface water on the central whorl leaf did not condense from the atmosphere (Sree *et al.*, 1994), but was exuded on to the leaf surface (Sivaramakrishnan *et al.*, 1994).

The strong association between shoot fly attack and leaf surface water, coupled with the knowledge that the surface water was an exudate, suggested the possibility that shoot fly attack would be affected by the plant water status. This was shown to be the case (Soman *et al.*, 1994) and further experiments have shown that shoot fly attack in the field can be substantially reduced by soil water management (K.F. Nwanze, personal communication).

14.8 Current status of knowledge

Recent progress in modelling leaf wetness in relation to plant diseases reflects an adequate understanding of the physical principles for evaporation of surface water from leaves. The number of reports of tests comparing observed and predicted wetness periods from multi-layer canopy evaporation models is limited, but it is apparent that simulation of realistic evaporation rates are possible in situations ranging from films of water to discrete drops. In spite of this, the complexity of multi-layer canopy models has severely restricted their adoption for application to plant disease problems. The simplified approach of Zhang and Gillespie (1990) to predict only the persistence of large drops deep in the canopy may help to overcome this problem.

For practical disease prediction schemes, it should be borne in mind that there is often an inoculum source at the base of the canopy, in which case wetness on the (diseased) lower leaves would relate to sporulation. Wetness persistence on (healthy) leaves near the top of the canopy is relevant to infection, especially since infection of upper leaves is likely to have a significant effect on crop productivity.

Physically based models for estimating dew have been shown to work well in a wide range of environments, but their use is commonly restricted by the availability of hourly data on cloud cover, wind speed and humidity. Empirical models therefore still have a role to fill, and dew records from judiciously placed wetness sensors offer a good alternative when recording equipment is available.

The most serious limitation to prediction of wetness from rain is the difficulty of estimating interception in canopies. This is especially true for crops on which water accumulates as discrete drops, where the interception process can be dominated by rainfall intensity and wind speed. Examples in this chapter demonstrate the size of this problem, but our current knowledge is inadequate to quantify the effects sufficiently well to exploit our ability to estimate evaporation of surface water from canopies. There is also a need for information on the distribution of intercepted water at different heights in the canopy. These comments have similar relevance to wetness estimates from multi-layer models, single-layer models and wetness sensors, since wetness duration depends on the amount of intercepted water as well as the evaporation rate.

Over the past 10 years numerous reports have been published of the response of different fungal pathogens to wetness. The majority of these were for infection, but there is an increasing number for sporulation, particularly for pathogens producing airborne spores. There are few examples, however, where knowledge of pathogen response to wetness and models to estimate wetness periods have been combined, either for practical disease management, or for improved understanding of the host–pathogen–environment system.

Reports on the response of insects to surface water are few, but those that exist provide excellent examples of adaptation to the environment in semi-arid tropical conditions. Such adaptation may be more widespread than is currently known and, where survival depends on moisture, this is likely to be valuable for integrated pest management.

References

Alderman, S.C., Lacy, M.L. and Everts, K.L. (1985) Influence of interruptions of dew period on numbers of lesions produced on onion by *Botrytis squamosa*. *Phytopathology* **75**, 808–810.

Barr, A. and Gillespie, T.J. (1987) Maximum wetness duration for water drops on leaves in the field. *Agric. For. Meteorol.* **41**, 267–274.

Bashi, E. and Rotem, J. (1974) Adaptation of four pathogens to semi-arid habitats as conditioned by penetration rate and germinating spore survival. *Phytopathology* **64**, 1035–1039.

Beard, J.S. (1962) Rainfall interception by grass. *J.S.A. Forestry Assoc.* **42**, 12–15.

Beck, S.D. (1980) *Insect Photoperiodism*, 2nd Edition. Academic Press, New York.

Beckman, P.M. and Payne, G.A. (1982) External growth, penetration and development of *Cercospora zeae-maydis* in corn leaves. *Phytopathology* **72**, 810–815.

Blum, A. (1963) The penetration and development of the sorghum shootfly in susceptible sorghum plants (in Hebrew). *Hassadeh* **44**, 23–25.

Brain, P. and Butler DR. (1985) A model of drop size distribution for a system with evaporation. *Plant Cell Environ.* **8**, 247–252.

Burgy, R.H. and Pomeroy, C.R. (1958) Interception losses in grassy vegetation. *Trans. Am. Geophys. Union* **39**, 1095–1100.

Butler, D.R. (1980) Dew and thermal lag: measurements and an estimate of wetness duration on cocoa pods. *Q. J. R. Meteorol. Soc.* **106**, 539–550.

Butler, D.R. (1983) Microclimate, rainfall interception and wetness persistence within a cereal canopy. PhD Thesis, University of Bristol, UK.

Butler, D.R. (1985) The energy balance of water drops on a leaf surface. *Boundary Layer Meteorol.* **32**, 337–350.

Butler, D.R. (1986) Evaporation from rain drops in a cereal canopy: a simulation. *Boundary Layer Meteorol.* **36**, 39–51.

Butler, D.R. and Huband, N.D.S. (1984) Throughfall and stem-flow in wheat. *Agric. For. Meteorol.* **35**, 329–338.

Butler, D.R. and Jadhav, D.R. (1991) Requirements of leaf wetness and temperature for infection of groundnut by rust. *Plant Pathol.* **40**, 395–400.

Butler, D.R. and King, S.C. (1982) An inexpensive balance for measuring throughfall in cereals. *Agric. Meteorol.* **27**, 233–240.

Butler, D.R., Wadia, K.D.R. and Jadhav, D.R. (1994) Effects of leaf wetness and temperature on late leaf spot infection of groundnut. *Plant Pathol.* **43**, 112–120.

Butler, D.R., Wadia, K.D.R. and Reddy, R.K. (1995) Effects of humidity, leaf wetness, temperature and light on conidial production by *Phaeoisariopsis personata* on groundnut. *Plant Pathol.* **44**, 662–674.

Carisse, O. and Kushalappa, A.C. (1990) Development of an infection model for *Cercospora carotae* on carrot based on temperature and leaf wetness duration. *Phytopathology* **80**, 1233-1238.

Carisse, O. and Kushalappa, A.C. (1992) Influence of interrupted wet periods, relative humidity and temperature on infection of carrots by *Cercospora carotae*. *Phytopathology* **82**, 602–606.

Carnegie, S.F. (1980) Aerial dispersal of the potato gangrene pathogen *Phoma exigua* var. *foveata*. *Ann. Appl. Biol.* **94**, 165–173.

Clark, O.R. (1940) Interception of rainfall by prairie grasses, weeds and certain crops plants. *Ecol. Monogr.* **10**, 243–277.

Clayton, C.N. (1942) The germination of fungal spores in relation to controlled humidity. *Phytopathology* **32**, 921–943.

Cole, J.S. (1971) Sporulation of powdery mildews, particularly *Erysiphe cichoracearum* DC. on tobacco. In: *Ecology of Leaf Surface Microorganisms* (eds T.F. Preece and C.H. Dickinson). Academic Press, New York, pp. 323–337.

Couturier, D.E. and Ripley, E.A. (1973) Rainfall interception in mixed grass prairie. *Can. J. Plant Sci.* **53**, 659-663.

Davis, D.R. and Hughes, J.E. (1970) A new approach to recording the wetness parameter by the use of electrical resistance sensors. *Plant Disease Reporter* **54**, 474–479.

Dennis, J.I. (1987) Temperature and wet period conditions for infection by *Puccinia striiformis* f. sp. *tritici* race 104E137A. *Trans. Br. Mycol. Soc.* **88**, 119–121.

Dillard, H.R. (1989) Effect of temperature, wetness duration, and inoculum density on infection and lesion development of *Colletotrichum coccodes* on tomato fruit. *Phytopathology* **79**, 1063–1066.

Everdingen, E. van. (1926) Het verband tusschen de weergesteldheid en de aardappelziekte (*Phytophthora infestans*). *Tijdschrift Plantenzeikten* **32**, 129–140.

Figueroa, L., Fitt, B.D.L., Welham, S.J. *et al.* (1995) Early development of light leaf spot (*Pyrenopeziza brassicae*) on winter oilseed rape (*Brassica napus*) in relation to temperature and leaf wetness. *Plant Pathol.* **44**, 641–654.

Fitt, B.D.L., McCartney, H.A. and Walklate, P.J. (1989) The role of rain in dispersal of pathogen inoculum. *Annu. Rev. Phytopathol.* **27**, 241–270.

Fogg, G.E. (1947) Quantitative studies of wetting of leaves by water. *Proc. R. Soc.* **134**, 503–522.

Gillespie, T.J. and Barr, A. (1984) Adaptation of a dew estimation scheme to a new crop and site. *Agric. For. Meteorol.* **31**, 289–295.

Glynne, M.D. (1953) Production of spores by *Cercosporella herpotrichoides*. *Trans. Br. Mycol. Soc.* **36**, 46–51.

Goos, R.D. and Tschirch, M. (1963) Greenhouse studies on the *Cercospora* leaf spot of banana. *Trans. Br. Mycol. Soc.* **46**, 321–330.

Griffiths, E. and Ao, H.C. (1976) Dispersal of *Septoria nodorum* spores and spread of glume blotch of wheat in the field. *Trans. Br. Mycol. Soc.* **67**, 413–418.

Grove, G.G., Madden, L.V., Ellis, M.A. *et al.* (1985) Influence of temperature and wetness duration on infection of immature strawberry fruit by *Phytophthora cactorum*. *Phytopathology* **75**, 165–169.

Häckel, H. (1974) Eine elektrische Methode zur Messung der Blattbenetzungsdauer unmittelbar am Blatt. *Agric. Meteorol.* **13**, 91–103.

Hall, R.L. (1985) Further interception studies of heather using a wet-surface weighing lysimeter system. *J. Hydrol.* **81**, 193–210.

Harrison, J.G. and Lowe, R. (1989) Effects of humidity and air speed on sporulation of *Phytophthora infestans* on potato leaves. *Plant Pathol.* **38**, 585–591.

Hirst, J.M. (1957) A simplified surface wetness recorder. *Plant Pathol.* **6**, 57–61.

Hirst, J.M. and Steadman, O.J. (1963) Dry liberation of fungus spores by raindrops. *J. Gen. Microbiol.* **33**, 335–344.

Hirst, J.M., Sotrey, I.F., Ward, W.C. et al. (1955) The origin of apple scab epidemics in the Wisbech area in 1953 and 1954. *Plant Pathol.* **4**, 91–96.

Holloway, P.J. (1969) The effects of superficial wax on leaf wettability. *Ann. Appl. Biol.* **63**, 145–153.

Holloway, P.J. (1970) Surface factors affecting the wetting of leaves. *Pestic. Sci.* **1**, 156–163.

Huband, N.D.S. and Butler, D.R. (1984) A comparison of wetness sensors for use with computer or microprocessor systems designed for disease forecasting. *Proc. Br. Crop Prot. Conf. Pests Dis.* **2**, 633–638.

Huber, L. (1988) Modèle de simulation de la durée d'humectation des feuilles au champ après une pluie. Cas d'un couvert homogène. *Agronomie* **8**, 675–684.

Huber, L. (1992) Déterminisme et mesure de la durée d'humectation en vue de la protection des plantes. *Agronomie* **12**, 281–295.

Huber, L. and Gillespie, T.J. (1992) Modeling leaf wetness in relation to plant disease epidemiology. *Annu. Rev. Phytopathol.* **30**, 553–577.

Huber, L and Itier, B. (1990) Leaf wetness duration in a field bean canopy. *Agric. For. Meteorol.* **51**, 281-292.

Huber, L., Flura, D., Durand, B. *et al.* (1991) Etude microméteorologique de la durée d'humectation des grapes de vigne. *EPPO Bull.* **21**, 441–449.

Koizumi, K. (1952) Water content and the hormone center for pupation of hibernating larvae of *Chilo simplex* Butler. *Annotationes Zool. Jpn* **25**, 156–167.

Manners, J.G. (1982) *Principles of Plant Pathology*. Cambridge University Press, Cambridge.

Manners, J.G. and Hossain, S.M.M. (1963) Effects of temperature and humidity on conidial germination in *Erysiphe graminis*. *Trans. Br. Mycol. Soc.* **46**, 225–234.

McCartney, H.A. and Lacey, M.E. (1990) The production and release of ascospores of *Pyrenopeziza brassicae* on oilseed rape. *Plant Pathol.* **39**, 17–32.

Merriam, R.A. (1961) Surface water storage on annual rye grass. *J. Geophys. Res.* **66**, 1833–1838.

Mills, W.D. (1944) Efficient use of sulfur dusts and sprays during rain to control apple scab. *Cornell Univ. Ext. Bull.* **630**, 1-4.

Miranda, R.A.C. and Butler, D.R. (1986) Interception of rainfall in a hedgerow apple orchard. *J. Hydrol.* **87**, 245–253.

Monteith, J.L. and Butler, D.R. (1979) Dew and thermal lag: a model for cocoa pods. *Q. J. R. Meteorol. Soc.* **105**, 207–215.

Nene, Y.L. and Reddy, M.V. (1987) Chickpea diseases and their control. In: *The Chickpea* (eds M.C. Saxena and K.B. Singh). C.A.B. International, Oxon, UK, pp. 233–270.

Nicholson, R.L. and Moraes, W.B.C. (1980) Survival of *Colletotrichum graminicola*: importance of the spore matrix. *Phytopathology* **70**, 255–261.

Norman, J.M. (1982) Simulation of microclimates. In: *Biometeorology in Integrated Pest Management* (eds J. Hatfield and I. Thomason). Academic Press, New York, pp. 65–99.

Norman, J.M. and Campbell, G. (1983) Application of a plant environment model to problems in irrigation. In: *Advances in Irrigation* (ed. D. Hillel). Academic Press, New York, pp. 155–189.

Nwanze, K.F., Reddy, Y.V.R. and Soman, P. (1990) The role of leaf surface wetness in larval behaviour of the sorghum shoot fly, *Atherigona soccata*. *Entomol. Exp. Appl.* **56**, 187-195.

Okuda, T. (1988) Effect of artificial wetting and rainfall on the larval diapause of a stem borer, *Busseola fusca* in western Kenya. *Entomol. Exp. Appl.* **48**, 263–267.

Okuda, T. (1990) Significance of water contact as a factor terminating larval diapause in a stem borer, *Busseola fusca*. *Entomol. Exp. Appl.* **57**, 151–155.

Pedro, M.J. and Gillespie, T.J. (1982) Estimating dew duration. II -Utilising standard weather station data. *Agric. Meteorol.* **25**, 297–310.

Raina, A.K. (1981) Movement, feeding behaviour and growth of larvae of the sorghum shoot fly, *Atherigona soccata*. *Ins. Sci. Appl.* **2**, 77–81.

Rathaiah, Y. (1977) Stomatal tropism of *Cercospora beticola* in sugar beet. *Phytopathology* **67**, 358–362.

Reddy, A.S. and Chippendale, G.M. (1973) Water involvement in diapause and the resumption of morphogenesis of the southwestern corn borer, *Diatraea grandiosella*. *Entomol. Exp. Appl.* **16**, 445–454.

Reuveni, R., Cohen, Y. and Rotem, J. (1971) Sporulation of *Erysiphe cichoracearum* as influenced by conditions favoring photosynthesis in the host. *Isr. J. Bot.* **20**, 78–83.

Rivnay, E. (1960) Field crop pests in the Near East. (Pests of graminaceous and leguminous crops and stored products.) *Hassadeh* (in Hebrew). *Monogr. Biol.* **47**, 25–27.

Rotem, J., Cohen, Y. and Bashi E. (1978) Host and environmental influences on sporulation in vivo. *Annu . Rev. Phytopathol.* **16**, 83–101.

Royle, D.J and Butler, D.R. (1986) Epidemiological significance of liquid water in crop canopies and its role in disease forecasting. In: *Water, Fungi and Plants* (eds P.G. Ayres and L. Boddy). Cambridge University Press, Cambridge, pp. 139–156.

Rudgard, S.A. (1986) Witches' broom disease on cocoa in Rondonia, Brazil: basidiocarp production on detached brooms in the field. *Plant Pathol.* **35**, 434–442.

Rutter, A.J. (1975) The hydrological cycle in vegetation. In: *Vegetation and the Atmosphere*. Vol. I. (ed. J.L. Monteith). Academic Press, London, pp. 111–154.

Savary, S. and Janeau, J.L. (1986) Rain-induced dispersal in *Puccinia arachidis*, studied by means of a rainfall simulator. *Netherlands J. Plant Pathol.* **92**, 163-174.

Schein, R.D. (1964) Comments on the moisture requirements of fungus germination. *Phytopathology* **54**, 1472.

Scherm, H. and van Bruggen, A.H.C. (1993) Sensitivity of simulation dew duration to meteorological variations in different climatic regions of California. *Agric. For. Meteorol.* **66**, 229–245.

Shaw, M.W. (1991) Interacting effects of interrupted humid periods and light on infection of wheat leaves by *Mycosphaerella graminicola* (*Septoria tritici*). *Plant Pathol.* **40**, 595–607.

Sivaramakrishnan, S., Soman, P., Nwanze, K.F *et al.* (1994) Resistance in sorghum to shoot fly *Atherigona soccata*: Evidence for the source of leaf surface wetness. *Ann. Appl. Biol.* **124**, 215–218.

Smith, D.H. and Crosby, F.L. (1973) Aerobiology of two peanut leafspot fungi. *Phytopathology* **63**, 703–707.

Soman, P., Nwanze, K.F., Laryea, K.B. *et al.* (1994) Leaf surface wetness in sorghum and resistance to shoot fly, *Atherigona soccata*: role of soil and plant water potentials. *Ann. Appl. Biol.* **124**, 97–108.

Sosa-Alvarez, M., Madden, L.V. and Ellis, M.A. (1995) Effects of temperature and wetness duration on sporulation of *Botrytis cinerea* on strawberry leaf residues. *Phytopathology* **79**, 609–615.

Sree, P.S., Nwanze, K.F., Butler, D.R. *et al.* (1994) Biophysical factors associated with resistance in sorghum to the shoot fly, *Atherigona soccata*: morphological characteristics of the central whorl leaf. *Ann. Appl. Biol.* **125**, 467–476.

Steadmam, O.J. (1980) Dispersal of spores of *Rhyncosporium secalis* from a dry surface by water drops. *Trans. Br. Mycol. Soc.* **75,** 339–340.

Stephan, J. and Mildebrath, E. (1938) Registrierende Taumessungen. *Bioklimatische Beiträge* **5,** 34–39.

Sutton, J.C., Gillespie, T.J. and James, T.D.W. (1988) Electronic monitoring and use of microprocessors in the field. In: *Experimental Techniques in Plant Disease Epidemiology* (eds J. Kranz and J. Rotem). Springer-Verlag, Berlin, pp. 99–113.

Thompson, N. (1981) The duration of leaf wetness. *Meteorological Magazine* **110,** 1–12.

Voglino, P. (1904) Intorno allo sviluppo e parassitismo delle *Septoria graminum* Desm. e *glumarum* Pass. *Ann. R. Acad. Torino* **46,** 259–282.

Wadia, K.D.R. and Butler, D.R. (1994) Infection efficiency of *Phaeoisariopsis personata* and the influence of different wetness patterns on germ tube growth of the pathogen. *Plant Pathol.* **43,** 802–812.

Ward, S.E. (1988) Wetness monitoring in crops. MSc thesis, University of Bristol, UK.

Ward, S.V. and Manners, J.G. (1974) Environmental effects on the quantity and viability of conidia produced by *Erysiphe graminis*. *Trans. Br. Mycol. Soc.* **62,** 119–128.

Weihong, L. and Goudriaan, J. (1991) Analysis of leaf wetness in the rice crop caused by dew formation — a simulation study. *EPPO Bull.* **21,** 551–559.

Weiss, A., Lukens, D.L. and Norman, J.M. (1989) Leaf wetness in dry beans under semi-arid conditions. *Agric. For. Meteorol.* **48,** 149–162.

Weissling, T.J. and Giblin-Davis, R.M. (1993) Water loss dynamics and humidity preference of *Rhynchophorus cruentatus* (Coleoptera: Curculionidae) adults. *Environ. Entomol.* **22,** 93–98.

Williamson, B., Ducan, G.H., Harrison, J.G. *et al.* (1995) Effect of humidity on infection of rose petals by dry-inoculated conidia of *Botrytis cinerea*. *Mycol. Res.* **99,** 1303–1310.

Winters, R. and Small, C.G. (1934) An automatic moisture recording device. *Phytopathology* **24,** 284–288.

Zhang, Y. and Gillespie, T.J. (1990) Estimating maximum water droplet wetness duration on crops from nearby weather station data. *Agric. For. Meteorol.* **51,** 145–158.

15

Surface wetness and pollutant deposition

J. N. Cape

15.1 Introduction

When viewed from the perspective of an atmospheric chemist, pollutant deposition is simply a process by which material is removed from the atmosphere. The removed material, however, is deposited at the ground, and for most terrestrial surfaces that means not the soil, but leaf surfaces. If the deposition process involves precipitation, contact times with vegetation may be relatively short, before rainwater passes through the vegetation canopy and reaches the soil. Snow may accumulate over a period on vegetation, before melting, more or less quickly, and being deposited on soil. Cloud water, however, may persist on canopies, if deposition rates are too slow to saturate the canopy holding capacity (usually between 1 and 2 mm rain equivalent), and evaporation rates are small. 'Dry' deposition, that is the turbulent deposition of gases and particles to the ground, also involves deposition to plant surfaces. In this case, the chemical and physical structure of the vegetation surface may control the rate of dry deposition, in so far as the rate of deposition is determined by accumulation, and equilibration of surfaces with the gas phase. For water-soluble gases, vegetation surfaces may represent an inert substrate for the interaction of the gas with surface water, or may play an active role in influencing chemical processes on the leaf surface.

Interactions between the leaf surface and the atmosphere occur across a very wide range of temporal and spatial scales. Time scales range from seconds, for the transit of a raindrop through a plant canopy, to decades, if one considers the hydrological influence of a growing forest. At the smallest spatial scales, at a molecular level, a leaf surface may have an enormous specific area (i.e. effective area relative to ground area). Electron micrographs (see e.g. Chapter 2) illustrate the complex wax structures that may be found on the surfaces of many leaves, and the heterogeneous nature of such structures across a leaf. Within the layer of air closest to a leaf surface, where gas transport is achieved by diffusion alone, these complex structures can provide ratios of wax surface to volume of air (S/V) very much greater than 1 m^{-1}. At the much larger scale of the whole leaf within a plant canopy, the leaf area relative to the ground, even assuming a smooth surface, is still large; a leaf area index (LAI) of 5, which would not be unusual for a plant canopy, would give a two-sided leaf area ratio to the ground of 10 m^2 (m^2 ground)$^{-1}$. The volume of air within such a forest canopy, with height 10 m and volume 10 m^3

Plant Cuticles, edited by G. Kerstiens.

(m^2 ground)$^{-1}$ would be characterized by S/V of 1 m^{-1}. At both the leaf and the canopy level there is also a wide variety of surfaces present for interaction with the atmosphere. Water on a leaf may exist as a thin film trapped in surface structures, as discrete droplets, or as a continuous liquid layer which overlays the detailed fine structure of the leaf surface. Within the canopy there will be areas with greater or lesser affinities for liquid water (e.g. bark and needles in a conifer forest), and with greater or lesser exposure to wind, light and canopy drip.

15.2 What is 'wetness'?

In rain or cloud a vegetation canopy is covered in a more or less continuous aqueous film, but surfaces may be 'wet', in the sense of allowing chemical reactions of dissolved materials, without a visible liquid film. Dew formation, i.e. the condensation of water vapour on a leaf surface which is cooler than the surrounding air, may lead to droplets or to a continuous liquid film (see Chapter 14). However, plant surfaces exposed to the atmosphere are continuously exposed to water vapour, so that the potential for the formation of molecular films is always present. At low relative humidities the amount of water adsorbed on a leaf surface may be too small to permit significant amounts of material to dissolve, but as humidity increases, hygroscopic particles begin to deliquesce and form 'puddles' of aqueous solution. As humidity increases further, these 'puddles' may grow and coalesce to give a film which permits a limited amount of ion transport. Similarly, amounts of adsorbed water may grow to the point where non-hygroscopic salts begin to dissolve, and absorption of soluble gases from the surrounding air leads to formation of ionic solutions. This phenomenon has been observed in measurements of surface conductivity in the laboratory (Burkhardt and Eiden, 1994), in response to added ammonia vapour, and is commonly seen also in field measurements.

15.2.1 *Field measurements*

Surface wetness is often measured using an interlaced network of electrical conductors on an insulating plate; as water accumulates, ion transport permits an electric current to flow. Such detectors are used to detect rainfall, but also respond to changes in humidity. Figure 15.1 shows the characteristic response of a resistance-based detector mounted in a forest canopy. Between dry conditions (resistance > 75 MΩ) and liquid water on the detector (resistance $<< 10$ MΩ), there is a region where resistance decreases below the 'dry' value, but remains well above the 'wet' value. The relationship between resistance and relative humidity is not simple, as it depends to some extent on the substrate used for the detector, and on the amounts of soluble salts deposited from the atmosphere on the detector, or left behind after evaporation of the last rain. Leaf surfaces can be wet, in the absence of rain, for long periods; Burkhardt and Eiden (1994) observed rain wetting for 28% of time in a spruce forest in summer 1992, and a further 40% of time when the electrical conductance of spruce needle surfaces was proportional to relative humidity. In a Scots pine forest in central Scotland, a wetness sensor registered liquid water for 25% of time over 6 months in summer, and an additional 30% of time when the sensor showed a measurable electrical conductance (Figure 15.1; Cape *et al.*, 1995). In a deciduous forest in Canada, a similar type of wetness sensor registered 'wet' for 50 (± 15)% of the time during summer (Fuentes *et al.*, 1992).

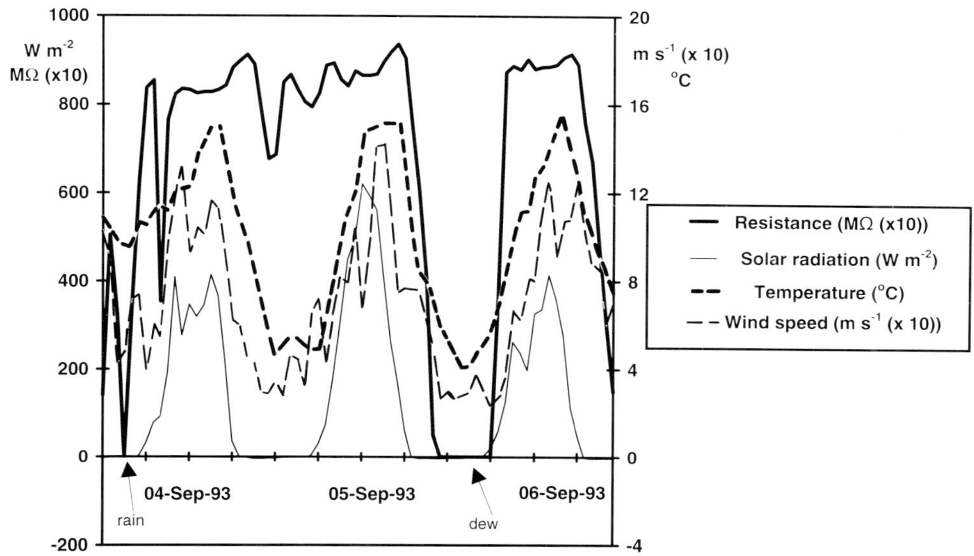

Figure 15.1. *Measurements of surface wetness (electrical resistance of an interlaced conductor), solar radiation, air temperature and wind speed at canopy height in a Scots pine forest in central Scotland (Cape* et al., *1995). A short rain shower early on 4 September brought the resistance of the wetness sensor close to zero, then resistance gradually increased through the day to the 'dry' value of 90 MΩ . A small decrease the following night (4–5 September), correlated with air temperature fluctuations, may have been caused by slight dew formation. After a clear, sunny day on 5 September, the clear sky led to rapid surface cooling, and marked cooling of the air as well, to 4°C. The low wind speed (< 0.4 m s^{-1}) promoted dew formation, seen as a 'zero' resistance for the wetness sensor from 23:00 on 5 September until shortly after sunrise on 6 September.*

15.2.2 *Chemistry in thin water films*

Although canopy surfaces may be regarded as 'wet' for long periods, the chemistry of thin water films may be very different from that which would occur in bulk aqueous solution. Effective concentrations of dissolved salts or gases may be very large, and ionic interactions need to be taken into account. An example of the differences between reactions in dilute solution and reactions in surface aqueous films may be deduced from measurements of activity coefficients in concentrated solutions of sodium chloride, in which the solubility of sulphur dioxide (SO_2) is greatly reduced relative to pure water. At a concentration of 6 M NaCl the solubility of SO_2 was approximately halved (Millero *et al.*, 1989). The non-ideality of chemical processes in water films is particularly important as surface water evaporates, and solutes are concentrated to the point where recrystallization may occur.

Laboratory measurements may be able to provide indications of the behaviour of surface water films, but are usually made in closed systems, where the relative humidity (r.h.), or even the total amount of available water, is strictly controlled (e.g. Benner *et al.*, 1992). The atmosphere is an open system, and water vapour will deposit on surface films in response to the solution of salts and soluble gases, to maintain solution compositions which are in equilibrium with the water vapour content of the surrounding air. A 'steady-state', as often produced in laboratory systems, may rarely exist in the field.

15.3 Liquid water on vegetation surfaces

The interaction of fog, cloud and rainwater with vegetation surfaces is at a larger scale, typical of the macroscopic structures of leaves, such as veins or trichomes. Characteristic droplet sizes range from around 10 μm, for fog and cloud droplets, up to several hundred μm for raindrops. The capture and retention of water droplets by vegetation surfaces depends on several factors: droplet size, velocity and surface tension, and leaf chemistry and surface roughness. Large droplets, falling under gravity or wind-driven, may have sufficient kinetic energy relative to droplet surface energy that they fragment on hitting a solid surface (Cape, 1988). Even relatively small drops, if driven by the wind, may similarly fragment. Whether a slow-moving drop sticks to a leaf or bounces off depends on both the chemical composition of the drop (in so far as the surface tension of the droplet may be affected) and the structure and composition of the leaf surface. The interaction between water droplets and the surface may be measured by the droplet contact angle. This interaction has been studied in some detail in terms of formulating droplets for the effective application of agricultural pesticides, by controlling droplet surface tension. Theories to explain wettability have been developed and tested in the laboratory (e.g. Adam, 1963; Warburton, 1966; Watanabe and Yamaguchi, 1992). Leaf wettability is affected by air pollutants, and has been used as a measure of the effects of air pollution on leaves (e.g. Jagels, 1994). The dependence on leaf surface structure has also been studied, both in terms of surface wax chemistry (Holloway, 1969) and physical structures such as leaf ridges or trichomes (Holloway, 1970). Structures have been shown to be important even in the absence of different surface waxes; carbon replicas of leaf surfaces of different species may have very different contact angles as a result of different structures (Challen, 1960). The contact angle between water and the surface determines not only the initial interaction between water droplets and the surface, but also the behaviour of the droplet once captured. Contact angles of less than 90° lead to spreading of droplets as leaves are shaken by the wind, and eventual amalgamation to form a continuous liquid film. Angles greater than 90° promote coverage of the surface by isolated droplets.

15.3.1 *Contact angles and water retention*

The contact angle determines the shape of the water drop on the surface, so also determines the contact area, i.e. the area of leaf wetted by the droplet. Figure 15.2 shows the variation in contact area as a function of contact angle for a fixed droplet volume. A contact angle of 60° increases contact area by 63% relative to a hemispherical droplet. This simple geometrical relationship has been shown to hold in practice (Brewer *et al.*, 1991).The actual area of contact between the droplet and the leaf surface may be less than this, if pockets of air are trapped beneath a droplet in the fine-scale structures of the leaf surface. Trapping of air is responsible for very large contact angles on some hairy leaves, where droplets may not be in contact with the cuticle surface, but suspended above it. Contact angles can be measured in the laboratory under controlled conditions, but this can only give an indication of behaviour in the field, where leaves may be subject to mechanical agitation, or to

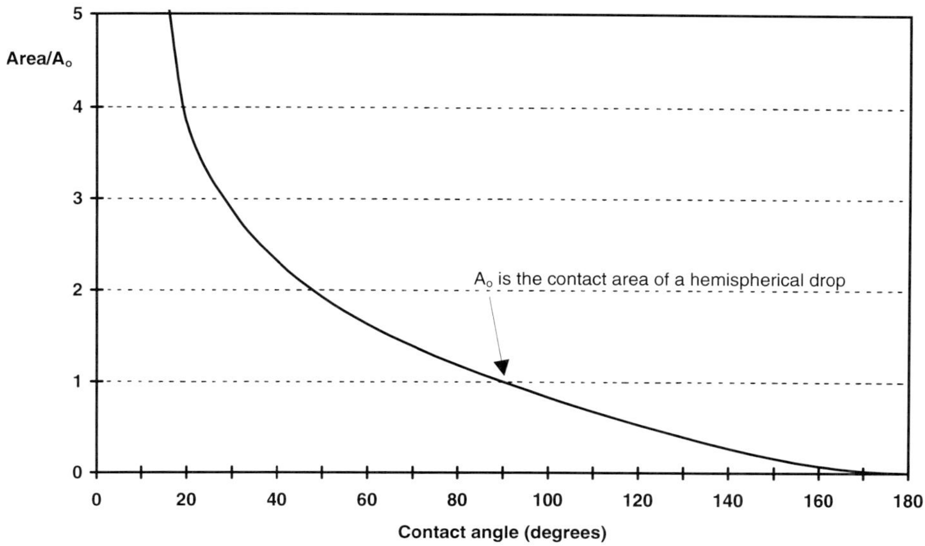

Figure 15.2. *Theoretical dependence of the contact area of a water droplet of fixed volume on the contact angle between the droplet and a solid surface. The area is expressed relative to the contact area of a hemispherical droplet with contact angle 90°.*

the impact of further droplets. A large contact angle may lead to the retention of large numbers of droplets in still conditions, but little retention when shaken; for example, *Triticum* leaves with a contact angle of 130° retained only 1.5 mg water cm^{-2} leaf surface, while *Sinapis* leaves under the same conditions, with a contact angle of 72° retained 5.12 mg cm^{-2} (Fogg, 1947). Trichomes can promote such water-repellent behaviour, or can actively retain water droplets (Brewer *et al.*, 1991; Monson *et al.*, 1992). Mechanical agitation, as well as causing non-wetting drops to be displaced, may also lead to the amalgamation of wetting drops to form a continuous liquid film. In this case, water may drain from the leaf tip or edge until a uniform thin aqueous film is produced. Subsequent addition of water to the film, which is more likely to retain droplets than a 'dry' solid surface, will lead to further drainage, maintaining a fairly constant liquid volume on the leaf. At the canopy level, this represents the canopy holding capacity, or interception volume, which is typically equivalent to 1–2 mm rainfall (see Chapter 14).

15.3.2 *Spatial heterogeneity*

Within the canopy, however, and even across the surface of an individual leaf, the properties that affect water retention may vary markedly. Different regions of a single pine needle, for example, show very different contact angles (Leyton and Juniper, 1963). The distribution of particles and droplets marked with a fluorescent dye, when applied experimentally to spruce needles, has been shown to be very non-uniform. Accumulation of particles and droplets was seen in regions around the stomata (Burkhardt *et al.*, 1995; Jagels, 1991). The reasons for this are not known, but may be related to differences in the wax structures around stomata on spruce needles, improving retention close to stomata, or to mechanisms that promote diffusion and deposition of fine droplets and particles close to the stomata. Within a forest canopy, different age-classes of conifer needle on the same branch show marked differences, with needles becoming more wettable with age (Cape, 1983; Cape *et al.*, 1989), and there may be even greater heterogeneity, if senescing leaves and bark are included. There are also very large differences among plant species, related to amounts and structures of surface waxes, and to the underlying leaf structure (Holloway, 1969).

15.3.3 *Temporal heterogeneity*

In addition to the spatial heterogeneity, there is a marked temporal variation. On a time scale of hours, changes in leaf turgor can affect the contact angle (Fogg, 1944), although this response may not be universal (Weiss, 1988). Contact angles change through the year, as new leaves are formed, grow and senesce, and, over longer time scales, as canopy architecture changes. The wettability of a leaf, or a plant canopy, is not static, and may therefore influence rates of pollutant deposition in different ways through the year.

15.4 Pollutant deposition

15.4.1 *Wet deposition*

Measurements of wet deposition below plant canopies, as throughfall and stemflow, give an indication of some of the processes that operate on leaf surfaces. Deposition of many ions, and of organic solutes, is often greatly enhanced in throughfall, although some ions in rain may be retained by the plant canopy, and utilized as nutrients by the plant itself, or by surface microorganisms. There has been considerable debate as to the origin of the material in throughfall in excess of input in rain, sometimes referred to as 'net throughfall'. How much represents the wash-off of dry-deposited material, either particles or gases, and how much represents leaching from the canopy? Several techniques have been developed in an attempt to answer this question, but for most ions a quantitative conclusion is still uncertain. For many metal ions, the canopy is thought to be a net source of material (e.g. K^+, Mn^{2+}), while other ions are not thought to participate in exchange processes to any great extent (e.g. Na^+, Cl^-, SO_4^{2-}). Ions such as Ca^{2+} and Mg^{2+} have both airborne and canopy sources, and separation of these quantitatively has proved difficult. There has been some success in predicting deposition of Ca^{2+} in throughfall in relation to the period of dry weather before a rain event, on the assumption that the Ca^{2+} released from the canopy comes from the solution of particles deposited during the dry period (Lovett and Lindberg, 1984). Measurements of dew collected on inert plates in Arkansas, and compared with dry-deposited material on nights without dew formation, showed that most of the Ca^{2+}, Mg^{2+}, K^+,Na^+, Cl^- and SO_4^{2-} in dew came from dissolution of dry-deposited particles (Wagner *et al.*, 1992). However, on leaf surfaces it is also possible that acidity in the falling rain, or produced by the solution of acidic gases in dew, displaces Ca^{2+} ions from ion exchange sites accessible to liquid water at the leaf surface, which have been slowly recharged from inside the leaf during the preceding dry period.

15.4.2 *Dry deposition of water-soluble gases*

The wetness of a plant canopy does not greatly affect the process of wet deposition, but wettability may dictate the pattern of drying of the canopy after rain. Deposition of gaseous pollutants to leaf surfaces may, however, be strongly dependent on the wetness of the surface. Measurements of the dry deposition of pollutant gases in the field operate at the canopy level, and therefore integrate the spatial heterogeneities discussed above. Nevertheless, clear patterns of behaviour in relation to surface wetness have been observed, notably for water-soluble gases; for example, the deposition of SO_2 to dew on wheat was observed to be a self-limiting process as saturation of the dew layer occurred (Fowler and Unsworth, 1979). Pollutant uptake by dew has also been modelled (Chameides, 1987; Wesely *et al.*, 1990), taking account of the restricted liquid volume available for the dissolution of water-soluble gases.

Field measurements of the deposition of gases from the atmosphere to vegetation and soil at the earth's surface not only integrate over the heterogeneity of the canopy, but include a number of physical processes which have to be unravelled before the controlling nature of the vegetation surface can be assessed. Deposition flux data are often expressed, by analogy with an electrical circuit, in terms of resistances to transport from the turbulent atmospheric boundary layer to the ultimate sink for the gas in question. Three independent series resistances can be identified (Fowler, 1978): (i) aerodynamic resistance, r_a, which is determined by the turbulence of the atmosphere, and is related to wind speed and the roughness of the vegetation; (ii) r_b, or laminar sub-layer resistance, which describes the process of molecular diffusion through the layer of still air close to a leaf surface, and is also a function of wind speed and leaf shape; (iii) the surface resistance, which describes the affinity of the surface for the gas. The surface resistance is made up from several parallel resistances, representing different pathways or sinks for the gas. Uptake by stomata usually represents a perfect sink for molecules, but stomata occupy only a small proportion of most leaf surfaces. The cuticle, or surface wax on the cuticle, has the largest relative area for potential deposition. When stomata are closed, as deduced from measurements of the exchange of CO_2 or water vapour between the atmosphere and the plant surface, estimates of the cuticular surface resistance to gas deposition can be made from field data. If stomata are not fully closed, however, stomatal uptake may still dominate the deposition pathway if surface resistances are high (cf. Chapter 5).

15.4.3 *Surface resistance*

For some gases, notably nitric acid vapour (HNO_3), even the apparently inert surfaces of leaves act as a good sink, and HNO_3 is removed from the atmosphere by leaf surfaces as fast as the atmospheric resistances ($r_a + r_b$) will allow. Natural vegetation also seems to have a negligible surface resistance to the deposition of ammonia (NH_3) from the atmosphere (Duyzer *et al.*, 1994; Sutton *et al.*, 1993). In general, however, dry leaf surfaces are not an effective sink for inorganic gases, because of their non-ionic nature. If wetted, however, plant canopy surfaces can be excellent sinks for water-soluble gases. Long-term field measurements of dry deposition of SO_2 to a range of vegetation types in the Netherlands showed that surface resistances were zero, within the measurement uncertainty, except in very dry weather (r.h. $< 60\%$) (Erisman *et al.*, 1993, 1994), and that surface resistance decreased towards zero as r.h. increased from 30 to 95% (Erisman and Wyers, 1993). These studies have led to parametrizations of surface resistance in terms of r.h. (Erisman *et al.*, 1994), which have been used in modelling dry deposition to the countryside (Draaijers and Erisman, 1993). Similar patterns have been seen for NH_3. The surface resistance for deposition of SO_2 to wet surfaces need not be close to zero, however. Measurements of deposition rates above a Scots pine forest in

Scotland showed that deposition to a wet surface at night (with stomata closed) was not necessarily greater than to a dry surface (Fowler and Cape, 1983). This result was interpreted in terms of the equilibration of the falling rain with SO_2, so that further uptake by the water layer was slow. Further deposition and solution of SO_2 requires neutralization of acidity generated by dissolved SO_2. Acidity may be neutralized by the solution of previously dry-deposited alkaline particles (e.g. $CaCO_3$), or even by ion exchange at the leaf surface, where H^+ ions are replaced by Ca^{2+} or K^+. In areas where NH_3 and SO_2 concentrations are similar in magnitude, acidity can be neutralized by gaseous NH_3, leading to continuous large deposition rates, and an effective surface resistance close to zero. The same argument, but for generation of hydroxyl ions, applies to the solution of NH_3, but in this case carbon dioxide in the atmosphere may act as a partial neutralizing agent.

15.5 Illustrative model of uptake of soluble gases by a water film

The dependence of deposition rates to wet surfaces on the presence of other gases, and of further reactions at the leaf surface, is best illustrated by means of an example. In the following model a uniform thin film of water is assumed to exist on a plant canopy, which plays no part in the process other than to control the atmospheric turbulence such that the maximum deposition velocity (v_d) for any gas is 10 mm s^{-1}. This is equivalent to setting ($r_a + r_b$) to 100 s m^{-1} and the surface resistance to zero. Actual deposition rates of less than 10 mm s^{-1} represent a measurable surface resistance in addition to the atmospheric terms. For the purposes of the model, the surface has no stomata, so that the only sink or source for gases is the atmosphere. As the model starts with a layer of pure water, the best analogy with nature is that of dew formation on an inert surface. Chemical equilibria and rate constants included in the model are shown in Table 15.1.

15.5.1 *Deposition velocity*

Figure 15.3a shows the effective deposition velocity for SO_2 (1 ppbV) (1 ppbV = 1 part in 10^9 by volume) to a water layer 0.1 mm thick, for a range of different NH_3 concentrations, and for a model with or without oxidation of S^{IV} (i.e. $SO_2(aq)$, HSO_3^-, SO_3^{2-}) to sulphate. For the simplest case, with no oxidation and very low

Table 15.1. *Conditions imposed on the model of gas exchange with a thin water film*

Temperature	20°C
Carbon dioxide concentration	340 ppmV
Thickness of water film	0.1 mm (0.05–0.2 mm)
Maximum deposition velocity	10 mm s^{-1}
Oxidation rate: $S^{IV} \rightarrow SO_4^{2-}$	1.35×10^{-3} s^{-1} (Benner *et al.*, 1992)
Henry's Law constant SO_2	1.5×10^{-5} M Pa^{-1} (1.5 M atm^{-1})
Henry's Law constant NH_3	9.14×10^{-4} M Pa^{-1} (92.6 M atm^{-1})
Henry's Law constant CO_2	3.79×10^{-7} M Pa^{-1} (3.84×10^{-2} M atm^{-1})
Dissociation constants S^{IV}	1.48×10^{-2} M; 7×10^{-8} M
Dissociation constants NH_4^+	1.71×10^{-5} M
Dissociation constants $[H_2CO_3]$	4.17×10^{-7} M; 4.79×10^{-11} M

The rate of transfer of a gas across the air–liquid interface is assumed to be fast relative to the rate at which gas is transferred to/from the atmosphere (i.e. maximum rate of surface exchange = 10 mm s^{-1}). The model was run using the *Facsimile* program (AEA Technology plc).

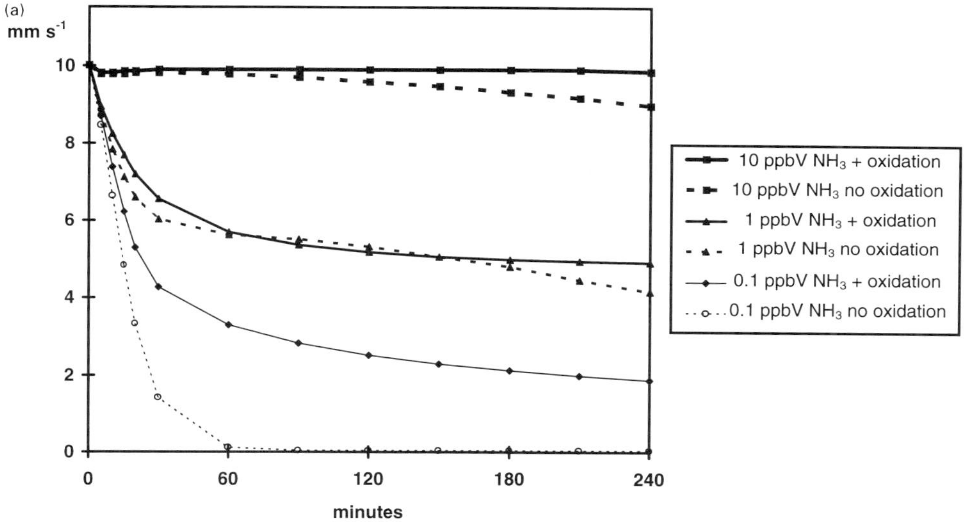

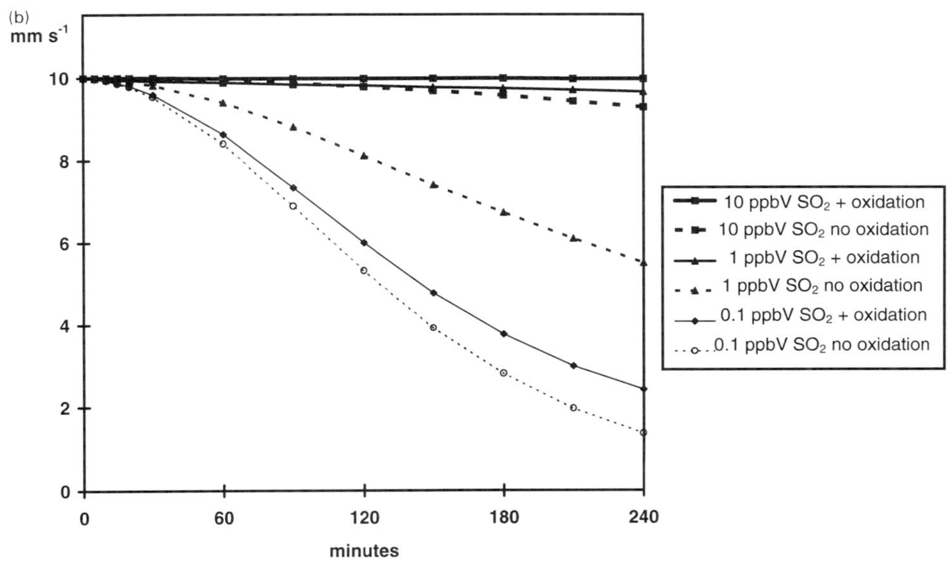

Figure 15.3. *(a) Deposition velocity (mm s^{-1}) of SO_2 (1 ppbV) to a water film 0.1 mm thick as a function of NH_3 concentration and oxidation in the aqueous phase ($S^{IV} \rightarrow SO_4^{2-}$). The maximum deposition velocity was set at 10 mm s^{-1}. Other conditions are in Table 15.1. (b) Deposition velocity (mm s^{-1}) of NH_3 (1 ppbV) to a water film 0.1 mm thick as a function of SO_2 concentration and oxidation in the aqueous phase ($S^{IV} \rightarrow SO_4^{2-}$). The maximum deposition velocity was set at 10 mm s^{-1}.*

NH_3 concentrations (0.01 ppbV), equilibrium is achieved in about 1 h, and no further deposition occurs. As NH_3 concentrations increase, with no aqueous-phase oxidation, the rate of deposition increases (dashed lines) until, with NH_3:SO_2 ratios of 10:1, the rate of transfer to the surface is almost at the maximum rate, and both S^{IV} and NH_3 are absorbed by the aqueous film. Saturation of the film is not achieved over time scales of several hours. Oxidation of S^{IV} does not change the picture greatly, except that uptake rates are enhanced slightly. There is constant production of NH_4^+ and SO_4^{2-} ions in the aqueous film, at a rate determined by the atmospheric transfer rate (i.e. surface resistance is zero). Figure 15.3b shows a similar set of deposition velocities for NH_3, and it is immediately obvious that equilibrium and saturation of the aqueous film take much longer. Deposition rates, even at very low SO_2 concentrations, are still substantial after several hours. Even when SO_2:NH_3 ratios are 1:1, if oxidation of S^{IV} occurs, NH_3 is deposited as fast as possible, within the constraint imposed by atmospheric transport. The comparison of these two scenarios illustrates why surface resistances for NH_3 are generally smaller than for SO_2. Atmospherically limited uptake of SO_2 depends on the presence of large concentrations of NH_3, but the converse does not apply. It should also be noted that the transition from atmospherically limited to surface-limited deposition occurs at concentrations that are widely found in the atmosphere over time scales of minutes to hours.

15.5.2 *Composition of surface solution*

Figure 15.4a–e shows the development of the composition of the aqueous film with time, starting with pure water equilibrated with atmospheric CO_2. Figure 15.4a shows the increase in NH_4^+ concentration for a constant 1 ppbV NH_3 under different SO_2 concentrations and with or without oxidation of S^{IV}. The solid straight line marks the atmospherically limited rate of increase, and even after 4 h concentrations are only approaching 60 μM. Smaller SO_2 concentrations and no oxidation both have the effect of reducing NH_3 uptake, and the rate of accumulation of NH_4^+. The reason can be seen clearly in Figure 15.4b, which shows the development of solution pH under the same conditions. At low SO_2 concentrations and in the absence of oxidation, pH increases above the CO_2 equilibrium value of 5.6. Conversely, large SO_2 concentrations and oxidation of S^{IV} lead to a marked reduction in pH, as low as pH 4.0 after 4 h, so that there is no surface resistance to further NH_3 uptake. The anions to balance the production of NH_4^+ and H^+ in solution come from S^{IV}, and also from sulphate when oxidation occurs. Figure 15.4c shows that in the absence of oxidation, NH_4^+ concentrations are matched by S^{IV}, up to 60 μM. When oxidation occurs, however, the production of acidity limits further dissociation of S^{IV}, and less is taken up by the solution, leading to S^{IV} concentrations (now almost all as HSO_3^-) of only a few μM. Figure 15.4d shows the rate of sulphate production when oxidation is allowed to occur. In the presence of 1 ppbV NH_3 the rate of sulphate production is linear, reflecting the constant deposition velocity (not always at the maximum rate, cf. Figure 15.3a). The relative production of NH_4^+ and SO_4^{2-} ions in the film is shown in Figure 15.4e. When the gases are present in equal concentrations, the solution composition approximates $(NH_4)_2SO_4$ after a few minutes, and even when the NH_3:SO_2 ratio is 10:1 the ionic composition is biased towards a 2:1 mole ratio. In this case the balancing anions are HSO_3^- and HCO_3^-.

15.5.3 *Effect of different water film thicknesses*

These illustrations have all been for a constant film thickness of 0.1 mm. The thickness of the water film dictates the capacity of the surface to absorb pollutant

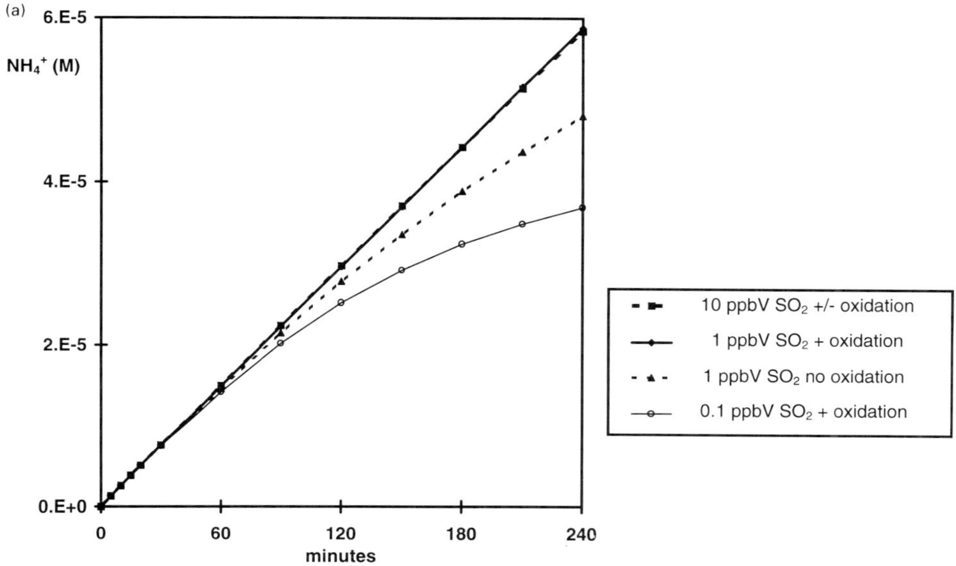

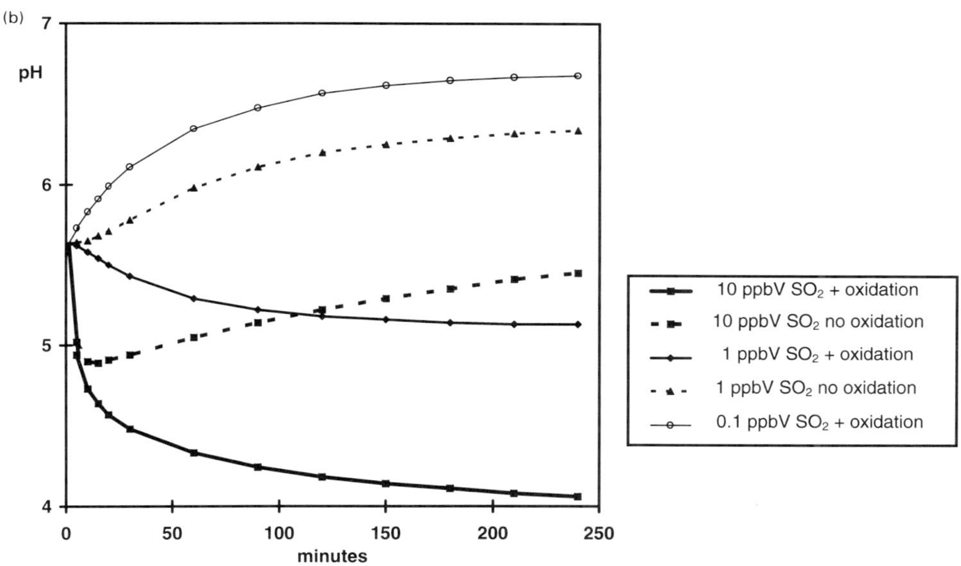

Figure 15.4. *Change in solution concentration of a water film (0.1 mm thick) exposed to 1 ppbV NH_3 and a range of SO_2 concentrations, with and without oxidation of S^{IV} to SO_4^{2-} in the aqueous phase. The maximum deposition velocity (rate of transfer from the atmosphere to the surface) was set at 10 mm s^{-1}. Other model conditions are given in Table 15.1. (a) Concentration of NH_4^+ ions (M). (b) pH. (c) Concentration of S^{IV} ($SO_2(aq)$ + HSO_3^- + SO_3^{2-}) (M). (d) Concentration of SO_4^{2-} (M) after oxidation. (e) Ammonium:sulphate molar ratio in aqueous film with oxidation of S^{IV}.*

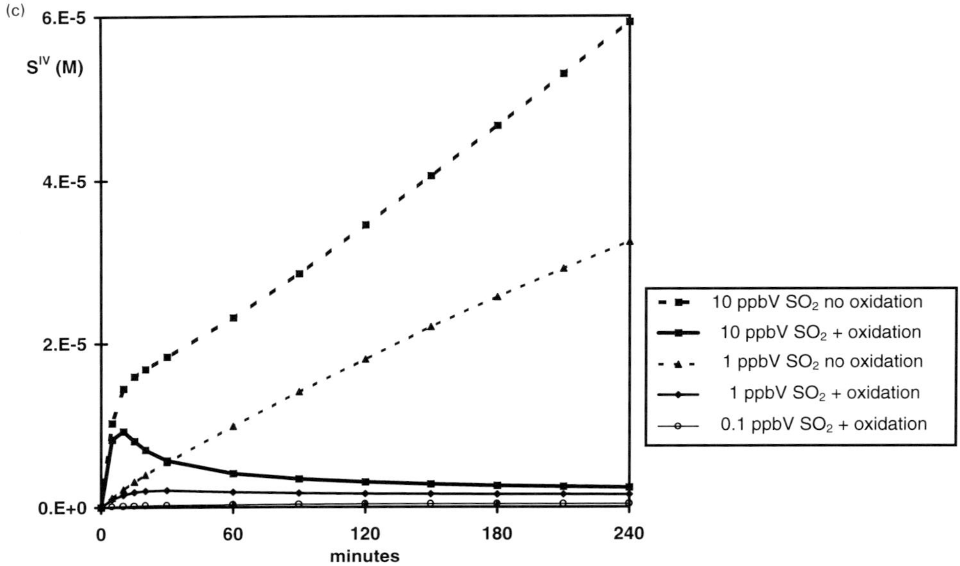

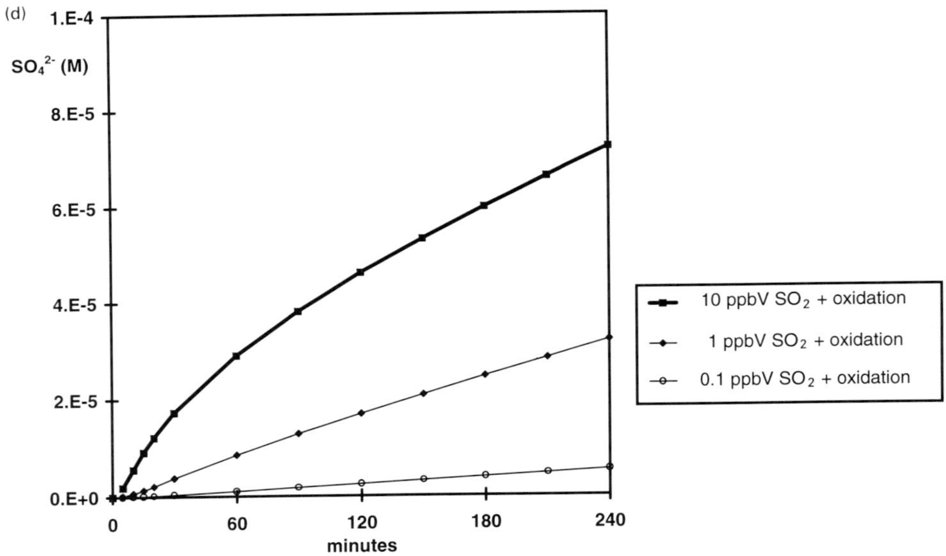

Figure 15.4. *Continued.*

gases; halving or doubling the film thickness changes the deposition velocity accordingly. The effect is much greater for SO_2 than for NH_3 (compare Figure 15.5a and b, noting the change in scale of the vertical axis in b). This differential dependence on film thickness goes some way to interpreting the much greater apparent dependency of SO_2 deposition rates on r. h., compared with NH_3.

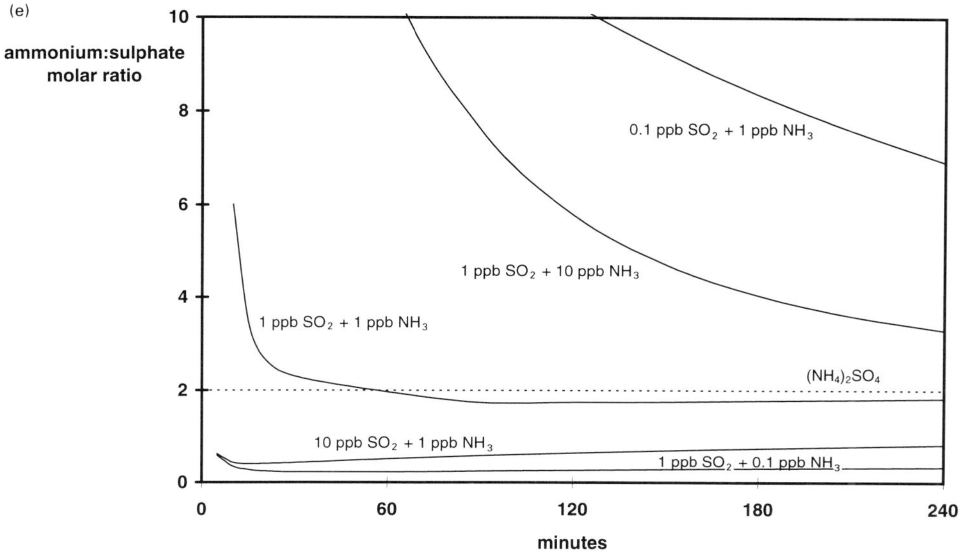

Figure 15.4. *Continued.*

The oxidation rate chosen here is derived from laboratory measurements in thin films (Benner *et al.*, 1992) and is rapid relative to many measurements derived from laboratory studies of bulk solutions. It has been suggested that S^{IV} oxidation by molecular oxygen can occur rapidly at water surfaces, through oxidation of HSO_3^- (Adema and Heeres, 1995). The rate used here (1.35×10^{-3} s^{-1}) is first-order in HSO_3^-, and was derived from experiments with 1 ppmV SO_2. If anything, this may underestimate the heterogeneous reaction rates at lower gas-phase concentrations. This rate also exceeds the modelled rate of oxidation of S^{IV} by ozone in solution, which is relatively slow because ozone is not very water-soluble. However, the reaction rate is similar to that obtained for Mn-catalysed oxidation, modelled to represent conditions on leaf surfaces where Mn^{2+} concentrations may be several micromolar (Arends and Eenkhoorn, 1990). Rates of oxidation of S^{IV} in surface films may also be affected by photochemical reactions, as happens in water droplets (Faust and Allen, 1992; Warneck, 1992).

15.5.4 *Comparison of model with real leaf surfaces*

Leaf surfaces are not just inert substrates on which chemistry can happen. There are many interactions that influence the composition of surface water, from the solution of particles on the surface (Hutchinson and Adams, 1987), to the exchange of cations (Mecklenburg *et al.*, 1966). Cation exchange, and solution of alkaline particles, may act as efficient buffers of acidity produced as S^{IV} is oxidized, promoting further SO_2 uptake by maintaining a high pH. Buffering may also be achieved by organic acids in the aqueous layer, either deposited from the atmosphere, or produced by the leaf. Eventually, however, the cation exchange capacity will be exceeded, and acidity will accumulate. The presence of buffering capacity serves to delay the production of acidity in solutions on leaf surfaces. Rates of cation transport through leaf cuticles are slow (Scherbatskoy and Tyree, 1990),

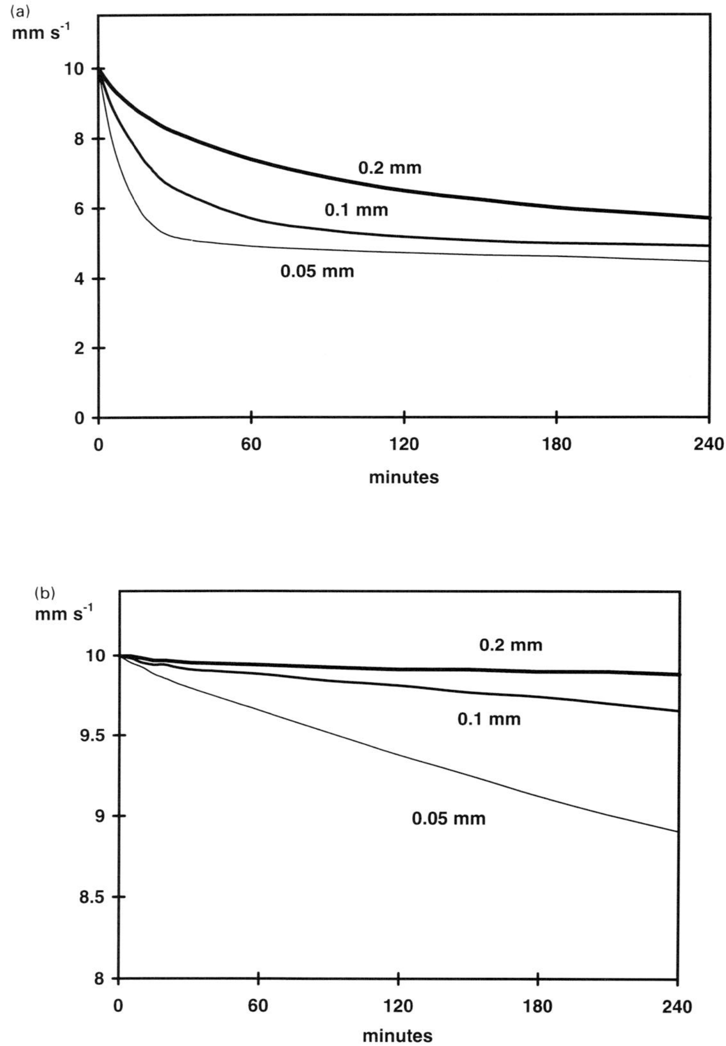

Figure 15.5. *Change in deposition velocity as a function of water film thickness, with* SO_2 *and* NH_3 *concentrations both at 1 ppbV, and oxidation of* S^{IV} *in the aqueous film. Other conditions as in Table 15.1. (a) Sulphur dioxide. (b) Ammonia.*

and not sufficient to provide continuous buffering at rates commensurate with atmospheric deposition and oxidation (Lauver *et al.*, 1992). The cation buffering capacity of the leaf surface is probably only regenerated during dry periods, by exchange with apoplast solutions. However, the effective buffering capacity of the leaf surface in dry periods is probably dominated by the capture of alkaline particles derived from soil (Reiners *et al.*, 1986). In the presence of surfactants, ion transport rates may be enhanced (Cantliffe and Wilcox, 1972), but pollutant ionic surfactants (e.g. Na salts of detergents in sea-spray) may themselves act as weak acids and buffer acidity, reducing the extent of cation exchange (Paoletti *et al.*, 1989).

15.6 Dry deposition of ozone to wet surfaces

The deposition of gases that are not very soluble in water is also influenced by the presence of surface water. This influence may be as simple as a reduction in uptake of gases through stomata, because of an additional water film across the stomatal opening. Reductions in photosynthetic rates of wetted leaf surfaces have been attributed to the increased diffusion resistance of the surface film (Brewer and Smith, 1994). Similarly, reductions in the deposition of ozone to wetted plant canopies have been attributed to the covering of the major surface sink in the stomata (Fuentes *et al.*, 1992). However, surface wetness may also influence the rate of deposition to the external surfaces of leaves. In general, surface water may have little detectable effect on surface resistance to ozone deposition, but measurements over deciduous forest (Fuentes *et al.*, 1992) and over grape vines (Grantz *et al.*,1995) have shown that surface wetness can greatly enhance ozone deposition rates. For the deciduous forest, the mechanism appears to be the reaction of ozone with organic solutes in the water film, derived from the leaf surface. Experimental wetting of red maple (*Acer rubrum*) leaves enhanced ozone deposition, whereas wetting poplar (*Populus deltoides tricocarpa*) leaves did not. Ozone deposition to leaf extracts of red maple was also enhanced, leading to the conclusion that a solute, presumed to be organic but as yet not identified, was involved (Fuentes and Gillespie, 1992; Fuentes *et al.*, 1994). Nevertheless, ozone deposition to plant canopies is still dominated by stomatal uptake and reaction inside the leaf.

15.7 Evaporation of surface water

Pollutants can be deposited to wet leaf surfaces and can accumulate in solution, but what happens as the surface dries, and water evaporates? For gases such as ozone, which are presumed to have reacted with an organic solute, the process is not reversible. Oxidation of dissolved SO_2 is also not reversible on evaporation, but dissolved NH_3 and SO_2 can, in principle, be released from the aqueous phase as water evaporates. This process has been observed in the field as upward fluxes of gases from drying canopies (Erisman and Wyers, 1993; Fowler and Cape, 1983), but the process, as for deposition, is not linear. Evaporation of water leads to a concentration of non-volatile solutes, and the consequent increase in acidity may lead to displacement of SO_2 from solution faster than predicted from a loss of liquid volume (Cape, 1988). The process of evaporation of water droplets on surfaces has been modelled (Bourges and Shanahan, 1993; Brain and Butler, 1985). Although the droplet volume decreases, the area of contact with the leaf does not necessarily recede. The process of evaporation of droplets from leaf surfaces can be observed when making contact angle measurements; contact areas seldom decrease, and often increase as droplets evaporate. In theory, evaporation can occur even during simultaneous deposition of cloud droplets (Unsworth, 1984), leading to persistent exposure of leaf surfaces to high solute concentrations. Although concentration of solutes during evaporation has been observed in the field, the pattern does not always correspond to theory (Milne *et al.*, 1990).

The concentration of ionic solutes as droplets dry can produce injury to leaves, presumably by generating large ionic concentration gradients which encourage ion penetration through the cuticle (Heller *et al.*, 1995). However, at high concentrations, as noted earlier, simple extrapolation of dilute aqueous chemistry does not apply. Not only dissociation equilibria, but also cation exchange equilibria are likely to be affected in ways that are not readily accessible to simple models.

15.8 Conclusion

When leaf and other plant surfaces are wetted by dew, rain or by microscopic layers of water associated with surface salts, chemical reactions that do not occur in dry conditions become possible. The surface water, even if not a continuous liquid film, may cause particles to dissolve, and may enhance the exchange of cations between surface salts and the underlying leaf surface. Organic substances may also be brought into solution. These can originate from inside the leaf, or from microbial or insect activity on the leaf surface. Great spatial heterogeneity may be found across a leaf, and throughout a plant canopy, either as a result of structural or physiological differences in the vegetation, or from earlier atmospheric deposition of particles or water drops. Such heterogeneity contributes to the great variability in the pattern of water deposition below a plant canopy, and in the chemical composition of throughfall. The wetted surface also provides a site for the removal of gases from the atmosphere, either by irreversible reaction, or reversible solution. Rates of removal of gases which form H^+ or OH^- ions in solution, such as SO_2 and NH_3, are controlled by the absolute and relative concentrations of such gases, and by the ions available from surface salts. Overall uptake rates will be limited by the large-scale structure of the canopy and by wind speed, which together determine the maximum transport rate of gases from the atmosphere to the earth's surface. For pollutant gas concentrations typical of the atmosphere, between 0.1 and 10 ppbV, equilibration with thin films of water on leaf surfaces may take minutes to hours. True atmosphere/surface chemical equilibrium is seldom likely to be achieved in the continuously varying pattern of temperature, relative humidity and gas concentrations which characterize the atmospheric boundary layer. However, relatively simple parametrizations of the surface resistance to pollutant deposition have been derived from a limited range of field measurements, as functions of surface wetness and relative humidity. Long-term monitoring of dry deposition rates at a wider range of field sites, with different patterns of gas concentrations, different vegetation types and different climates, will lead to further parametrizations which may one day be sufficient to provide reliable estimates of dry deposition of trace gases across a wide range of vegetation types and pollution climates.

References

Adam, N.K. (1963) Principles of water repellency. In: *Waterproofing and Water-Repellency*, (ed. J.L. Moilliet). Elsevier, Amsterdam, pp.1–23.

Adema, E.H. and Heeres, P. (1995) Dry deposition of sulphur dioxide and ammonia on wet surfaces and the surface oxidation kinetics of bisulphate. *Atmos. Environ.* **29**, 1091–1103.

Arends, B.G. and Eenkhoorn, S. (1990) The influence of manganese leached from plants on dew chemistry. *Environ. Tech.* **11**, 181–188.

Benner, W.H., Ogorevc, B. and Novakov, T. (1992) Oxidation of SO_2 in thin water films containing NH_3. *Atmos. Environ.* **26A**, 1713–1723.

Bourges, C. and Shanahan, M.E.R. (1993) Influence of evaporation on the contact angle of water drops. *Comptes Rendus Acad. Sci. III* **316**, 311–316

Brain, P. and Butler, D.R. (1985) A model of drop size distribution for a system with evaporation. *Plant Cell Environ.* **8**, 247–252.

Brewer, C.A. and Smith, W.K. (1994) Influence of simulated dewfall on photosynthesis and yield in soybean isolines (*Glycine max* [L] Merr cv Williams) with different trichome densities. *Int. J. Plant Sci.* **155**, 460–466.

Brewer, C.A., Smith, W.K. and Vogelman, T.C. (1991) Functional interaction between leaf trichomes, leaf wettability and the optical properties of water droplets. *Plant Cell Environ.* **14**, 955–962.

Burkhardt, J. and Eiden, R. (1994) Thin water films on coniferous needles. *Atmos. Environ.* **28**, 2001–2011.

Burkhardt, J., Peters, K. and Crossley, A. (1995) The presence of structural surface waxes on coniferous needles affects the pattern of dry deposition of fine particles. *J. Exp. Bot.* **46**, 823–831.

Cantliffe, D.J. and Wilcox, G.E. (1972) Effect of surfactant on ion penetration through leaf wax and a wax model. *J. Am. Soc. Hort. Sci.* **97**, 360–363.

Cape, J.N. (1983) Contact angles of water droplets on needles of Scots pine (*Pinus sylvestris*) growing in polluted atmospheres. *New Phytol.* **93**, 293–299.

Cape, J.N. (1988) Chemical interactions between cloud droplets and trees. In: *Acid Deposition at High Elevation Sites* (eds M.H. Unsworth and D. Fowler). Kluwer Academic Publishers, Dordrecht, pp. 639–649.

Cape, J.N., Paterson, I.S. and Wolfenden, J. (1989) Regional variation in surface properties of Norway spruce and Scots pine needles in relation to forest decline. *Environ. Pollut.* **58**, 325–342.

Cape, J.N., Sheppard, L.J., Binnie, J. *et al.* (1995) Throughfall deposition of ammonium and sulphate during ammonia fumigation of a Scots pine forest. *Water Air Soil Pollut.* **85**, 2247–2252.

Challen, S.B. (1960). The contribution of surface characters to the wettability of leaves. *J. Pharm. Pharmacol.* **12**, 307–311.

Chameides, W.L. (1987) Acid dew and the role of chemistry in the dry deposition of reactive gases to wetted surfaces. *J. Geophys. Res.* **92D**, 11 895–11 908.

Draaijers, G.P.J. and Erisman, J.W. (1993) Atmospheric sulphur deposition to forest stands — throughfall estimates compared to estimates from inference. *Atmos. Environ.* **27A**, 43–55.

Duyzer, J.H., Verhagen, H.L.M., Weststrate, J.H. *et al.* (1994) The dry deposition of ammonia onto a Douglas fir forest in the Netherlands. *Atmos. Environ.* **28**, 1241–1253.

Erisman, J.W., Wyers, G.P. (1993) Continuous measurements of surface exchange of SO_2 and NH_3 — implications for their possible interaction in the deposition process. *Atmos. Environ.* **27**, 1937–1949.

Erisman, J.W., Versluis, A.H., Verplanke, T.A.J.W. *et al.* (1993) Monitoring the dry deposition of SO_2 in the Netherlands — results for grassland and heather vegetation. *Atmos. Environ.* **27A**, 1153–1161.

Erisman, J.W., van Elzakker, B.G., Mennen, M.G. *et al.* (1994) The Elspeetche Veld experiment on surface exchange of trace gases — summary of results. *Atmos. Environ.* **28**, 487–496.

Faust, B.C. and Allen, J.M. (1992) Aqueous-phase photochemical sources of peroxyl radicals and singlet molecular oxygen in clouds and fog. *J. Geophys. Res.* **97D**, 12 913–12 926.

Fogg, W.H. (1944) Diurnal fluctuation in a physical property of leaf cuticle. *Nature* **154**, 515.

Fogg, W.H. (1947) Quantitative studies on the wetting of leaves by water. *Proc. R. Soc. B* **134**, 503–522.

Fowler, D. (1978) Dry deposition of SO_2 on agricultural crops. *Atmos. Environ.* **12**, 369–373.

Fowler, D. and Cape, J.N. (1983) Dry deposition of SO_2 onto a Scots pine forest. In: *Precipitation Scavenging, Dry Deposition and Resuspension.* (eds H.R. Pruppacher, R.G. Semonin and W.G.N. Slinn). Elsevier, Amsterdam, pp. 763–774.

Fowler, D. and Unsworth, M.H. (1979) Turbulent transfer of sulphur dioxide to a wheat crop. *Q. J. R. Meteorol. Soc.* **105**, 767–783.

Fuentes, J.D. and Gillespie, T.J. (1992) A gas exchange system to study the effects of leaf surface wetness on the deposition of ozone. *Atmos. Environ.* **26A**, 1165–1173.

Fuentes, J.D., Gillespie, T.J., den Hartog, G. *et al.* (1992) Ozone deposition onto a deciduous forest during dry and wet conditions. *Agric. For. Meteorol.* **62**, 1–18.

Fuentes, J.D., Gillespie, T.J. and Bunce, N.J.(1994) Effects of foliage wetness on the dry deposition of ozone onto Red Maple and Poplar leaves. *Water Air Soil Pollut.* **74**, 189–210.

Grantz, D.A., Zhang, X.J., Massman, W.J. *et. al.* (1995) Effects of stomatal conductance and surface wetness on ozone deposition in field-grown grape. *Atmos. Environ.* **29**, 3189–3199.

Heller, L.I, Shaw, A.J. and Jacobson, J.S. (1995) Exposure of red spruce seedlings to acid mist: importance of droplet composition just prior to drying periods. *New Phytol.* **129**, 55–61.

Holloway, P.J. (1969) The effects of superficial wax on leaf wettability. *J. Appl. Biol.* **63**, 145–153.

Holloway, P.J. (1970) Surface factors affecting the wetting of leaves. *Pestic. Sci.* **1**, 156–163.

Hutchinson, T.C. and Adams, C.M. (1987) Comparative abilities of leaf surfaces to neutralize acidic raindrops I. The influence of calcium nutrition and charcoal-filtered air. *New Phytol.* **106,** 169-183.

Jagels, R. (1991) Biophysical aspects of fog deposition on the needles of three conifers. *J. Exp. Bot.* **42,** 757-763.

Jagels, R. (1994) Leaf wettability as a measure of air pollutant effects. In: *Air Pollutants and the Leaf Cuticle* (eds K.E. Percy, J.N. Cape, R. Jagels and C.J. Simpson). Springer-Verlag, Berlin, pp. 97-105.

Lauver, T.L., McCune, D.C., Shaff, J.E. *et al.* (1992) The use of ion-selective microelectrodes for measuring calcium and hydrogen ion transfer between foliar surfaces and simulated rain solutions. *New Phytol.* **121,** 179-185.

Leyton, L. and Juniper, B.E. (1963) Cuticle structure and water relations of pine needles. *Nature* **198,** 770-771.

Lovett, G.M. and Lindberg, S.E. (1984) Dry deposition and canopy exchange in a mixed oak forest as determined by analysis of throughfall. *J. Appl. Ecol.* **21,** 1013-1027.

Mecklenburg, R.A., Tukey, H.B. Jr and Morgan, J.V. (1966) A mechanism for the leaching of calcium from foliage. *Plant Physiol.* **41,** 610-613.

Millero, F.J., Hershey, J.P., Johnson, G. *et al.* (1989) The solubility of SO_2 and the dissociation of H_2SO_3 in NaCl solutions. *J. Atmos. Chem.* **8,** 377-389.

Milne, R., Crossley, A. and Henderson, C. (1990) Field measurements of the acidity of cloudwater deposited on polypropylene surfaces and Sitka spruce shoots. *Trees* **4,** 205-210.

Monson, R.W.K., Grant, M.C., Jaeger, C.H. *et al.* (1992) Morphological causes for the retention of precipitation in the crowns of alpine plants. *Environ. Exp. Bot.* **32,** 319-327.

Paoletti, E., Gellini, R. and Barbolani, R. (1989) Effects of acid fog and detergents on foliar leaching of cations. *Water Air Soil Pollut.* **45,** 49-61.

Reiners, W.A., Olson, R.K., Howard, L. *et al.* (1986) Ion migration from interiors to outer surfaces of balsam fir needles during dry interstorm periods. *Environ. Exp. Bot.* **26,** 227-231.

Scherbatskoy, T. and Tyree, M.T. (1990) Kinetics of exchange of ions between artificial precipitation and maple leaf surfaces. *New Phytol.* **114,** 703-712.

Sutton, M.A., Fowler, D. and Moncrieff, J.B. (1993) The exchange of atmospheric ammonia with vegetated surfaces. I: Unfertilized vegetation. *Q. J. R. Meteorol. Soc.* **119,** 1023-1045.

Unsworth, M.H. (1984) Evaporation from forests in cloud enhances the effects of acid deposition. *Nature* **312,** 262-264.

Wagner, G.H., Steele, K.F. and Peden, M.E. (1992) Dew and frost chemistry at a midcontinent site, United States. *J. Geophys. Res.* **97D,** 20 591-20 597.

Warburton, F.L. (1963) The effect of structure on waterproofing. In: *Waterproofing and Water-Repellency* (ed. J.L. Moilliet). Elsevier, Amsterdam, pp.24-51.

Warneck, P. (1992) Chemistry and photochemistry in atmospheric water drops. *Ber. Bunsenges. Phys. Chem.* **96,** 454-460.

Watanabe, T. and Yamaguchi, I. (1992) Studies on wetting phenomena on plant leaf surfaces. 3: A retention model for droplets on solid surfaces. *Pestic. Sci.* **34,** 273-279.

Weiss, A. (1988) Contact angle of water droplets in relation to leaf water potential. *Agric. For. Meteorol.* **43,** 251-259.

Wesely, M.L., Sisterson, D.L. and Jastrow, J.D. (1990) Observations of the chemical-properties of dew on vegetation that affect the dry deposition of SO_2. *J. Geophys. Res.* **95,** 7501-7514.

16

Interactions between epiphytic microbes and deposited compounds

S. Huttunen

16.1 Introduction

Foliar associated microorganisms can be classified as epiphytes living on plant surfaces (Blakeman, 1981; Dickinson and Preece, 1976), endophytes inhabiting internal plant tissues without causing apparent injury to their host (Caroll, 1986; Petrini, 1986, 1991; Petrini *et al.*, 1992) and plant pathogens causing visible injury to the host plant (Agrios, 1988). Defence responses of plants to pathogens have been recently reviewed by Kombrink and Somssich (1995).

Interactions between microbes and deposited compounds are mainly determined by infection behaviour and factors affecting successful growth and colonization of microbes. Deposited air pollutants affect microbial growth on the surface (phyllosphere) and wettability via epicuticular waxes (EW). The annual change of needle wettability in evergreens is the higher, the more northerly the location, and it shows significant relationships to long-term climatic variables (Turunen and Huttunen, 1996). Phyllospheric saprophytes can have an antagonistic effect on foliar pathogens. Atmospheric pollutants decrease phyllospheric saprophytes, for example yeasts, but filamentous fungi seem to be tolerant to them (Magan and McLeod, 1991). Disruption in the balance between antagonistic phyllospheric saprophytes and pathogens may result in a change in pathogen activity and disease development. The presence of surface microorganisms affects leaf wettability (Schreiber, 1996) and buffering of compounds deposited in surface water films. Both effects affect deposition rate. Therefore, we have two feedback cycles that intersect with each other: (a) deposition of air pollutants–phyllosphere–wettability and buffering–deposition rates; (b) deposition of air pollutants–EW–wettability–deposition rates.

In this chapter, I will consider interactions between microbes and deposited compounds, starting from deposition aspects and then concentrating on some examples.

Plant Cuticles, edited by G. Kerstiens.

16.2 Pollutant deposition

16.2.1 *Acidifying compounds*

Oxides of nitrogen or sulphur deposited on leaf and stem surfaces may dissociate in the water film and/or react with the cuticular wax components and result in damaged cuticular surfaces. Solubility of NO in aqueous media varies with temperature, as NO is more soluble at low temperatures than at high temperatures (Stephen and Stephen, 1963). This reduction in solubility as the temperature rises has important implications for plants growing at low temperatures, especially as the rates of conversion of NO and NO_2 are reduced at low temperatures.

Aerial algae react positively to increased atmospheric nitrogen deposition. A Swedish inventory revealed that the green alga *Pleurococcus vulgaris* on young Norway spruces (*Picea abies*) was most abundant in Blekinge, Skåne, Halland and Bohuslän, where total nitrogen deposition is over 20 kg ha^{-1} $year^{-1}$ or more (Göransson, 1988; Figure 16.1). Bråkenhielm and Qinghong (1995) monitored green algae on Norway spruce needles for 4 years. The algal colony thickness and colonization rate on needles increased from north to south and from east to west, as did the length of the growing season and the amount of atmospheric nitrogen and sulphur deposition. The algal colonization rate increased under the impact of just a few kilograms of nitrogen deposition.

Donelly and Dowding (1994) found the algal cell coverage of *Pinus sylvestris* needles to be up to 80% at some sites in Ireland. The microflora on the needle surfaces, which was characterized by algae and fungi, was more pronounced on pines close to a fur farm (< 100 m) than on the control trees (> 1000 m) according to Bäck *et al.* (1996b).

The role of surface water in modifying pollutant flux rates to the leaf surface is crucial (see Chapter 15). Thin water films are present on spruce needles at relative humidity (r.h.) values as low as 50%, despite the hydrophobic nature of the surfaces (Burkhardt and Eiden, 1994). Dew droplets collected from Scots pine needles showed significantly higher ion concentrations than those taken from Norway spruce trees. This can be explained by the micromorphology of the needle surface. The higher microscale roughness of the wax layer of the pine causes a more efficient flux of atmospheric aerosol particles compared to the spruce needle surface. Dew on coniferous needles was shown to be capable of maintaining pH values below 3.0 for several hours (Burkhardt and Eiden, 1990).

Plant surfaces are normally somewhat acid and negatively charged. Certain plants resist cellular pH change, even when exposed regularly to simulated acid rain down to pH 3.0, by pumping the extra acidity into their vacuoles. The extent and nature of buffering is an inherited characteristic.The buffering capacity of leaves and needles and their surfaces varies due to the season, the species and the site (Huttunen *et al.*, 1981a). In fumigation experiments with SO_2, NO_2 and O_3, the leaf surface pH was significantly decreased by fumigation in all the species investigated by Steubing *et al.* (1989). The neutralizing capacity of *Allium ursinum* leaf surfaces was elevated by SO_2, NO_2 and O_3 fumigation (Steubing *et al.*, 1989). Hutchinson *et al.* (1986) found the neutralizing capacity against acid deposition to be related to wettability rather than susceptibility to foliar injury. Hutchinson and Adams (1987) found high neutralizing capacities against acid droplets in herbaceous species. Surface wettability and the speed with which acid droplets dried out were important in determining the neutralizing ability of leaf surfaces.

The leaf surface pH varies somewhat less in conifers than in broad-leaved plants. The surface pH values of pine and spruce needles were studied in groups with acid

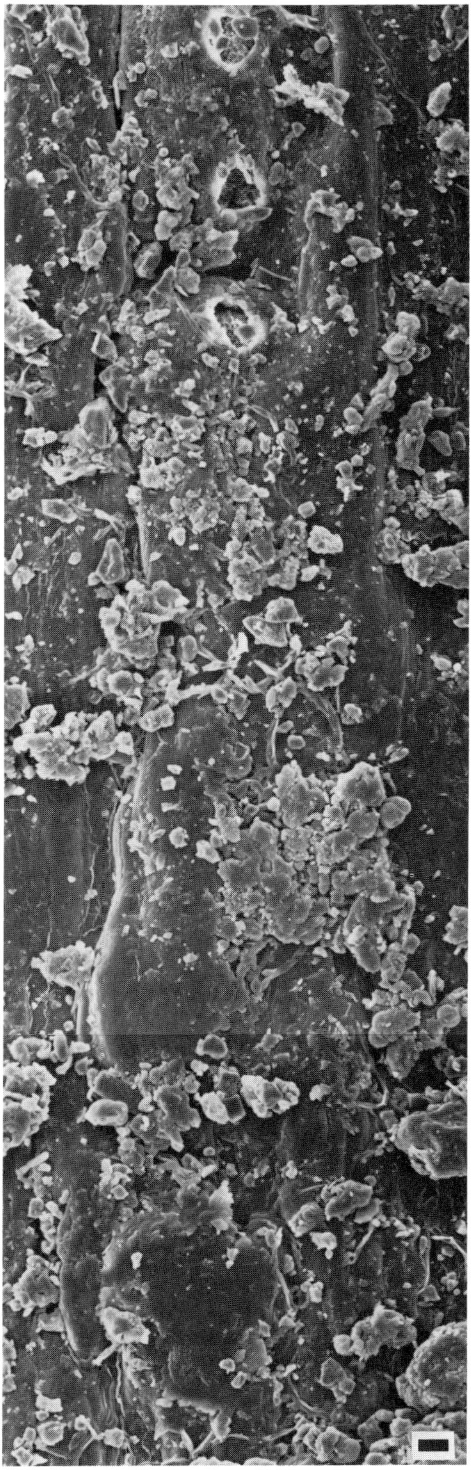

Figure 16.1. *Scots pine (*Pinus sylvestris *L.) needle surface with autotrophic algal associations; previous year's needles on a traffic-polluted site in southern Sweden. Heterotrophic fungi seem to colonize well among algae. (Bar = 1 μm.)*

rain treatment at pH 3.0 and pH 4.0, with clean water and without treatment (Bäck *et al.*, 1996a; Huttunen *et al.*, 1990). The needle samples for pH measurements were collected monthly from August 1986 to October 1987 and again in May and August 1989. Seasonal changes were observable in the needle surface pH values, with a gradual increase from August to mid-October (Huttunen and Laine, 1987; Huttunen *et al.*, 1990; Turunen and Huttunen, 1989). The summertime surface pH was higher (pH 4.6–4.7) in young developing needles than in older needles (3.9–4.0). *Picea abies* had summertime pH values between 4.0 and 4.5 and wintertime values between 4.4 and 4.8. In *Pinus sylvestris* (Scots pine) the summertime values varied between 3.8 and 4.8, depending on needle age, and the wintertime values were between 4.4 and 4.9. The mean leaf surface pH was 4.6 in pine and 4.3 in spruce needles. A significant difference between treatments was found in previous year's pine needles between the clean water and pH 3.0 treatments. Neutral water treatment increased surface pH significantly. A slight delay in surface pH changes in autumn was observed in acid-treated seedlings (Turunen and Huttunen, 1989). The southern provenances of Scots pine and Norway spruce needles tended to have significantly lower surface pH values. The surface pH values tended to increase with needle age (Table 16.1). The surface pH values of different broad-leaved tree species varied from pH 3.3 to 5.7 and in needles from 3.8 to 5.0 (Table 16.2).

Increasing acidity of irrigation water did not affect the development of rust, *Melampsoriudium betulinum*, but nitrogen in the acidified water resulted in a prolongation of seedling growth, and the leaves hence became senescent later in the autumn and the rust grew for a longer time (Vuorinen, 1992). Epiphytic microfungi on leaves and needles are usually reduced by acid treatment, and needle and leaf endophytes are negatively affected by acid treatments (Helander *et al.*, 1993a,b; Ranta, 1990).

Delayed development of the epistomatal waxes and four types of deformed stomatal complex were found in a simulated acid rain experiment in Kevo (Turunen and Huttunen, 1991). An acid rain-induced delay of wax synthesis was observed in Scots pine and Norway spruce seedlings. This may, together with the Ca leaching observed in young needles, predispose and modify these needles for epiphytic and endophytic fungi or pathogens (Huttunen *et al.*, 1989; Turunen *et al.*, 1995). Kirkwood *et al.* (1989) found *Aureobasidium pullulans*, *Botrytis cinerea* and *Schlerophoma pithyophila* to grow best at low pH (2.5–3.0). In the Kevo area the *Aureobasidium/Hormonema* complex was most common (Helander, 1994b).

The direct effect of leaf surface pH and buffering capacity on microbial ecology may be associated with the pH optimum of the infection, microbial growth and nutrition, or the phenological coincidence of favourable surface pH. The decreasing leaf surface pH values may limit favourable infection or growth or alter the surface architecture by decaying waxes. The changes in surface roughness can favour spore landing and promote the growth of some pathogenic fungi. The coincidence of pH changes and changes in buffering capacity or surface roughness can cause favourable or unfavourable conditions for phylloplanic microbes. Organic acids may play a significant role on wetted surfaces (Figures 16.2–16.5).

16.2.2 *Ozone*

A flux of ozone to plant surfaces results in its decomposition at the surface and its uptake through the cuticle. The deposition velocity is much higher than cuticular ozone permeance at equivalent ozone concentrations (cf. Section 5.3). Due to the ozone decay in the cuticle, ozone permeance is lower in thick than in thin cuticles

Table 16.1. *Surface pH of Norway spruce* (Picea abies) *and Scots pine* (Pinus sylvestris) *needles in simulated acid rain experiment (observed significant differences)*

Scots pine current needles
Abaxial surfaces of southern provenances more acid than surfaces of northern provenances

Scots pine previous year's needles
Water-treated: significantly higher pH than in pH 3.0 treated needles

Norway spruce 2-year-old needles
Abaxial and adaxial surfaces of southern provenances (mean pH 4.7) more acid than northern provenances (mean pH 5.2)

Norway spruce 3-year-old needles
Abaxial surfaces of southern provenances (mean pH 4.8) more acid than northern provenances (mean pH 5.1)

Table 16.2. *The pH value of needle and leaf surfaces in August 1986 (means ± standard error)*

	Botanical gardens		Park trees	
Species	Adaxial surface	Abaxial surface	Adaxial surface	Abaxial surface
Acer negundo	4.3±0.06	3.3±0.03	3.7±0.05	3.5±0.05
Alnus glutinosa	4.3±0.05	4.0±0.03	4.9±0.05	4.6±0.04
A. incana	5.1±0.04	5.2±0.05	5.0±0.04	4.9±0.05
A. rugosa v. *americana*	5.1±0.04	4.6±0.07		
Betula nana	4.7±0.06	4.5±0.05		
B. pubescens	4.8±0.05	4.8±0.06	4.2±0.03	4.0±0.05
Caragana arborescens	5.3±0.06	5.7±0.05	5.1±0.05	5.6±0.06
Cornus alba v. *sibirica*	5.7±0.05	5.5±0.03	5.2±0.06	5.3±0.04
Larix laricina	4.6±0.05	4.5±0.06		
L. laricina v. *alaskensis*	4.3±0.04	4.2±0.04		
L. sibirica	4.4±0.04	4.2±0.06	4.0±0.09	3.8±0.08
Lonicera tatarica	5.4±0.06	5.4±0.04	4.9±0.08	5.3±0.07
Malus baccata	4.6±0.06	4.3±0.10		
M. x *robusta*			4.5±0.12	4.9±0.09
Picea abies	4.3±0.05	4.4±0.07	4.2±0.11	4.2±0.09
P. mariana	4.2±0.07	4.2±0.08		
Pinus banksiana	4.1±0.05	4.0±0.08		
P. cembra	4.7±0.07	4.6±0.08	4.2±0.07	4.1±0.03
P. contorta v. *latifolia*	4.8±0.11	4.8±0.14		
P. mugo:				
(1)	4.0±0.13	3.9±0.14	4.4±0.08	4.4±0.07
(2)	4.3±0.12	4.1±0.08		
P. silvestris	4.8±0.09	4.4±0.12	4.3±0.21	4.4±0.11
Populus balsamifera	4.9±0.07	4.9±0.06	4.4±0.06	4.7±0.07
P. tremula x *tremuloides*	5.2±0.10	5.3±0.10		
P. tremula			4.9±0.05	5.5±0.04
Prunus padus	5.1±0.06	5.3±0.05	5.1±0.04	5.3±0.08
Sorbus aucubaria:				
(1)	5.3±0.06	5.4±0.07	5.2±0.09	5.2±0.10
(2)	5.0±0.09	5.3±0.06		
Thuja occidentalis	5.0±0.07	4.9±0.03	4.9±0.06	4.8±0.05

Pinus mugo: (1) provenance Kankaanpää, Niinisalo; (2) provenance Karstula.
Sorbus aucuparia: (1) provenance Oulu, Järvitie; (2) provenance Oulu, Oritkari.

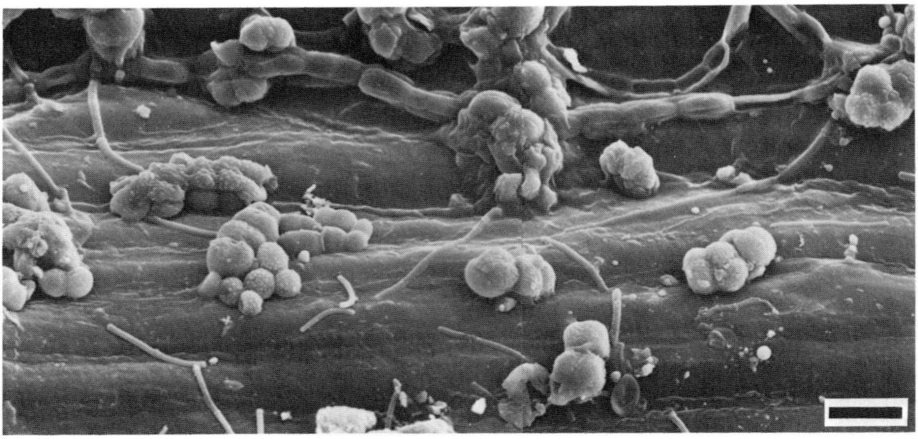

Figure 16.2. Lophodermium seditosum *on Scots pine (*Pinus sylvestris *L.) needle surface. Previous year's needles on a traffic-polluted site in southern Sweden. (Bar = 1 μm.)*

(Kerstiens and Lendzian, 1989a). The ozone deposition velocity increases along with moisture (Fuentes *et al.*, 1994). The plant defence systems induced by ozone seem to be similar to the pathogen-induced defence responses in plants (Kangasjärvi *et al.*, 1994). As in various stresses, the induction of a general defence system is indicated by an accumulation and increase in activity of pathogenesis-related proteins, polyamines and several other enzymes of oxygen radical scavenging and ethylene synthesis (Foyer *et al.*, 1994; Pääkkönen, 1996). This opens up many questions concerning potential interactions. The formation of active oxygen species is most probably responsible for the plasma membrane lipid peroxidation that occurs after ozone treatment (Chevrier *et al.*, 1990). During the early stages of pathogen–plant interactions (Devlin and Gustine, 1992; Schwacke and Hager, 1992) or exposure to ultraviolet (UV) light, generation of superoxide radicals takes place in the apoplast. These may trigger defence responses in the cells of ozone-exposed plants (Coleman *et al.*, 1992; Jones and Coleman, 1991).

A range of responses take place in the interactions between pathogens and ozone (Costonis and Sinclair, 1972; Winner *et al.*, 1987). The effects on the host plant, on the pathogen or on both may lead to stimulations or inhibitions. Ozone exposure may decrease the vitality of trees and predispose them to pathogens, insects and diseases (Chappelka and Chevone, 1992). On the other hand, inhibitory effects of ozone against microorganisms and fungi have also been reported. Coleman *et al.* (1987) found that urediospore production was strongly affected by ozone treatment, cottonwood genotype and leaf age, but no interactions between these factors occurred. Jones *et al.* (1994) found that the growth of leaf spot fungus on the eastern cottonwood *Populus deltoides* was unaffected by O_3 and that O_3 exposure of the living plant reduced decomposition rates of leaf litter. The phenolic deposition process associated with ozone injury may have been the primary cause.

Moderately exposed (combinations of SO_2, NO_2 and O_3) Scots pine, Norway spruce and pubescent birch seedlings were studied in open-top chambers under climatic circumstances resembling those of southern Sweden today. The pine seedlings had higher needle, wood and root dry weight in the non-filtered (ambient air) group than in the $O_3 + NO_2$ or $O_3 + SO_2$ treated groups. O_3 and SO_2 caused a delay in EW synthesis (Figure 16.6). Birch leaves exhibited earlier senescence after

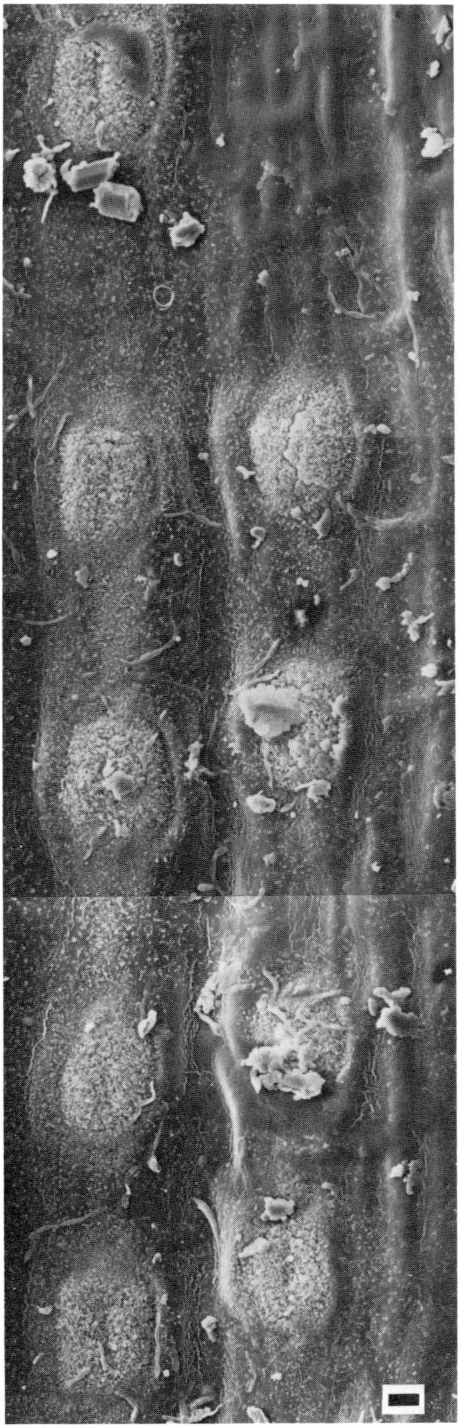

Figure 16.3. *Current year Norway spruce (*Picea abies *L. Karst.) needles from southern Sweden. Starting growth of* Chrysomyxa ledi *and small holes caused by* Lachniidae *herbivory (samples collected in November). (Bar = 1 μm.)*

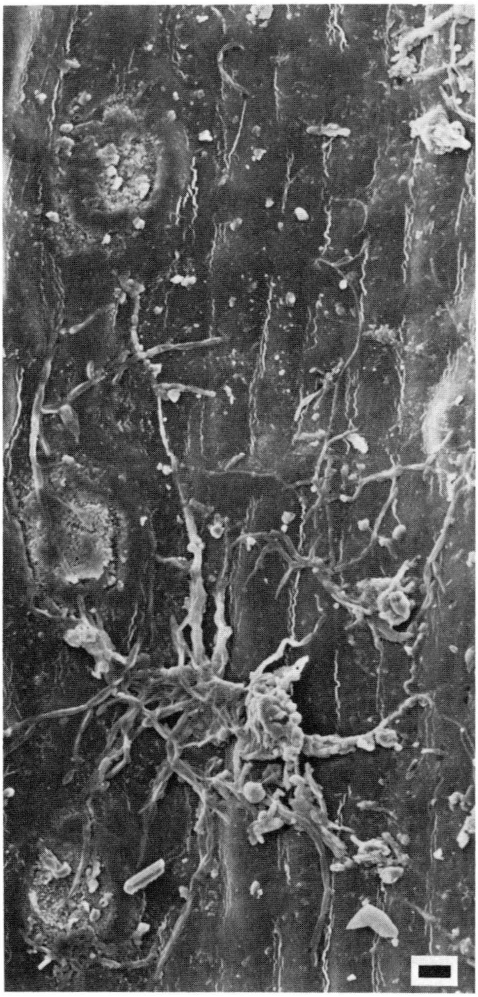

Figure 16.4. Chrysomyxa ledi *on previous year's Norway spruce needles in southern Sweden (samples collected in November). (Bar = 1 μm.)*

O_3 and NO_2 and $O_3 + NO_2 + SO_2$ treatments (Bäck *et al.*, 1996a). Delayed needle development and early senescence of birch leaves may have effects on host-pathogen interactions, as has been earlier indicated in the case of *Lophodermella sulcigena* (Jalkanen *et al.*, 1981) and *Melampsora betulinum* (Helander, 1994a). The water permeance of non-isolated and isolated plant cuticles was not affected by realistic ozone exposure (Kerstiens and Lendzian, 1989b).

16.2.3 *Other pollutants*

Studies on the kinetics of exchange of metal ions between artificial precipitation and sugar maple (*Acer saccharum*) leaf surfaces (Scherbatskoy and Tyree, 1990) indicated that the mechanisms of exchange differed depending on the ion species, the solution pH had significant effects on the rate of exchange, and the rates of ion diffusion through cuticles were small. Copper and lead were rapidly adsorbed, zinc was unchanged, and potassium, calcium and magnesium were released from the leaf surface.

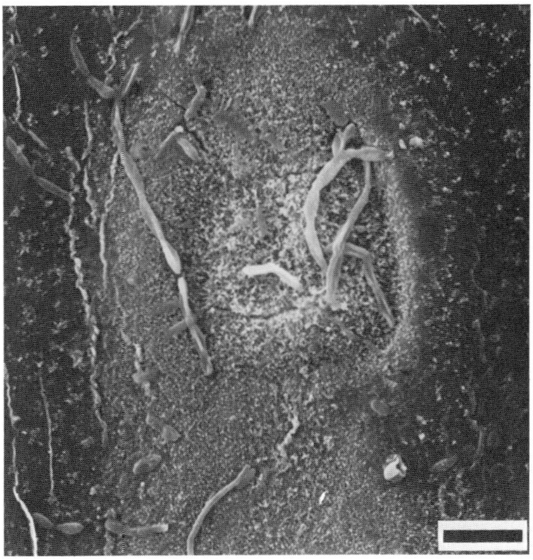

Figure 16.5. Chrysomyxa ledi *on previous year's Norway spruce needles. (Bar = 1 μm.)*

Bewley (1981) studied the number and species composition of the phylloplanic microflora of Corsican pine (*Pinus nigra* var. *maritima*) saplings which were uncontaminated, exposed to zinc, lead and cadmium from a smelter, or artificially dosed with zinc, lead and cadmium oxides. There was little difference in the total numbers of most groups of microorganisms isolated from each site. A higher percentage of metal-tolerant isolates were found in polluted samples. Artificially dosed plants also supported greater numbers of *Aureobasidium pullulans* and orange and pink chromogenic bacteria. These particular organisms were also among the most metal-tolerant fungi and bacteria *in vitro*. Helander (1994b) found in gradient analyses around smelters a reduced number of *Cenangium ferruginosum*-infected Scots pine needles. *Hormonema*-infected needles were not affected by the distance. The *Hormonema* strain isolated near the smelters had a higher metal tolerance than the strain isolated from a less polluted area. In the experimental part of the study, abundance of *Hormonema* on CuNi-treated trees and of *C. ferruginosum* on pH 3.0-treated needles was significantly reduced. Turunen *et al.* (M. Turunen, S. Huttunen, K.E. Percy and C.K. McLaughlin, unpublished data) analysed the same experiment and found the epicuticular characteristics to vary more between the needle age classes than between the treatments. During the first year of the needles' life span, the epistomatal waxes were reduced by 47% and the droplet contact angles by 9°. The effect of the treatment on EW was significant for diesters, the proportion of which was lowest in needles treated with Cu and Ni and highest in needles treated with pH 3.1 acid rain. Metal-treated needles had significantly less wax tubes than the control needles. In 1994, significant treatment effects on droplet contact angles were observed in the youngest and oldest needle classes. The different responses of the two endophyte species studied and the self-sufficient response of the host leave possible interactions open. In the smelter area (M. Turunen, S. Huttunen, K.E. Percy, C.K. McLaughlin and J. Lamppu, unpublished data), the annual wax erosion rate and wettability were found to increase towards the smelter. The proportion of dehydroabietic acid in waxes also increased towards the smelter, which may or may not have had effects on *Cenangium ferruginosum* abundance.

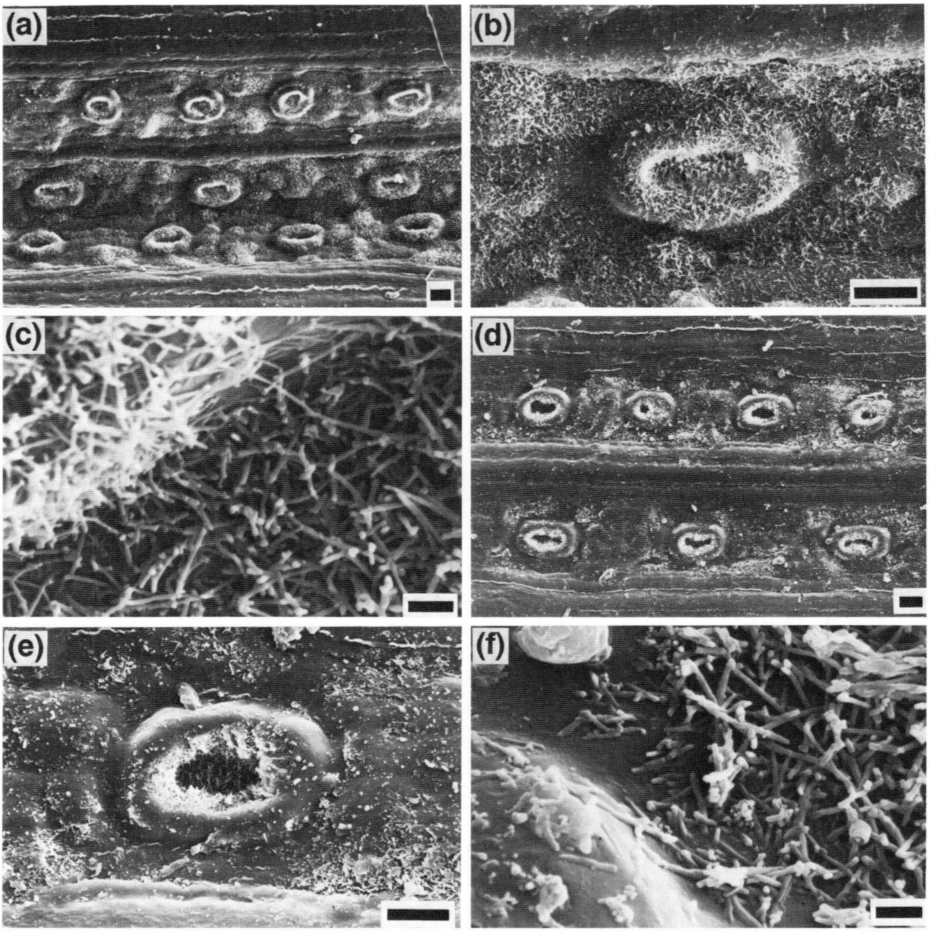

Figure 16.6. *Current needle surfaces of northern Finnish pine seedlings at the end of October 1992. (a–c). Ambient air in an open-top chamber climate, developmental status of epicuticular waxes (EW): less developed than on outside plots without chambers. (d–f) O_3- and SO_2-treated pine seedlings (same provenance). Development of wax further delayed, changes in the architecture of small tubes, tube fusions. (Bars (a,b,d,e) = 10 µm, (c,f) = 1 µm.)*

Helander (1994a) found some inhibition of birch rust infection on the abaxial surface of the leaf of pH 4.0 and pH 3.0 irrigated birches, and a decreased number of fungal colonies on the leaf surfaces (Helander and Rantio-Lehtimäki, 1990) by intensive short-term heavy metal treatment ($CuSO_4$ and $NiSO_4$). On the other hand, metals in combination with acid water (pH 3.0) produced marked rust infection. A lowering of the pH level of a medium containing copper reduced its toxicity to fungi (Gadd, 1993).

Because plant cuticular surfaces carry a net negative ionic charge, anionic surfactants, sulphates and nitrates tend to decay the surfaces (Chamel *et al.*, 1992). Richard *et al.* (1996) found that the surfactant dodecyl benzene sulphonate, at equal concentrations, was accumulated to a greater extent in the EW of needles in synthetic seawater than in distilled water. This raises the level of pollution in plants growing on seashores.The surfactant mainly accumulated in the EW of the needles. Alterations in the EW structure were also observed. Pollution by surfactants may modify the chemical composition and the related structure of the EW on conifer needles (Riederer *et al.*, 1995)

It has been demonstrated that a large accumulation of organic micropollutants occurs in the cuticular waxes (Granier and Chevreuil, 1992; Reischl *et al.*, 1989; Schönherr and Riederer, 1989). Oxidation of airborne C_2-chlorocarbons into trichloroacetic (TCA) and dichloroacetic acid (DCA) results in an accumulation of TCA and DCA in conifer cuticles and needles. The absolute values differ in the course of the season, depending upon the meteorological conditions; after sunny and dry periods they are relatively high, being up to 300 $\mu g\ kg^{-1}$ needles (Frank, 1991; Frank *et al.*, 1989). The TCA concentration in conifer needles in boreal forests ranged from 3 to 126 $\mu g\ kg^{-1}$ (Frank *et al.*, 1992). Several herbicides (e.g. TCA) have been found to generate active oxygen species either by direct involvement in radical production or by inhibition of biosynthetic pathways (e.g. in wax biosynthesis; Franich and Wells, 1980). The mode of action of some herbicides (e.g. diphenyl esters) is based on the ability to induce abnormal accumulation of photosensitizing tetrapyrroles, especially photoporphyrin. Natural photosensitizers induce oxidative damage in the light and the damage may be enhanced by cuticular degradation. The membrane damage caused by the fungal photosensitizer cercosporin provides nutrients for fungal growth and sporulation in the host (Foyer *et al.*, 1994). Herbicides cause strong inhibition of alkane accumulation, but this is depressed in the presence of UV-B (Tevini and Steinmüller, 1987).

16.3 Microbial ecology

Phyllosphere researchers are interested in the synergistic effects of nutritional and inhibitory substances in water droplets and films on leaf surface (Fokkema, 1991). Tukey (1971) studied the substances that can be leached with water from the leaves. The differences in the sugar and amino acid contents of leachates are usually small. Chelating agents, terpenoids, phenolic compounds and substances associated with wax may all play major roles in the ecology of the phyllosphere.

Recent observations demonstrate that bacteria may change the nutrient situation and release toxic substances. The same is true of the yeasts known as killer yeasts (Andrews and Hirano, 1991). Bacteria and other epiphyllic microorganisms can increase leaf wettability via the production of biosurfactants. Surface-active *Pseudomonas* spp. strains had an effect on the wettability of *Triticum aestivum* cv. Minaret leaves (Bunster *et al.*, 1989). Schreiber (1996) found evidence of pH dependence of wetting due to epiphyllic microorganisms on older needle surfaces of grand fir (*Abies grandis*).

The nitrogen-fixing bacteria (Huttunen *et al.*, 1981b) are among the ones most sensitive to air pollution. They have been isolated from leaf surfaces of many plants (Jones, 1970; Last and Warren, 1972; Tenberge and Peveling, 1991), and they are of significance for the global biogeochemical cycles. Nitrogenase activity (acetylene reduction activity) in lichens with nitrogen-fixing bacteria was significantly decreased after 4 months of simulated acid rain at pH 3.0 (Kytöviita and Crittenden, 1994). One year after the discontinuation of the treatments, lichens treated at pH 3.0 showed no clear sign of recovery.

16.4 Tree leaf surfaces

Results support the conjecture that a major factor constraining the population size of microorganisms on leaves is the number and quality of protected sites on the leaf surface (Fokkema, 1991). The effects of air pollutants on plant surfaces are often more pronounced in long-living evergreens, and microbial interactions are thus more likely in ageing surfaces. The dominant mycoflora populations on needle surfaces of Scots pine and Norway and Sitka spruce (*Picea sitchensis*) included *Aureobasidium pullulans*, *Schlerophoma pithyophila*, *Epicoccum nigrum* and *Cladosporium* spp., which were markedly reduced by sulphur dioxide treatment. *A. pullulans*, *Botrytis cinerea* and S. *pithyophila* grew best at pH 2.5–3.0 (Kirkwood *et al.*, 1989). The crust of microorganisms on the surface of spruce needles occurs as a green to brown crumbling layer. Most of the algae are unicellular green algae belonging to the *Chlorococcales*. Most of the fungi are thin filamentous fungi belonging to the *Ascomycetes*. The colonization by microorganisms starts at the end of the first season of a new needle in late autumn.The thickness of the layer of epiphytes and the portion of the covered needle surface increased with age from less than 30 μm to up to 130 μm and from less than 10% to 70–80 % (Peveling *et al.*, 1992).

Heterotrophic fungi seem to have at least three different sources of nutrients. First, they penetrate the algal cells forming haustoria, known as organs of nutrient absorption. Second, the intrahyphal hyphae described in endophytic fungi of spruce needles and in epiphytes suggest parasitism among the epiphytic fungi. Third, fungi may utilize dead organic material, indicating saprotrophic nutrition. Many observations give evidence to suggest that algae, fungi and, to a certain extent, also bacteria live in symbiotic to parasitic associations on the needle (Peveling *et al.*, 1992). There are also indications that some of these epiphytes may damage the outer epidermal wall beneath them (Tenberge and Peveling, 1991). The quantity of the photosynthetically active radiation usable by the needle has been shown to be reduced following colonization by epiphytes.

In conclusion, algae and fungi mainly live in symbiotic relationships on the needle without causing siginificant damage to the epidermal cell wall. However, for trees already weakened in health, the effect of these microorganisms on their vitality may be more important, especially in view of the cumulative effect of the reduced photosynthetic rate in covered needles (Peveling *et al.*, 1992). Petrini (1991) points out that air pollutants apparently also affect species composition and the frequency of colonization of endophytes. Fenn *et al.* (1989) found that chronic exposure to ambient ozone or SO_2 in polluted areas may affect phyllospheric fungal communities, while short-term exposures may not significantly disturb phyllospheric fungi. Possible pollution interaction pathways may also be related to leaf development (Figure 16.6). Kowalski (1982) found the following species to play the main role in causing needle cast in Polish polluted young Scots pine stands: *Cenangium ferruginosum*, *Lophodermium seditosum*, *Namemacyclus minor* and *Schlerophoma pithyophila*. The leaves of beech (*Fagus sylvatica*) in Europe are mainly colonized by *Apiognominia errabunda* (Sieber and Hugentobler, 1987); for instance, forest health surveys in England (Innes and Boswell, 1988) revealed an increase of beech crown dieback with an increasing distance from the east coast and in the more rugged areas. It is also greater in areas with higher rainfall. It is not easy to find any general phenomenon related to air pollution. On the other hand, beech might be one of the species that have interactions in their phyllosphere due to nutrient-rich and humid circumstances (Figures 16.7–16.9).

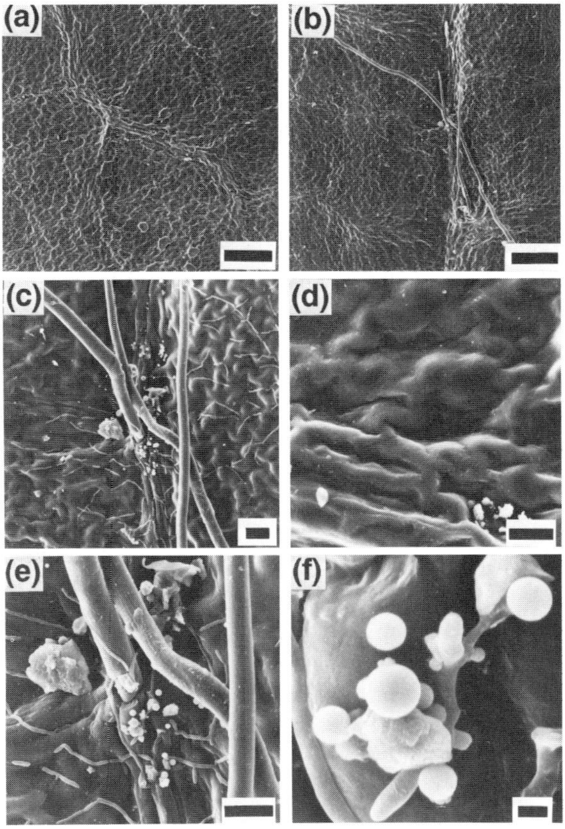

Figure 16.7. *Microbial ecology of green* Fagus sylvatica *leaf surfaces in June in northern England. (a) Adaxial surface, (b,c) orientation of hyphae, (d–f) spore germination on protected sites. (Bars (a,b) = 100 μm, (c–e) = 10 μm, (f) = 1 μm.)*

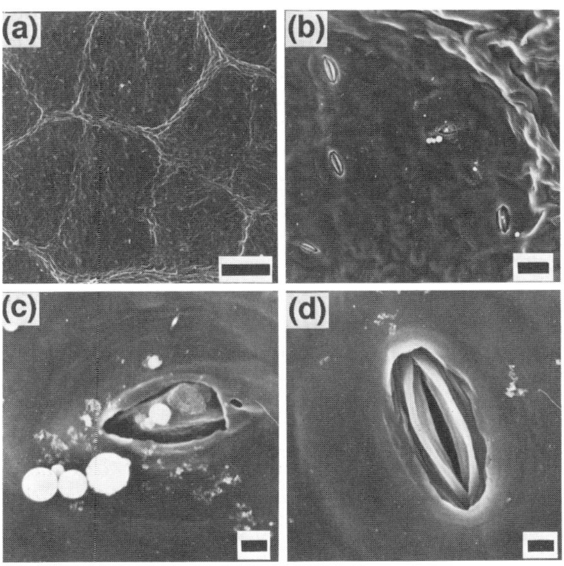

Figure 16.8. *Microbial ecology of green* Fagus sylvatica *leaf surfaces in June in northern England, abaxial surface. (a) Epidermal structure, (b) further surface architecture, (c) spore landing on a protected stomata-associated site, (d) stomata. (Bars (a) = 100 μm, (b) = 10 μm, (c,d) = 1 μm.)*

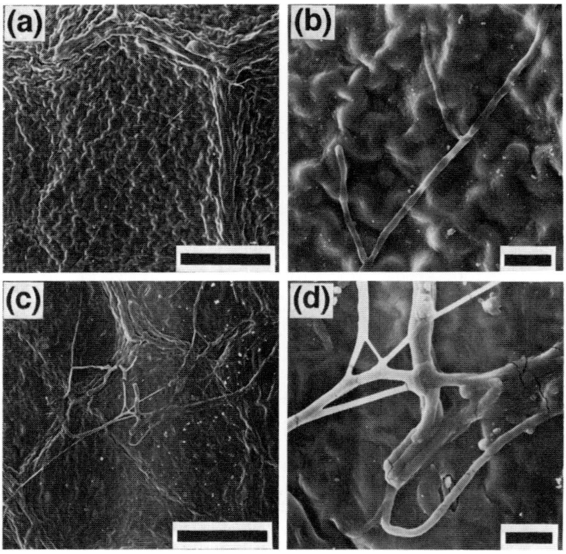

Figure 16.9. *Microbial ecology of a necrotic area of a* Fagus sylvatica *leaf surface in June in northern England. (a,b) Adaxial surface, (c,d) abaxial surface. (Bars (a,c) = 100 μm, (b,d) = 10 μm.)*

The fungal communities of green, symptomless needles and leaves can be considered to be in an ecological species equilibrium as long as no senescence process starts. Senescing tissues gradually allow the establishment of a new community composed of saprobes. Saprophytic phyllophanic fungi also act as litter decomposers of plants after leaf fall (Canavesi, 1987).

References

Agrios, G.N. (1988) *Plant Pathology*. Academic Press, London.

Andrews, J.H. and Hirano, S.S. (1991) *Microbial Ecology of Leaves*. Brock/Springer-Verlag, Berlin.

Bäck, J., Huttunen, S. and Roitto, M. (1996a) Responses of tree seedlings to combined exposure to moderately elevated concentrations of SO_2, NO_2 and O_3. In: *Air Pollution and Multiple Stresses*, Proceedings of the 16th International Meeting for Specialists in Air Pollution Effects on Forest Ecosystems (eds R.M. Cox, K.E. Percy, K.F. Jensen and C.J. Simpson). Canadian Forest Service, Fredericton, Canada, pp. 113–120.

Bäck, J., Turunen, M., Ferm, A. *et al.* (1996b) Scots pine needle structures and epiphytic microflora under heavy ammonia deposition from fur farming. *Water Air Soil Pollut.* (in press).

Bewley, R.J.F. (1981) Effect of heavy metal pollution on the microflora of pine needles. *Holarct. Ecol.* **4**, 215–220.

Blakeman, J.P. (1981) *Microbial Ecology of the Phylloplane*. Academic Press, London.

Bråkenhielm, S. and Quinghong, L. (1995) Spatial and temporal variability of algal and lichen epiphytes on trees in relation to pollutant deposition in Sweden. *Water Air Soil Pollut.* **79**, 61–74.

Bunster, L., Fokkema, N.J. and Schippers, B. (1989) Effect of surface-active *Pseudomonas* spp. on leaf wettability. *Appl. Environ. Microbiol.* **55**, 1340–1345.

Burkhardt, J. and Eiden, R. (1990) The ion concentration of dew condensed on Norway spruce (*Picea abies* L. Karst.) and Scots pine needles. *Trees* **4**, 22–26.

Burkhardt, J. and Eiden, R. (1994) Thin water films on coniferous needles. *Atmos. Environ.* **28**, 2001–2017.

Canavesi, F. (1987) *Beziehungen zwischen endophytischen Pilzen von Abies alba Mill. und den Pilzen der Nadelstreue*. Dissertation. ETH no. 8325, Swiss Federal Institute of Technology, Zurich.

Caroll, G.C. (1986) The biology of endophytism in plants with particular reference to woody perennials. In: *Microbiology of the Phyllosphere* (eds N.J. Fokkema and J. van den Heuvel). Cambridge University Press, Cambridge, pp. 205–222.

Chamel, A., Gambonnet, B. and Coret, J. (1992) Effects of two ethoxylated nonylphenols on sorption and penetration of (14C) isoproturon through isolated plant cuticles. *Plant Physiol. Biochem.* **30(6),** 713–721.

Chappelka, A.H. and Chevone, B.I. (1992) Tree responses to ozone. In: *Surface Level Ozone Exposures and their Effects on Vegetation* (ed. A.S. Lefohn). Lewis Publishers, Inc., Chelsea, pp. 271–323.

Chevrier, N., Chung, Y.S. and Sarhan, F. (1990) Oxidative damages and repair in Euglena gracilis exposed to ozone II. Membrane permeability and uptake of metabolites. *Plant Cell Physiol.* **31,** 987–992.

Coleman, J.S., Jones, C.G. and Smith, W.H. (1987) The effect of ozone on cottonwood-leaf rust interactions: independence of abiotic stress, genotype, and leaf ontogeny. *Can. J. Bot.* **65,** 949–953.

Coleman, J.S., Jones, C.G. and Krischik, V.A. (1992) Phytocentric and axploiter perspectives of phytopathology. *Adv. Plant Pathol.* **8,** 149–195.

Costonis, A.C. and Sinclair, W.A. (1972) Susceptibility of healthy and ozone-injured needles of *Pinus strobus* to invasion by *Lophodermium pinastri* and *Aureobasidium pullulans. Eur. J. For. Pathol.* **2,** 65–73.

Devlin, W.S. and Gustine, D.L. (1992) Involvement of the oxidative burst in phytoalexin accumulation and the hypersensitive reaction. *Plant Physiol.* **100,** 1189–1195.

Dickinson, C.H. and Preece, T.F. (1976) *Microbiology of Aerial Plant Surfaces*. Academic Press. London.

Donnelly, A. and Dowding, P. (1994) A comparison of epicuticular wax of *Pinus sylvestris* needles from three sites in Ireland. In: *Air Pollutants and the Leaf Cuticle,* NATO ASI Series G 36 (eds K. Percy, J.N. Cape, R. Jagels and C.J. Simpson). Springer-Verlag, Berlin, pp. 351–358.

Fenn, M.E., Dunn, P.H. and Durall, D.M. (1989) Effects of ozone and sulfur dioxide on phyllosphere fungi from three tree species. *Appl. Environ. Microbiol.* **55,** 412–418.

Fokkema, N.J. (1991) The phyllosphere as an ecologically neglected milieu: a plant pathologist's point of view. In: *Microbial Ecology of Leaves* (eds J.H. Andrews and S.S. Hirano). Brock/Springer-Verlag, Berlin, pp.3–8.

Foyer, C.H., Lelandais, M. and Kunerst, K.J. (1994) Photo-oxidative stress in plants. *Physiol. Plant.* **92,** 696–717.

Franich, R.A. and Wells, L.G. (1980) Inhibition of *Pinus radiata* primary needle epicuticular wax biosynthesis by trichloroacetate. *J. Exp. Bot.* **31,** 829–838.

Frank, H. (1991) Airborne chlorocarbons, photooxidants, and forest decline. *Ambio* **20,** 13–18.

Frank, H., Vital, J. and Frank, W. (1989) Oxidation of airborne C2-chlorocarbons to trichloroacetic and dichloroacetic acid. *Fresenius Z. Anal. Chem.* **333,** 713.

Frank, H., Scholl, H., Sutinen, S. *et al.* (1992) Trichloroacetic acid in conifer needles in Finland. *Ann. Bot. Fennici* **29,** 263–268.

Fuentes, J.D., Gillespie, T.J. and Bunce, N.J. (1994) Effects of foliage wetness on the dry deposition of ozone onto red maple and poplar leaves. *Water Air Soil Pollut.* **74,** 189–210.

Gadd, G.M. (1993) Interactions of fungi with toxic metals. *New Phytol.* **124,** 25–60.

Göransson, A. (1988) Luftalger och lavar indikerar luftföroreningar. *Naturvårdsverket Rapport* 3562, Laboratoriet för Miljökontroll **15,** 1–17.

Granier, L. and Chevreuil, M. (1992) On the use of tree leaves as bioindicators of the contamination of air by organochlorines in France. *Water Air Soil Pollut.* **64,** 575–584.

Helander, M.L. (1994a) Effects of heavy metals and acid irrigation on the birch rust, *Melampsoridium betulinum. Eur. J. For. Pathol.* **24,** 171–180.

Helander, M.L. (1994b) *Responses of Foliar Microfungi to Air Pollution*. Reports from the Department of Biology, University of Turku, No. 40.

Helander, M.L. and Rantio-Lehtimäki, A. (1990) Effects of watering and simulated acid rain on quantity of phyllosphere fungi of birch leaves. *Microb. Ecol.* **19,** 119–125.

Helander, M.L., Neuvonen, S., Sieber, T. *et al.* (1993a) Simulated acid rain affects birch leaf endophyte populations. *Microb. Ecol.* **26,** 227–234.

Helander, M.L., Ranta, H. and Neuvonen, S. (1993b) Responses of phyllosphere microfungi to simulated sulphuric and nitric acid deposition. *Mycol. Res.* **97,** 533–537.

Hutchinson, T.C. and Adams, C.M. (1987) Comparative abilities of leaf surface to neutralize acidic rain drops. I. The influence of calcium nutrition and charcoal filtered air. *New Phytol.* **106,** 169–183.

Hutchinson, T.C., Adams, C.M. and Gaber, B.A. (1986) Neutralization of acidic raindrops on leaves of agricultural crop and boreal forest species. *Water Air Soil Pollut.* **31,** 475–484.

Huttunen, S. and Laine, K. (1987) Effects of acid precipitation on the winter hardiness and cold tolerance of Scots pine and Norway spruce seedlings. In: *Acid Rain: Scientific and Technical Advances* (eds R. Perry, R.M. Harrison, J.N.B. Bell and J.N. Lester). Selper Ltd, London, pp. 614–617.

Huttunen, S., Karenlampi, L. and Kolari, K. (1981a) Changes in osmotic potential and some related physiological variables in needles of polluted Norway spruce (*Picea abies* L. Karst.). *Ann. Bot. Fennici* **18,** 63–71.

Huttunen, S., Karhu, M. and Kallio, S. (1981b) The effect of air pollution on transplanted mosses. *Silva Fennica* **15,** 495–504.

Huttunen, S., Turunen, M. and Reinikainen, J. (1989) Studies on Scots pine (*Pinus sylvestris* L.) and Norway spruce (*Picea abies* L. Karst.) needle cuticles. *Ann. Sci. For.* **46** (suppl.), 553–556s.

Huttunen, S., Reinikainen, J. and Turunen, M. (1990) Wintering response of conifers to acid rain treatment under northern conditions. In: *Acidification in Finland* (eds P. Kauppi, P. Anttila and K. Kenttåmies). Springer-Verlag, Berlin, pp. 607–633.

Innes, J.L. and Boswell, R.C. (1988) Forest health surveys 1987. *For. Comm. Bull.* **79.**

Jalkanen, R., Huttunen, S. and Väisänen S. (1981) The wax structure of the developing needles of Pinus sylvestris progenies infected by *Lophodermella sulcigena. Silva Fennica* **15,** 377–381.

Jones, K. (1970) Nitrogen fixation in the phyllosphere of the Douglas fir, *Pseudotsuga douglasii. Ann. Bot.* **34,** 230–244.

Jones, C.G. and Coleman, J.S. (1991) Plant stress and insect herbivoiry: Towards an integrated perspective. In: *Response of Plants to Multiple Stresses* (eds H.A. Mooney, W.E Winner and E.J Pell). Academic Press, San Diego, pp. 249–279.

Jones, C.G., Coleman, J.S. and Findlay, S. (1994) Effects of ozone on interactions between plants, consumers and decomposers. In: *Plant Responses to the Gaseous Environment* (eds R.G. Alscher and A.R. Wellburn). Chapman & Hall, London, pp. 339–364.

Kangasjärvi, J., Talvinen, J., Utriainen, M. *et al.* (1994) Plant defence systems induced by ozone. *Plant Cell Environ.* **17,** 783–794.

Kerstiens, G. and Lendzian, K.J. (1989a) Interactions between ozone and plant cuticles. I. Ozone deposition and permeability. *New Phytol.* **112,** 13–19.

Kerstiens, G. and Lendzian, K.J. (1989b) Interactions between ozone and plant cuticles II. Water permeability. *New Phytol.* **112,** 21–27.

Kirkwood, I.A., Magan, N. and McLeod, A.R. (1989) Effect of atmospheric sulphur dioxide on the phylloplane mycoflora of conifer needles. In: *Air Pollution and Forest Decline*, Proc. 14th Int. Meeting for Specialists in Air Pollution Effects on Forest Ecosystems IUFRO P 2.05 (eds J. Bucher and I. Bucher-Wallin), Interlaken, Switzerland, 2-8 October 1988. EAFV, Birmensdorf, pp. 457–459.

Kombrink, E. and Somssich, I.E. (1995) Defence responses of plants to pathogen. *Adv. Bot. Res.* **21,** 1–30.

Kowalski, T. (1982) Fungi infecting *Pinus sylvestris* needles of various ages. *Eur. J. For. Pathol.* **12,** 182–190.

Kytöviita, M.M and Crittenden, P.D. (1994) Effects of simulated acid rain on nitrogenase activity (asetylene reduction) in the lichen *Stereocaulon paschale* (L.) Hoffm., with special reference to nutritional aspects. *New Phytol.* **128,** 263–271.

Last, F.T and Warren, R.C. (1972) Non-parasitic microbes colonizing green leaves: their form and functions. *Endeavour* **31,** 143–150.

Magan, N. and McLeod, A.R. (1991) Effects of atmospheric pollutants on phyllosphere microbial communities. In: *Microbial Ecology of Leaves* (eds J.H. Andrews and S.S. Hirano). Brock/Springer-Verlag, Berlin, pp.379–400.

Pääkkönen, E. (1996) *Ozone-Complex Risk for Birch*. Kuopio University Publications C. Natural and Environmental Sciences **42.**

Petrini, O. (1986) Taxonomy of endophytic fungi of aerial plant tissues. In: *Microbiology of the Phyllosphere* (eds N.J. Fokkema and J. van den Heuvel). Cambridge University Press, Cambridge, pp.175–187.

Petrini, O. (1991) Fungal endophytes of tree leaves. In: *Microbial Ecology of Leaves* (eds J.H. Andrews and S.S. Hirano). Brock/Springer-Verlag, Berlin, pp. 179–197.

Petrini, O., Sieber, T.N., Toti, L. *et al.* (1992) Ecology, metabolite production, and substrate utilization in endophytic fungi. *Natural Toxins* **1,** 185–196.

Peveling, E., Burg, H. and Tenberge, K.B. (1992) Epiphytic algae and fungi on spruce needles. *Symbiosis* **12,** 173–187.

Ranta, H.M. (1990) Effect of simulated acid rain on quantity of epiphytic microfungi on Scots pine (*Pinus sylvestris* L.) needles. *Environ. Pollut.* **67,** 349–359.

Reischl, A., Reissinger, M. and Hutzinger, O. (1989) Organic micropollutants and plants. In: *Ecological Studies,* Vol. 77 (eds E.D. Schulze, O.L. Lange and R. Oren). Springer-Verlag, Berlin, pp. 193–209.

Richard, B., Grieu, P., Badot, P.M. *et al.* (1996) Influence of marine salts on the localisation and accumulation of surfactant in the needles of *Pinus halepensis* Mill. *Water Air Soil Pollut.* (in press).

Riederer, M., Badot, P.M., Garrec, J.P. *et al.* (1995) The plant cuticle as an interface between leaves and air-borne pollutants. *EUROSILVA Contribution to Forest Tree Physiology,* Dourdan, France, 7–10 November 1994. INRA, Paris (Les Collogues no. 76), pp.101–118.

Scherbatskoy, T. and Tyree, M.T. (1990) Kinetics of exchange of ions between artificial precipitation and maple leaf surfaces. *New Phytol.* **114,** 703–712.

Schönherr, J. and Riederer, M. (1989) Foliar penetration and accumulation of organic chemicals in plant cuticles. *Rev. Environ. Contam. Toxicol.* **108,** 1–70.

Schreiber, L. (1996) Wetting of the upper needle surface of *Abies grandis*: influence of pH, wax chemistry and epiphyllic microflora on contact angles. *Plant Cell Environ.* **19,** 455–463.

Schwacke, R. and Hager, A. (1992) Fungal elicitors induce a transient release of active oxygen species from cultured spruce cells that is dependent on Ca^{2+} and protein kinase activity. *Planta* **187,** 136–141.

Sieber, T.N. and Hugentobler, C. (1987) Endophytische Pilze in Blättern und Ästen gesunder und geschädigter Buchen (*Fagus sylvatica L.*). *Eur. J. For. Pathol.* **17,** 411–425.

Stephen, H. and Stephen, T. (1963) *Solubilities of Inorganic and Organic Compounds,* Vol. 1, Part 1. Macmillan, New York.

Steubing, L., Fangmeier, R., Both, R. *et al.* (1989) Effects of SO_2, NO_2 and O_3 on population development and morphological and physiological parameters of native herb layer species in a beech forest. *Environ. Pollut.* **58,** 281–302.

Tenberge, K.B. and Peveling, E. (1991) Strukturelle Veränderungen an Epidermiszellen von Fichtennadeln bei Besiedlung mit Epibionten. *Allgemeine Forst Zeitschrift* **15,** 762–765.

Tevini, M. and Steinmuller, D. (1987) Influence of light, UV-B radiation, and herbicides on wax biosynthesis of cucumber seedlings. *J. Plant Physiol.* **131,** 111–121.

Tukey, H.B. (1971) Leaching of substances from plants. In: *Ecology of Leaf Surface Micro-organisms* (eds T.F. Preece and C.H. Dickinson). Academic Press, London, pp. 67–80

Turunen, M. (1996) *Responses of Scots pine needle surfaces to air pollutants.* Doctoral Thesis. Acta Universitatis Ouluensis A 277.

Turunen, M. and Huttunen, S. (1989) The pH and structure of the needle surface of Scots pine and Norway spruce seedlings after acid precipitation. In: *Air Pollution and Forest Decline,* Proc. 14th Int. Meeting for Specialists in Air Pollution Effects on Forest Ecosystems, IUFRO P 2.05 (eds J.B. Bucher and I. Bucher-Wallin), Interlaken, Switzerland, 2–8 October 1988. EAFV, Birmensdorf, pp. 535–537.

Turunen, M. and Huttunen, S. (1991) Effects of simulated acid rain on the epicuticular wax of Scots pine needles under northerly conditions. *Can. J. Bot.* **69,** 412–419.

Turunen, M. and Huttunen, S. (1996) Scots pine needle surfaces on radial transects across the north boreal area of Finnish Lapland and the Koln Peninsular of Russia. *Environ. Pollut.* **50** (in press).

Turunen, M., Huttunen, S. and Jaana Back (1995) Acid-rain-induced changes in cuticles and Ca distribution in Scots pine and Norway spruce seedlings. *Can. J. For. Res.* **25,** 1313–1325.

Winner, W.E., Cotter, I.S., Powers, H.R. Jr *et al.* (1987) Screening loblolly pine seedling responses to SO_2 and O_3: Analysis of families differing in resistance to fusiform rust disease. *Environ. Pollut.* **47,** 205–220.

Vuorinen, M. (1992) Effects of acid rain treatments on the birch rust, *Melampsoridium betulinum* and birch, *Betula pendula*. *Eur. J. For. Pathol.* **22,** 183–187.

17

Plant cuticles under challenge

J. Grace and P.R. van Gardingen

17.1 Introduction

As we have seen in the preceding chapters, the aerial surfaces of land plants are covered by a cuticular membrane (CM) that provides the necessary barrier against uncontrolled evaporation and against the ingress of gaseous and aqueous pollutants. The chemical and mechanical properties of this membrane are remarkable, as cuticle must be both water-impermeable and resist weathering, abrasive damage and attack by pathogens and insect herbivores. There are many occasions in the natural environment when the integrity of the cuticle is challenged. The purpose of this chapter is to consider the circumstances where this can happen, especially in the light of what has been said in previous chapters on the structure of cuticle, and what is known about the weathering of plant surfaces.

17.2 The structure of plant cuticle in relation to its function

The structure of cuticle has been discussed by Juniper and Jeffree (1983), Schönherr (1982), and in Chapter 2, yet is still not fully understood. The reticulate inner region of cuticle is composed of the polyester cutin (and/or cutan) in which polysaccharides including pectin are embedded. This polysaccharide-rich polymer matrix may perhaps absorb mechanical strain caused by flexure and serve as a cushion between cell walls and the outer region of the cuticle. Most authors suggest that the resistance to the diffusion of water vapour resides mainly within a thin layer close to the outer surface of the cuticle (Becker *et al.*, 1986; Geyer and Schönherr, 1990; Lendzian and Kerstiens, 1991; Riederer and Schneider, 1990; Schmidt *et al.*, 1989; Schönherr, 1982).

A superficial layer of wax, the epicuticular layer, is extruded on the cuticle surface although no-one has convincingly demonstrated a pathway for transport of this material or its precursors to the surface of the leaf from below. This layer develops a microcrystalline appearance, as does any wax surface if left unpolished (Jeffree *et al.*, 1975), and scatters light, giving the surface a pale matt appearance, described as a 'bloom'.

It is thought that the cuticular permeance to water vapour is controlled by the presence of soluble cuticular lipids (SCL) in the CM. Some studies have noted a

Plant Cuticles, edited by G. Kerstiens.

relationship between the amount of SCL and cuticular permeability, but the functional role of SCL was initially obscure (Geyer and Schönherr, 1990). Attempts to correlate cuticular properties with the chemical composition were not successful, indicating that a simple chemical model of cuticular permeability would not be satisfactory (Riederer and Schneider, 1990). It now seems likely that cuticular permeability to water vapour is best explained by the combined effects of the amount and chemical composition of SCL and their physical arrangement in the cuticle (Riederer and Schneider, 1990; Schreiber *et al.*, 1995; see Chapter 4).

In some studies, there have been indications that the epicuticular waxes (EW) may act as an important barrier for water vapour in mesophytic species. Polishing or brushing of the leaf surface to displace the wax increases the water loss from leaves through an apparent increase in cuticular conductance (Hall, 1966; Hall and Jones, 1961; Pieniazec, 1944; Pitcairn *et al.*, 1986). Similar conclusions are obtained from the observation that glaucous varieties of *Brassica oleracea* display lower transpiration rates than non-glaucous types (Denna, 1970). In contrast, in many other species it is difficult or impossible to influence water loss by gentle rubbing treatments (Grace *et al.*, 1975; Rees and Grace, 1981), presumably because the diffusion barrier resides within the CM. Some of the difficulty in interpreting results from these studies arises from the possibility that measurements have been influenced by partial stomatal opening (van Gardingen and Grace, 1992; van Gardingen *et al.*, 1989; see Chapter 5). Partial opening may, in some studies, be induced by damage to the stomatal apparatus during rubbing treatments (Pitcairn *et al.*, 1986).

The role and relative importance of EW in limiting water loss from leaves is unclear from the existing literature and needs to be re-examined. However, the undisputed importance of the EW seems to lie in that it (i) influences the short-wave reflectance and hence the leaf energy balance (see Chapter 7); this may be important in survival for species growing in hot environments such as deserts, and in mountain environments where total solar radiation and the ultraviolet component may be very high; (ii) confers a hydrophobic property to the leaf surface, thus protecting the surface to some extent from the chemical ingredients of rain and cloud water, and making the surface less available for uptake by inorganic pollutants that may be present as an aqueous solution in acid rain or mist (see Chapter 15). It should, however, be noted that this only applies to transfers from *aqueous solutions*: if we are also concerned with gaseous dry deposition, then the wax presents a surface area for uptake many times greater than the projected leaf area. It may also have importance in relation to infection by pathogens (see Chapter 14). In addition, the CM has useful biomechanical properties; isolated cuticle is rather tough in relation to its strength, and pieces may be manipulated with forceps with no apparent damage.

17.3 The nature of weathering

Contrary to popular belief, wax surfaces are not inert. There are now many instances where it has been shown that EW is modified as a result of rather mild treatments of rain (Baker and Hunt, 1986), aromatic hydrocarbons (Sauter and Pambor, 1989) and acid rain (Adams *et al.*, 1990; Percy and Baker, 1990; Turunen and Huttunen, 1990, 1991) as well as by ozone, SO_2 and other more severe treatments (Cape and Percy, 1993; Cape *et al.*, 1995; Crossley and Fowler, 1986; Günthardt-Goerg *et al.*, 1994; Percy *et al.*, 1992; Turunen and Huttunen 1991). For water vapour, it has been shown that a layer of paraffin wax suspended in humid air ad- and absorbs up to 80 μg H_2O cm^{-2} in 1 h, far more than has been observed for materials such as brass, steel and polytetrafluoroethane (Dixon and Grace, 1982; van Gardingen and Grace, 1992). The underlying mechanisms for any of these interactions are poorly understood (Riederer, 1989).

There is also mechanical weathering. When leaves are taken from glasshouse-grown plants, or from young plants in sheltered environments, the leaf surface is said to be 'pristine'. The pristine surface shows an unbroken covering of cuticle, and often has crystalline EW. If pristine material is exposed to turbulent wind in a wind tunnel, the leaves undergo mechanical strain and abrasion through mutual contact. Over periods of hours and days, the pristine surface is altered in three ways: (i) the EW is smeared, smoothed and to some extent redistributed; (ii) the EW and possibly the cuticle proper (CP) is cracked; (iii) the epidermis is broken and small lesions are formed, often not visible to the naked eye (Grace, 1974; MacKerron, 1976; Thompson 1974; van Gardingen *et al.*, 1991; Wilson, 1984). Over this period water loss from the leaves increases in parallel with the damage to the leaf surface. Part of this increase can be attributed to an increase in cuticular transpiration, but part is almost certainly due to failure of the stomata to close or even to an increase in stomatal opening brought about by the change in turgor relations of the epidermis following puncturing of some of the epidermal cells. Some regions of the leaf may look 'bruised' and brown lesions are apparent.

Similar damage is evident in scanning electron microscope (SEM) images of plant material taken from gardens, fields or forests (Crossley and Fowler, 1986; Hoad *et al.*, 1992, 1994). Such damage is not always visible with the naked eye, but can be visualized using a dye tracer such as Coriphosphine O, which binds to any pectin exposed to the surface (Jeffree *et al.*, 1994). More usually, an integrated index of surface weathering is obtained from measuring contact angles made with water droplets placed on the surface (Cape, 1983), although this may misrepresent the whole story as it emphasizes the overall removal of surface waxes but does not provide information about the important local hotspots of damage. A gradual decrease in contact angle is observed over months and years, and the rate of decline is influenced by chemical treatment of the leaf surface by acid mists and pollutant gases (Cape, 1994). Part of the decline is likely to be caused by the accumulation of hydrophilic materials on the surface.

Weathering in the natural environment may also include impaction of abrasive particles that become embedded in the cuticle (Armbrust *et al.*, 1974), and chemical weathering from dry and wet deposition. The importance of acid deposition and chemical weathering may have been exaggerated in recent studies. Although the EW layer is susceptible, it is well-known that the CM is resistant to very strong acid treatment; indeed, cuticle is known to be one of the most resistant natural materials and a common way to prepare cuticles for examination has been to treat leaves in Jeffrey's solution (chromic and nitric acid) so that all the leaf is digested except for the cuticle.

Weathering is not uniform over the leaf. The aerodynamic properties of any individual leaf are complex. Variations in topography, venation and serrations of leaf margins are all likely to influence the transition from laminar to turbulent flow, and hence the rate at which materials are deposited. Also, parts of the leaf that stand proud are exposed to abrasive damage from neighbouring leaves (Hoad *et al.*, 1992), and this may be especially apparent during the stages of leaf unfolding. Distal regions of the leaf move in arcs during natural fluttering, and so may undergo greater mechanical stress through collisions than do proximal regions. In graminaceous leaves there is often a whiplash effect, whereby extreme mechanical damage occurs resulting in necrosis of the leaf tip (Grace, 1974). Pubescent leaves are more likely to be damaged than glabrous leaves following brushing, polishing, and wind exposure through damage to trichomes and their basal epidermal cells when the trichomes are torn off.

17.4 Can damaged cuticles be repaired?

When EW is rubbed off in glaucous genera like *Chrysanthemum* or *Eucalyptus* there is recovery in the appearance of the wax surface over periods varying from a few hours to days (Hallam, 1967). In the grass *Festuca arundinacea*, Pitcairn *et al.* (1986) subjected the cuticle to brief disturbance by either polishing or abrading with emery cloth. It was the more gentle treatment of polishing that caused the greatest increase in water loss from the cuticle. Partial repair appeared to occur in several days, as was also found by Wilson (1984) and Pitcairn *et al.* (1986). It is, however, difficult to say what fraction of that water loss and subsequent apparent recovery was genuinely cuticular, as we cannot exclude the possibility that the polishing elicited a stomatal opening response.

Our understanding of possible mechanisms of physical recovery following damage to the cuticle is based on the conceptual model of cuticular function described above. The amount and structural conformation of SCL in the CM are known to be dynamic (Geyer and Schönherr, 1990). This suggests that processes of repair may be possible in mature or fully expanded leaves. This is supported by the recent work of Schreiber *et al.* (1996) who observed that the waxy transport barrier of cuticles is a dynamic system. They suggest that since wax molecules are mobile, physical healing of cuticles is possible. These authors suggest that this process of healing may be of ecological significance in evergreen species such as conifers.

When leaf surfaces are wounded or penetrated, the exposed internal tissues take on a new role as the impermeable barrier. Mitotic activity is stimulated and a wound periderm is produced. Suberin and lignin may be deposited on the outer walls of such a periderm, and the synthesis of cutin has been reported at such sites (Lipetz, 1970), although more usually the suberization serves to isolate the wounded region from water supply, leading to a brown, necrotic region. In pine needles these appear as flecks or spots, and are quite common.

17.5 Interactions between cuticular damage and pollution stress

When the leaf surface is damaged, extra pathways for the ingress of pollutant materials are created and uptake of pollutant materials may proceed at a greatly accelerated rate. Jeffree *et al.* (1994) applied droplets of ^{35}S-labelled sulphuric acid to leaf surfaces of beech (*Fagus sylvatica* L.), taken from either exposed or sheltered sides of a garden hedge. The mean uptake of sulphur, measured after the droplets had been allowed to dry, was fivefold higher in the exposed than the sheltered leaf. The distribution of uptake rate over the leaf surface was very variable, but concentrated near the distal region of the leaf where damage was more concentrated. Autoradiography of leaves that had been covered with ^{35}S-sulphuric acid showed several areas on the surface where uptake was especially high, believed to be the result of intense abrasion from other leaves or perhaps buds. Similar results were obtained in the earlier study of Hoad *et al.* (1992) showing significantly higher levels of uptake of sulphate by leaves of *F. sylvatica* from an exposed site in the

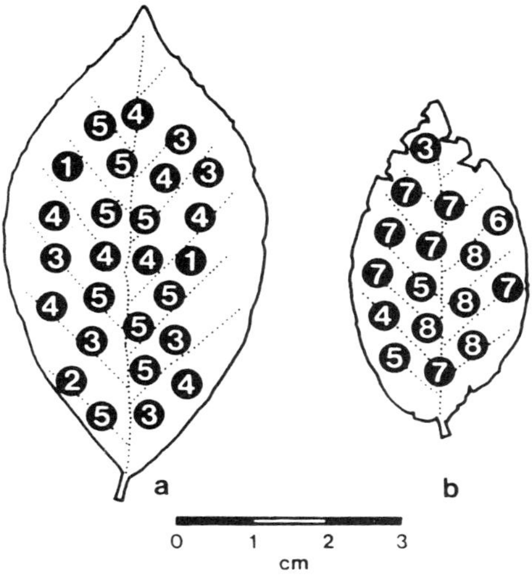

Figure 17.1. Distribution of ^{35}S-sulphate uptake via the adaxial surface of Fagus sylvatica *leaves grown in (a) sheltered and (b) exposed sites at Dunslair Heights, Peebleshire, UK. The sample positions are indicated by the circles. Numbers within the circles represent the range of ^{35}S-sulphate uptake from a 2.5 mM pH 3.0 solution of sulphuric acid/ammonium sulphate. (Reproduced from Hoad* et al. *(1992) Effects of wind and abrasion on cuticular integrity in* Fagus sylvatica *and consequences for transfer of pollutants through leaf surfaces.* Agric. Ecosyst. Environ. **42**, *275-289, with permission from Elsevier Science - NL.)*
Key:

Code number	Uptake range mol (SO_4^{2-}) m^{-2}
1	$0 \rightarrow 5 \times 10^{-8}$
2	$5 \times 10^{-8} \rightarrow 10^{-7}$
3	$10^{-7} \rightarrow 5 \times 10^{-7}$
4	$5 \times 10^{-7} \rightarrow 10^{-6}$
5	$10^{-6} \rightarrow 5 \times 10^{-6}$
6	$5 \times 10^{-6} \rightarrow 10^{-5}$
7	$10^{-5} \rightarrow 5 \times 10^{-5}$
8	$5 \times 10^{-5} \rightarrow 10^{-4}$

Scottish uplands when compared with leaves from a sheltered site at the same location (Figure 17.1).

If the cuticle is intact, it has been shown that acidic deposition has no effect on the leaf. Hoad *et al.* (1994) sprayed leaves of 3-year old birch and beech with simulated acid rain at pH 5 or 3, as well as with water as a control. The leaves were of three types: (i) from plants that were well separated and exposed in a field so that they received direct wind; (ii) from plants placed close together so that their leaves clashed and abraded each other in wind, known as the indirect wind treatment; and (iii) from plants kept in a sheltered place, also outdoors. In the case of the sheltered plants, the acid rain had no effect on the number of lesions that developed nor on the cuticular conductance (Figure 17.2). However, wind treatments displayed an enhancement in lesions and cuticular conductance, particularly with the indirect wind treatment. The most likely interpretation of this result, is that the acid precipitation enters through cracks in the cuticle of leaves that have already been formed during exposure to wind, thus causing hydrolysis of the polysaccharide materials underneath the cuticle, and perhaps hydraulic isolation of the cuticle. Thus acidic deposition may act by exacerbating the effect of the small crack and causing physiologically significant effects. There are some other observations that are consistent with the idea that acid deposition does not harm conifer leaves unless the cuticle has been damaged. Eamus *et al.* (1989) found that treatment of red

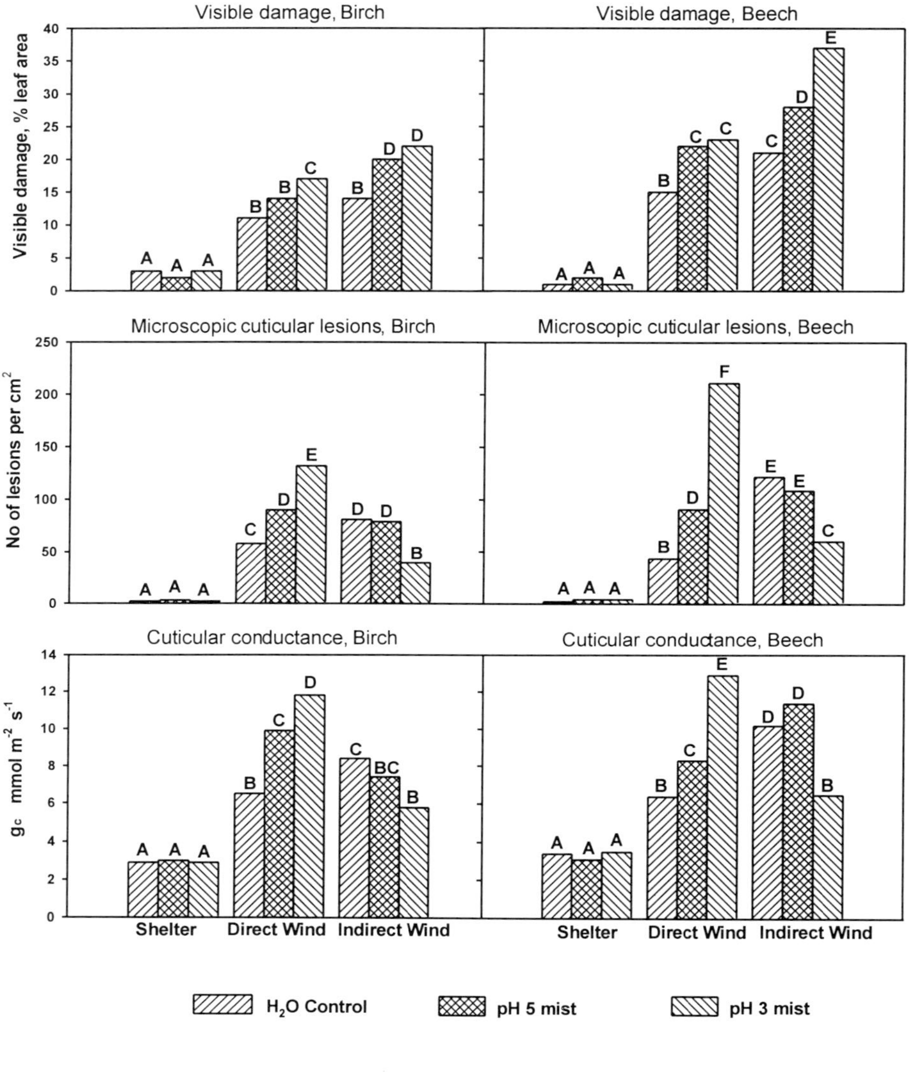

Figure 17.2. *The effects of a combination of acid mist and wind treatment applied to 3-year-old birch and beech plants, on visible leaf damage, the numbers of microscopic cuticular lesions and the cuticular conductance (g_c). Different letters above each column indicate differences significant at* $p \leq 0.05$. *(Reproduced with permission from Hoad* et al. *(1994) Effects of wind and simulated acid mist on leaf cuticles. In:* Air Pollutants and the Leaf Cuticle, *NATO ASI Series (eds Percy* et al.*). Springer-Verlag, Heidelberg, pp. 225–236, with permission from Springer-Verlag.)*

spruce, *Picea rubens,* with acid mist had no effect on surface conductance. Werkman (1993) found that fumigation of *Picea sitchensis* with 140 ppb ozone had no effect unless it followed exposure to high wind speed for 120 h.

17.6 Tree-lines and other challenging environments

The altitudinal limit to the distribution of trees occurs wherever the mean temperature of the warmest month is about 10°C (Daubenmire, 1954; Grace, 1977; Wardle, 1965). The immediate cause of failure of trees in the alpine environment is often said to be mortality of leaves in winter caused by water stress, referred to in the early German literature as 'Frosttrocknis' or frost-drought, and occurring because soil water is unavailable (frozen or cold and thus viscous) whilst cuticular transpiration proceeds at a finite rate (Tranquillini, 1979; Wardle, 1971). Many physiologists have advanced the idea that where the summer is very short or very cool (as at the tree-line), the developmental process of the cuticle is incomplete, thus exposing the leaves to frost-drought in the winter. The term 'cuticular ripening' has been used to describe the hypothetical process whereby a cuticle passes from an incompetent juvenile stage to the point of being fully functional (Wardle, 1971). The data supporting this hypothesis are well known, and incorporated into a more general theory of tree-line by Tranquillini (1979). The critical observations are as follows. Lange and Schulze (1966) measured the thickness of cuticle over the summer at an elevational transect that included the alpine tree-line. They found that the thickness increased long after needle expansion had ceased. They also found that the leaves from high altitudes did indeed have thinner cuticles. Baig and Tranquillini (1976) reported cuticular thickness in *Picea abies* and *Pinus cembra* near the tree-line in Austria. Both the cuticle and the cutinized cell wall layer declined in thickness with altitude, and below a thickness of 0.5 μm the minimum transpiration rate was markedly increased. However, authors of both these older studies relied on cuticular measurements made from light microscopy, and additional work with more refined techniques is required to achieve an understanding of cuticular development in an ecological context. In *Clivia miniata* chemical and ultrastructural changes in the cuticle occur after leaf expansion and are accompanied by an increase in cuticular thickness and mass (Riederer and Schönherr, 1988).

More recent work tends to refute the hypothesis that imperfect or incomplete cuticular development is the cause of death of foliage as a result of winter desiccation. Studies by Grace (1990), Hadley and Smith (1983, 1986), Holtmeier (1980) and Marchand and Chabot (1978) suggest that the increase in minimum transpiration that definitely does occur over the course of the winter, is not the result of thin or poorly developed cuticles, but of cuticular damage by abrasion during winter storms. Cuticular conductances are normally so low (about 1 mmol m^{-2} s^{-1}) that small abrasions cause increases that double or triple the winter conductance, and are therefore of great importance in the water balance. Grace (1990) isolated cuticles of *Pinus sylvestris* from an altitudinal gradient in Scotland using a mixture of 10% nitric and 10% chromic acid, and showed that cuticular mass per area of leaf declined only slightly with altitude, although the minimal surface conductance increased considerably. He concluded that the increase in transpiration was the result of cuticular lesions and possibly incomplete stomatal closure rather than lack of cuticular development. Both McCracken *et al.* (1985) working in Switzerland and New Zealand, and James *et al.* (1994) in Scotland showed that the water potential of the foliage did not decline as the winter progressed, as required by Tranquillini's hypothesis.

The montane environment may be especially challenging as there are unique combinations of physical factors, the most important of which appear to be the high winds and short season for growth. However, there are other cases that deserve attention but that have been somewhat neglected. In the coastal environment the wind speeds are often high, and sometimes locally exacerbated by uneven topography. At high wind speeds, sand particles become airborne, and at about 7 m s^{-1} 'white-horses' appear on the surface of the water so that sea-spray becomes an important ecological factor. During storms, sea-spray is deposited over many kilometers inland, and subsequent browning of foliage occurs as a result of the toxic effect of chloride ions. Patterns of deposition follow lines of flow, and the characteristic wind-shaping of trees near the coast is usually caused by such deposition. Maritime plants such as *Hippophae*, *Elymus* and *Ammophila* all have a considerable development of EW, so that seawater forms globules and runs off, thick cuticles, and well-developed sclerophylly. It is, of course, dangerous to link structure and function without carrying out experimental work case-by-case, but it seems likely that maritime plants, of all species, may have undergone the strong selection pressure for cuticular development. There are some manmade environments that also deserve attention. Roadside plants are exposed to intermittent mechanical excitation whenever vehicles pass at speed. On motorways, the violent agitation coupled with the deposition of gases and particles from vehicle exhausts may lead to lethal doses of pollutant materials (Colwill *et al.*, 1979) and roadside plants often show a 'scorched' appearance similar to coastal damage following storms. In cold countries the use of salt as a de-icer is likely to have accentuated this effect. It is well-known that maritime species have appeared in the floras of motorway verges far from the coast.

17.7 The need for more research

Our knowledge of the form and function of cuticles has progressed through research involving approaches including ultrastructural, physico-chemical and physiological investigations combined with direct measurement of the transport properties of cuticles. The utility of these approaches and their results are well illustrated throughout the rest of this volume. In this chapter, there has been a change of emphasis to consider the ecological significance of plant cuticles. We find that whilst existing research addresses many ecological issues, there are, however, several questions that, as yet, remain unanswered.

- What is the relationship between cuticular composition, structure and cuticular function?
- What is the ecological significance and function of EW, and how far does it influence cuticular permeability?
- Do leaves repair their cuticles after damage, or is this an artefact of experimental treatments?
- What is the significance of damage to the cuticle for uptake of pollutants and agrochemicals?

- What is the role of cuticles in determining timberline and influencing the survival of plants in hostile environments?

It is likely that these and other questions will continue to challenge researchers seeking to apply their understanding of cuticular form and function to the study of ecological processes.

Acknowledgements

We would like to thank Drs C.E. Jeffree and G. Kerstiens for their helpful comments on drafts of this manuscript.

References

Adams, M.B., Caporn, S.J.M. and Hutchinson, T.C. (1990) Crystal occurrence and wax disruption on leaf surfaces treated with simulated acid rain. *New Phytol.* **114,** 147–158.

Armbrust, D.V., Paulsen, G.M. and Ellis, R. (1974) Physiological responses of wind and sandblast-damaged winter wheat plants. *Agron. J.* **66,** 421–423.

Baig, M.N. and Tranquillini, W. (1976) Studies on upper timberline: morphology and anatomy of Norway spruce (*Picea abies*) and stone pine (*Pinus cembra*) needles from various habitat conditions. *Can. J. Bot.* **54,** 1622–1632.

Baker, E.A. and Hunt, G.M. (1986) Erosion of waxes from leaf surfaces by simulated rain. *New Phytol.* **102,** 161–173.

Becker, M., Kerstiens, G. and Schönherr, J. (1986) Water permeability of plant cuticles: permeance, diffusion and partition coefficients. *Trees* **1,** 54–60.

Cape, J.N. (1983). Contact angles of water droplets on needles of Scots Pine (*Pinus sylvestris*) growing in polluted atmospheres. *New Phytol.* **93,** 293–299.

Cape, J.N. (1994) Evaluation of pollutant critical levels from leaf surface characteristics. In: *Air Pollutants and the Leaf Cuticle*, NATO ASI series (eds K.E. Percy, N. Cape, R. Jagels and C.J.Simpson). Springer-Verlag, Heidelberg, pp.123–138.

Cape, J.N. and Percy, K.E. (1993) Environmental influences on the development of spruce needle cuticles. *New Phytol.* **125,** 787–799.

Cape, J.N., Sheppard, L.J. and Binnie, J. (1995) Leaf surface-properties of Norway spruce needles exposed to sulfur-dioxide and ozone in an open-air fumigation system at Liphook. *Plant Cell Environ.* **18,** 285–289.

Colwill, D.W., Thompson, J.R. and Rutter, A.J. (1979) *The Impact of Road Traffic on Plants.* Transport and Road Research Laboratory, Berkshire.

Crossley, A. and Fowler, D. (1986). The weathering of Scots pine epicuticular wax in polluted and clean air. *New Phytol.* **103,** 207–218.

Daubenmire, R. (1954) Alpine timberlines in the Americas and their interpretation. *Butler Univ. Bot. Studies* **2,** 119–136.

Denna, D.W. (1970) Transpiration and the waxy bloom in *Brassica oleracea* L. *Aust. J. Biol. Sci.* **23,** 27–31.

Dixon, M. and Grace, J. (1982) Water uptake by some chamber materials. *Plant Cell Environ.* **5,** 323–327.

Eamus, D. Leith, I. and Fowler, D. (1989) Water relations of red spruce seedlings treated with acid mist. *Tree Physiol.* **5,** 387–397.

Geyer, U. and Schönherr, J. (1990). The effect of the environment on the permeability and composition of *Citrus* leaf cuticles.1. Water permeability of isolated cuticular membranes. *Planta* **180,** 147–153.

Grace, J. (1974) The effect of wind on grasses 1. Cuticular and stomatal transpiration. *J. Exp. Bot.* **25,** 542–551.

Grace, J. (1977) *Plant Response to Wind*. Academic Press, London.

Grace, J. (1990) Cuticular water loss unlikely to explain tree-line in Scotland. *Oecologia* **84,** 64–68.

Grace, J., Malcolm, D.C. and Bradbury, I. (1975) The effect of wind and humidity on leaf diffusive resistance in Sitka spruce seedlings. *J. Appl. Ecol.* **12,** 931–940.

Günthardt-Goerg, M.S., Keller, T. and Matyssek, R. (1994) Environmental effects on Norway Spruce needle wax. *Eur. J. For. Pathol.* **24,** 92–111.

Hadley, J.L. and Smith, W.K. (1983) Influence of wind exposure on needle desiccation and mortality for timberline conifers in Wyoming, USA. *Arctic Alp. Res.* **15,** 127–135.

Hadley, J.L. and Smith, W.K.(1986) Wind effects on needles of timberline conifers. *Ecology* **67,** 12–19.

Hall, D.M. (1966) A study of the surface wax deposits on apple fruit. *Aust. J. Biol. Sci.* **19,** 1017–1025.

Hall, D.M. and Jones, R.L.(1961) Physiological significance of surface wax on leaves. *Nature* **191,** 95–96.

Hallam, N.D. (1967) An electron microscope study of the leaf waxes of the genus *Eucalyptus* L'Heritier. PhD thesis, University of Melbourne.

Hoad, S.P., Jeffree, C.E. and Grace, J. (1992) Effects of wind and abrasion on cuticular integrity in *Fagus sylvatica* and consequences for transfer of pollutants through leaf surfaces. *Agric. Ecosyst. Environ.* **42,** 275–289.

Hoad, S.P., Jeffree, C.E. and Grace, J. (1994) Effects of wind and simulated acid mist on leaf cuticles. In: *Air Pollutants and the Leaf Cuticle*, NATO ASI series (eds K.E. Percy, N. Cape, R. Jagels and C.J. Simpson). Springer-Verlag, Heidelberg, pp. 225–236.

Holtmeier, F.K. (1980) Influence of wind on tree physiognomy at the upper timberline in the Colorado Front Range. In: *Mountain Environments and Subalpine Tree Growth* (eds U.Beneke and M.R. Davis). Forest Research Institute, Wellington, pp. 247–261.

James, J., Grace, J. and Hoad, S. (1994) Growth and photosynthesis of *Pinus sylvestris* at its altitudinal limit in Scotland. *J. Ecol.* **82,** 297–306.

Jeffree, C.E., Baker, E.A. and Holloway, P.J. (1975) Ultrastructure and recrystallization of plant epicuticular waxes. *New Phytol.* **75,** 539–549.

Jeffree, C.E., Grace, J. and Hoad, S.P. (1994) Spatial distribution of sulphate uptake by wind-damaged beech leaves. In: *Air Pollutants and the Leaf Cuticle*, NATO ASI series (eds K.E. Percy, N. Cape, R. Jagels and C.J.Simpson). Springer-Verlag, Heidelberg, pp. 183–193.

Juniper B.E. and Jeffree C.E. (1983) *Plant Surfaces*. Edward Arnold, London.

Lange, O.L. and Schulze, E.-D. (1966) Untersuchungen über die Dickenentwicklung der kutikularen Zellwandschichten bei der Fichtennadel. *Forstwiss. Z.* **85,** 27–38.

Lendzian, K.J. and Kerstiens, G. (1991) Gas transport in plant cuticles. *Rev. Environ. Contam. Toxicol.* **121,** 66–128.

Lipetz, J. (1970) Wound healing in higher plants. *Int. Rev. Cytol.* **27,** 1–28.

MacKerron, D.K.L. (1976) Wind damage to the surface of strawberry leaves. *Ann. Bot.* **40,** 351–354.

Marchand, P.J. and Chabot, B.F. (1978) Winter water relations of tree line plant species on Mount Washington, New Hampshire. *Arct. Alp. Res.* **10,** 105–116.

McCracken, I.J., Wardle, P., Beneke, U. *et al.* (1985) Winter water relations of tree foliage at timberline in New Zealand and Switzerland. *Eidgenössische Anstalt für das forstliche Versuchswesen* **270,** 85–93.

Percy, K.E. and Baker, E.A. (1990) Effects of simulated acid rain on epicuticular wax production, morphology, chemical composition and on cuticular membrane thickness in 2 clones of Sitka Spruce *Picea sitchensis* (Bong) Carr. *New Phytol.* **116,** 79–87.

Percy, K.E., Jensen, K.F. and McQuattie, C.J. (1992) Effects of ozone and acidic fog on red spruce needle epicuticular wax production, chemical composition, cuticular membrane ultrastructure and needle wettability. *New Phytol.* **122,** 71–80.

Pieniazec, S.A. (1944) Physical characters of the skin in relation to apple fruit transpiration. *Plant Physiol.* **19,** 529–536.

Pitcairn, C.E.R. and Grace, J. (1984) The effect of wind on provenances of *Molinia caerulea* L. *Ann. Bot.* **54,** 135–143.

Pitcairn, C.E.R., Jeffree, C.E. and Grace, J. (1986) The influence of polishing and abrasion on the diffusion conductance of the leaf surface of *Festuca arundinacea* Schreb. *Plant Cell Environ.* **9,** 191–196.

Rees, D.J. and Grace, J. (1981) The effect of wind and shaking on the water relations of *Pinus contorta*. *Physiol. Plant.* **51,** 222–228.

Riederer, M. (1989), The cuticles of conifers: structure, composition and transport properties. In: *Forest Decline and Air Pollution. A Study of Spruce (Picea abies) on Acid Soils* (eds E.D. Schulze, O.L. Lange and R. Oren). Springer-Verlag, Berlin, pp. 155–192.

Riederer, M. and Schneider, G. (1990) The effect of the environment on the permeability and composition of *Citrus* leaf cuticles. II. Composition of soluble cuticular lipids and correlation with transport properties. *Planta* **180,** 154–165.

Riederer, M. and Schönherr, J. (1988) Development of plant cuticles, fine structure and cutin composition of *Clivia miniata* Reg. leaves. *Planta* **174,** 127–138.

Sauter, J.J. and Pambor, L. (1989) The dramatic corrosive effect of road side exposure and of aromatic hydrocarbons on the epistomatal wax crystalloids in spruce and fir and its significance for the Waldsterben. *Eur. J. For. Pathol.* **19,** 370–378.

Schmidt, H.W., Mérida, T. and Schönherr, J. (1989) Water permeability and fine-structure of cuticular membranes isolated enzymatically from leaves of *Clivia miniata* Reg. *Z. Pflanzenphysiol.* **105,** 41–51.

Schönherr, J. (1982) Resistance of plant surfaces to water loss: transport properties of cutin, suberin and related lipids. In: *Encyclopedia of Plant Physiology 12B Water relations and carbon assimilation* (eds O.L. Lange, P.S. Nobel, C.B. Osmond and H. Ziegler). Springer-Verlag, Berlin, pp. 153–179.

Schreiber, L., Bach, S., Kirsch, T. *et al.* (1995) A simple photometric device analyzing cuticular transport physiology — surfactant effect on permeability of isolated cuticular membranes of *Prunus laurocerasus* L. *J. Exp. Bot.* **46,** 1915–1921.

Schreiber, L., Kirsch, T. and Riederer, M. (1996) Transport properties of cuticular waxes of *Fagus sylvatica* L. and *Picea abies* (L.) Karst — estimation of size-selectivity and tortuosity from diffusion coefficients of aliphatic molecules. *Planta* **198,** 104–109.

Thompson, J.R. (1974) The effect of wind on grasses 2. Mechanical abrasion in *Festuca arundinacea* (Schreb.). *J. Exp. Bot.* **25,** 117–122

Tranquillini, W. (1979) *Physiological Ecology of Alpine Timberline*. Springer-Verlag, Berlin.

Turunen, M. and Huttunen, S. (1990) A review of the response of epicuticular wax of conifer needles to air pollution. *J. Environ. Qual.* **19,** 35–45.

Turunen, M. and Huttunen, S. (1991) Effect of simulated acid rain on the epicuticular wax of Scots pine needles under northerly conditions. *Can. J. Bot.* **69,** 412–419.

van Gardingen, P.R. and Grace, J. (1992) Vapour pressure deficit response of cuticular conductance in intact leaves of *Fagus sylvatica* L. *J. Exp. Bot.* **43,** 1293–1299.

van Gardingen, P.R., Jeffree, C.E. and Grace, J. (1989) Variation in stomatal aperture in leaves of *Avena fatua* L. observed by low-temperature scanning electron microscopy. *Plant Cell Environ.* **12,** 887–898.

van Gardingen, P.R., Grace, J. and Jeffree, C.E. (1991). Abrasive damage by wind to the needle surfaces of *Picea sitchensis* (Bong.) Carr. and *Pinus sylvestris* L. *Plant Cell Environ.* **14,** 185–193.

Wardle, P. (1965) A comparison of alpine timberlines in New Zealand and North America. *N.Z. J. Bot.* **3,** 113–135.

Wardle, P. (1971) An explanation for alpine timberline. *N.Z. J. Bot.* **9,** 371–402.

Werkman, B.R. (1993) Effects of acid mist, ozone and wind on Norway spruce. PhD thesis, University of Edinburgh.

Wilson, J. (1984) Microscopic features of wind damage to leaves of *Acer pseudoplatanus* L. *Ann. Bot.* **53,** 73–82.

Subject index

Species index

References to more than one species per genus have been combined as *Genus* spp.